PROCEEDINGS OF THE THIRD INTERNATIONAL SYMPOSIUM ON TRICHOPTERA

Perugia
July 28–August 2, 1980

SERIES ENTOMOLOGICA

EDITORS

E. SCHIMITSCHEK & K. A. SPENCER

VOLUME 20

DR W. JUNK PUBLISHERS
THE HAGUE · BOSTON · LONDON 1981

PROCEEDINGS OF THE THIRD INTERNATIONAL SYMPOSIUM ON TRICHOPTERA

Perugia, July 28–August 2, 1980

Edited by

G. P. MORETTI

DR W. JUNK PUBLISHERS
THE HAGUE · BOSTON · LONDON 1981

Distributors:

for the United States and Canada
Kluwer Boston, Inc.
190 Old Derby Street
Hingham, MA 02043
USA

for all other countries
Kluwer Academic Publishers Group
Distribution Center
P.O. Box 322
3300 AH Dordrecht
The Netherlands

Library of Congress Cataloging in Publication Data

International Symposium on Trichoptera (3rd: 1980: Perugia, Italy)
  Proceedings of the Third International Symposium on Trichoptera, Perugia, July 28-
August 2, 1980.

  (Series entomologica; v. 20)
  1. Caddis-flies – Congresses.   I. Moretti,   G.P. (Giampaolo P.)   II. Title.
III. Series.
QL516.I57   1980   595.7'45   81–3688CR2
                                         AACR2

ISBN-13:978-94-009-8643-5   e-ISBN-13:978-94-009-8641-1
DOI: 10.1007/978-94-009-8641-1

COVER DESIGN: ALESSANDRO SENSIDONI AND MAX VELTHUIJS

CONTENTS

PREFACE

G.P. MORETTI

The Triennial Symposium of Trichoptera would seem to have become a regular event on the calendar. Initiated by Prof. Malicky at Lunz in Austria in 1974, they continued at Reading in England in 1977 (Convenor: Dr M.I. Crichton), the last, this year, took place in Perugia, Italy (Convenor: Prof. G.P. Moretti) and the next will be hosted by Dr J.C. Morse in Clemson, U.S.A. in 1983.

The most outstanding points of the *3rd International Symposium on Trichoptera* held at Perugia from July 28 to August 2, 1980 were *1)* the high number of participants; *2)* the extent, scientific interest and coverage of the papers presented and *3)* the warmth and immediate contact which drew everyone together from the first moment.

Twenty-one nations (Australia, Austria, Belgium, Bulgaria, Canada, Czechoslavakia, Denmark, France, Germany-F.G.R., Germany-G.D.R., Hungary, Iceland, Italy, Mexico, Netherlands, Poland, Spain, Sweden, Switzerland, United Kingdom, United States) were represented by a total of 63 trichopterologists who presented 54 papers and 8 posters during in 8 sessions chaired by M.I. Crichton, H. Malicky, A. Nielsen, O. Flint, L. Botosaneanu, F. Vaillant, G.B. Wiggins and J.C. Morse. The scientific value of the contributions reflected the advanced level reached in their investigations, as well as the diversity of the biological trends. Papers dealt with the morphology, ultrastructure, systematics, taxonomy, phylogenesis, palaeontology, zoogeography, ecology, population dynamics, distribution, pollution, classification, biological cycles, sex ratios, histology, cytology, biochemistry, physiology and parasitology of both larvae and adult Trichoptera. New taxa were described and systematic catalogues were proposed by countries and regions where there was previously a paucity of information on Trichopteran fauna. Two notices of particular importance were announced. One, that Dr L.W. Higler would appreciate receiving any zoogeographical paper on Trichoptera published in any part of the world, so that Fischer's precious and indispensable *Trichopterorum Catalogus* may be kept up-to-date; the other, that Prof. H. Malicky pointed out the utility and value of supplying him with any information for onward transmission to trichopterologists through that mine of interesting news, his *Trichoptera Newsletter.*

The great variety, value and sheer number of papers, as well as the fact that some institutes presented a group of contributions, has created problems for the editorial staff and the publishers (JUNK), who kindly agreed to allocating more pages than originally agreed to and, also, to reducing the bibliographical references to the brief form used by many American journals.

The programme was enriched by two field trips to typical Umbrian biotopes; one to Lake Trasimene, the largest laminar lake in Italy; the second, to the waters of the Sibillini Mounts, the highest and most extensive orohydrographic system in the Umbrian-Marches Apennines. Typical Limnal,

Rhitral and Crenal Trichoptera were collected on both occasions and the lighter moments of 'scientists at work' were caught and documented photographically by our technician, Leo Marini. *Sericostoma italicum* Moretti, which are endemic to Central Italy, were collected in the characteristic irrigated meadows ("marcite") of Norcia and another Central Italian term, *Rhyacophila italica*, from our breedings tanks which are cared for by Giorgina Vignaroli, was given to the participants as a memento of their visit.

I should like to conclude by offering my hearty thanks to my irreplaceable collaborators at the Institute of Zoology and Hydrobiology and to admit that without their unflagging support I should never have been able to carry out all the work which the 3rd symposium entailed. I am particularly grateful to Fernanda Cianficconi for her valuable help, Judy Dale who has assisted me in preparing this volume, Carla Cassioli – the institute secretary – Elio Aisa for his help in obtaining financial support and Maria Vittoria Di Giovanni for the organization of the social programme.*

I cannot close without a word of appreciation to Hans Malicky and Ian Crichton, Lazaire Botosaneanu and Krassimir Kumansky who acted as signposts during my meanders through current Trichoptera systematics by supplying me with data and specimens, as well as convalidating my research, thereby steering me away from the pitfalls innate to all catalogues.

St Francis of Assisi's Day   *Perugia*
                                                                October 4, 1980

---

*I also thank Clara Bicchierai for the careful revision of the proofs.

LIST OF PARTICIPANTS

AISA, Prof. Elio, Istituto di Zoologia, Facoltà di Scienze, Università degli Studi, Via Elce di Sotto, 06100 Perugia, Italy.

ANDERSEN, Dr Trond, Museum of Zoology N-5014 Bergen University, Norway.

BADCOCK, Dr Ruth M., Department of Biology, The University, Keele, Staffordshire, ST5 5BG, England.

BICCHIERAI, Dr M. Clara, Istituto di Zoologia, Facoltà di Scienze, Università degli Studi, Via Elce di Sotto, 06100 Perugia, Italy.

BOTOSANEANU, Dr Lazare, Instituut voor Taxonomische Zoölogie, Universiteit von Amsterdam, Plantage Middenlaan 64, 1018 DH Amsterdam, The Netherlands.

BOURNAUD, Dr Michel, Départment de Biologie Animale et Zoologie, Université Claude Bernard, Lyon 1, 43 Boulevard du 11 Novembre 1918, F-69621, Villeurbanne, France.

BOUVET, Di Yvette, Department de Biologie Animale et Zoologie, Université Claude Bernard, Lyon 1, 43 Boulevard du 11 Novembre 1918, F-69621 Villeurbanne, France.

BUENO-SORIA, Dr Joaquin, Instituto de Biologia UNAM 70-153 México 20 D.F.

BURKHARDT, Mr Rüdiger, AuBenstelle Künanz-Haus des Zoologischen Instituts der Universitat Gieben Hoherodskopf, D-6479 Schotten 12, Germany.

CANTON, Mr S.P., Harner-White Ecological Consultants, 4901 E. Dry Creek Road, Littleton, Colorado, 80122, U.S.A.

CIANFICCONI, Prof. Fernanda, Istituto di Zoologia, Facoltà di Scienze, Università degli Studi, Via Elce di Sotto, 06100 Perugia, Italy.

CORALLINI, Dr Carla, Istituto di Zoologia, Facoltà di Scienze, Università degli Studi, Via Elce di Sotto, 06100 Perugia, Italy.

CRICHTON, Dr M. Ian, Department of Zoology, The University, Whiteknights, Reading, RG6 2AJ, England.

DENIS, Dr Christian, Laboratoire de Biologie Animale, Faculté des Sciences, Université de Rennes, B.P. 25A, F-35031 Rennes Cedex, France.

DI GIOVANNI, Prof. M. Vittoria, Istituto di Zoologia, Facoltà di Scienze, Università degli Studi, Via Elce di Sotto, 06100 Perugia, Italy.

FLANNAGAN, Mr John F., Freshwater Institute, 501 University Crescent, Winnipeg, Manitoba, R3T 2N6, Canada.

FLINT, Dr Oliver S. Jr., Department of Entomology, National Museum of Natural History, Smithsonian Institution, Washington, D.C. 20560, U.S.A.

FLORIN, Dr Janett, Haldenstrasse 2a, CH- 9302, Kronbuhl, Switzerland.

FRIEDLANDER, Dr Michael, Entomologisches Institut – ETH Zentrum, CH- 8092 Zurich, Switzerland.

GARCIA DE JALON, Dr Ing Montes Diego, Departamento de Biologia y Entomologia ETST Montes, Ciudad Universitaria, Madrid-3, Spain.

GATTAPONI, Dr Probo, Istituto di Zoologia, Facoltà di Scienze, Università degli Studi, Via Elce di Sotto, 06100 Perugia, Italy.

GÍSLASON, Dr Gisli M., Institute of Biology, University of Iceland, Grensasvegur 12, 108 Reykjavik, Iceland.

GIUDICELLI, Prof. Jean, Laboratoire de Biologie Animale, Ecologie, Faculté des Sciences de Saint-Jerôme, Rue Poincaré, 13397 Marseille Cedex 4, France.

HARRISON, Mr Jeremy, Department of Biology, The University, Keele, Staffordshire, ST5 5BG, England.

HIGLER, Dr Lambertus W.G., P.C. Hooftlaan 1, P.O. Box 634, Zeist, The Netherlands.

KISS, Dr Otto I., Cifrakapu u. 120 1/5, H-3300 Eger, Hungary.

KUMANSKI, Dr Krassimir, Bulgarian Academy of Science, National Natural History Museum, Boulevard Russki 1, BG-1000 Sofia, Bulgaria.

McELRAVY, Dr Eric P., Division of Entomology, Parasitology, 218 Wellman Hall, University of California, Berkeley California 94720, U.S.A.

MALICKY, Dr Hans, Biologische Station Lunz, A-3293 Lunz, Austria.

MARLIER, Dr Georges, Institut Royal des Sciences Naturelles, Avenue Montjoie 229, 1180 Bruxelles, Belgium.

MEARELLI, Dr Mario, Istituto di Idrobiologia e Pescicoltura, Facoltà di Scienze, Via Elce di Sotto, 06100 Perugia, Italy.

MORETTI, Prof. Giampaolo, Istituto di Zoologia, Facoltà di Scienze, Via Elce di Sotto, 06100 Perugia, Italy.

MORSE, Dr John C., Department of Entomology, Fisheries and Wildlife Clemson University, Clemson, South Carolina 29631, U.S.A.

NEBOISS, Dr Arturs, Department of Entomology, National Museum of Victoria Annexe, 71 Victoria Crescent, Abbotsford, Victoria 3067, Australia.

NIELSEN, Dr Anker, Zoological Museum, Universitetsparken 15, DK-2100 København, Denmark.

NIMMO, Dr Andrew P., Department of Entomology, University of Alberta, Edmonton, Alberta T6G 2E3, Canada.

NOVAK, Dr Karel, Institute of Entomology, Czechoslovak Academy of Sciences 7., Vinicnà CS 128 00 Praha 2, Czechoslovakia.

OTTO, Dr Christian, Ecology Building, Helgonav. 5, S-223 62 Lund, Sweden.

PIRISINU, Dr Quirico, Istituto di Zoologia, Facoltà di Scienze, Via Elce di Sotto, 06100 Perugia, Italy.

PRAT, Dr Narcis, Fac. Biologia, Departamento Ecologia, Universidad de Barcelona, Granvia, 585 Barcelona- 7, Spain.

PUIG, Dr M.A., Fac. Biologia, Departamento Ecologia, Universidad de Barcelona, Granvia, 585-Barcelona-7, Spain.

RESH, Prof. Vincent H., Department of Entomology, University of California, Berkeley, California 94720, U.S.A.

SIEGENTHALER, Dr Claudine, Chemin du Frene 4, CH-1004 Lausanne, Switzerland.

SMITH, Prof. Stamford, Department of Biology, Central Washington, Univ. Ellensburg, Washington 98926, U.S.A.

SOLEM, Dr John O., Kongelige Norske Videnskabers Selskab Museet, Universitet i Trondheim, N-7000 Trondheim, Norway.

SPINELLI, Dr Giuliana, Istituto di Zoologia, Facoltà di Scienze, Via Elce di Sotto, 06100 Perugia, Italy.

STANFORD, Dr Jack A., University of Montana Biological Station East Shore, Flathead Lake, Bigfork, Montana 59911, U.S.A.

STATZNER, Dr Bernhard, Zoologisches Institut I, Kornblumenstr. 13, Postfach 6380, D-75 Karlsrube, 1, FR Germany.

SVENSSON, Dr Björn S., Ecology Building, Helgonav. 5, S-223 62 Lund, Sweden.

SYKORA, Prof. Jan, Graduate School of Public Health, University of Pittsburgh, Pittsburgh, Pa 15261, U.S.A.

TACHET, Dr Henri, Département de Biologie Animale et Zoologie, Université Claude Bernard, Lyon 1, 43 Boulevard du 11 Novembre 1918, F-69621 Villeurbanne, France.

TATICCHI, Dr Maria I., Istituto di Idrobiologia e Pescicoltuta, Facoltà di Scienze, Via Elce di Sotto 06100 Perugia, Italy.

TERRA, Dr Luiz, Estacao Aquicola-4481 Vila do Conde Codex, Portugal.

TOMASZEWSKI, Prof Cezary, University od Lódź. Department of Evolutionary Biology u. Benache 12/16, 90–237 Lódź, Poland.

TYSSE, Mr Asmund, Museum of Zoology N-5014 Bergen University, Norway.

VAILLANT, Prof. Francois, Allée de Pont Croissant, Montbonnot, 38330 Saint-Ismier, France.

WARD, Dr James V., Department of Zoology and Entomology, Colorado State University, Fort Collins, Colorado 80523, U.S.A.

WICHARD, Dr Wilfried, Siebengebirgsstr. 221 D-5300 Bonn 3, F.R.G.

WIGGINS, Prof. Glenn B., Department of Entomology, Royal Ontario Museum, 100 Queen's Park, Toronto, Ontario M5S 2C6, Canada.

WILKINSON, Miss Bridget, Is Manor Drive, Southgate, London N.14 S.J.H., England.

WILLIAMS, Prof. Dudley, Division of Life Sciences, Scarborough College, University of Toronto, West Hill, Toronto, Ontario M1C 1A4, Canada.

WILLIAMS, Mrs N.E., Division of Life Sciences, Scarborough College, University of Toronto, West Hill, Toronto, Ontario M1C 1A4, Canada.

ZINTL, Dr Heribert Grossherzogin, Maria-Anna-Weg 16a, D-8172 Lenggries, Germany.

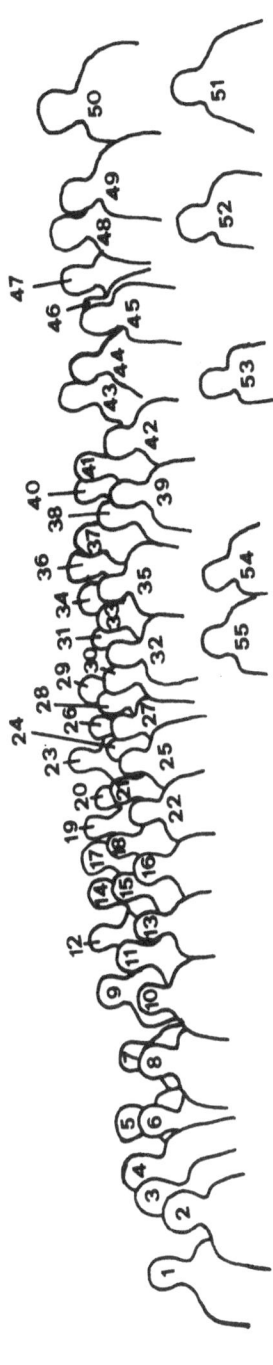

Symposium photograph taken in a corner of the Roman mosaic floor (II sec. A.D.) that depicts "Orpheus taming the beasts", at the entrance to the Institute of Zoology, Perugia University, July 30, 1980: 1. J.C. MORSE, 2. L. TERRA, 3. B.S. SVENSSON, 4. A. TYSSE, 5. M.A. PUIG, 6. D. GARCIA DE JALON, 7. I. BAUTISTA, 8. T. ANDERSEN, 9. K. KUMANSKI, 10. J.O. SOLEM, 11. J. BUENO SORIA, 12. A. NIMMO, 13. C. SIEGENTHALER, 14. J.D. HARRISON, 15. E.P. McELRAVY, 16. M. BOURNAUD, 17. N. PRAT, 18. J. FLANNAGAN, 19. A. NEBOISS, 20. O. FLINT, 21. V.H. RESH, 22. L. BOTOSANEANU, 23. H. ZINTL, 24. R. BADCOCK, 25. A. NIELSEN, 26. C. DENIS, 27. G.P. MORETTI, 28. W. WICHARD, 29. G.M. GISLASON, 30. F. VAILLANT, 31. G.B. WIGGINS, 32. G. SPINELLI, 33. G. FERRARA, 34. A. SENSIDONI, 35. M.I. CRICHTON, 36. H. MALICKY, 37. I.V. WARD, 38. C. TOMASZEWSKI, 39. F. CIANFICCONI, 40. S. SMITH, 41. C. OTTO, 42. C. CORALLINI-SORCETTI, 43. B. STATZNER, 44. D. WILLIAMS, 45. M.C. BICCHIERAI, 46. N.E. WILLIAMS, 47. J. GIUDICELLI, 48. H. TACHET, 49. B. WILKINSON, 50. J. FLORIN, 51. L. HIGLER, 52. P. GATTAPONI, 53. M.V. DI GIOVANNI, 54. Y. BOUVET, 55. M. FRIEDLANDER. Not in photograph: E. AISA, R. BURKHARDT, G. MARLIER, M. MEARELLI, Q. PIRISINU, J. SYKORA, M.I. TATICCHI. Not present, but submitted papers: P. CANTON STEVEN, G. COBB, D.B. FISHER, F.S. GIANOTTI, P. GUERRIERI, O. KISS, K. NOVAK, R. PADILLA, S. RIVERA, I.A. STANFORD, M.I. TORT, I.S. WEAVER III, H. WOLDA.

LIST OF POSTERS PRESENTED AT THE SYMPOSIUM

Trichoptera from the Baltic Amber.
  L. BOTOSANEANU and W. WICHARD.

Trichoptères du groupe de *Stenophylax*.
  Y. BOUVET

Modifications de la morphologie larvaire de *Sericostoma galeatum* au cours du dévelopment.
  C. DENIS.

Observations morphologiques sur le labre des larves de certains Leptoceridae.
  G. MARLIER.

Further ultrastructural observations on the Trichoptera androconia.
  G. P. MORETTI and M.C. BICCHIERAI.

Total lipid content and fatty acid composition in *Hydropsiche dissimulata* Kum. Bots. and *H. pellucidula* Curt. larvae fed with different diets.
  G.P. MORETTI; F. CIANFICCONI; F. FEDERICI and G.F. TROVARELLI.

Larvae of the *Rhyacophila*.
  S. SMITH.

Taxonomie et ripartition des Hydropsychidae dans le Rhöne (entre Lyon et Genève).
  H. TACHET and M. BOURNAUD.

A TRIBUTE TO HERBERT H. ROSS, 1908–1978

J.C. MORSE

Clemson University, U.S.A.

North American caddisfly workers owe much to their predecessors – men such as Herman Hagen, Nathan Banks, and Cornelius Betten. Yet Herbert Ross stands in a very special position in the annals of trichopterology. His studies in caddisfly systematics form the modern foundation for comprehension of the Nearctic fauna. His book, *The Caddis Flies, or Trichoptera, of Illinois,* though published 36 years ago, is still the essential desk reference for initiating the identification of North American genera and species, particularly for the adult forms. In this regard, his contributions on our side of the Atlantic might readily compare in significance with those of Robert MacLachlan on this side.

In another respect, however, Professor Ross consistently sought to comprehend the order on a world scale, especially through studies of its evolution and its historical patterns of distribution. These interests are reflected in such landmark publications as *Evolution and Classification of the Mountain Caddisflies* (1956), 'The evolution and past dispersal of the Trichoptera' (1967), and, indeed, the paper he presented to this body 3 years ago in Reading, 'The present distribution of components of the Sericostomatidae *sens. lat.*' (1978). In this way, he has provided fertile ground for all of us in our interpretations of comparative morphology, ecology, physiology, behavior, or other such interests pursued with Trichoptera faunas anywhere in the world.

For these reasons, it is most appropriate that we remember him today in this special manner.

Herbert Holdsworth Ross was born on March 3, 1908, in Leeds, England. In 1912 his family emigrated to Canada, fortunately after having been unsuccessful in booking a passage on the ill-fated ship *Titanic.* He grew up in a suburb of Vancouver, British Columbia. In 1922, at the age of only 14, he entered the University of British Columbia, graduating in 1927 with a Bachelor of Science degree in agriculture. He completed his Master of Science (1929) and Doctor of Philosophy (1933) degrees in entomology at the University of Illinois in Urbana. His career at the Illinois Natural History Survey in Urbana spanned more than 41 years from 1927 through 1969. From 1947 through 1969 he also held a joint appointment with the University of Illinois as a professor of entomology. Retiring from the University and from the Survey in 1969, he assumed a teaching and research professorship at the University of Georgia at Athens until retiring again in 1975. Even following his second retirement, he continued to lecture, to travel extensively, and to conduct research on caddisfly systematics until his death in Athens, Georgia, on November 2, 1978.

He will be remembered by entomologists throughout the world for his volume, *A Textbook of Entomology* (1948, 1956, 1965), translated into several non-English languages, the fourth edition of which he was completing at the time of his death. He also compiled the remarkable record of having published over 220 papers, books, and chapters of books on such diverse orders as Orthoptera, Plecoptera, Hemiptera, Homoptera, Megaloptera, Neuroptera, Diptera, and Hymenoptera in addition to Trichoptera. Besides his accomplishments with the caddisflies he was a widely recognized authority on North American Symphyta (Hymenoptera; sawflies) and Cicadellidae (Homoptera; leafhoppers).

Professor Ross was a synthesizer of ideas from across many of the traditionally isolated taxonomic specialties, publishing two books on evolution and one on systematics (*A Synthesis of Evolutionary Theory* in 1962, *Understanding Evolution* in 1966, and *Biological Systematics* in 1974). These works and many others demonstrate that he was able to integrate effectively significant, modern concepts in geology, paleontology, microbiology, botany, and zoology with theories in evolution, community ecology, biogeography, and systematics.

On a more personal note, I would say that Professor Ross was also an extraordinary teacher, especially in daily informal sessions, collectively with his students, sharing a pot of coffee. In these and in frequent individual consultations, Dr Ross' intense curiosity and probing questions regarding the student's research always left a young scientist with a renewed sense of excitement that he and his research were important enough to have been considered seriously by such a noted authority and that such significant issues were implied by his investigations.

Yet Dr Ross was probably the humblest and most gentle man I have ever known, capable of leading a student to realize his own error in some particular conclusion and then of praising him genuinely for the discovery.

As a highly productive entomological research scientist, as a synthesizer of ideas from across many disciplines of natural history, as a farsighted leader in championing the causes of systematics, as the father of modern trichopterology in North America and of phylogenetic trichopterology on a world scale, as a revered mentor and a cherished friend, Herbert Ross will be loved and respected for many years to come.

OPENING SPEECH

G. DOZZA
The Chancellor, Perugia University.

Mr Chairman, Ladies and Gentlemen, Participants,

It is with great pleasure that I welcome you on behalf of Perugia University.

The number, the renown, as well as the worldwide support from twenty-one countries, the interesting breakdown of the proposed subject for discussion. All these are reasons for satisfaction at the Perugia Athenaeum, particularly to the Director of the Institute of Zoology, Professor Giampaolo Moretti, to whom, at this point, we should like to reiterate our appreciation and gratitude.

The topic of this conference is, for all of you, a constant research reference that evolves daily and that restates the vital rôle of Trichoptera as biological indicators of the degree of water pollution.

Valuable contributions on this particular aspect of trichopterology have been provided by Professor Moretti's Institute within the C.N.R. projects.

Awareness of our deteriorating environment is something that concerns us all and the dangerous pollution levels reached cause us to ponder the vital importance of research that aims at the conservation of our natural environment and the improvement of the quality of life.

The outstanding scientific programme and the itinerary Professor Moretti has prepared with particular care will allow you to make two journeys of special importance. One to Lake Trasimene where the Institute of Hydrobiology works and where Trichoptera play their part in the food chain of still and running waters; the other to Norcia where the more interesting Trichoptera of the Italian fauna are found, as it is here that the Mediterranean and the Northern species meet.

To all of you, a hearty welcome to Perugia and our best wishes for a successful Symposium.

Perugia
July 28, 1980

OPENING SPEECH

G.P. MORETTI
Chairman and Director,
Institute of Zoology

Chancellor, Ladies and Gentlemen, Participants,

I hope you will be lenient with the far from perfect English in which I shall do my best to welcome you to the *Third International Symposium on Trichoptera.*

After the flattering speech of the Chancellor, it falls to me to say what a pleasure and honour it is to have you all, dear colleagues, with us today. This Third Symposium is about to take place in the Science Faculty of Perugia University, in this ancient, bare basilica, which has been transformed into a teaching hall, and which we shall shortly leave to go to a classroom full of far-from-comfortable benches, old maps and subsidiary educational apparatus, where we can get our heads down and drown ourselves in Trichoptera confabulations.

I hope this will not cause you to think back with regret to the elegant inaugurations in the Hotel at Lunz and the modern facilities at the University of Reading and the efficient organization at both.

We shall do our best to compensate for our somewhat chilly scholastic biotope by escaping to nature's wonderland on two occasions: the first time to dreamy, historical Lake Trasimene, the fourth largest lake in Italy, and the second time to wander upstream along the River Nera, a typical Appennine rhithron, so that we may collect interesting mountain reophyl species in the *'chiare, fresche, dolci acque'* so dear to Petrarca's heart.

An especially warm welcome to Malicky and Crichton who gathered trichopterologists from all over the world in a way that I can in no way imitate and that links all of us in one family within the precincts of the language of our field to exchange ideas and findings.

Your coming here is a long-awaited reward, because the many degree theses on Trichoptera that I have inflicted on my students have earned me the name of monodidactic teacher, and I have, also, been accused of flirting with the ridiculous because of my fidelity to the Trichoptera. Could that be the reason the madman of the Tarocch cards is depicted brandishing an entomologist's net in his hand?

The many participants at this symposium, as you can see, have come from far and wide and the papers about to be presented cover a wealth of subjects of particular interest. This means that the pathway opened by Malicky in 1974, followed by Crichton in 1977 and continued by myself, the speaker, today, is recognized as being valid and on the track and is destined to go ahead into the future.

At this point I should like to publically thank Malicky for having proposed Perugia as the seat of the Third International Symposium after the meeting at Reading. I, also, thank him in the name of the Chancellor and the Dean of the Faculty of Sciences, who is unfortunately unable to be with us owing to examinations at Rome, my collaborators and, yours humbly, the Chairman.

I should once again like to offer my thanks to the Chancellor, as well as to the President of the Azienda di Turismo, The Italian National Council for Research, Prof. Livia Tonolli, Prof. Sandro Ruffo and Prof. Pierfrancesco Ghetti, who have encouraged this symposium with substantial financial support.

As I mentioned previously, this programme is rich in scientific findings and I confess I find myself very moved seeing the names and faces of the most erudite trichopterologists whom I always think of as my true teachers, the teachers who put me on the track of the studies I have pursued for long years, although university commitments have at times made it impossible to dedicate myself to these studies as fully and continuously as I should have wished. I am one of your pupils, despite the fact that I have one foot on the threshold of retirement. An elderly scolar, perhaps a little presumptuous, who would doubtlessly do better to speak less and work harder for our insect friends.

I have been fortunate in having several young collaborators who have fallen in love with this field of investigations and who, I am certain, will contribute to filling the many gaps in our knowledge on the Trichoptera of Italy and other parts of the world that I have not been able, or have not known how to, close. Allow me to thank them with conviction and faith in their future contribution.

The streaming of you, trichopterologist friends, from water-sheds of the world to Perugia, although this has meant considerable personal sacrifice, has been due to your enthusiasm. For this my heartfelt thanks and, also, for the unforgettable collaboration that you have so willingly given to help make this Third Symposium as memorable as the previous two.

With these brief remarks, I declare the *Third International Symposium on Trichoptera* open.

Chancellor, Ladies and *Gentlemen, Participants,* I thank you.

St Nazario's Day   *Perugia*
                                                                      July 28, 1980

A REPORT ON GREGARINES OF TRICHOPTERA IN THE INTESTINES OF *SALMO FARIO* L.

E. AISA AND P. GUERRIERI

SUMMARY

During a research on parasites in the fish populations of Central Italy, *Asterophora hydropsyches* Baudoin sporadins were found in the intestines of the *Salmo fario* L. specimens from the upper reaches of the River Nera and its tributary, the Vigi, in the Apennines. The parasite passes to the fish with the infested Trichoptera larvae on which it feeds and it is then liberated in the intestines where it survives and may even grow to a large size.

INTRODUCTION

Gregarines are well-known endoparasites which are found in the digestive tract, coelom and other cavities, of invertebrates, particularly arthropods and annelids. Although there have been reports on this parasite in the gut content of Trichoptera larvae for some time, a survey of the literature has not revealed any reports on its presence in fish intestines. However, during an investigation on the fish parasites of Central Italy, gregarines were found in the intestines of *Salmo fario* L. specimens.

MATERIALS AND METHODS

Thirty-two specimens of *Salmo fario*, 25 from the River Nera and 7 from its tributary, the Vigi, between 3 and 5 years of age, in a normal state of development and nutrition were used for this investigation.

A pathological examination was carried out on all fish and a stereomicroscope examination on the gills, abdominal organs, intestinal walls and gut contents. Light microscope examination was carried out on gut content samples immediately, and after staining with Lugol. Further samples were flotated and examined both fresh and after Lugol staining. Samples of intestinal walls were examined histologically for the presence of gregarines and alterations caused by parasites.

RESULTS

There were no contents in the stomach; intestinal contents were very scarse or minimal and consisted of mucous-like material with traces of food. This suggests that the fish were caught some time after digesting their last meal. One to 3 empty Trichoptera cases and chitinous larva fragments were found in the anterior and median intestines of two specimens from the Nera River and two from the Vigi tributary (total 4).

Gregarines and trematodes were observed in the intestinal wall and contents at stereomicroscopy. There were two to five gregarines at the sporadin stage lying on the superficial mucosa in the anterior and median intestines of three fish which contained larva cases. Morphological and systematic examinations showed them to be *Asterophora hydropsyches* Baudoin in an advanced stage of development. They were characterized by a nucleus with several nucleoli, well marked septum division, squat protomerites that were sometimes as wide as the deutomerites. The somatometric values were T.L. mμ 76, 00–138, 25; P.L. mμ 23, 50–58, 75; D.L. mμ 50, 50–82, 75; P.W. mμ 30, 25–50, 00; D.W. mμ 30, 25–50, 25.

Four fish from the Nera and 3 from the Vigi (total 7), including all those with gregarines, had one to four adult trematodes of the *Crepidostomum metoecus* Braun species in the anterior and median intestines.

After flotation, microscopic studies revealed two cephaline remains and trematod eggs in fish infested by parasites. No cellular or tissue alterations were seen in histological sections taken from various tracts of the intestinal wall or the area affected by gregarines, but atrophy and erosion of the superficial epithelial mucosa cells associated with congestion of the vessels was sometimes observed at the sites of trematod infestation.

DISCUSSION

The presence of both gregarines and the cases of Trichoptera larvae in the intestines of the fish examined suggests that the endoparasitic infestation is due to ingestion of the larva. The finding of *A. hydropsyches* in 48% of the *Hydropsyche pellucidula* larvae from the Metauro River which is in the same geographical area of the Central Apennines, by Moretti et al. (1978) during a research on Trichoptera larvae parasites confirm this supposition.

The presence of living gregarines after digestion of Trichoptera larvae testifies both to their resistance to the digestive juices and their considerable capacity to survive in the digestive tract of *S. fario*. The size of the sporadins, also, demonstrates their capacity to develop in the intestines of these fish. Some are, probably, liberated from the larva intestines in the cephaline stage during the digestion of the larva in the fish digestive tract and they further develop to become mature individuals. The finding of the sporadin stage and the absence of other forms would seem to exclude a biological cycle in the intestines of *S. fario*. The fact that the SY3-49YC form was not found may be attributed to the interruption of the gregarine biological cycle when the fish were killed, or to environmental differences in the intestines of Trichoptera larvae and fish. Moreover, it has been observed (Vivier et al., 1963; Durchon and Vivier, 1964; Schrevel, 1971) that, in some cases, the host's hormonal state influences the mating of gamonts. As the intestinal lumen of the fish is far larger than that of the larvae, dispersion of gamonts would be favoured rather than their meeting.

The degree of infestation was moderate and this may be due to the non-survival and the consequent loss of individual parasites, or even to modest infestation in the larvae themselves. This is in agreement with the research carried out by Moretti et al. (1978) who counted one to 15 *A. hydropsyches* and *Globulocephalus hydropsyches* Baudoin specimens in infested *H. pellucidula* larvae. The simultaneous presence of *A. hydropsyches* and *C. metoecus* must mean that these two parasites are able to cohabit.

Our data show that gregarines may survive in the intestine of *S. fario*, but are insufficient to reach conclusions on their development. At the moment, it is not possible to state whether the ecological communication between the niches of *A. hydropsyches* and *S. fario* leads to a state of accidental parasitism as well as to pseudoparasitism. Future experimental research will be carried out on this aspect.

CONCLUSIONS

The *Asterophora hydropsyches* Baudoin sporadins found in the anterior and median intestines of *Salmo fario* L. specimens from the upper reaches of the River Nera and its tributary, the Vigi, in the Central Italian Apennines, are probably due to the fish feeding on Trichoptera larvae infested by this parasite. Once in the fish intestines the parasites are well tolerated and they not only survive, but may grow to a quite large size. As the results of this investigation do not give adequate proof of this being a case of accidental parasitism, it will, for the time being, be classified as pseudoparasitism.

REFERENCES

Bykhovskaya-Pavlovskaya, I.E. et al., Opredelitel' parazitov presnovodnykh ryb SSSR, Izdatel'stvo Akademii Nauk SSSR, Moskva-Leningrad, 1962 (Key to Parasites of Freshwater Fish of the U.S.S.R., Israel Program for Scientific Translation, pp. 576–577, Jerusalem, 1964).
Baudoin, I., Ann. Stat. Biol., 2: 15–160, 1967.
Durchon, M. and Vivier, E., Ann. Endocr. (Paris), 25: 43–48, 1964.
Geus, A., Sporentierchen, Sporozoa Die Gregarinida der land- und süsswasserbewohnenden Arthropoden Mitteleuropas, Die Tierwelt Deutschlands, 57: 3–608, ed. G. Fischer, Jena, 1969.
Grassé, P.P., Traité de Zoologie, I, 2, Sous-embranchement des Sporozoaires, pp. 545–797, ed. Masson, Paris, 1953.
Greel, K.G., Protozoology, Springer, Berlin, Heidelberg, New York, 1973.
Kudo, R.R., Protozoology, C. Thomas, Springfield, Illinois, 1971.
Moretti, G.P. and Corallini Sorcetti, C., Boll. Zool., 43: 69–73, 1976.
———; Corallini Sorcetti, C. and Montini, R., Riv. di Idrobiol., 17: 3–26, 1978.
Schrevel, J., Protistologica, 7: 101–130, 1971.
Vivier, E.; Schrevel, J. and Henneré, E., Arch. Zool. exp. gén., 102: 231–238, 1963.

PIGMENTS AND PRESERVATION IN CADDIS LARVAE AND PUPAE

A progress report

R.M. BADCOCK

SUMMARY

The vivid green colour in the haemolymph of certain caddis larvae and pupae, which is most intense in the prepupa and pupa, is very difficult to preserve after death. In *Rhyacophila dorsalis* (Curt.) it is thought to be due to a bilin-type, blue-green zoochrome and a xanthophyll. Possible methods of fixation and preservation to retain the colour are discussed.

INTRODUCTION

If the title of this paper had been submitted at the time of the Symposium instead of several months in advance, it would have been called 'An aspect of colour and its preservation in caddis larvae and pupae: a progress report'. It is very much a progress report, with some areas still to be tackled and many points unresolved or having only partial answers. The preservation of natural colour after death is often difficult and the problem is not confined to Trichoptera or even to insects. Only a small facet of it is discussed here.

During work on the Hydropsychidae, it was found that well grown larvae of *Cheumatopsyche lepida* (Pict.) could often be spotted in the field by their vivid green colour, that this colour persisted in the pupa and in the newly emerged imago but quickly faded after death and was destroyed by most of the methods of preserving structure. A similar green was present in many larvae and pupae of *Rhyacophila dorsalis* (Curt.), although the vividness varied somewhat. Greenness was also found in certain polycentropodid larvae e.g. *Cyrnus* and *Holocentropus dubius* (Ramb.) and in some hydroptilids, such as certain species of *Hydroptila*.

Microscopic observation showed that the greatest intensity of the green colour was in the haemolymph and was shifted about the body according to how the animal was contracting and where the blood was accumulating. Sometimes the head would be unusually green, then the blood could be watched ebbing away and accumulating in another part. The colour showed most clearly in the abdomen. The green colour developed as the larva aged, early instars were not green and the greatest intensity of colour was found in well-grown larvae — especially the prepupa — and in the pupa.

Some attempt has been made to determine the nature of the substances involved but the full answer has not yet been obtained. Needham (1974, p. 4) points out that the term pigment usually means 'a particulate and insoluble coloured material or even a mixture of such materials, used as a paint for external application'. He regards the terms 'biochrome' and 'zoochrome' as more scientific and preferable, a biochrome being defined as 'a specific chemical substance with a coloured molecule, synthesised by living organisms' and a zoochrome as 'any such chrome found in the bodies of animals'. Needham concludes that the term pigment 'seems most usefully applied only to materials of solid or indefinite physical states, to mixtures of chromes and to unanalysed and unknown coloured materials'. So 'pigment' may be used for zoochromes in solid or semi-solid form, within or outside cells but it is not considered suitable for blood zoochromes which are mostly in solution, often in the plasma, or where the chromes are dissolved in lipid in intracellular vacuoles.

In the present study, it was found that extraction of haemolymph with micro-syringes yielded very small quantities and more was eventually obtained from larvae, prepupae and pupae of *Rhyacophila dorsalis* by ligaturing anteriorly and posteriorly (unnecessary in the pupa) to prevent release of gut contents, making a slit through the abdominal integument without piercing the gut, suspending the organisms, by the ligature thread, in the tubes of a hand centrifuge and centrifuging slowly. The investigation into the nature of the colour was undertaken using *Rhyacophila dorsalis*, as it was common and the only green species readily available in adequate quantity at the time.

Examination of the blood led to the conclusion that the colour was probably mostly in solution in the haemolymph. Some greenish cells were seen and one wondered whether solution might have come about through lysis. However Needham (1974, p. 103) states that 'In most animals, some types of blood cells ingest experimentally injected dyes (Verne, 1926, p. 222; Wigglesworth, 1959) and presumably they normally ingest coloured metabolic products'. Probably the zoochrome is genuinely in solution in the haemolymph and some cells may have ingested the colour.

After extraction from the organisms, haemolymph required for further physical and chemical study was kept under nitrogen in a stoppered tube in the dark in a refrigerator, since work on preservation of colour in the larvae had shown that it was rapidly destroyed by oxidation in light. After trials, methanol was selected as a suitable solvent for extraction of the zoochrome in order to study its absorption spectrum. This spectrum showed a very pronounced shoulder at approximately 440 nm, with a lesser hump between 470 and 480 nm. Now an intense Soret absorption peak around 410 nm, or 400–420 nm, is attributed by Needham (1974, 1978) to porphyrins collectively and it is pointed out that the exact position varies according to conditions and the molecular structures under consideration. Vuillaume (1975) refers to 'dipyrryl méthènes (appelés aussi propentdyopents)' showing maximum main absorption between 400 and 450 mμ. Needham (1974) states that bilins have only one peak in the visible range and it is very flat-topped. From this, it is thought that we are dealing with a porphyrin substance, probably one of the bilins or verdohaems which gives a bluish green tint to the blood, a hue which becomes a vivid green when mingled with a carotenoid pigment, probably a yellow xanthophyll, to which the lesser peak around 475 nm is attributed.

The precise identity of the blue-green bilin-type zoochrome is still uncertain. A positive Gmelin test (which should show a series of colour changes on oxidation with fuming nitric acid) was not obtained but this may have been due to faulty extraction or to the zoochrome having decomposed, as the blood had been stored. Needham (1978) refers to the reactivity of bilins and this would account for the difficulty in preserving the colour. However, while some bilins give positive Gmelin tests, others do not (Fox, 1976). It has been observed that on keeping, the blood becomes yellower, also that larvae, on preserving, tend first to lose their vivid green colour, becoming a more yellow green, then yellowish and finally colourless. This is attributed to the bilin-type zoochrome being much less stable than the yellow xanthophyll. In an attempt at silica gel chromotography, it was found difficult to get the blue green pigment to move up the column. In paper chromatography of haemolymph from *Rhyacophila dorsalis,* far greater intensities of colour were obtained using prepupae and pupae than were obtained with larvae. It is feasible that the bilin may have a function in respiratory metabolism during the pupal phase but until more about the precise nature of the zoochrome is known, further speculation is not profitable. Vuillaume (1975) states that the biological role of the prophyrins in insects is even less well known than that of other pigments but that they seem to intervene in metabolism and, in particular in respiration, excretion and perhaps even in photoreception. Bilins are known to be degradation products of haemoproteins; haemoglobin as such is not known in either Trichoptera or Lepidoptera, nevertheless, some of its components may be produced or derived from food and may be functional. Schmidt and Young (1971) reported a bile pigment, probably of bilitriene type and also a yellow carotenoid pigment in the larval haemolymph of the tortricid moth, *Corystoneura* spp.

FIXATION AND PRESERVATION OF CADDIS LARVAE

Most methods currently in use for fixing and preserving caddis larvae result in loss of natural colour, especially of green. While this does not matter for the study of structure, for museum display and for reference collections it may be desirable to retain the colour.

Wiggins (1977) suggests boiling water for fast killing of larvae before fixing. However I find that heat rapidly destroys the green colour and while his recommended fixatives, Kahle's fluid, 80% ethyl or isopropyl alcohol, are excellent for structure, green colour is lost. Kahle's fluid contains ethyl alcohol, besides unbuffered formalin and glacial acetic acid. Wiggins advises subsequent transference to 80% alcohol for storage but comments that it removes abdominal colours retained in Kahle's fluid.

In order to preserve colour, it is desirable to allow the appropriate fixative to kill the larvae, although they can be fixed in a more relaxed condition if narcotised first with carbon dioxide or with a little propylene phenoxetol in the water (Steedman, 1976, p. 91) but must be removed as soon as immobile, for propylene phenoxetol helps to remove colour after death.

Experiments with a variety of fixatives have shown that there can be an osmotic problem, especially for larvae with gills. Too strong a fixative causes gills to shrivel. For hydropsychid and *Rhyacophila* larvae, $2\frac{1}{2}$% formalin (i.e. 1% formaldehyde) buffered to pH 7.4 with sodium acetate ($\simeq 1$%) is recommended. This retains the green colour and gills are fixed in a natural condition. As $2\frac{1}{2}$% formalin is about the minimum strength that will give adequate fixation, care should be taken not to dilute it needlessly on adding the larvae and the fluid should be changed. Acid formalin rapidly removes colour, so it is essential for the formalin to be buffered to pH 7.4 if the green colour is to be retained. I am grateful to Dr H.F. Steedman for discussing changes in the pH of formalin with me before his work was published (1976). Fixation time is critical. For well-grown larvae and pupae of *Rhyacophila dorsalis* and hydropsychids, not less than eighteen hours and not more than twenty-four hours is recommended. For smaller forms such as hydroptilids, the time must be reduced or too much colour is lost. It is a matter of achieving a balance between adequate fixation and too long an exposure to the fixative which results in loss of colour. During fixation the material must be in the dark. The addition of propylene glycol, which should help the penetration of formaldehyde and reduce brittleness in the material (Steedman, 1976) was abandoned because it increased fading of colour.

The next process is to rinse the material twice in glass distilled water, again buffered to pH 7.4 with sodium acetate. Initially, when the water was not buffered, marked loss of colour occurred during rinsing and it was found that even double glass distilled water could be quite acidic.

Earlier it was thought that p-toluene sulphonic acid, recommended by Adams, Flerchinger and Steedman (1976) for ctenophores, might be a suitable fixative. As their acid solution resulted in loss of colour, a neutral solution was obtained using 2 g p-toluene sulphonic acid and 1 g sodium acetate, with caustic soda ($\simeq 0.5$%) to give a solution of pH 7.0 when made up to 100 ml with distilled water. Material fixed in this solution for 24 hours suffered scarcely any colour loss and the method seemed promising. The larvae were rinsed in alkaline distilled water and stored in liquid paraffin. However after a few weeks the material was found to be in poor condition and to have faded, so these modifications of the p-toluene sulphonic acid method are not advisable. At present, alkaline formalin is the most satisfactory fixative known for retention of the green colour.

Currently, preservation of colour during storage is the main difficulty. The blue-green bilins are very readily oxidised, especially in light, and colour is leached in most organic solvents, hence dehydration using alcohols or acetone has not been successful. For preservation in fluid, the most satisfactory technique so far has been to transfer the material from alkaline distilled water to a tube full of liquid paraffin and keep it in the dark. At first, propylene phenoxetol and p-chlorophenoxetol were added as bactericide and fungicide respectively but they caused rapid loss of colour. Plain liquid paraffin seems to be a possible storage fluid in the short term but its long term efficacy is not yet known and the material has to be kept in the dark. In daylight, the colour fades in two days, so larvae in liquid paraffin cannot be displayed for any length of time and can only be examined briefly if the colour is to be retained. Evacuating the paraffin with a vacuum pump and replacing any air with nitrogen might help but this would be too inconvenient a technique for many zoologists to use. The light wave lengths operative in loss of colour are not yet known. Since glass should filter out most of the ultraviolet light, it may not be U.V.. It may be possible to provide a storage fluid or container which filters out the relevant wavelengths.

Freeze drying was tried as an alternative to fluid storage. The larvae were fixed in buffered formalin and rinsed as before. It was necessary to use glass distilled water buffered to pH 7.4 with sodium acetate, although Harris (1976) advises triple-glass distilled water to remove any solutes which might be precipitated. The material was frozen at $-20°C$, as recommended by Harris. Freezing in liquid nitrogen was less satisfactory, causing some larvae to break. Colour retention with freeze drying was good and has persisted in the short term with larvae kept in air in the dark. In daylight, fading is slower in freeze dried larvae kept dry in glass tubes than in larvae stored in liquid paraffin, nevertheless some fading occurred after a week in daylight and it was progressive.

It is now hoped to investigate the nature of the light operative in the photochemical oxidations and the possibility of protecting larvae and pupae by embedding them in, or coating them with, an appropriate plastic or resin.

Colour photography was unlikely to give a sufficiently accurate record of the colour changes and degree of fading under different processes unless the films were always developed by the same person under identical conditions, so the colours were monitored using the Methuen handbook of Colour (Kornerup and Wancher, 1967). This provides abbreviated charts based on the Munsell notation which gives figures for hue, value and chroma. The Methuen scales can be translated into Munsell notation. However the work is not very critical, as the Methuen scale is incomplete compared with the Munsell and the colour squares are only small, also colours should always be compared under strictly controlled, constant conditions, whereas daylight was used. Nevertheless records were kept of the colours of larvae and pupae after the various processes. Owing to variations between larvae, the ranges of colour after each process were considered to be less meaningful than following the colour of an individual through the various processes and timings. As the notation does not convey much without the colour charts for reference, details are not given here.

Your comments on methods of improving the preservation of colour in caddis larvae would certainly be appreciated, for it seems to be a very difficult problem and one which I have not yet succeeded in solving satisfactorily.

ACKNOWLEDGEMENTS

I acknowledge with gratitude much helpful discussion and advice from Dr H.F. Steedman over the years and before his work was published in 1976. I am also grateful to my technician, Mr R. Genn, for assistance in collecting large numbers of larvae and pupae of *Rhyacophila dorsalis*, for extracting haemolymph from them and for paper chromatography. I thank Dr G. Jones for extracting the zoochromes and obtaining their absorption spectra, also for attempting the Gmelin reaction and I thank Mrs. H. Cable for freeze drying the material.

REFERENCES

Adams, H.R., Flerchinger, A.P. and Steedman, H.F., *In* Zooplankton Fixation and Preservation, ed. H.F. Steedman, pp 270–271, The Unesco Press, Paris, 1976.

Fox, D.L., Animal Biochromes and Structural Colours. University of California Press, Berkeley, Los Angeles, London, 1976.

Harris, R.H., *In* Zooplankton Fixation and Preservation, ed. H.F. Steedman, pp 97–99, The Unesco Press, Paris, 1976.

Kornerup, A. and Wanscher, J.H., Methuen Handbook of Colour. 2nd Edition, pp 243, Methuen, London, 1967.

Needham, A.E., The Significance of Zoochromes. Springer-Verlag, Berlin, Heidleberg, New York, 1974.

Needham, A.E., *In* Biochemistry of Insects, ed. M. Rockstein, pp 233–305, Academic Press, New York and London, 1978.

Schmidt, F.H. and Young, C.L., J. Insect Physiol. 17: 843–855, 1971.

Steedman, H.F., *In* Zooplankton Fixation and Preservation, ed. H.F. Steedman, pp 87–94 and pp 103–105, The Unesco Press, Paris, 1976.

Verne, J., Les Pigments dan l'Organisme Animale, Gaston Doin et Cie, Paris, 1926.

Vuillaume, M., *In* Traité de Zoologie, Anatomie, Systematique, Biologie, VIII Fasc. III, ed. P.P. Grassé, pp 77–184, Paris, 1975.

Wiggins, G.B., Larvae of the North American Caddis Fly Genera (Trichoptera), University of Toronto Press, Toronto and Buffalo, 1977.

Wigglesworth, V.B., Control of Growth and Form, Cornell Univ. Press, Ithaca N.Y.; Oxford University Press, London, 1959.

DISCUSSION

Moretti: Do you think the green colour in fat bodies is conserved by copper chloride?

Badcock: I have not tried this and do not know. Thank you for mentioning it. What strength of copper chloride do you suggest using?

Moretti: I use a 1% solution.

10

ORDO TRICHOPTERA ET HOMO INSAPIENS

L. BOTOSANEANU

SUMMARY

The author's aim is to summarize most of the published information, as well as information received from colleagues, concerning altered, endangered, or vulnerable habitats of Trichoptera; caddis-fly taxa or populations which are — locally or generally — vulnerable, endangered, or already destroyed by human activities.

"L'HOMME EST LA CRÉATURE LA PLUS SAGE ET INTELLIGENTE". QUI A DIT ÇA ? L'HOMME!"
J.S. Lec (*"Pensées en broussaille"*).

Le premier devoir professionnel de chaque biologiste est, pour moi, l'activité protectionniste dans les domaines de sa compétence. Je donne ici un aperçu de la situation en ce qui concerne les trichoptères et leurs habitats; j'utilise les données de la littérature, mais aussi des informations inédites fournies par des collègues. Les documents sont extrêmement lacunaires: on ne sait pratiquement *rien* pour d'énormes territoires: l'hécatombe des habitats, des populations et des espèces se déroulera là-bas dans le silence le plus total. — Je commence par exposer les faits concernant les habitats des trichoptères (groupés par Crenal, Rhithral, Potamal et Limnal). Je continue par des exemples d'espèces localement ou même généralement menacées, en voie de disparition ou anéanties (évidemment, ces aspects sont souvent indissociables de ceux concernant les habitats).

La menace pesant sur les habitats souvent ponctuels, isolés, vulnérables, du Crenal, est particulièrement grave dans les zones basses, à agriculture ± intensive, et dans celles sémi-arides ou arides, soit qu'il s'agisse de faunules endémiques, soit qu'il est question de populations périphériques ± isolées. Quelques exemples seront fournis par mon expérience personnelle. — Sur les rives occidentales de la Mer Morte il y a un petit nombre de sources d'eau douce ou saumâtre, qui représentent de véritables oasis dans ce 'désert hydrobiologique' et qui sont peuplées par une remarquable faunule de trichoptères soit endémiques, soit à distribution limitée et à populations fort éparses: *Hydroptila hirra* Mos., *H. adana* Mos., *Ithytrichia dovporiana* Bots., *Chimarra lejea* Mos., *Tinodes negevianus* Bots. & Gasith, *Setodes alalus* Mos.

(Botosaneanu, 1973); or, certaines de ces sources (par exemple celles d'Ein Feshkha) sont très sérieusement menacées par des aménagements en relation avec les loisirs et le tourisme. — Il y a des années, il m'avait été possible de sauver d'une destruction imminente les sources de Corbii Ciungi, situées à 100 m. d'alt. dans la Plaine agricole de Valachie, Roumanie (Botosaneanu and Negrea, 1961); ces sources hébergent, entre autres, des populations très isolées, périphériques, relictaires, de: *Ernodes articularis* Pict., *Adicella filicornis* Pict., *Lype reducta* Hag., *Lithax obscurus* Hag., *Notidobia ciliaris* L. — L'unique localité à *Plectrocnemia conspersa* Curt. de la province roumaine de Dobroudja est la source frontale du petit ruisseau Casimcea (Botosaneanu and coll., 1959): je crois que cette enclave crénique a été anéantie ces dernières années. — En Provence, dans la zone de Crau, il subsiste un groupe de sources à faune de trichoptères bien diversifiée, incluant les endémiques *Agapetus cravensis* Giudicelli et *Hydroptila giudicellorum* Bots.; ce sont les sources de l'Etang du Comte, sur lesquelles des recherches sont entreprises dans le laboratoire de J. Giudicelli; ces sources sont menacées de subir une brutale modification du milieu. — H. Malicky m'a fourni les renseignements suivants: au cours des 30 dernières années absolument toutes les sources du Mt. Pentadactylos (Chypre) ont été captées et murées: il n'y a actuellement plus de faune dans les sources et ruisseaux de ce massif, et s'il y a eu des trichoptères intéressants, ils ont péri. — C'est d'ailleurs aussi le cas pour les basses régions d'Europe septentrionale: voir plus loin ce qui s'est passé dans certaines sources du Danemark. — Si la situation du Crenal des hauts massifs montagneux d'Europe reste encore bonne, celle des sources des moyennes montagnes devient souvent préoccupante; cette situation est reflétée dans le travail de Moretti and Cianficconi (1975) dédié aux trichoptères des sources de la chaîne des Appennins: la capture progressive des sources par l'homme représente un péril imminent pour des crénobiontes endémiques comme *Drusus improvisus* McL., *Drusus* sp. 1 et *Drusus* sp. 2.

La situation du Rhithral des zones montagneuses et surtout de haute montagne (par exemple dans les pays alpins d'Europe) est encore en général bonne. Mais dans les zones de piémont, dans les larges vallées très peuplées, parfois même dans les moyennes montagnes, la situation de cours d'eau appartenant, par exemple, au hyporhithral, se dégrade repidement si elle n'est pas déjà catastrophale. — Un document saisissant est celui apporté par Vaillant (1970) sur l'Isère à Grenoble; en 1946–47 l'auteur y trouve un certain nombre d'espèces; en 1957 il y avait encore des larves de *Rhyacophila dorsalis* Curt., celles d'*Allogamus auricollis* Pict., étant très nombreuses; en 1960 et plus tard, seule *R. dorsalis* se maintenait; c'est la pollution chimique qui a anéanti la faune, à Grenoble d'abord, puis progressivement en amont. — Dans les zones de basse altitude, tous les cours d'eau sont souvent affectés; l'exemple publié le plus impressionnant est peut-être celui des ruisseaux et des petites rivières de Jutlande (Danemark) que rapporte Nielsen (1976a). Cet auteur brosse un tableau très pessimiste: en quelques décennies les ruisseaux et les petites rivières de Jutlande sont devenus parmi les plus pollués sur le plan mondial, ce qui est la conséquence de l'installation de centaines de fermes salmonicoles déversant dans les cours d'eau une énorme quantité de déjections; Nielsen a été le témoin de la destruction pratiquement totale d'une faune jadis riche, comprenant *Oligoplectrum maculatum* Fourcr., *Ecclisopteryx dalecarlica* Kol., plusieurs autres limnephilides; *Hydropsyche* (surtout *angustipennis* Curt.) est favorisé par un certain taux de pollution organique, pour disparaître à son tour. – Une situation similaire est celle des ruisseaux d'un pays comme la Hollande. – Déjà au cours des années 30 et 40, Ross (1944) avait remarqué la régression de certains habitats rhithriques par suite de l'extension de l'agriculture, la disparition dans certaines zones de l'Illinois d'espèces qui y étaient antérieurement présentes en masse; comment la situation a-t-elle évoluée ces 4 dernières décennies? – La situation dans les ruisseaux des petites îles est souvent grave; H. Malicky (i.l.) considère que dans certaines îles grecques l'utilisation massive d'insecticides contre la mouche de l'olive pourrait mettre en péril l'existence même d'espèces comme *Tinodes megalopompos* Mal., *T. peterressli* Mal. ou *Adicella dionisos* Mal., à aréal restreint à 3–4 ruisseaux seulement. Similaire est la situation observée par moi-même à l'Isla de Pinos, au S. de Cuba: certains des ruisseaux à faune éradiquée par l'utilisation chaotique d'insecticides dans les plantations d'agrumes, tandis que dans un fort petit nombre de ruisselets directement menacés, subsiste encore une faune originale-endémiques et espèces à aréal limité, comme *Ochrothrichia islenia* Bots. ou *Oecetis maspeluda* Bots. (Botosaneanu, 1979). Flint (1964) suppose, à son tour, que des Leptoceridae, par exemple, ont été gravement affectés par l'utilisation d'insecticides à Puerto Rico. – Un facteur considéré par plusieurs auteurs comme étant de première importance pour la dégradation des habitats et des faunes des ruisseaux, est le déboisement des rives et la liquidation de la végétation ripale. Edington & Hildrew (1973) signalent, dans de telles cironstances, le remplacement de *Diplectrona felix* McL. par *Hydropsyche* dans des ruisseaux du Northumberland. L'enorme importance de ce facteur est evoquée par Morse (1976), qui souligne son impact sur la diversité de la faune d'insectes aquatiques dans la Caroline du Sud. J'avais été très surpris de ne capturer, à Haïti, que des Hydroptilidae sur de très beaux cours d'eau de montagne, tandis que dans les montagnes boisées de la province orientale de Cuba, des cours d'eau similaires avaient fourni une faune beaucoup plus riche et diversifiée; maintenant je sais que c'est le résultat du déboisement total à Haïti, avec d'incalculables conséquences négatives sur le régime thermique, la photosynthèse, l'aspect et la distribution des niches écologiques, les ressources trophiques, des cours d'eau.

C'est probablement dans le Potamal des régions densément peuplées et industrielles que la situation est la plus désastreuse; et encore est-elle connue de façon très fragmentaire: car que sait-on de la situation des trichoptères du Guadalquivir, de la Seine ou de la Loire, du Dniepr ou de la Volga, du Missisipi-Missouri ou du Nile inférieur? – Les aspects positifs sont rares: c'est le cas du Danube, c'est dans une certaine mesure le cas des rivières des pays alpins ou de moyenne montagne, à grande capacité d'autoépuration, et c'est probablement encore le cas de nombre de rivières et de fleuves d'Asie, d'Afrique, d'Amérique du Sud – qui attendent toujours leur trichoptérologue. – Pour l'Europe Centrale, la solide comparaison que Novák (1977) a heureusement pu réaliser entre les résultats obtenus par F. Klapálek en Bohême à la limite entre le XIXe et le XXe siècles, et ceux de ses propres recherches, a abouti à un triste bilan: Klapálek connaissait 37 espèces de trichoptères dans la Vltava à Prague, il en reste 16; il signalait du cours moyen d'Elbe 33 espèces, dont 19 seulement ont été retrouvées par Novák (v. détails plus loin). − La Dimbovitza à Bucarest (Roumaine) hébergeait encore jusqu'en 1955 *Hydropsyche dissimulata* Kùm. and Bots., *Agraylea sexmaculata* Curt., des *Triaenodes*, des *Limnephilus*: elle y est maintenant réduite à l'état azoïque. − Dans le Rhône supérieur (entre Schaffhausen et Basel), il y avait vers la fin du XIXe siècle (Ris, 1896) une trichoptérofaune d'une 'fabuleuse richesse': ca. 24 espèces sont mentionnées, la plupart 'massenhaft'; j'aimerais savoir quelle est la situation actuelle − est-ce que *Sericostoma flavicorne* Schneid., *Micrasema setiferum* Pict., *Ceraclea aurea* Pict., *Triaenodes conspersus* Ramb., *Chimarra marginata* L. *Agapetus ochripes* Curt., *Hydroptila rheni* Ris, *Ithytrichia lamellaris* Eat., se sont maintenus? Je vais faire référence, un peu plus loin, à la situation actuelle dans le Rhône inférieur. − Aussi incroyable que ceci puisse paraître, il est apparemment impossible de reconstituer, même incomplètement, ce qu'était la faune du Rhin moyen et inférieur il y a 80 ou 100 ans (il y avait certainement ca. 40 espèces de trichoptères!); mais pour la situation actuelle, elle est fort simple: W. Wichard me fait savoir qu'à Bonn il y a une seule espèce qui survit, mais en quantités astronomiques: c'est *Hydropsyche contubernalis* McL; au début de ce siècle *H. tobiasi* Mal. (= l'*exocellata* de Tobias) était abondante dans le Rhin Allemand, où elle n'a plus été retrouvée depuis (inf. H. Malicky). − W. Tobias me communique qu'une seule espèce survit dans le Main à Mainz: *Ecnomus tenellus* Ramb. (forcé d'adopter un mode de vie madicole). − G.P. Moretti (intervention au Ie Symp. sur les trichoptères), constate que dans le Tibre les espèces les plus résistantes à la pollution sont *Hydropsyche dissimulata* Kum. and Bots., *Hydroptila angulata* Mos., *Ecnomus tenellus* Ramb. − Le Rock River dans l'Illinois était même pendant les années 30−40 une rivière assez polluée; elle est actuellement très déteriorée par l'urbanisation et l'industrialisation; c'est une des rares rivières américaines pour lesquelles des données comparatives existent (Resh, 1976): *Athripsodes mentieus* Walk. qui avait été l'espèce la plus abondante dans des captures de 1924 à 1927, avait complètement disparu en 1971−72, ayant été remplacée par *A. transversus* Hag., très rare dans les collections plus anciennes. − L'importance des 'vieux bras' et des zones innondables des fleuves européens pour le maintien d'une partie de leur faune, est certaine. Je détiens de MM. Dumont et Rivier (C.T.G.R.E.F., Aix-en-Provence) les renseignements suivants, significatifs pour le Rhône inférieur: dans le Rhône canalisé à Donzére, il n'y a plus que 3 espèces qui se maintiennent; par contre, dans le Vieux-Rhône 16 espèces forment une faune de trichoptères encore assez vigoureuse; l'influence des eaux du Vieux-Rhône se fait encore ressentir à la confluence aval avec le Canal: 8 espèces. Il est probable que les anciens bras du Rhin allemand on été quelque peu épargnés par l'hécatombe: peut-être hébergent-ils encore quelques-uns des vrais potamobiontes du fleuve? Des notes signalant l'importance des 'vieux bras' m'ont été envoyées par H. Malicky (sur la March, Autriche) et par I.D. Wallace (pour la Tamise). Pour le Danube inférieur, la lutte barbare menée ces dernières décennies contre la zone inondable, aura certainement des conséquences néfastes pour le peuplement du fleuve.

14

Abordons le domaine des eaux stagnantes. Les eaux temporaires sont extrêmement périclitées dans de vastes régions agricoles et à dense population, en Europe et peut-être ailleurs. Hiley (1978) discute ce sujet et évoque le sort de 11 espèces de *Limnephilus* et *Grammotaulius* qu'il trouve adaptées à ces eaux en Angleterre. Dans les plaines d'Europe Orientale, toutes les populations d'eau temporaire des trichoptères, disparaissent avant même d'avoir pu faire l'objet d'une étude. Wichard (1979) considère aussi les espèces des eaux temporaires comme étant particulièrement menacées dans le Land Nordrhein-Westfalen. – Presque rien de vraiment concret ne peut être affirmé sur la situation des trichoptères des marécages et des tourbières; Wichard (1979) évoque cette situation comme étant la plus préoccupante dans le Nordrhein-Westfalen. Je me demande si, après la liquidation presque totale des marécages de Huleh (au N. du lac Kinneret, en Israël) il subsiste encore quelque chose de leur faune caractéristique et très incomplètement documentée (Botosaneanu, 1963), comprenant des espèces qui ont fort peu de chances de se retrouver ailleurs dans ce pays (*Triaenodes reuteri* McL, par exemple). – Novák (1977) constate que dans les étangs de Bohême plusieurs espèces communes du temps de F. Klapálek n'ont plus jamais été retrouvées depuis cet auteur, ou bien qu'elles sont devenues d'une extrême rareté (v. plus loin). – Mey and Tietze (1979) analysent l'effet de la pollution atmosphérique dans une région fort industrialisée d'Allemagne orientale, sur les trichoptères des eaux stagnantes. – La conclusion à laquelle aboutit Ujhélyi (1971) dans une étude bien documentée, est particulièrement déprimante: '. . . the Trichoptera fauna of Lake Balaton whose major representatives belong to the above family [Leptoceridae: surtout espèces d'*Athripsodes* et d'*Oecetis*] since 1966 almost completely disappeared'. – Aussi bien documentée l'étude de Resh (1976) portant sur le Lac Erie et sur sa faune (en partie endémique) d'*Athripsodes*; une comparaison rigoureuse des captures à la lumière, des années 30 et 60–70, montre que 4 espèces d'*Athripsodes* (*erullus* Ross, *erraticus* Milne, *saccus* Ross, *submacula* Walk.) ont été totalement extirpées du lac; 5 autres espèces plus tolérantes, du même genre, ont survécu, mais en populations beaucoup amoindries. – Les protectionnistes australiens ont mené une lutte acerbe pour sauver le lake Pedder de Tasmanie de sa destruction par inondation dans le cadre d'un projet hydroélectrique; je ne connais pas exactement la situation, mais il me semble que ces efforts ont échoué, et que ce lac a disparu, entraînant dans sa disparition aussi sa faune remarquable: 3 espèces connues uniquement de Pedder (le limnephilide *Archaeophylax vernalis* Neb.; les Kokiriidae *Taskiria mccubbini* Neb. et *Taskiropsyche lacustriis* Neb.), ainsi que le Hydrobiosinae *Austrochorema complexa* Jacquemart, le Triplectidinae *Westriplectes pedderensis* Neb., etc. (Neboiss, 1977).

Des documents sur l'extermination totale de taxa de trichoptères, sont, heureusement, fort rares; Nielsen (1976b) considère que deux sous-espèces d'*Apatania muliebris* McL (*intermedia* Nielsen et *nielseni* Schm.) qui habitaient en populations uniques deux sources du Danemark (Rold Kilde, Lille Blaakilde) peuvent être considérées comme extinctes; il y a controverse quant à la cause de cette situation: échantillonnage trop brutal, ou modification du milieu par suite de modifications dans la végétation terrestre (peut-être les deux?). – Par ailleurs, on dispose de documentation pour un assez grand nombre de cas d'espèces exterminées ou bien vulnérables et sérieusement menacées dans certaines parties de leur aréal. Plusieurs cas ont déjà été exposés dans les paragraphes précédents. Voici quelques cas concernant la faune européenne. – *Hydropsyche fulvipes* Curt., espèce de distribution très localisée (ponctuelle), est documentée de Belgique uniquement par 1 ♂ de la coll. Selys, capturé à Spa, il y a un siècle environ, n'ayant jamais été capturée ultérieurement (inf. G. Marlier); elle est très rare et localisée en Angleterre (Badcock, 1974); d'après mes renseignements, l'unique localité connue pour la Roumanie (sources à Greaca) a été gravement affectée par le dessèchement de la zone inondable du Danube. – Le cas assez mystérieux du gros limnephilide *Platyphylax frauenfeldi* Brau., a été évoqué par Malicky (1975); si l'on excepte 2 exemplaires capturés par A. Adlmannseder sur l'Inn, en Autriche, tous les (rares) exemplaires connus ont été capturés au siècle passé, à 'Vienne', sur la Mur, dans les Alpes de Styrie et d'Italie, et à 'Marseille' (! ?), apparemment surtout – ou uniquement – sur des grandes rivières: c'est donc peut-être un potamobionte ne supportant pas la pollution, et en voie avancéé d'extinction. – La présence en Roumanie de *Hydropsyche ornatula* McL. est documentée par un seul ♂ capturé en Transylvanie, sur le Mures, en 1922 (Botosaneanu and Schneider, 1978), et il est possible que cette espèce n'y ait pas survécu aux pressions anthropogènes. – Novák (1977) a pu démontrer que, depuis la fin du XIXe siècle, les espèces suivantes ont été complètement éliminées des grandes rivières de Bohême: *Orthotrichia angustella* McL., *Athripsodes leucophaeus* Ramb., *Oecetis tripunctata* Fbr., *Setodes viridis* Fourcr.; tandis que des espèces comme *Chimarra marginata* L., *Orthotrichia costalis* Curt., *Oxyethira falcata* Mort., *Allotrichia pallicornis* Eat., étaient éliminées des eaux d'autres types de la même province. – Le Danemark a déjà été évoqué ici; selon Svensson and Tjeder (1975) qui utilisent des indications de A. Nielsen, les espèces suivantes doivent être considérées parmi celles certainement disparues ou en voie de disparition de la faune danoise: *Glossosoma boltoni* Curt., *Agapetus ochripes* Curt., *Micrasema minimum* McL., *Oligoplectrum maculatum* Fourcr. – Des données obtenues en Angleterre (Badcock, 1976) permettent de tracer un tableau exact et sombre de la situation de quelques espèces d'*Hydropsyche*: *H. saxonica* McL. y était connue d'une seule localité, près Oxford, mais depuis 1955 elle n'a plus pu être retrouvée: travaux de construction, draguage et pollution du ruisseau, ont achevée cette population; *H. exocellata* Duf. n'a plus été capturée depuis 1901; la capture la plus récente de *H. guttata* Pict. (peut-être en réalité *H. bulgaromanorum* Mal.?) date de 1915; ces 2 dernières espèces étaient régulièrement capturées sur la Tamise à Londres au cours du siècle passé: on peut les considérer comme disparues d'Angleterre. – D'autres observations alarmantes me sont communiquées d'Angleterre par J.D. Wallace: *Adicella filicornis* Pict., à distribution très localisée, est menacée dans sa seule localité connue qui subsiste encore, un groupe de sources en Ecosse; *Leptocerus lusitanicus* McL., autrefois espèce très commune dans la Tamise entre Oxford et Reading, y est actuellement très localisée, et l'habitat de ses stades aquatiques (les racines submergées des *Salix* etc.) est en danger d'anéantissement par suite d'activités de loisir (l'espèce a disparu de la rivière proprement-dite,

16

elle se maintient dans les 'vieux bras' non navigables). – A ma connaissance, il existe actuellement seulement deux essais de dresser des listes de trichoptères menacés dans une région déterminée. Wichard (1979) publie une 'Liste Rouge des Trichoptères menacés dans le Land Nordrhein Westfalen'. Y sont considérées comme étant en danger de mort *Cyrnus insolutus* McL., *Hagenella clathrata* Kol., *Ironoquia dubia* Steph., *Limnephilus elegans* Curt., *Rhadicoleptus alpestris* Kol., comme sérieusement menacées 13 autres espèces, comme menacées 24 espèces (ca. 25% du total des espèces qui existent encore dans ce Land: mais il faut penser aussi aux espèces déjà disparues sans laisser de trace – celles du Rhin notamment!). – Pour l'Amérique du N., il y a la liste publiée par le First South Carolina Endangered Species Symposium (1976): une espèce – *Rhyacophila appalachia* Morse and Ross – est considérée comme étant en danger sur le plan *national*, car connue seulement de 2 ruisselets tributaires de la North Fork Little River et sujets à des pressions de la part de l'agriculture; 25 espèces sont 'rares' ou/et 'périphériques' dans la Caroline du S. (*Polycentropus carlsoni* Morse. par exemple, est connue d'une unique localité, un ruisselet de source). – Mentionnons aussi qu'une 'Species Conservation Monitoring Unit' a été récemment créée (Cambridge, U.K.), qui rédige le premier 'Invertebrate Red Data Book', dans lequel nos trichoptères trouveront certainement leur place.

Parfois, les répercussions des activités humaines sur les trichoptères ont une apparence (ou une facette) positive. De nombreux cas sont connus où la pollution organique des cours d'eau entraîne le développement exagéré de certaines espèces (surtout constructeurs de filets filtrateurs), mais ceci se fait toujours au dépens du reste de la faune. Un exemple fourni par M.M. Dumont et Rivier (C.T.G.R.E.F., Aix-en-Provence) est suggestif: pour un échantillonnage absolument similaire, on a récolté dans l'Onde (ruisseau oligotrophe) 32 larves du limnephilide *Allogamus auricollis* Pict., dans le Gyr (affluent de la Durance, à forte pollution organique) ca. 1560 exemplaires, et dans la Durance (fort polluée, en aval de Briançon) plus de 14.500 exemplaires! – Il y a aussi la tendance de certaines espèces plus opportunistes de coloniser des habitats aquatiques créées par l'homme, ce qui les sauve parfois de l'anéantissement. L'unique trichoptère qui m'est connu de Malta, *Tinodes maclachlani* Kimm., y habite des canaux d'irrigation; en Doubroudja j'ai pu remarquer que des *Hydroptila* et des *Hydropsyche* colonisent les canaux d'irrigation, en remplacement de sources détruites; Moretti and Cianficconi (1975) mentionnent le peuplement de nouveaux habitats crénicoles créées par les activités humaines dans les Appennins; Malicky (discussion au VIe Symp. d'Entomofaunistique en Europe Centrale) remarque que, dans les îles grecques calcaires, les ruisseaux à la faune la plus riche ne sont pas ceux à végétation intacte (et où le processus de formation du travertin crée des conditions défavorables à la plupart des trichoptères), mais dans ceux ayant subi de fortes modifications anthropogènes (destruction de la végétation, etc.).

J'adresse ici un appel à tous les trichoptérologues: nous devons nous sentir plus concernés par le sort de nos insectes, et faire preuve de plus de combativité pour les sauver. Pour les raisons exprimées par Morse (1976) la possibilité de faire protéger par les lois des espèces isolées d'insectes aquatiques est illusoire quand des objectifs économiques sont en jeu; ce qu'il *faut* faire, c'est choisir avec soin, dans les zones les plus vulnérables, des habitats encore intacts, avec la plus grande variété de conditions du milieu et la plus grande diversité spécifique, pour en faire des refuges protégés. A partir de tels refuges le repeuplement des habitats altérés pourrait être tenté (Moretti and Cianficconi, 1975; Fittkau and Reiss, Tagung Westdeutschen Mittglieder I.V.L., 1979). D'autre part, nous devrions tous montrer une plus grande préoccupation pour le rassemblement et le stockage de collections et de documents, avant que l'homme ne reste accompagné sur la terre uniquement par des espèces comme le cochon, les blattes, le Colorado, les rats ou *Erystalis.*

DISCUSSION

Malicky: Some further information: 1) the problem of deterioration of caddisflies in central Europe is worse in large rivers. Light traps, which operated by Prof. Kinzelbach in 1979 along the Rhine near Mainz, yielded enormous numbers of *Hydropsyche contubernalis* only; the other species (about 20–30) are disappeared. In particular, *Hydropsyche tobiasi*, which was known only from the Middle Rhine and which was abundant there at the beginning of the century, is probably extinct. 2) *Platyphylax frauenfeldi* was rare also earlier, as McLachlan has already stated, and it still is. But in the last years Dr. Ujhelyi has caught a few specimens in South Western Hungary. Badcock: I entirely agree with your main thesis about the danger to many species. However there is a small ray of light about *Hydropsyche saxonica.* Since suggesting that it might have become extinct in Great Britain after disappearing from the Oxford site, I have confirmed the identity of a specimen of *H. saxonica* in the Peacy collection which is now in the British Museum (Nat. Hist.). It came from an area of Cotswolds which is still unspoilt and may persist there.

REFERENCES

Badcock, R.M., In Proc. 1st. Int. Symp. Trichoptera, ed. H. Malicky, pp. 49–58, Junk The Hague, 1976.
Botosaneanu, L., Bull. Ent. de Pologne 33: 95–99, 1963.
Botosaneanu, L., Fragm. Entom. 9: 61–80, 1973.
Botosaneanu, L., Stud. Fauna Curaçao and other Caribbean Islands 59: 33–62, 1979.
Botosaneanu, L. and Negrea, S., Hydrobiologia 18: 199–218, 1961.
Botosaneanu, L. and Schneider, E.A., Studii si Communicari Muz. Brukenthal, St. Nat. 22: 307–326, 1978.
Botosaneanu, L. et al., Arch. f. Hydrobiol. 55: 30–51, 1959.
Edington, J.M. and Hildrew, A.G., Verh. Internat. Verein. Limnol. 18: 1549–1558, 1973.
Flint, O.S., Univ. P. Rico Agric. Exp. Sta. Techn. Paper 40: 1–80, 1964.
Hiley, P.D., In Proc. 2nd Int. Symp. Trichoptera, ed. J. Crichton, pp 297–301, Junk The Hague, 1978.
Malicky, H., Verh. d. VI. Int. Symp. über Entomofaunistik in Mitteleuropa, pp. 105–117, Junk The Hague, 1975.
Mey, W. and Tietze, F., Hercynia N.F. 16: 264–272, 1979.

Moretti, G.P. and Cianficconi, F., In Atti V Simposio Nazionale Conservazione Natura 2: 69–83, 1975.

Morse, J.C. In Proc. 1st. S. Carolina Endangered Species Symp. pp. 111–113, 1976.

Neboiss, A., Mem. Nat. Mus. Vict. 38: 1–208, 1977.

Nielsen, A., In Proc. 1st. Symp. Trichoptera, ed. H. Malicky, pp 159–161, Junk The Hague, 1976.

Nielsen, A., In Proc. 1st Int. Symp. Trichoptera, ed. H. Malicky, pp 163–165, Junk The Hague, 1976.

Novak, K., Verh. d. VI Int. Symp. über Entomofaunistik in Mitteleuropa, pp 119–123, Junk The Hague, 1977.

Ris, F., Mitth. d. Schweiz. Ent. Ges. 9: 423–431, 1897.

Resh, V., In Proc. 1st. Int. Symp. Trichoptera, ed. H. Malicky, pp 167–170, Junk The Hague, 1976.

Ross, H.H., Illinois Nat. Hist. Survey Bull. 23: 1–326, 1944.

Svensson, B. and Tjeder, B., Ent. Scand. 6: 261–274, 1975.

Ujhelyi, S. Folia Entom. Hung. 24: 119–137, 1971.

Vaillant, F., Trav. Lab. Hydrobiol. Grenoble 61: 17–32, 1970.

Wichard, W., Schriftenreihe d. Landesanstalt., Nordrhein-Westfalen 4: 65–66, 1979.

Status Report, in Proc. 1st. S. Carolina Endangered Species Symp. pp 46–51, 1976.

19

OBSERVATIONS MORPHOLOGIQUES ÉTHOLOGIQUES ET ÉCOLOGIQUES SUR *HYDROPTILA HIRRA* MOSELY (TRICHOPTERA : HYDROPTILIDAE)

L. BOTOSANEANU ET J. GIUDICELLI

SUMMARY

The first description of *Hydroptila hirra* Mosely came from S.W. Arabia, but it was subsequently found by us in Israel, Sinai, the Dead Sea Depression and the Aïr Mountains in the Republic of Niger. It is a relict inhabitant of permanent springs and streamlets in semi-desert environments. The larva has abdominal dorsal sclerites on segments I–VI that probably correspond to 'chloride epithelia'. The building behaviour is extremely peculiar as prior to pupation the larva starts to build a dome of fine and coarse sand, which strongly resembles the pupal case of Glossosomatids, around its typical *Hydroptila* case.

INTRODUCTION

Hydroptila hirra a été décrite (♂) par Mosely (1948) du S.O. de la Péninsule Arabique. Botosaneanu et Gasith (1971) mentionnent l'espèce dans des petits ruisseaux et des sources, au Sinaï, au bord de la Mer Morte et en Galilée supérieure; ils figurent la femelle. Botosaneanu (1973) mentionne à nouveau l'espèce, cette fois aussi dans d'autres stations de la Mer Morte; il en donne de nouvelles figures des genitalia et développe des considérations de biogéographie historique.

Le 8 Avril 1971, L.B. a pu faire – malheureusement en hâte et super-ficiellement – des observations et des prélèvements sur les stades aquatiques de cette espèce, dans une des localités du bord de la Mer Morte : le complexe de sources d'En Turabe (Ein Ghuweir). Les 5 et 6 Février 1978, J.G. a redécouvert l'espèce dans un petit ruisseau permanent du massif de l'Aïr (République du Niger); il a pu aussi faire des observations et des prélèvements sur les jeunes stades. Ceux-ci se montrant remarquables à plus d'un égard, nous avons décidé de combiner nos observations pour en tirer les notes que voici.

Les sources d'En Turabe font partie des émergences assez nombreuses, parfois impressionnantes par leur débit et remarquables par la constance de leurs paramètres abiotiques, qui s'égrennent le long de la rive occidentale de la Mer Morte, apportant une note inattendue de fraîcheur et de verdure entre deux milieux extrêmes: la Mer Morte, d'une part, le Désert de Juda (Midbar Yehuda), de l'autre.

Ces sources sont situées à 22 km à vol d'oiseau de l'extrémité nord de la Mer Morte, à une altitude de ca. −300 m. La nappe aquifère alimentant les sources est mise en charge par l'eau collectée sur les monts de Judée qui culminent à plus de 1000 mètres. L'eau sort au jour au bord de la mer. La végétation y est luxuriante, dominée par les joncs; elle forme des taches nettes de verdure qui permettent de repérer les sources d'assez loin. L'eau est tout à fait douce. (Fig. 1)

Fig. 1. Sources d'En Turabe

Le fond du complexe des sources est recouvert de galets, de gravier et de sable grossier. L'écoulement forme des ruisselets assez peu individualisés où une lame d'eau de quelques centimètres seulement recouvre les galets et le sable. Ailleurs, dans les endroits pratiquement stagnants, entre les ruisselets, les algues filamenteuses se développent en abondance. L'insolation de l'eau est nettement réduite par le couvert végétal.

Sur les faces supérieure et inférieure des pierres vit une riche communauté d'Invertébrés composée, entre autres, de Triclades, de Gastéropodes, d'Ephéméroptères, de Chironomides.

En avril, on a récolté trois espèces de Trichoptères constituant d'abondantes populations sur les galets et graviers:

— *Setodes alalus* Mos. représentée par des stades jeunes.
— *Chimarra lejea* Mos. représentée par des nymphes.
— *Hydroptila hirra* représentée par des nymphes, des praepupae, des larves ayant commencé à préparer les constructions en vue de la nymphose; il était alors possible de voir les très nombreuses constructions, rappelant étrangement celles des nymphes de Glossosomatidae, fixées aux faces *inférieures* des petites pierres dans les ruisseaux.

Au centre du massif de l'Aïr, juste au sud de l'oasis de Tîmia, à 18° 08′ N et 8° 46′ E et à 1200 m d'altitude, un ruisseau permanent coule dans une petite vallée constituant un affluent du kori Tîmia*. C'est un des très rares habitats permanents d'eau courante de l'Aïr, important massif sub-saharien rappelant le Hoggar algérien et à climat sub-désertique (entre 100 et 125 mm de précipitations par an à Tîmia). L'eau émerge sous forme d'une assez grande source, au contact d'une coulée de basalte avec le socle granitique (Fig. 2). La nappe aquifère est alimentée par l'eau qui ruisselle, à la saison des pluies (Juillet à Octobre), sur l'Adrar Egalah (1871 m).

Fig. 2. Source et ruisseau de Tîmia (habitat du premier type)

*Dans la langue tamasheq, parlée par les touaregs de l'Aïr, 'kori' est synonyme de 'oued' ou 'wadi'.

paramètres hydrologiques.

(mesures réalisées dans le ruisseau, en Février, en saison sèche)

    débit: 4 l/sec.

    température: 24°C à 09 heures, 26°C à 17 heures

    pH: 7,7 – calcium: 24 mg/l – magnésium: 8 mg/l – chlore: 7 mg/l

Deux types distincts d'habitats ont été constatés. Il s'agit d'abord de secteurs où le fond du ruisseau est constitué de basalte avec peu d'éléments déposés (quelques cailloux); dans le deuxième type d'habitat, le fond est constitué de galets et de graviers, le courant étant un peu plus rapide (environ 50 cm/sec.) que dans l'habitat précédent. Dans les habitats du premier type, la faune est assez clairsemée et les *Hydroptila hirra* construisent des logettes nymphales, assez fragiles, en sable. Dans les habitats du deuxième type, le substrat est entièrement recouvert d'organismes, les algues filamenteuses et les invertébrés formant un revêtement de 0,5 cm d'épaisseur*; les *Hydroptila hirra* n'ont pas été observées avec des logettes nymphales minérales. Au mois de Février, *H. hirra* était représentée par des larves à divers stades, des praepupae, des nymphes, des imagos.

CONSIDERATIONS BIOGEOGRAPHIQUES

La découverte d'*Hydroptila hirra* dans le massif de l'Aïr est intéressante. Elle montre que son caractère d'espèce autrefois expansive, souligné par Botosaneanu (1973, p. 72), est encore plus accentué. Elle montre aussi que l'espèce a pu coloniser l'Afrique (tout comme *Chimarra lejea* Mos. et *Setodes alalus* Mos.) à une époque où l'Arabie n'était pas encore séparée du continent africain (*op. cit.*, 71–72); il est probable qu'elle était largement répandue dans l'ensemble des massifs sahariens, il y a environ 10.000 ans, alors que sur cette région régnait un climat tropical humide. Ce Trichoptère représente donc une espèce relicte, peuplant actuellement des habitats isolés et précaires dans un environnement semi-désertique accentué.

OBSERVATIONS MORPHOLOGIQUES

Des observations sur les exemplaires recueillis à En Turabe et à Tîmia montrent que la larve au stade V de *H. hirra* présente, sur chacun des dorsa abdominaux I - VI, un sclérite allongé transversalement à bords irréguliers, à zone périphérique plus marquée, et dépourvu de pores et de soies. Ces sclérites abdominaux ne sont jamais absents, mais parfois assez indistincts (Fig. 3).

Une mise au point s'impose sur la présence des sclérites abdominaux chez

---

*Invertébrés dominants (récolte au filet Surber sur 1/20 m²)

Dugesia subtentaculata (Draparnaud) (Triclade)	: 1.080 individus
Allonais paraguayensis Michaelsen (Oligochète)	: 580 individus
Simulium ruficorne Macquart (Dipt. Simuliide)	: 1.770 individus
Rheotanytarsus fuscus Freeman (Dipt. Chironomide)	: 970 individus
Caliophrys sp. (Dipt. Muscide)	: 125 individus
Hydroptila hirra Mosely (Trichoptère)	: 1.050 individus

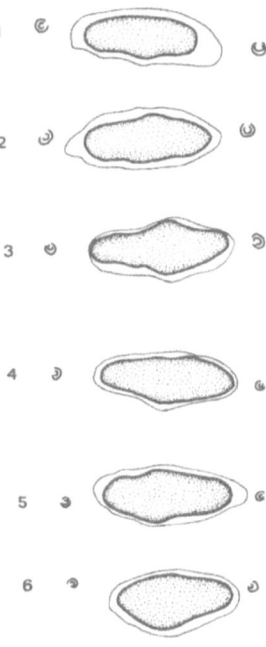

Fig. 3. Sclérites des dorsa abdominaux I à VI de la larve (exemplaire de Tîmia)

les larves d'*Hydroptila* au stade V (pour toutes les jeunes larves d'Hydroptilides l'existence des tergites abdominaux était bien connue). En fait, dans toute la littérature avant 1948, et souvent même après, on ne trouve aucune mention de ces tergites abdominaux, sauf, bien sûr, sur le segment IX. Pratiquement sans exception, dans les clés de détermination et dans les descriptions, les larves d'*Hydroptila* sont considérées comme dépourvues de sclérites abdominaux.

Nielsen (1948) est le premier à avoir décrit chez *H. femoralis* (= *tineoides*) des sclérites sur les segments I à VI. Il les considère comme des 'cuticular plates' correspondant à l'ouverture d'une glande, ou plutôt obstruant cette ouverture. Des glandes unicellulaires géantes sont présentes, semble-t-il, chez la plupart des larves d'Hydroptilides au stade V; chez *Hydroptila*, les deux cellules glandulaires de chaque segment abdominal 'touch each other and their cuticular plates, which are larger in proportion to the cell, have become fused as to form a single approximately elliptic transversely placed plate'. Les figures de sclérites données par Nielsen sont identiques aux nôtres, sauf pour le sclérite du segment I qui est plus petit et parfaitement rond.

Dans leur description de la larve d'*Hydroptila acuta* Mos., Jacquemart et Coineau (1962) ne mentionnent pas ces sclérites; ils figurent un premier segment abdominal sans sclérite, mais un deuxième segment abdominal sur lequel un sclérite est certainement présent.

La larve de *Hydroptila aegyptia* Ulm., décrite par Moretti et coll. (1978),

25

présente aussi des tergites abdominaux sur les segments I à VI; les auteurs considèrent que ces plaques représentent un organe 'de type glandulaire', sans plus de précision.

C'est Wichard (1974) qui s'est référé pour la première fois à l'existence de 'chloride epithelia' sur l'abdomen des larves d'Hydroptilides, ces champs épaissis (correspondant aux sclérites que nous venons de mentionner) jouant un rôle dans l'osmorégulation par absorption des ions du milieu aquatique. L'étude histophysiologique de Wichard semble convaincante. Wiggins (1977) a observé ces sclérites chez une *Hydroptila* sp. de l'Ontario, ainsi que chez d'autres Hydroptilides, et il a adopté le point de vue de Wichard.

N'étant pas histophysiologistes, nous n'avons pas de point de vue personnel à apporter concernant la fonction de ces formations. Mais nous insistons sur le fait qu'il s'agit de vrais sclérites abdominaux qui, bien que plus petits et plus pâles, sont très probablement homologues des grands sclérites abdominaux, parfois complexes et bien pigmentés, des *Stactobia*, *Alisotrichia*, *Leucotrichia* et autres genres d'Hydroptilides. Nous nous demandons si, chez ceux-ci, ces sclérites correspondent vraiment à des 'chloride epithelia' ou à des glandes (ceci nous semblant hautement improbable!). Il est possible, d'autre part, que la distribution de ces sclérites sur les segments abdominaux, leur forme et leur coloration puissent fournir des caractères permettant de distinguer les larves au stade V de diverses espèces d'*Hydroptila*.

OBSERVATIONS ETHOLOGIQUES

En Avril 1971, L.B. avait été frappé de voir, placées sur les surfaces inférieures des petites pierres à En Turabe, de nombreuses constructions en gros grains de sable ressemblant visiblement aux logettes nymphales des Glossosomatides, mais abritant des fourreaux typiques d'*Hydroptila*. Ces constructions minérales (à l'intérieur desquelles les fourreaux proprement dits jouent, pour ainsi dire, le rôle de cocons) ont l'aspect de dômes; elles sont le plus souvent assez fragiles et ne résistent pas aux efforts faits pour les décrocher du substrat. Cependant, un certain nombre de logettes de solidité supérieure ont été récoltées et une de celles-ci a pu être assez bien conservée (Figs. 4, 5).

Les grains de sable les plus gros (dépassant souvent en dimensions le fourreau larvaire et demandant sans doute un très gros effort pour leur manipulation) sont apparemment toujours attachés à l'une des tranches du fourreau, surtout à la tranche supérieure et aux deux extrémités (c'est le cas de 7 grains de sable grossier dans la Fig. 4). Il est probable qu'après avoir fini d'attacher ces gros grains formant la 'charpente' de la logette nymphale, la larve les réunit entre eux à l'aide de grains plus fins; dans l'échantillon il y a de nombreux fragments de logette (Fig. 6) qui montrent des grains de sable solidement rassemblés à l'aide de faisceaux de soie d'aspect et de solidité diverses. Tout ceci montre qu'il s'agit de la construction de véritables logettes et nullement d'une simple fixation de quelques grains de sable au fourreau larvaire.

26

Fig. 4. Logette nymphale, renversée pour montrer à l'intérieur le fourreau de la larve qui se prépare pour la nymphose; des fragments de la logette, constitués de grains de sable plus petits, se sont sans doute détachés de la 'charpente' de gros grains (Source d'En Turabe).

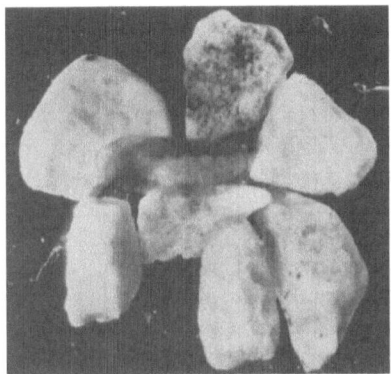

Fig. 5. Logette nymphale, renversée pour montrer à l'intérieur le fourreau de la larve.

Au terme de l'activité constructrices, le fourreau nymphal à l'intérieur de la logette est porté sur la tranche et – dans les cas typiques – fixé au substrat à l'aide de deux pédicelles sécrétés (comme chez les autres *Hydroptila*).

Nous ne savons pas si la larve attache son fourreau au substrat *avant* de commencer l'élaboration de la logette nymphale, ou bien à une certaine étape de la construction de celle-ci. Il nous est malheureusement aussi impossible de dire comment se réalise exactement la fixation de la logette nymphale au substrat: il semble que ce soit quelque chose d'assez lâche et il est même possible que l'ensemble ne soit fixé que par les pédicelles du fourreau proprement dit.

Les observations faites à Tîmia montrent qu'il y a une assez grande labilité dans le comportement constructeur de la larve d'*Hydroptila hirra*.

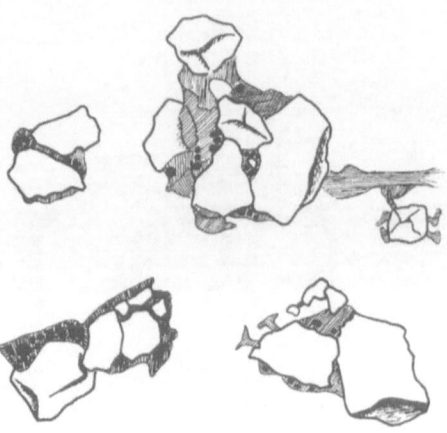

Fig. 6. Fragments divers de logettes nymphales (source d'En Turabe). En blanc et en gris: les grains de sablei' en fines hachures: la soie sécrétée; en noir: espaces entre les faisceaux de soie sécrétée. Même grossissement que pour fig. 4.

Dans les habitats du premier type, les larves construisent de vraies logettes nymphales minérales, mais celles-ci sont extrêmement fragiles. Dans les habitats du deuxième type, sans doute à cause de l'énorme densité du peuplement animal et végétal, on n'observe plus de telles constructions, mais tout au plus quelques grains de sable attachés aux fourreaux.

Peut-être en relation directe avec ce que nous venons de mentionner, à En Turabe les deux pédicelles adhésifs du fourreau proprement dit sont normalement indistincts, difficiles à distinguer des faisceaux de filaments utilisés pour le rattachement des grains de sable de la logette. Par contre, dans la population de Tîmia, ces pédicelles adhésifs sont normalement très bien développés et typiquement constitués.

Il y a labilité aussi dans les matériaux utilisés pour la construction des fourreaux larvaires. A En Turabe on a observé, pour des fourreaux de taille égale, l'utilisation parfois uniquement d'algues à filaments extrêmement fins, parfois uniquement d'algues à filaments nettement plus robustes et parfois pas d'algues du tout. A Tîmia, des matériaux divers sont aussi souvent utilisés, par exemple des tubes de *Rheotanytarsus* (Chironomides).

Enfin, il faut insister sur l'intérêt que présente le comportement constructeur de *Hydroptila hirra* sous l'aspect phylogénétique, car c'est un des éléments spectaculaires qui donnent une idée de l'étroite parentée entre Hydroptilides et Glossosomatides.

NOTE

Au cours de l'impression de ce travail, nous avons pris connaissance de la publication de S. Jacquemart 'Un Trichoptère nouveau de l'Aïr: Hydroptila aïrensis sp. n. (Hydroptilide)' (Bull. Inst. r. Sci. nat. Belg., 52, 13: 1–5,

1980). *Hydroptila aïrensis* Jacquemart est évidemment un synonyme de *H. hirra* Mos.

DISCUSSION

Moretti: Quelle est l'alimentation de la larve de *Hydroptila hirra*?

Giudicelli: La larve se nourrit d'algues vertes filamenteuses; elle les utilise aussi pour contruire son étui.

Wichard: Some observations on the osmoregulation of Hydroptilidae in repect to the sclerotized fields on the abdomen of *Hydroptila hirra* Mosely.

The osmoregulatory absorption structures are not identical in all Hydroptilidae. We may observe anal papillae (e.g. *Ithytrichia*) and very small chloride epithelia (cell complexes) surrounded by well sclerotized fields similar to the abdominal fields of normal epithelia in Hydroptilidae (e.g. *Stactobia*).

If *Hydroptila hirra* does not possess such small cell complexes within the strongly sclerotized fields, they may absorb ions from the intestine, as a primitive form of osmoregulation.

BIBLIOGRAPHIE

Botosaneanu, L., Fragm. Entom. 9: 61–80, 1973.
Botosaneanu, L. and Gasith, A., Israel J. Zool. 20: 89–129, 1971.
Jacquemart, S. and Coineau, J., Bull. Inst. roy. Sc.nat. Belg. 38: 1–81, 1962.
Moretti, G.P., Tucciarelli, F. and Cruccolini, E., Riv. Idrobiol. 17: 27–84, 1978.
Mosely, M.A., British Museum (N.H.) 1: 67–85, 1948.
Nielsen, A.K. Danske Vidensk Selsk. Skr. 5: 1–200, 1948.
Wichard, W., In Proc. 1st Int. Symp. on Trichoptera. ed. H. Malicky pp 171–177, Junk, The Hague, 1974.
Wiggins, G.B., Larvae of the North American Caddisfly Genera. University of Toronto Press, 1977.

TRICHOPTERA FROM THE BALTIC AMBER

L. BOTOSANEANU AND W. WICHARD

SUMMARY

Caddis flies in Baltic amber are common. Since Ulmer's extensive study — Die Trichopteren des Baltischen Bernsteins (1912) — 152 species belonging to 56 genera have been described. Our aim is to call attention again to the Baltic amber caddis flies and to reawaken interest in fossil Trichoptera.

Fig. 1. Triplectides (Leptoceridae) from the Baltic amber.

32

OBSERVATIONS ON THE LONGITUDINAL DISTRIBUTION OF TRICHOPTERA LARVAE IN A STREAM AT ZEMPOALA MEXICO, MEXICO

J. BUENO-SORIA, J. PADILLA AND M. RIVERA

SUMMARY

The distribution and abundance of the larvae of ten Trichoptera species including the Glossosomatidae, Helicopsychidae, Hydropsychidae, Lepidostomatidae, Leptoceridae, Limnephilidae, Philopotamidae and Rhyacophilidae families were investigated. The main factors that determined the longitudinal distribution of the different species along the stream – the stream velocity, substratum type and temporal and spatial changes – are correlated with the developmental stages of the larvae. *Culotila* sp. 1 was the most abundant species along the stream; however, *Nectopsyche lahontanensis*, *Hesperophylax magnus* and *Clistoronia graniculata* reflected the influence of biotic and physical factors in their longitudinal distribution.

INTRODUCTION

Both laboratory experimental and field study reports on the distribution of aquatic insects have mainly focused their attention on the rôle played by physical factors (Hynes, 1976; Cummins, 1975) and, also, in some species, on the effect of drift on the distribution (Townsend and Hildrew, 1976).

The purpose of this paper is to show the longitudinal distribution of the larvae of different species of Trichoptera in a small stream and to correlate this distribution to certain physical and biotic environmental factors. The factors we chose as those most likely to exercise the greatest influence on distribution were water temperature, substratum type and stream velocity. This study was carried out on the borders of the Morelos, Mexico and Distrito Federal States, 65 k south of Mexico city (Fig. 1) in the southernmost part of Transvolcanic belt. It is a mountainous area all of which stands at more than 2,800 m above sea level. From here, the slopes descend southward to the Cuernavaca Valley. The annual rainfall is between 1200 and 1500 mm.

DESCRIPTION OF THE COLLECTING AREA

The study area is located along a small stream which originates from several small springs in the forest and which is approximately 1 km from its source at

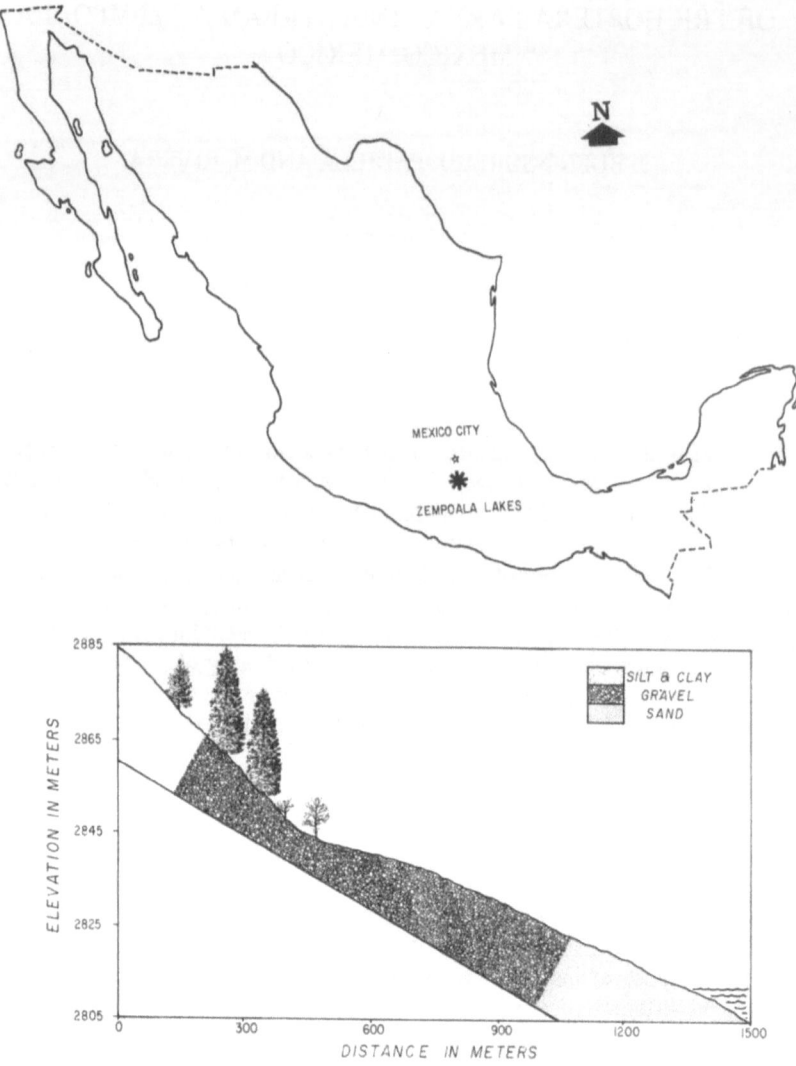

Fig. 1. View of the small stream from the side showing three types of substratum.

2882 m to its lowest point, where it flows into a small pond at an altitude of 2820 m. For purposes of investigation we divided this stream into three zones: (a) the forest area, (b) the forest-valley area, and (c) the valley area (Fig. 1). In the forest zone the substratum of the stream is made up of a great quantity of organic material from the trees and over this there are sand and stones. The very slow velocity of the stream averages 11 cm/sec and this favours the settling of organic matter and other abiotic elements of small weight.

The second zone we called the forest-valley area, it includes Stations 3—8

34

and constitutes the longest stretch of the stream. At Station 3 the vegetation which borders the stream is primarily *Senecio* sp.. Because of the varying sizes of the stones, as well as the scattered aquatic vegetation (Umbelliferae, *Hydrocotila ranunculoides*, Cruciferae, *Rorippa Nasturtium-Aquaticum* and Lemnaceae, *Lemna gibba*), the substratum in this area is slightly more varied. The stream velocity is faster and at some stations it reaches 24–96 cm/sec. There is no evident modification of the substratum in any part of this zone and the substratum measurements obtained were constant.

The third zone, where we located the last sampling station, corresponds to the valley. It is characterized by having an absence of trees and bushes along the banks and patches of acquatic plants scatter the sandy substratum. The stream velocity is 34.6 cm/sec. and the flow laminate, it is, therefore, slower than the second zone.

METHODS

We used a 30 cm surber sampler for collecting the larvae by the stratified sampling technique described by Cummins (1962). Three bank-centre-bank areas were chosen at each collecting station. The stones were checked *in situ* and all the larvae obtained were placed in 80% ethyl alcohol for identification.

RESULTS

The longitudinal distribution of the ten species in nine genera is shown in Fig. 2, where it will be seen that *Lepidostoma delongi* and *Lepidostoma* sp. 1 had a similar distribution pattern, as they both inhabited Stations 1 and 2 where the current velocity is slow and the detritus abundant and where the forest provides heavily shaded areas most of the day and large quantities of organic material. Despite the fact that both *Lepidostroma* species showed a preference for the same habitat throughout the year, we observed a marked difference in their numbers, *L. delongi* being more abundant at Station 1 and *Lepidostroma* sp. at Station 2.

Culoptila sp. was the most abundant species in the stream and the highest population density was registered at Station 2, a station characterized by the stoney substratum required by this species.

Helicopsyche mexicana and *Wormaldia planae* were found in the upper part of the slow speed stream. The first predominated in the detritus area and the second where the substratum was more rocky.

Diplectrona chiapensis and *Atopsyche hidalgoi* did not form aggregations in space, as did other species, so they were collected in small quantities at each sampling station.

The distribution pattern and feeding habits of *D. chiapensis* and *A. hidalgoi* are closely related. *D. chiapensis* requires a stoney substratum and a high current speed for the formation of its food-catching net, whereas *A. hidalgoi* is a free living predator that is found throughout the whole length of the stream.

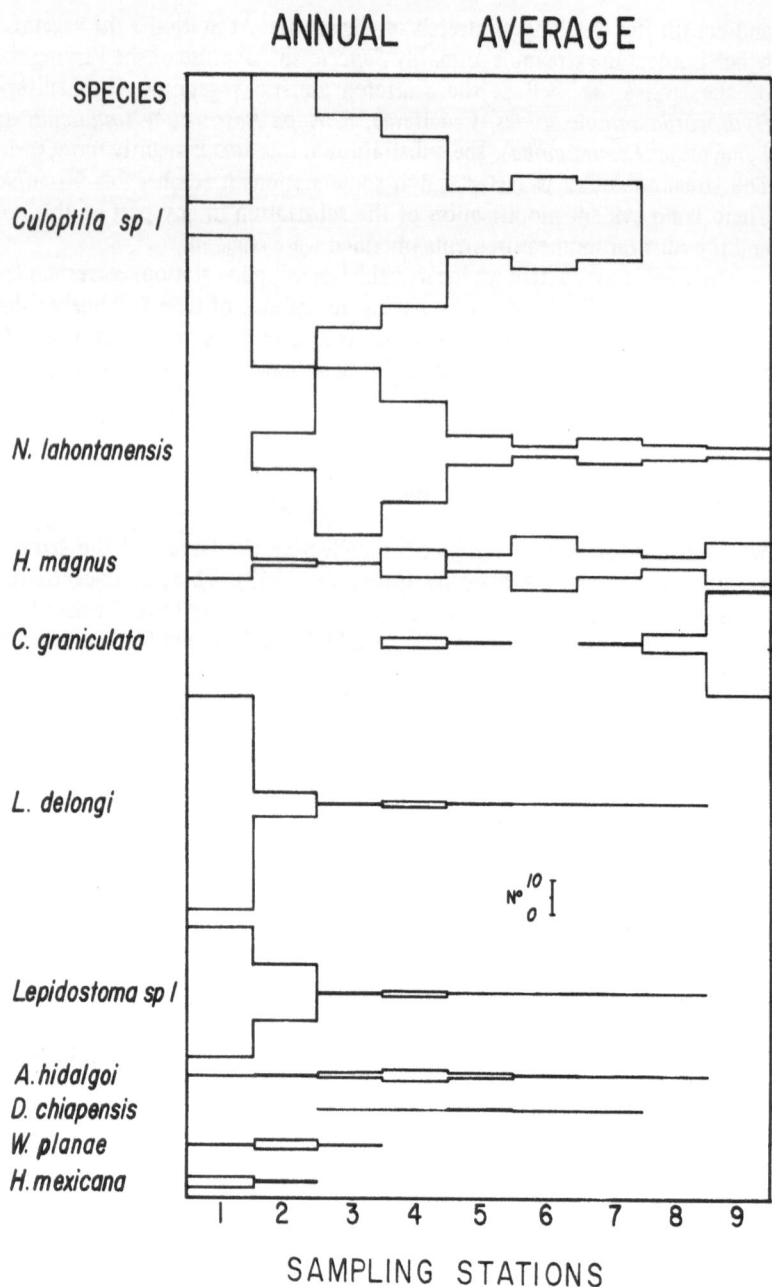

ANNUAL   AVERAGE

SPECIES

Culoptila sp I

N. lahontanensis

H. magnus

C. graniculata

L. delongi

Lepidostoma sp I

A. hidalgoi
D. chiapensis
W. planae
H. mexicana

SAMPLING STATIONS

Fig. 2. Average number for the 10 species at each sampling station.

Hesperophylax magnus, *Clistoronia graniculata* and *Nectopsyche lahontanensis* were gathered from aquatic plants found in all the length of the stream (Fig. 3), but there was a marked difference in the individual numbers of each species at each of the sampling stations. *N. lahontanensis*

36

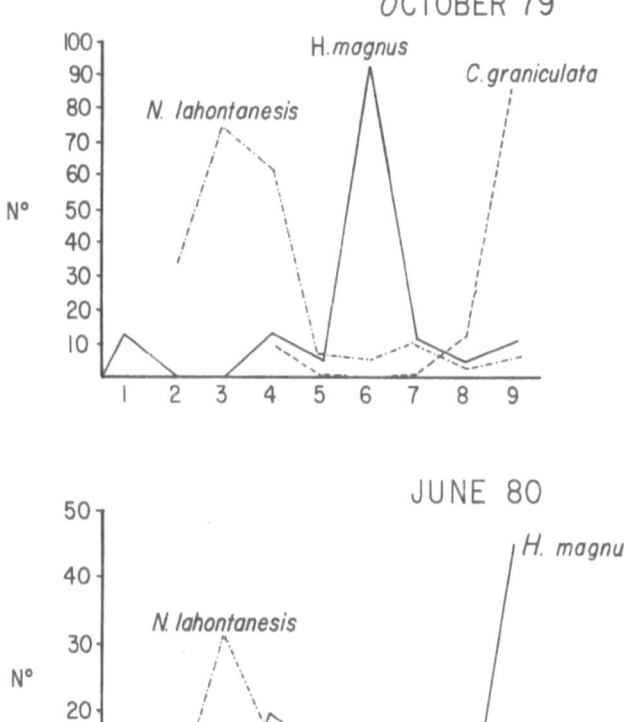

Fig. 3. Number of *Nectopsyche lahontanensis*, *Hesperophylax magnus* and *Clistoronia graniculata* collected with surber sampler each month in Lagunas de Zempoala.

larvae were much more abundant in the wooded part of the stream, where Stations 5 and 6 were situated, here they occupied a mossy substratum in a fast flowing area. *H. magnus* populations were more dense towards the centre of the stream (Stations 5 and 6), an environment not favoured by *N. lahontanensis*. Finally, the distribution of *C. graniculata* was clearly towards the pond, therefore the highest density was found at Station 9, the last sampling station.

Despite the fact that *H. magnus* and *C. graniculata* were gathered from the same plants, their time-space relationships were different. *N. lahontanensis*, *H. magnus* and *C. graniculata*, whose optimal habitats were at Stations 3, 6 and 9 respectively were most abundant in October, 1979. By June, 1980 the maximum density incidence for each species had changed and this was

37

particularly true of *H. magnus* and *C. garniculata*; the last was not collected at Station 9, the last sampling station, as the pupae were localized downstream where the stream flows into the pond and are, therefore, not shown on the graph. *N. lahontanensis* was present in its optimal habitat at Station 3, although the numbers were lower than in October 1979.

In conclusion, when the flow, substratum and food supply are taken into consideration, it will be seen that there is a certain relationship between the species. Although *Culoptila* sp. 1 was the most abundant, it did not interfer with species living nearby; perhaps because of the different habitats or ecological niches they occupied. On the other hand, the species with a wider distribution range, were not abundant in the optimal habitats of other species, for example the species *H. magnus* was mainly present at Stations 4, 6 and 9 where *W. planae* and *H. mexicana,* which were confined to the first three sampling stations, were very rare. *C. graniculata* was particularly concentrated between the last two stations.

Based on these observations, it can be concluded that, in this case, the longitudinal distribution of the Trichoptera larvae was modified by time-space changes related to the developmental stage of the larvae, as well as to the type of substratum needed for its food supply and protection, the current velocity and, in some cases, the water temperature.

ACKNOWLEDGEMENTS

I am much indebted to Dr Flint, Jr for his assistance in preparing the manuscript; M.S. Manuel Guzman for preparing the figures and M.S. and to Silvia Santiago and William Lopez-Forment for reviewing my English Translation.

REFERENCES

Cummins K.W., Am. Midl. Nat. 67: 477–504, 1962.
————., In River Ecology, ed. A.B. Whitton, p. 711, University of California Press, 1975.
Hynes, H.B.N. In the Ecology of Running Waters, 3rd edition, p. 555, University of Toronto Press, 1976.
Townsend, C.R. and Hildrew, A.G., J. Anim. Ecol. 45: 759–772, 1976.

DISCUSSION

Statzner: How do you relate your results to the ideas of Illies and Botosaneanu on the longitudinal section of running waters?
Bueno-Soria: On the basis of the preliminary study, we do not find a confirmation of the Illies-Botosaneanu system. We have, also, noticed that the behaviour of the adults effects the distribution of the larvae.

EMERGENCE OF TRICHOPTERA FROM TROUT CREEK, COLORADO, USA

S.P. CANTON AND J.V. WARD

SUMMARY

A total of 21 species of adult Trichoptera were collected over two years with black-light trap and sweep-net from a site on Trout Creek, a mountain stream in northwestern Colorado. The adult flight period was from mid-June to early October. A total of 7863 organisms were collected over the two years representing 8 families, 16 genera, and 21 species. Two species (*Lepidostoma moneka* and *Brachycentrus americanus*) accounted for 81% of the total numbers of adults. Seasonal emergence patterns exhibited three peaks in both numbers of adults and numbers of species collected. The early summer group (mid-June to mid-July) was represented by 4 species comprising 5% of total numbers. The mid-summer group (mid-July to mid-August) of 10 species accounted for 90% of total numbers. The late summer-early fall group (mid-August to early October) of 7 species comprised 5% of the total numbers of adults. Three species (*L. moneka*, *Wormaldia gabriella*, *Neotrichia okapa*) are new records for Colorado. Sex ratios for abundant species ranged from 34% females (*B. americanus*) to over 65% females (*Glossosoma ventrale*). Season appeared to have no effect on sex ratios.

INTRODUCTION

The caddisflies of Colorado are poorly known. Most published data are restricted to descriptions of new species, or are otherwise limited in scope (Denning 1947, 1949; Ross 1946, 1956; Flint and Herrmann 1976). Relatively comprehensive accounts of Colorado Trichoptera are limited to Dodds and Hisaw (1925), Mecom (1972), and Ward (this volume).

The purpose of this paper is to present data on Trichoptera emergence patterns collected over a two-year period from a Colorado mountain stream. This investigation was part of a larger study on the effects of coal surface mining on stream benthos (Canton and Ward 1978).

SITE DESCRIPTION

Trout Creek runs roughly north to south beginning in the Little Flat Tops area of Routt National Forest and flows into the Yampa River west of Milner, Colorado, in the upper Colorado River Basin. The stream traverses a mixture of forested and farmed land in the vicinity of the study area. The site selected for the emergence study was not influenced by mining activities. The altitude of the study site is 2260 m a.s.l. The substratum is rubble-cobble with gravel and sand underneath. Riparian vegetation is comprised primarily of willows (*Salix* sp.) and alders (*Alnus tenuifolia*), which shade approximately 50% of the stream. See Canton and Ward (1978) for detailed descriptions of physico-chemical conditions.

MATERIALS AND METHODS

Field collections of adult Trichoptera commenced with the loss of snow and ice cover (mid-April) and continued until the first nights of heavy frost (mid-October). Sampling in 1977 was conducted with a portable black-light and by sweep-netting the vegetation. In 1978, a more efficient black-light trap was used in conjunction with sweep-netting, enabling capture of many more adults than the previous year. The black-light was placed next to the stream and left over night every 10–14 days. Captured adults were placed in vials with 70% ethanol. Trichoptera were identified using Ross (1944, 1946, 1956), Denning (1956), and Nimmo (1971). The Hydroptilidae were identified by Dr R.L. Blickle, University of New Hampshire. Dr D.G. Denning, Moraga, California, verified identifications and provided specific names.

RESULTS AND DISCUSSION

Although intensive sampling began in mid-April, caddisfly adults were not collected until mid-June (Fig. 1). The mountain streams of Colorado are subject to extreme increases in discharge each spring as the result of runoff from snowmelt at the higher elevations. Trichoptera emergence in Trout Creek appeared to be timed to the receding runoff. Unusually high snowpack extended runoff in 1978 and delayed the emergence of the first caddisfly species (*Oligophlebodes minuta*) from the second to the fourth week in June (Table I). Peak emergence occurred in late July–early August with > 3200 adult *Lepidostoma moneka* collected 23–24 July, 1978, and > 2000 adult *Brachycentrus americanus* collected on 1–2 August, 1978.

A total of 7863 adult Trichoptera was collected from the site over two years representing 8 families, 16 genera and 21 species. The seasonal emergence pattern exhibited three peaks in total numbers of organisms and

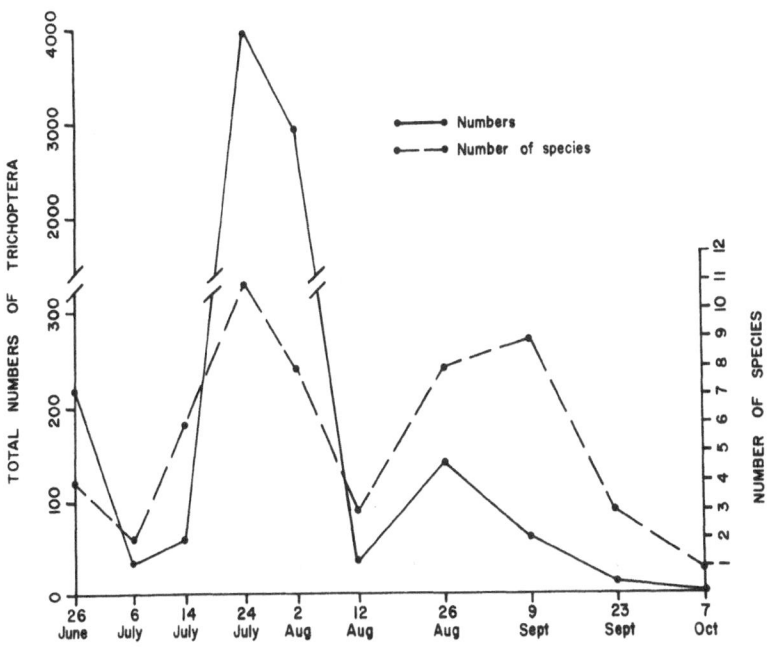

Fig. 1. Seasonal occurrence of adult Trichoptera from Trout Creek, Colorado, 1978, showing total numbers collected and number of species.

Table I. The adult occurrence of early summer emerging species of Trichoptera from Trout Creek, Colorado, during 1977 (o) and 1978 (x). Each month is divided into four weeks.

	June				July			
	1	2	3	4	1	2	3	4
Limnephilidae								
Oligophlebodes minuta (Banks)		o	o	o				
				x	x	x		
Hesperophylax occidentalis (Banks)				x	x			
Hydropsychidae								
Arctopsyche grandis (Banks)		o						
Leptoceridae								
Oecetis avara (Banks)				o				x

number of species collected (Fig. 1). The early summer group of 4 species which emerged from mid-June to mid-July (Table I) accounted for 5% of the total adults collected. *O. minuta* was the first and most abundant caddisfly to emerge during this period. *Arctopsyche grandis* adults were only collected during 1977. This species has a two-year life cycle (Smith 1968) and apparently the Trout Creek population is synchronous. The mid-summer group which emerged from mid-July through mid-August

41

Table II. The adult occurrence of mid-summer emerging species of Trichoptera from Trout Creek, Colorado, during 1977 (o) and 1978 (x). Each month is divided into four weeks.

	June				July				August				September			
	1	2	3	4	1	2	3	4	1	2	3	4	1	2	3	4
Limnephilidae																
Anabolia bimaculata (Walker)									x							
Hydropsychidae																
Hydropsyche cockerelli Banks				o	o	o x	o x	o x	o x	o						
Hydropsyche oslari Banks		o				o x	x	x	x							
Hydropsyche occidentalis Banks						o		x								
Lepidostomatidae																
Lepidostoma moneka Denning						o x	x	x	x	x	x	x	x	x		
Lepidostoma knowltoni Ross						o		x	x	x	x	x	x	x		
Brachycentridae																
Brachycentrus americanus (Banks)						o	o	o x	o x	o x	o x	o x	x	x		
Glossosomatidae																
Glossosoma ventrale Banks						o x	o x	o x	o x	o x			x	x		
Hydroptilidae																
Ochrotrichia logana Ross				x	x	o		x	x	x	x	x	x	x		
Agraylea multipunctata Curtis						o		x				x				

Table III. The adult occurrence of late summer-early autumn emerging species of Trichoptera from Trout Creek, Colorado, during 1977 (o) and 1978 (x). Months are divided into four weeks.

	July				August				September				October			
	1	2	3	4	1	2	3	4	1	2	3	4	1	2	3	4
Limnephilidae																
Limnephilus externus Hagen										x						
Dicosmoecus atripes (Hagen)								x	o x	x	. x	x				
Onocosmoecus unicolor (Banks)								x	o x	x	x	x				
Hesperophylax consimilus (Banks)								x	x	x	x					
Glossosomatidae																
Agapetus boulderensis Milne					o	o	o	o	o x	o x	o x	o x	o x	o		
Philopotamidae																
Wormaldia gabriella (Banks)						o				x						
Hydroptilidae																
Neotrichia okapa Ross								x								

43

(Table II) contained the largest numbers of organisms (90% of total) and species (10). The most abundant species were *L. moneka*, *B. americanus*, *Glossosoma ventrale*, and *Hydropsyche* spp., the immatures of which are common in the riffle sections of Trout Creek (Canton and Ward 1978). The emergence of many species including *Hydropsyche occidentalis*, *Agraylea multipunctata*, and *Ochrotrichia logana* was delayed in 1978 apparently due to the high and extended runoff. The late summer-early fall group of 7 species which emerged from mid-August to early October (Table III) accounted for 5% of the total numbers. *Agapetus boulderensis*, *Dicosmoecus atripes* and *Onocosmoecus unicolor* were the most abundant species in this group both years. *A. boulderensis* emerged one month earlier in 1977 than in 1978. *O. unicolor* also emerged in early fall in an Oregon stream (Anderson and Wold 1972).

Although adult collections were made from only one site on one stream, three species (*L. moneka*, *Wormaldia gabriella*, and *Neotrichia okapa*) were new published records for the state (Denning, pers. comm.; Nimmo 1974; Blickle 1979), which points to the need for additional study on Trichoptera in Colorado.

The emergence period of Trout Creek Trichoptera was relatively short when compared to other streams in North America (Anderson and Wold 1972; Resh et al., 1975; Flannagan 1977). However, temporal segregation of congeneric and confamilial species did occur. Emergence of *Hesperophylax occidentalis* and *H. consimilis* was separated by two months. The limnephilids *Anabolia bimaculata* and *Limnephilus externus* emerged 9 weeks apart. Although the flight periods of the glossosomatids *G. ventrale* and *A. boulderensis* overlapped, they did not coexist in the stream (Canton and Ward 1978). However, some closely related species (notably *Hydropsyche* spp.) exhibited widely overlapping emergence periods.

Sex ratios calculated for the abundant species ranged from 34% females for *B. americanus* to over 65% females for *G. ventrale*. Over the emergence period, the sex ratios remained constant for each species except *O. unicolor*, which commenced with 50% males and terminated with 100% males. *B. americanus* maintained a ratio of only 34% females over 2 months. The predominance of males is rare in caddisflies (Resh et al., 1975) and may relate to differential affinities to black-light of the sexes in this species. *G. ventrale* averaged 67% females over 9 weeks, a ratio similar to that found for another species of *Glossosoma* in an Oregon stream (Anderson and Wold 1972). *L. moneka*, the most abundant species, maintained a ratio of 50% females over an emergence period of 2 months. *O. minuta* also had a ratio of 50% females for its three-week emergence period. Season appeared to have no effect on the sex ratio of the Trichoptera collected in Trout Creek.

ACKNOWLEDGEMENTS

The research was supported in part by a National Science Foundation Energy Traineeship awarded to S.P. Canton, and by the U.S. Environmental

Protection Agency, Environmental Research Laboratory – Duluth, Research Grant No. R803950. The authors are most grateful to Drs R.L. Blickle and D.G. Denning for taxonomic assistance.

REFERENCES

Anderson, N.H. and Wold, J.L., Can. Ent. 104: 189–201, 1972.

Blickle, R.L., N.H. Agric. Exper. Sta. Bull. 509, Durham, 1979.

Canton, S.P., Ward, J.V., U.S. EPA Environ. Res. Lab. Duluth, 1978.

Denning, D.G., Bull. Brook. Ent. Soc. 42: 145–148, 1947.

————., Amer. Midl. Nat. 42: 112–122, 1949.

————., In The Aquatic Insects of California, ed. R.L. Usinger, pp. 237–270, Univ. Calif. Press, Berkeley, 1956.

Dodds, G.S. and Hisaw, F.L., Ecology 6: 380–390, 1925.

Flannagan, J.F., In Proc. 2nd Int. Symp. Trichopt. ed. M.I., Crichton. pp. 183–197, W. Junk, The Hague, 1977.

Flint, O.S. and Herrmann, S.J., Ann. Ent. Soc. Amer. 69: 894–898, 1976.

Mecom, J.O., Hydrobiologia 40: 151–176, 1972.

Nimmo, A.P., Quaest. ent. 7: 3–234, 1971.

————., Quaest. ent. 10: 315–349, 1974.

Resh, V.H., Haag, K.H. and Neff, S.E., Environ. Ent. 4: 241–253, 1975.

Ross, H.H., Bull. Ill. Nat. Hist. Surv. 23: 1–326, 1944.

————., Ann. Ent. Soc. Amer. 39: 265–291, 1946.

————., Evolution and Classification of the Mountain Caddis Flies. Univ. Ill. Press. Urbana, 1956.

Smith, S.D., Pan-Pac. Ent. 44: 102–112, 1968.

DISCUSSION

Moretti: What is the voltine pattern of the species?

Ward: *Arctopsyche grandis* is semivoltine. All other species seem to be univoltine.

Jalon: Have you found the continuous collecting at the same station to have any influence of the insect population?

Ward: I do not believe that any depletion has occurred. Does anyone have any definitive data on this matter?

Resh: Extensive sampling of *Ceraclea ancylus* did not result in population size reduction in the following generation (Resh 1977, Freshwat. Biol.), but destructive sampling of a rare population may result in depletion and density underestimates during the same generation (Lamberti and Resh, 1979, Envir. Entomol.).

FURTHER OBSERVATIONS ON LIMNEPHILID LIFE HISTORIES, BASED ON THE ROTHAMSTED INSECT SURVEY

M.I. CRICHTON AND D.B. FISHER

SUMMARY

It is proposed that the following eleven species of British limnephilids belong to the category having a short flight period, without a diapause, in spring and summer, sometimes extending into autumn: *Apatania muliebris* McL., *A. wallengreni* McL., *Drusus annulatus* (STEPH.), *Ecclisopteryx guttulata* (PICT.), *Limnephilus elegans* CURT., *L. extricatus* McL., *L. luridus* CURT., *Nemotaulius punctatolineatus* (RETZ.), *Rhadicoleptus alpestris* (KOL.), *Micropterna lateralis* (STEPH.) and *Hydatophylax infumatus* (McL.). A revised scheme of classification of flight periods for British Limnephilidae is put forward.

INTRODUCTION

The study of caddisflies of the family Limnephilidae by Crichton (1971) recorded captures from light traps of the Rothamsted Insect Survey over the years 1964–68. Using these records, with published and other information on life histories, a classification of flight periods for limnephilids in Britain was proposed as follows:

(1) an extended period, normally with a diapause, from spring through summer into autumn;

(2) a shorter period, without a diapause, in spring and summer, sometimes extending into autumn;

(3) a short period, without a diapause, in autumn.

The second period was the least well defined, with several species placed there provisionally. Since 1968, further information from the survey, and elsewhere, has made it possible to reconsider this group of species.

LIGHT TRAP SITES AND RECORDS

A standard Rothamsted light trap (Williams, 1948), using a 200 w tungsten filament lamp, was in continuous operation at each site. These sites form

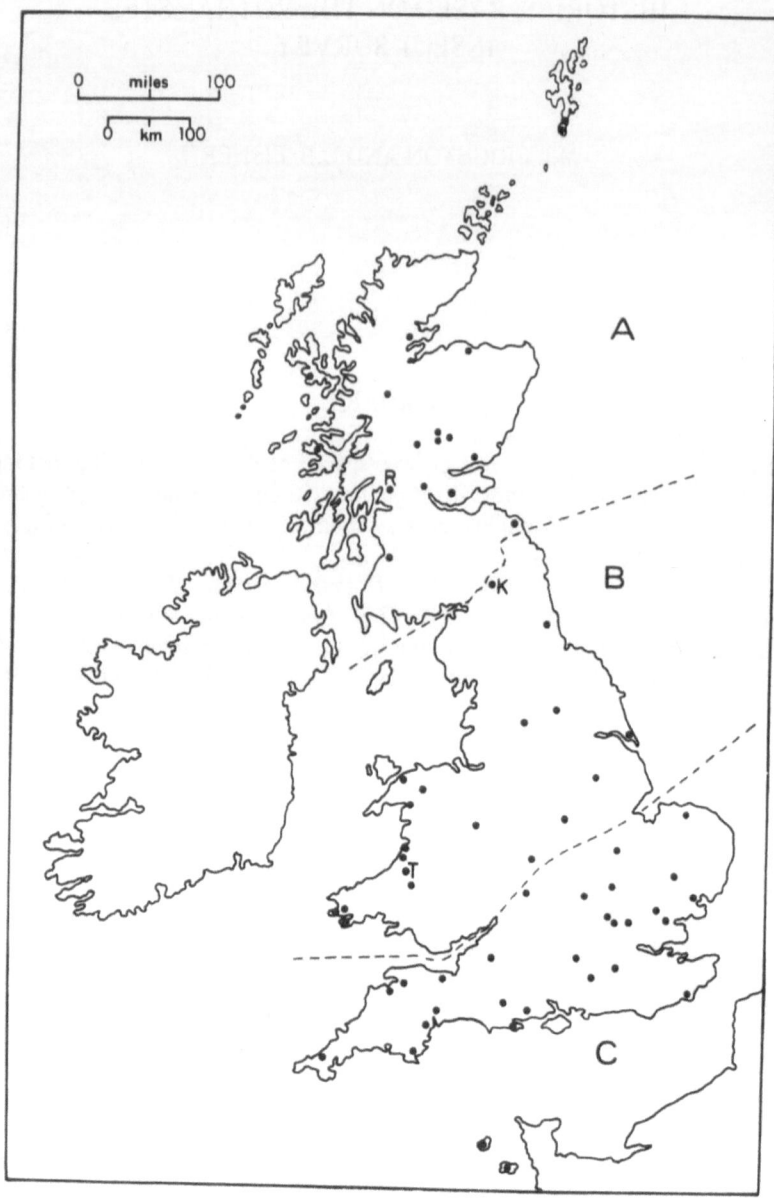

Fig. 1. Map of the British Isles showing the location of light trap sites. A. Scotland; B. Wales and Northern England; C. Southern England; K. Kielder; R. Rowardennan; T. Tregaron.

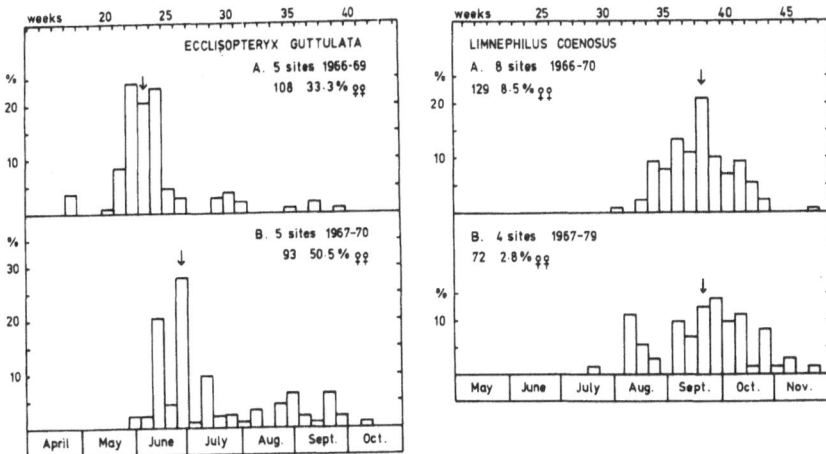

Fig. 2. Weekly catches of *Ecclisopteryx guttulata* and *Limnephilus coenosus* expressed as percentages of the total catch for each area. The arrows indicate median weeks. Total catch and percentage of females are given for each area.

part of the Rothamsted Insect Survey, which was set up in 1964 to monitor changes in insect populations, particularly of Macrolepidoptera. A few sites have been in operation continuously since the beginning, while others have operated for shorter periods. At the end of 1979 there were 165 traps in operation out of a total of 411 traps listed in the survey. Caddisflies have been recorded, for varying periods of time, from 72 sites. Details of these sites and the treatment of catches are given in Crichton (1971) and in Crichton, Fisher and Woiwod (1978). Two additional sites used in this study are No. 331, Tregaron, grid reference SN 687618, and No. 375, St Abb's Head, grid reference NT 912678. Because these two sites, and also one at Rowardennan, are referred to in the text and in Figs. 3 and 4, they are indicated separately in Fig. 1. This figure gives the location of the sites used in the present study, and also defines the three areas: A. Scotland; B. Wales and Northern England; C. Southern England.

The numbered weeks in the histograms are those of the Rothamsted Insect Survey which uses a year of exactly 52 weeks by omitting the nights of 29 February and 31 December.

RESULTS

The second category of limnephilids, with a spring and summer flight period, proposed by Crichton (1971), included the following six species:

Apatania wallengreni, Drusus annulatus, Limnephilus stigma, Rhadicoleptus alpestris, Potamophylax rotundipennis and *Micropterna lateralis.*

The following nine species were also included provisionally:

Apatania auricula, A. muliebris, Limnephilus elegans, L. extricatus,

49

Nemotaulius punctatolineatus, Potamophylax cingulatus, P. latipennis, Mesophylax impunctatus and *Hydatophylax infumatus.*

Consideration will be given first to the species now proposed for this second category. They are indicated in Table I, which lists British Limnephilidae in the order of genera and species given by Botosaneanu and Malicky (1978) in the Trichoptera section of the Limnofauna Europaea.

Apatania muliebris has been shown by Elliott (1971), in his study of its life history in the English Lake District, to have a short flight period in May and June.

Apatania wallengreni was caught at Rowardennan, on the shores of Loch Lomond, from April to June, with a pronounced peak early in May (Fig. 3). This is the earliest peak recorded for any British limnephilid. No other site in the survey yielded this species, but it has recently been recorded by Morrison (1978) at the same time of year, from the Lake of Menteith only 20 km from Loch Lomond.

The records of *Drusus annulatus* (Fig. 4) extend from May to the end of October, with the median week in August for Scotland, and in September for England and Wales. Gower (1973), in his detailed study of its life cycle

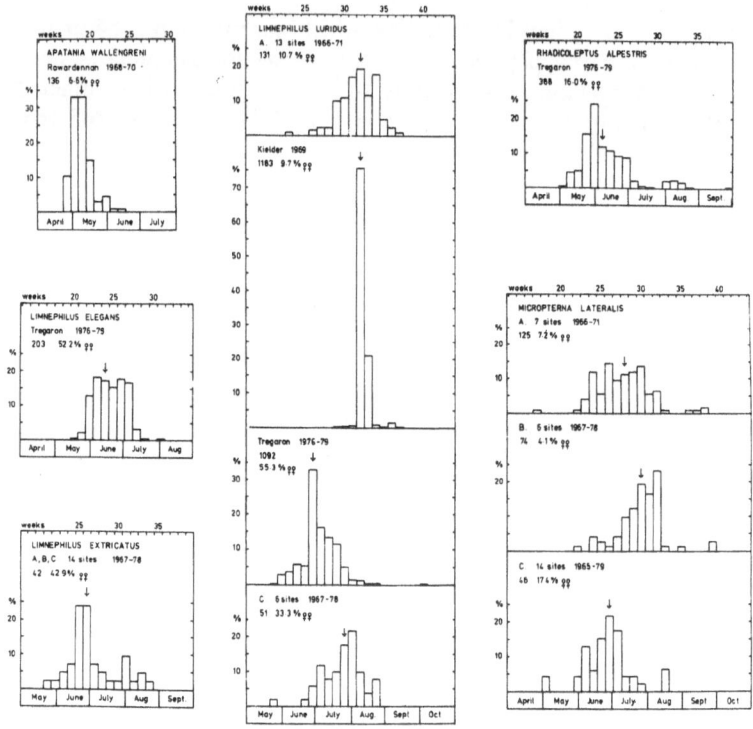

Fig. 3. Weekly catches of *Apatania wallengreni, Limnephilus elegans, L. extricatus, L. luridus, Rhadicoleptus alpestris* and *Micropterna lateralis* expressed as percentages of the total catch for each area or site. The arrows indicate median weeks. Total catch and percentage of females are given for each area or site.

50

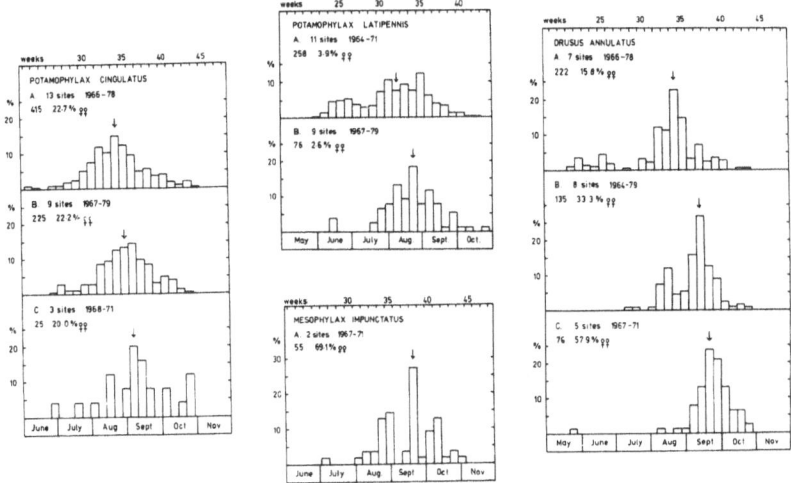

Fig. 4. Weekly catches of *Potamophylax cingulatus, P. latipennis, Mesophylax impunctatus* and *Drusus annulatus* expressed as percentages of the total catch for each area. The arrows indicate median weeks. Total catch and percentage of females are given for each area.

and larval growth in a Welsh mountain stream at 400 m above sea level, showed that the flight period extended from May to September with peak numbers in July. The later median weeks in Fig. 4 may result from the lower altitudes of the light traps in the survey. Although these records extend into autumn there is no evidence of an adult diapause, and at individual sites the flight period was comparatively short. It should be noted that in watercress beds in southern England, where the water from springs or bore holes had a constant temperature of about 10°C., *Drusus annulatus* adults could be found throughout the year, but most abundantly in April and September, so that it was considered as bivoltine under these special circumstances (Gower, 1965, 1973).

Ecclisopteryx guttulata (Fig. 2) was caught in greatest numbers in June, with much smaller catches in later months, so it is classed as a summer-flying species.

Limnephilus binotatus (formerly *L. xanthodes* McL.) was not found in the survey, but the scattered records in Britain (Kidd, 1963; Pelham-Clinton, 1964) suggest that it should be placed provisionally in the second category.

The catches of *Limnephilus elegans* in the light trap at Tregaron in Wales (Fig. 3) were concentrated in June, with much smaller numbers in May and July, thus giving a well defined summer peak. Few specimens were caught at other sites. The smaller total catch of *Limnephilus extricatus* from all areas (Fig. 3) showed a similar pattern of summer activity. This resembles that given by Svensson (1972) for the same species from a trap near Lund in Sweden, where, however, the peak period was somewhat later.

Limnephilus luridus (Fig. 3) had a short summer flight period, which was particularly well marked in the large catches from Kielder in

51

Northumbria, and also from Tregaron. Hiley (1978) stated that his studies indicated that this species was one of those which had an ovarian diapause in England, but the four histograms from the survey give no evidence of the pattern of activity associated with a diapause.

Adults of *Nemotaulius punctatolineatus* were recorded at Hillerød in Denmark, from the end of May to the middle of July by Wesenberg-Lund (1910), in his detailed study of the species. The only confirmed British records, of one male on 2 July 1965 and of one female on 18 June 1967, are from Aviemore in Scotland (Pelham-Clinton, 1966, and *in litt.*). It is therefore suggested that it should be classed as a summer-flying species in Britain, as in Denmark.

There were large catches of *Rhadicoleptus alpestris* from Tregaron (Fig. 3) with a peak in early June. It was recorded in small numbers from only four other sites in England and Wales. Holmes (1963) noted that it was fairly common for a short period in early June at Malham Tarn in Yorkshire.

Micropterna lateralis (Fig. 3) was recorded mostly in June, July and August, from 27 sites in the survey, confirming it as a summer-flying species. The catches of Svensson (1972), from Sweden, were in June and July, with peak numbers in late June.

Only ten specimens of *Hydatophylax infumatus* were recorded in the survey, in June, July and August. Dittmar (1955) gave a flight period of about one month from the middle of June, in his study of a stream in Sauerland in Germany. It would seem that this species should be regarded as being active in summer in Britain.

Consideration will now be given to *Limnephilus stigma, Potamophylax cingulatus* and *P. latipennis,* which it is proposed should be transferred from the second to the first category. Only ten speciments of *Limnephilus stigma* were caught in the survey, all in August and September. The observations of Novak and Sehnal (1963) in Czechoslovakia, and of Hiley (1978) in England, indicate a summer diapause associated with temporary waters. Both *Potamophylax cingulatus* and *P. latipennis* (Fig. 4) had long flight periods, from early June to the end of October, with peak numbers in August or early September. These two species are placed in the first category, although there appears to be no evidence of an adult diapause and they do not exhibit the build-up of numbers in September normally associated with species in this group. As explained earlier, the first category is proposed for species with an extended flight period, normally but not essentially, with a diapause. It should be added that Svensson (1972) gave early September for peak numbers of *P. cingulatus* in Sweden, where at emergence the ovaries were maturing or fully developed and there was no imaginal quiescence.

Mesophylax impunctatus had been placed provisionally in the second category. Holmes (1963) recorded it as common at Malham Tarn from June to October, and the evidence from further catches in Scotland (Fig. 4) suggests now that this is an autumn species, in the third category. Another species which should now be regarded as autumnal, at least in northern Britain, is *Limnephilus coenosus* because peak numbers were caught in September (Fig. 2). Further south in Czechoslovakia the findings of Novak and Sehnal (1963) suggested that it was not an autumnal species, but yet they did not collect it in May and June with species which had an ovarian diapause.

To summarize these proposals, the 56 species of the family Limnephilidae known in Britain are listed in Table I, where 48 of them are placed in their appropriate categories. Eight species are unplaced because there was not enough information from British localities. It should be noted that comparison has been made with localities in Denmark and the south of Sweden, which are at about the same latitude as Scotland. No reference has been made to the several papers on flight times in northern Scandinavia and also in Iceland, because weather conditions and day-length are so different from those in Britain, and result in considerable differences in life histories.

SEX PROPORTIONS

On each histogram in Figs. 2, 3 and 4 the total catch and the percentage of females have been indicated. There are considerable variations in sex proportions, but in most species more males were caught in the light traps. Species where the proportion of females was strikingly low were *Apatania wallengreni, Limnephilus luridus* in Scotland and at Kielder (but not at Tregaron), *Rhadicoleptus alpestris, Potamophylax latipennis* and *Micropterna lateralis.* This preponderance of males probably results from their greater activity at night and their wider dispersal.

DISCUSSION

Nielsen: From my findings in Denmark, *Ironoquia dubia* should be included in the first category.

Crichton: I did not have enough information from the few British records to place this species, so I am glad to have this suggestion.

Botosaneanu: *Micropterna lateralis* belongs to your second category. Are you sure there is no imaginal diapause in this species?

Crichton: I cannot be sure, but I have no evidence of an imaginal diapause in Britain.

Denis: The species of the first group you have defined involve, as you mentioned, a diapause during the imaginal stage and the females are immature when they emerge. The species of the second group have no imaginal diapause, but a larval one and, when they emerge, the females are in vitellogenesis. The third group is more difficult to explain. The species seem to have neither larval nor imaginal diapause, but I am not sure that this is true of all of them.

Crichton: Thank you for your comments. They help to explain the extent and variety of diapause. I should point out that my first category normally included an adult diapause, but this is not essential.

Gislason: Trichoptera in Iceland do not undergo a diapause. The species that emerge early in the summer (May—June), with long flight periods, inhabit waters that are not frozen solid; whereas larvae of the species that emerge

Table I. Species list of British Limnephilidae, indicating those recorded in the Rothamsted Insect Survey and a classification into the following flight periods: 1. an extended period, normally with a diapause, from spring through summer into autumn; 2. a shorter period, without a diapause, in spring and summer and sometimes extending into autumn; 3. a short period, without a diapause, in autumn. (N. indicates species referred to by Novak and Sehnal, 1963.)

	Recorded in Survey	Flight period category		
		1	2	3
Ironoquia dubia (STEPH.)				
Apatania auricula (FORSSL.)				
A. muliebris McL.	+		+	
A. wallengreni McL.	+		+	
Drusus annulatus (STEPH.)	+		+	
Ecclisopteryx guttulata (PICT.)	+		+	
Limnephilus affinis CURT.	+	+ N		
L. auricula CURT.	+	+ N		
L. binotatus CURT.				?
L. bipunctatus CURT.	+	? N		
L. borealis (ZETT.)				
L. centralis CURT.	+	+ N		
L. coenosus CURT.	+	N		+
L. decipiens (KOL.)	+	N		
L. elegans CURT.	+		+	
L. extricatus McL.	+	N	+	
L. flavicornis (F.)	+	+ N		
L. fuscicornis (RAMB.)	+	N		
L. fuscinervis (ZETT.)				
L. griseus (L.)	+	+ N		
L. hirsutus (PICT.)	+	N		
L. ignavus McL.	+	? N		
L. lunatus CURT.	+	+ N		
L. luridus CURT.	+		+	
L. marmoratus CURT.	+	+		
L. nigriceps (ZETT.)				+ N
L. politus McL.	+			+ N
L. rhombicus (L.)	+	+ N		
L. sparsus CURT.	+	+ N		
L. stigma CURT.	+	+ N		
L. subcentralis (BRAUER)		N		
L. vittatus (F.)	+	+ N		
Colpotaulius incisus (CURT.)	+	+		
Grammotaulius nigropunctatus (RETZ.)	+	+ N		
G. nitidus (MÜLL.)				
Glyphotaelius pellucidus (RETZ.)	+	+ N		
Nemotaulius punctatolineatus (RETZ.)			+	
Anabolia nervosa (CURT.)	+			+
Phacopteryx brevipennis (CURT.)				
Rhadicoleptus alpestris (KOL.)	+		+	
Potamophylax cingulatus (STEPH.)	+	+		
P. latipennis (CURT.)	+	+		
P. rotundipennis (BRAUER)	+			
Halesus digitatus (SCHRANK)	+			+
H. radiatus (CURT.)	+			+
Melampophylax mucoreus (HAG.)	+			+
Enoicyla pusilla (BURM.)				+
Stenophylax permistus McL.	+	+		

	Recorded in Survey	Flight period category		
		1	2	3
S. vibex (CURT.)	+	+		
Micropterna lateralis (STEPH.)	+		+	
M. sequax McL.	+	+		
Mesophylax aspersus (RAMB.)				
M. impunctatus McL.	+			+
Allogamus auricollis (PICT.)	+			+
Hydatophylax infumatus (McL.)	+		+	
Chaetopteryx villosa (F.)	+			+

in mid-summer (late June or July) inhabit waters that freeze to the bottom, such as swamps, small pools etc. in winter.

Crichton: This is interesting and is to be expected with the severe winter and short summer in Iceland.

ACKNOWLEDGEMENTS

The Rothamsted Insect Survey is dependent on the goodwill of many people who undertake the daily task of dealing with the light trap catches. We are especially grateful to Joan Nicklen at Rothamsted who did so much in separating caddisflies from other insets. We are also very grateful to R.A. Jenkins of the Welsh Water Authority for allowing us to use his complete records of caddisflies from three years of trapping at Tregaron. During the major part of the study, one of the authors (D.B.F.) was supported by a grant from the Natural Environment Research Council.

REFERENCES

Botosaneanu, L. and Malicky, H., Limnofauna Europaea, ed. J. Illies, Stuttgart: Gustav Fisher, 1978.
Crichton, M.I., J. Zool. Lond. 163: 533–563, 1971.
———., Fisher, D. and Woiwod, I.P., Holarctic Ecology. 1: 31–45, 1978.
Dittmar, H., Arch. Hydrobiol. 50: 305–522, 1955.
Elliot, J.M., Entomologist's Gaz. 22: 245–251, 1971.
Gower, A.M., Entomologist's mon. Mag. 101: 133–141, 1965.
———., J. Ent. (A) 47: 191–199, 1973.
Hiley, P.D., In Proc. 2nd Int. Symp. Trichoptera, ed. M.I. Crichton, pp. 297–301, Junk The Hague, 1978.
Holmes, P.F., Proc. Leeds phil. lit. Soc. 9: 31–35, 1963.
Kidd, L.N., Entomologist. 96: 246–247, 1963.
Morrison, B., Entomologist's Rec. J. Var. 90: 38–41, 1978.
Novak, K. and Sehnal, F., Cas. csl. Spol. ent. 60: 68–80, 1963.
Pelham-Clinton, E.C., Entomologist 97: 67, 1964.
———., Entomologist's Gaz. 17: 5–8, 1966.
Svensson, B.W., Oikos. 23: 370–383, 1972.
Wesenberg-Lund, C., Int. Rev. ges. Hydrobiol. Hydrog. 3: 93–114, 1910.
Williams, C.B., Proc. R. ent. soc. Lond. (A) 23: 80–85, 1948.

ACTION DE LA PHOTOPÉRIODE SUR LA MATURATION GÉNITALE DES FEMELLES DE QUELQUES LIMNEPHILIDÉS

C. DENIS

SUMMARY

The delay between emergence and beginning of vitellogenesis varies according to the photoperiod to which the females are exposed and, also, depends on the photoperiod during larval life; first during instars one to four and then during the fifth.

INTRODUCTION

L'existence, chez plusieurs espèces de Limnéphilidés, d'une diapause imaginale en été, due aux longues périodes diurnes, est un fait bien connu depuis les travaux de Novak et Sehnal (1963 et 1964). Cette diapause se traduit par une inhibition de la maturation génitale des femelles. Par des travaux antérieurs (Denis, 1972, 1973 et 1978), nous avons montré que cette inhibition ne se produisait pas si les adultes étaient exposés à des jours plus courts ou si les larves étaient élevées à une photopériode longue.

Cependant chez les femelles, immatures à l'émergence, la vitellogénèse ne commence jamais avant un certain délai et qui varie beaucoup selon les cas. Il nous a donc paru intéressant de déterminer ce délai pour des femelles maintenues dans différentes conditions de photopériode et issues de larves élevées, elles aussi, à différentes photopériodes.

La présence ou l'absence d'une diapause dépend beaucoup des conditions que les animaux ont subies au cours des étapes antérieures de leur vie. Mais en général, c'est pendant une partie seulement de leur cycle que les animaux sont sensibles aux facteurs qui conditionnent la diapause. Nous avons donc entrepris de déterminer la ou les étapes de la vie des Limnéphilidés pendant lesquelles ces animaux étaient sensibles à l'action de la photopériode. Nous avons déjà mis en évidence (Denis 1974 et 1978) la sensibilité particulière des larves du cinquième stade à cette action, chez *Limnephilus rhombicus*.

Ce travail a porté sur quatre espèces de Limnéphilidés communes en Bretagne: *Limnephilus rhombicus* Linné, *L. lunatus* Curtis, *L. centralis* Curtis et *Glyphotealius pellucidus* Retzius.

Dans toutes nos expériences, les larves et les adultes ont toujours été maintenus à une température de 15° C.

Deux étapes successives marquent la formation des oeufs dans l'ovaire. La première ou prévitellogénèse est celle où un follicule s'organise autour de chaque ovocyte formé et où les premiers follicules constitués se disposent en une série linéaire, dans chaque ovariole et commencent à augmenter de volume. La seconde ou vitellogénèse est celle où les ovocytes se chargent de vitellus et où les follicules augmentent rapidement de volume. Chez les espèces de Limnéphilidés qui subissent une diapause imaginale, les femelles, à l'émergence, sont toujours immatures et leurs ovaires contiennent des follicules à un stade plus ou moins précoce de la prévitellogénèse. La diapause se situe toujours pendant la prévitellogénèse. Et le début de la vitellogénèse indique que les femelles ont rompu leur diapause.

Dans ce travail, nous avons donc cherché à déterminer, pour chaque cas expérimental, le moment où la vitellogénèse commençait. Deux méthodes permettent de déterminer ce moment avec une assez bonne approximation. C'est soit par l'observation des premiers accouplements qui coïncident avec le début de la vitellogénèse, soit par l'examen, après dissection, de l'état du développement génital, lorsque la présence de la vitellogénèse est devenue patente (le tégument abdominal devient plus souple et transparent), en sachant que cette étape dure de deux à trois semaines.

I – *Détermination du délai entre l'émergence et le début de la vitellogénèse dans différentes conditions de photopériode.*

Les larves utilisées dans ces expériences ont effectué leur développement dans les conditions suivantes:
— Soit à une photopériode de 12 heures. Dans la nature, les animaux subissent une photopériode moyenne (comprise entre 8 et 14 heures) pendant la presque totalité de leur vie larvaire.
— Soit à une photopériode de 18 heures. Les plus longues périodes diurnes estivales sont de 16 heures.
— Soit à une photopériode de 6 heures. Les plus courtes périodes diurnes hivernales sont de 8 heures.

Les adultes ont toujours été maintenus à photopériode constante. Les durées d'éclairement utilisées étaient respectivement de 6, 12, 14, 16 et 18 heures.

Chaque expérience a porté sur une quinzaine à une trentaine de femelles.

Les données obtenues sont présentées dans les tableaux I, II, III et IV, puis matérialisées par les diagrammes de la figure 1.

Résultats

Cas des animaux dont les larves se sont développées à 12 heures. Lorsque les adultes sont maintenus à photopériode courte (6 h.) ou moyenne (12 h.), le délai entre l'émergence et le début de la vitellogénèse est relativement court (3 à 6 semaines), du moins pour les trois espèces du genre *Limnephilus*. Ce délai s'accroît et atteint plusieurs mois pour des femelles maintenues à des photopériodes plus longues.

Tableaux I, II, III et *IV*. Délai entre l'émergence et le début de la vitellogénèse pour des femelles maintenues à différentes photopériodes et issues de larves élevées, elles aussi, à différentes photopériodes.

Tableau I

Photopériode Adultes	LIMNEPHILUS RHOMBICUS			
Larves	6 ou 12 h	14 h	16 h	18 h
18 h	1,5 à 2 semaines	1,5 à 3 semaines	2 à 3 semaines	3 à 4 semaines
12 h	3 à 6 semaines	1,5 à 2 mois	2,5 à 3 mois	3 à 4 mois
6 h	3 à 6 semaines			1,5 à 2,5 mois

Tableau II

Photopériode Adultes	LIMNEPHILUS CENTRALIS			
Larves	6 ou 12 h	14 h	16 h	18 h
18 h	1,5 à 2 semaines	2 à 4 semaines	3 à 4,5 semaines	2 à 2,5 mois
12 h	3,5 à 6 semaines	2 à 2,5 mois	3 à 3,5 mois	3 à 4 mois
6 h	3,5 à 6 semaines			2 à 2,5 mois

Tableau III

Photopériode Adultes	LIMNEPHILUS LUNATUS			
Larves	6 ou 12 h	14 h	16 h	18 h
18 h	1,5 à 4 semaines	3 à 4,5 semaines	4 à 7 semaines	2,5 à 3,5 mois
12 h	3 à 6 semaines	2 à 2,5 mois	3 à 4 mois	4,5 mois au moins
6 h	3 à 6 semaines		5 mois au moins	5,5 mois au moins

Tableau IV

Photopériode Adultes	GLYPHOTAELIUS PELLUCIDUS			
Larves	6 ou 12 h	14 h	16 h	18 h
18 h	3,5 à 5 semaines		1 à 2 mois	2 à 2,5 mois
12 h	2 à 3 mois			5 mois au moins
6 h	2 à 3 mois			3,5 à 4,5 mois

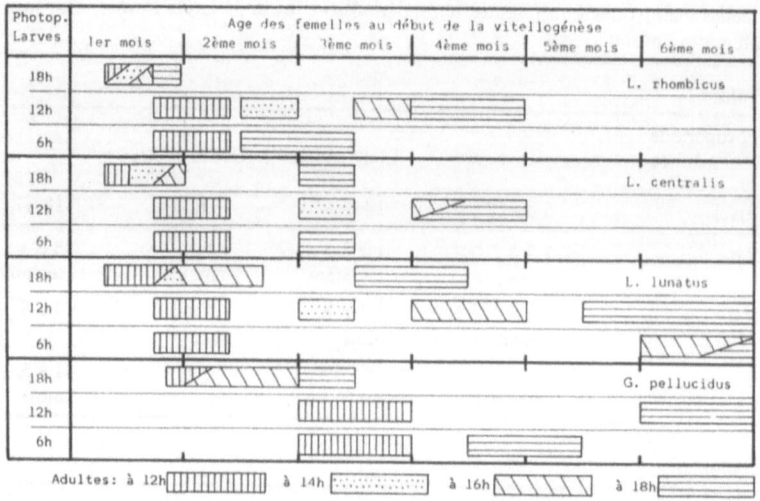

Fig. 1 – Diagrammes illustrant, pour les quatre espèces étudiées, la variabilité du délai entre l'émergence et le début de la vitellogénèse en fonction de la photopériode appliquée pendant la vie larvaire et durant le stade imaginal.

Donc à une photopériode courte ou moyenne, les femelles ont apparemment une maturation génitale directe, tandis qu' à une photopériode longue, elles subissent une diapause. Cette dernière commence à s'installer lorsque les périodes diurnes atteignent une durée de 14 heures. Aux latitudes de la France, ceci intervient vers le 20 avril. Les espèces étudiées ici émergeant au mieux en fin avril, mais surtout en mai, juin et même juillet, elles subissent donc obligatoirement une diapause.

La diapause peut être rompue malgré le maintien du facteur déclenchant; toutefois plus la photopériode est longue et plus la rupture de la diapause est retardée.

Dans le cas de *Glyphotaelius pellucidus,* le délai d'au moins deux mois entre l'émergence et le début de la vitellogénèse à photopériode courte ou moyenne ne permet pas d'affirmer que les femelles ont alors un dévellopement direct. L'état d'immaturité extrêmement poussé des femelles, à l'émergence (fig. 2), peut expliquer la longueur de ce délai, de même que le temps particulièrement long pour obtenir la rupture de la diapause à photopériode longue.

Cas des animaux dont les larves se sont développées à 18 heures. Nous constatons, par rapport aux animaux dont les larves se sont développées à 12 heures, une diminution générale et importante du délai entre l'émergence et le début de la vitellogénèse. Cependant ce délai demeure variable en fonction de la durée d'éclairement des adultes. Et il est toujours plus bref pour des femelles maintenues à photopériode courte ou moyenne que pour celles qui sont maintenues à photopériode longue. De plus ces délais sont,

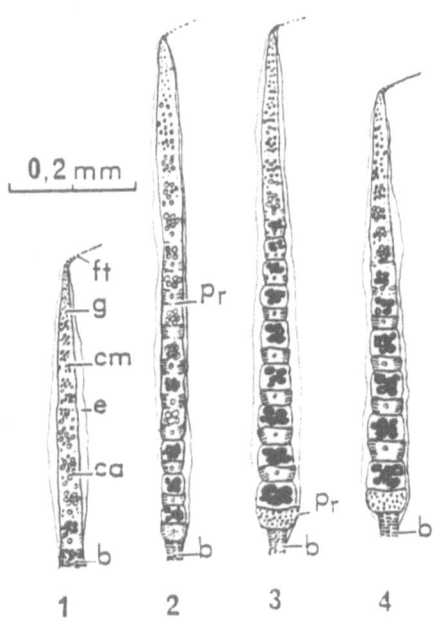

Fig. 2. Ovarioles et état de maturation des cellules sexuelles des quatre espèces de Limnéphilidés à la mue imaginale (d'après Le Lannic 1976). 1 – *Glyphotaelius pellucidus*, 2 – *Limnephilus lunatus*, 3 – *L' rhombicus* et 4 – *L' centralis*. b: bouchon du vitellarium, ca: complexes alignés dans la zone dilatée du vitellarium, cm: complexes mêlés dans la zone rétrécie du vitellarium, e: enveloppe épithéliale, ft: filament terminal, g: germarium, pr: follicule en prévitellogénèse.

d'une manière générale, très courts pour les femelles de *L. rhombicus;* ils sont nettement supérieurs pour celles des trois autres espèces lorsqu'elles sont maintenues à 18 heures.

Cas des animaux dont les larves se sont développées à 6 heures. Nous constatons, par rapport aux animaux dont les larves ont été élevées à 12 heures, une certaine diminution du délai entre l'émergence et le début de la vitellogénèse pour des femelles maintenues à photopériode longue chez *L. rhombicus, L. centralis* et, à un moindre degré, chez *G. pellucidus.* D'ailleurs, dans le cas de *L. centralis,* cette diminution est aussi importante que lorsque la vie larvaire se déroule à photopériode longue. Donc l'élevage des larves à une photopériode inférieure aux plus courtes existant aux latitudes de la France, diminue l'intensité de la diapause chez les trois espèces précédentes. Par contre, cela ne produit aucun effet sur les femelles de *L. lunatus.*

Remarque: Les tableaux I, II, III et IV montrent que dans des conditions données, les femelles d'une même espèce n'ont pas toutes exactement le même âge lorsqu'elles entrent en vitellogénèse. Ces écarts sont vraisemblablement dus, en partie, aux différences individuelles existant dans l'état d'immaturité de l'appareil génital.

II — *Mise en évidence des stades de la vie larvaire sensibles à l'action de la photopériode.*

Cette seconde partie des recherches a porté principalement sur *L. rhombicus* et *L. lunatus.* Nous avons toutefois effectué quelques expériences, à titre comparatif, sur les deux autres espèces. Les données obtenues sont présentées dans les tableaux V, VI, VII et VIII.

Tableau V. Délai entre l'émergence et le début de la vitellogénèse pour des femelles maintenues à différentes photopériodes et issues de larves ayant subi un changement de photopériode au cours de leur développement. Les données du tableau I sont rappelées à titre de comparaison avec les résultats des présentes expériences.

	Photopériode Adultes	LIMNEPHILUS RHOMBICUS		
	Larves	12 h	16 h	18 h
1	Tout à 18 h	1,5 à 2 semaines	2 à 3 semaines	3 à 4 semaines
2	(1 à 4) 12 h (5) 18 h		2,5 à 4 semaines	4 à 6 semaines
3	(1 à 4) 18 h (5) 12 h			2,5 à 3,5 mois
4	Tout à 12 h	3 à 6 semaines	2,5 à 3 mois	3 à 4 mois
5	(1 à 4) 12 h (5) 6 h			2,5 à 3,5 mois
6	Tout à 6 h	3 à 6 semaines		1,5 à 2,5 mois
7	(1 à 4) 6 h (5) 12 h		1,5 à 2 mois	1,5 à 2 mois
8	(1 à 4) 6 h (5) 18 h	1,5 à 2 semaines	2 à 3 semaines	4 à 6 semaines

Résultats

Cas de Limnephilus rhombicus (Tableau V). La comparaison entre les lignes 3 et 4 montre que l'exposition des quatre premiers stades à une photopériode de 18 heures ne modifie pas sensiblement la durée de la diapause. Cela a même un effet défavorable sur le déroulement du 5ème stade dont la durée s'accroît et pendant lequel la mortalité est alors élevée.

Par contre la comparaison entre les lignes 1, 2 et 3 montre qu'il est nécessaire et suffisant que le 5ème stade se déroule à une photopériode longue pour supprimer la diapause.

L'examen des lignes 4, 5, 6 et 7 indique que c'est en agissant sur les quatre premiers stades, qu'une photopériode courte diminue très nettement

62

la durée de la diapause. Il restera à déterminer si, pour obtenir un tel résultat, la photopériode courte doit agir sur l'ensemble de ces quatre stades ou sur certains seulement.

Enfin les actions conjuguées d'une photopériode courte sur les quatre premiers stades et d'une photopériode longue sur le cinquième (ligne 9) suppriment la diapause; mais la comparaison des données obtenues dans ce cas avec celles présentées aux lignes 1 et 2 semble indiquer que la combinaison de ces deux actions ne produit pas d'effets cumulés.

Tableau VI. Même légende que le tableau V. dt.5: début de 5ème stade; en fait, deux semaines sur un stade qui dure au moins deux mois.

	Photopériode Adultes Larves	LIMNEPHILUS CENTRALIS 12 h	16 h	18 h
1	Tout à 18 h	1,5 à 2 semaines	3 à 4,5 semaines	2 à 2,5 mois
2	(1 à 4) 12 h (5) 18 h	1,5 à 2 semaines		2 à 2,5 mois
3	(1 à 4) 18 h (5) 12 h	Forte perturbation. Animaux tous morts avant ou pendant nymphose		
4	Tout à 6 h	3,5 à 6 semaines		2 à 2,5 mois
5	(1 à dt. 5) 6 h (suite) 12 h			2 à 2,5 mois
6	(1 à dt. 5) 6 h (suite) 18 h		4 à 6 semaines	5 à 7 semaines

Cas le Limnephilus centralis (tableau VI). Nous constatons de nombreuses ressemblances avec le cas de *L. rhombicus*:

(a) Après exposition des larves du cinquième stade à une photopériode longue (ligne 2), le délai entre l'émergence et le début de la vitellogénèse est comparable à celui obtenu lorsque toute la vie larvaire se déroule à cette même photopériode.

(b) L'exposition des quatre premiers stades à une photopériode longue suivie d'une diminution de la durée d'éclairement (ligne 3) perturbe le développement et, semble-t-il, encore plus profondément que chez *L. rhombicus.*

(c) Après exposition des quatre premiers stades seulement à une photopériode courte (ligne 5), le délai entre l'émergence et le début de la vitellogénèse est diminué autant que dans le cas où toute la vie larvaire s'est déroulée à cette même photopériode (ligne 4).

Mais contrairement à *L. rhombicus,* chez *L. centralis* les actions conjuguées d'une photopériode courte sur les quatre premiers stades et d'une photopériode longue sur le cinquième (ligne 6), diminuent fortement le délai entre l'émergence et le début de la vitellogénèse. Il y a donc ici effets cumulés des deux actions, conduisant à une véritable suppression de la diapause.

Cas de Limnephilus lunatus (Tableau VII). La comparaison entre les lignes

Tableau VII. Même légende que pour le tableau V. dt.5: début du 5ème stade; 7 à 10 jours sur un stade qui dure de 30 à 40 jours.

	Photopériode Adultes		LIMNEPHILUS LUNATUS	
	Larves	12 h	16 h	18 h
1	Tout à 12 h	3 à 6 semaines	3 à 4 mois	4,5 mois au moins
2	(1 à 4) 12 h (5) 18 h		5 à 8 semaines	3 mois au moins
3	Tout à 18 h	1,5 à 4 semaines	4 à 7 semaines	2,5 à 3,5 mois
4	(1 à 4) 18 h (5) 12 h		3 à 7 semaines	1 à 2 mois
5	(1 à dt. 5) 18 h (suite) 12 h	1,5 à 2 semaines	2,5 à 7 semaines	3 à 7 semaines
6	(1 à 2) 12 h (3 à dt. 5) 18 h (suite) 12 h	1,5 à 2 semaines	3 à 6 semaines	6 à 8 semaines

2 et 3 montre que lorsqu'une photopériode longue agit sur le cinquième stade seulement, les délais entre l'émergence et le début de la vitellogénèse sont identiques à ceux obtenus après action de cette même photopériode sur l'ensemble de la vie larvaire. Mais ces délais sont les plus réduits après l'action d'une photopériode longue sur les quatre premiers stades, suivie d'une diminution de la durée d'éclairement (ligne 4). L'effet est particulièrement sensible sur des femelles maintenues à 18 heures. Les délais sont alors voisins de ceux les plus courts obtenus pour *L. rhombicus*.

Donc l'action d'une photopériode longue sur le cinquième stade diminue la durée de la diapause; mais pour supprimer véritablement cette dernière, il faut successivement qu'une photopériode longue agisse sur les quatre premiers stades puis qu'une diminution de la photopériode se produise au dernier stade larvaire. Et *L. lunatus* diffère donc profondément de *L. rhombicus* en ce qui concerne la détermination de la chronologie du développement génital des femelles.

Afin de préciser les conditions qui permettent une suppression de la diapause, nous avons effectué des expériences complémentaires. Dans le cas dont les résultats figurent à la ligne 5, les diminutions de la photopériode intervenaient sept à dix jours ou même douze à quinze jours après le dernière mue larvaire. (Dans nos expériences, le 5ème stade durait une trentaine à une quarantaine de jours.) Les expériences dont les résultats sont présentés aux lignes 5 et 6 montrent que pour supprimer la diapause, il suffit d'une part, que les troisième et quatrième stades se déroulent à photopériode longue et, d'autre part, que la diminution ultérieure de la photopériode intervienne durant la première moitié du dernier stade.

Cas de Glyphotaelius pellucidus (tableau VIII). Les quelques expériences réalisées sur cette espèce montrent que:

Tableau VIII. Même légende que pour le tableau V.

	Photopériode Adultes	GLYPHOTAELIUS PELLUCIDUS		
	Larves	12 h	16 h	18 h
1	Tout à 18 h	3,5 à 5 semaines	1 à 2 mois	2 à 3 mois
2	(1 à 4) 12 h (5) 18 h			2 à 3 mois
3	(1 à 4) 18 h (5) 12 h		2,5 à 3 mois	3 à 3,5 mois
4	Tout à 12 h	2 à 3 mois		5 mois au moins

(a) Après exposition des larves du cinquième stade à une photopériode longue (ligne 2), le délai entre l'émergence et le début de la vitellogénèse est aussi court que lorsque toute la vie larvaire se déroule à photopériode longue. Ce résultat rapproche le cas de *G. pellucidus* de ceux de *L. rhombicus* et de *L. centralis*.

(b) L'exposition des quatre premiers stades à une photopériode longue suivie d'une diminution de la photopériode (ligne 3) provoque une réduction de la durée de la diapause, ce qui rapproche un peu *G. pellucidus* de *L. lunatus;* mais l'effet est beaucoup moins important chez la première espèce que chez la seconde.

CONCLUSIONS

De cette étude qui montre combien les espèces réagissent différemment les unes des autres à l'action de la photopériode, nous retiendrons les points suivants:

1) Il est difficile de distinguer de façon rigoureuse les cas où la diapause est présente et ceux où elle est réellement supprimée. Il importe surtout de constater la grande variabilité dans la chronologie du développement génital des femelles en fonction des conditions de photopériode qu'elles subissent à l'état larvaire, comme à l'état adulte.

2) Chez les Limnéphilidés qui subissent une diapause imaginale sous l'effet des jours longs, la chronologie de la maturation génitale femelle est réglée de façon différente d'une espèce à l'autre. Cependant les résultats obtenus conduisent à penser que les mécanismes dont dépend cette chronologie présentent des éléments communs, mais aussi des différences. Les éléments communs sont:

— Chez les larves, l'existence de deux périodes sensibles à l'action de la photopériode. La première se situe durant les quatre premiers stades et la seconde correspond au cinquième stade.

— Chez les adultes, le fait de l'accroissement progressif du délai entre l'émergence et le début de la vitellogénèse en fonction de l'augmentation de la photopériode.

Les différences résident:

− Chez les larves, dans le fait que c'est l'une ou l'autre des deux périodes sensibles qui joue le rôle prépondérant dans le conditionnement des réactions des adultes. Dans le fait, également, qu'une même photopériode peut produire, au cours d'une période donnée, des conditionnements différents selon les espèces.

− Chez les adultes, dans le fait que l'augmentation progressive du délai entre l'émergence et le début de la vitellogénèse, en fonction de la photopériode, est plus ou moins rapide selon les espèces.

L'existence de ces mécanismes peut aider à comprendre pourquoi les espèces septentrionales, notamment celles qui vivent en Islande, ont un développement génital femelle direct (Gislason 1977 et 1978).

Mais la présence de ces mécanismes chez les espèces qui vivent aux latitudes moyennes devrait leur permettre d'avoir une maturation génitale directe et nous devons nous demander pourquoi elles subissent une diapause. L'assèchement temporaire de certains lieux où vivent ces Trichoptères n'est pas un argument suffisant, car ces derniers fréquentent également des cours d'eau et des étangs permanents. La diapause s'explique en partie par le fait que, dans la nature, la presque totalité du cinquième stade se déroule à l'époque des photopériodes moyennes. Mais si le cycle de *L. lunatus* commençait au début de l'été, les quatre premiers stades auraient lieu pendant les jours longs − la température élevée de l'eau et l'abondance de la nourriture favorisant une croissance rapide − et le cinquième stade se déroulerait pendant les jours courts. De telles conditions de développement conduiraient à une maturation génitale femelle directe. Mais le cinquième stade aurait alors lieu en hiver. Or c'est pendant la seconde moitié de ce stade que se produit la spermatogénèse, et Le Lannic (1976) a montré qu'une basse température perturbe fortement le déroulement de ce processus, surtout chez *L. lunatus*. D'autre part, dans la descendance d'animaux ayant effectué leur cycle sans diapause, nous avons constaté des anomalies du développement génital de quelques femelles; mais leurs causes restent à déterminer.

REFERENCES

Denis, C., Bull. Soc. Sci. Bretagne 47: 33−38, 1972.
─────., Bull. Soc. Sci. Bretagne 48: 197−207, 1973.
─────., Proc. of the 2nd Int. Symp. Trich., 109−115, 1977.
Gislason, M., Proc. of the 2nd Int. Symp. Trich., 135−146, 1977.
─────., Verh. Internat. Verein. Limnol. 20: 2622−2629, 1978.
Le Lannic, J., Thèse Dr 3ème Cycle, Rennes, 1976.
Novak, K. et Sehnal, F., Cas. csl. Spol. ent. 60: 68−80, 1963.
─────., ─────., Proc. XII Int. Congr. Ent. London, p. 434, 1964.

CHANGES IN THE PROFUNDAL TRICHOPTERA OF LAKE WINNIPEG 1928–32 TO 1969

J.F. FLANNAGAN AND D.G. COBB

SUMMARY

During the open-water season of 1969, the offshore benthos of Lake Winnipeg was sampled approximately monthly at up to 60 stations. Comparison of the Trichoptera collected during this investigation with the 1928–32 collections by Neave indicate that major qualitative and quantitative changes have occurred. *Molanna flavicornis* and *Phryganea cinerea* have both disappeared completely from the South Basin and both show distributions more restricted in 1969 than in the earlier survey. Although average densities appear to be reduced the maximum densities recorded in both surveys are similar.

Oecetis spp. imagines are recorded from the area by Neave from collections in 1927–32, however, he did not find larvae in his profundal benthic samples. In our study *Oecetis inconspicua* was absent from the main part of the North Basin, was relatively common in the Narrows, and was widespread and occurred in densities up to $180/m^2$ in the South Basin.

It seems likely that the changes in the Trichoptera fauna of the lake are related to the changed agricultural, industrial and/or domestic activities of man. However the effects of natural biological or physical phenomena cannot be discounted.

INTRODUCTION

Lake Winnipeg, which is a remnant of Glacial Lake Agassiz, is located in central Canada (Fig. 1). It is the 13th largest lake in the world, 7th largest in N. America (Table I). The lake is essentially isothermal during the open-water period and receives a very high nutrient loading from the rivers which enter it (Table I), especially those from the west and south. However, at least in the South Basin, primary production appears to be limited by turbidity, not nutrient supply.

Some major physical, chemical and biological changes have occurred in the Lake since the last extensive surveys of the benthos (Bajkov 1930; Neave 1932, 1933, 1934). Flannagan (1979) reported extensive reductions in the

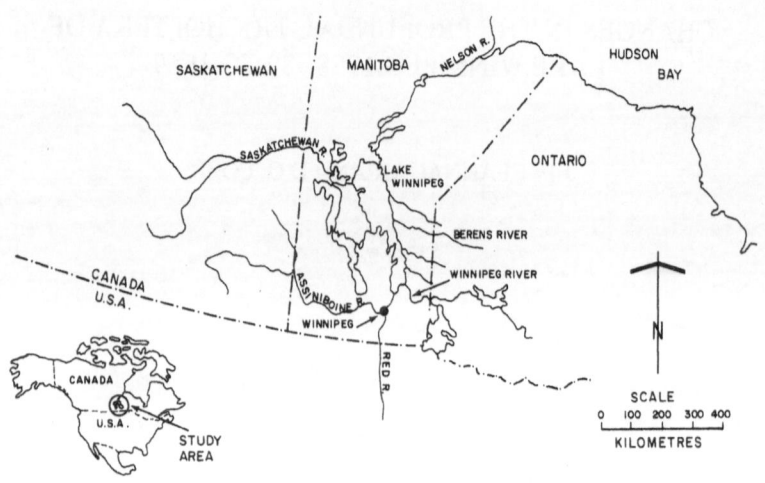

Fig. 1. The location of Lake Winnipeg, Manitoba, Canada.

Table I. Selected physical and chemical characteristics of Lake Winnipeg.*

Maximum length	436 km
Maximum breadth	111 km
Surface area	23 750 km^2
Mean depth	12.0 m
Maximum depth	36 m
Water renewal time	3.1 yrs
Secchi disc	
North Basin	0.5–2.6 m
South Basin	0.1–1.0 m
Dissolved oxygen	near saturation throughout the year at all depths
Nutrient loading	
P	5 000 tonnes/yr
N	62 000 tonnes/yr

*from Brunskill (1973); Brunskill, Schindler, Elliot and Campbell (1979); Brunskill et al. (1979, 80).

density of the mayflies (*Hexagenia rigida* and *H. limbata*) and a change in their relative abundance. These changes may be indirectly related to increases in the depth of the Lake (mean elevation increased from 216.7 meters above sea level in 1931 to about 218.4 m a.s.l. at the present time). The damming of the large, turbid Saskatchewan River, which flows into the North Basin of the Lake, has caused the suspended solids in the river and, subsequently, the North Basin of the Lake to be considerably reduced. The transparency of the waters of this basin has increased from < 1 m in Bajkov's (1930) time to up to 3 m in 1969. In contrast, it appears that the transparency of the South Basin has markedly decreased over the same period (Brunskill 1979a, b). In addition parts of the commercial fisheries on the lake were closed in the years 1969–72, because of mercury contamination, particularly from the Winnipeg River. Further, the Lake catchment basin contains a fairly large human population and it has been suggested that human activities may be hastening the eutrophication process in the lake (Jo-Anne Crowe, Manitoba Government, personal communication).

The above changes in Lake Winnipeg, the long interval since Neave's and Bajkov's survey, and the importance of the commercial fish catch from the Lake (over 10% of the total freshwater fishery catch in Canada, $\simeq 4{,}700$ metric tonnes/yr[1]) led to a joint survey, in 1969, of the limnology of the lake by staff of the Freshwater Institute. The present paper contains Trichoptera results from this survey and compares these results with those of Neave (1933, 34).

The Trichoptera of the Lake Winnipeg area are listed in Neave (1934). He also carried out an extensive study of the two deep-water species, *Phryganea cinerea* and *Molanna flavicornis* (Neave 1933). In this latter study Neave concluded that *P. cinerea* was distributed throughout the whole lake and was not limited by depth, while *M. flavicornis*, although occurring in the deepest parts of the lake and from the extreme south to the extreme north, was apparently limited in its distribution to areas in which coarse sand for case building was present.

METHODS

Lake Winnipeg was sampled approximately monthly at up to 60 offshore stations during the open-water season of 1969 (see Fig. 3 for station locations). At each station, attempts were made to take three 15 cm Birge-Ekman grabs, however, when the substrate was too hard or too coarse the samples were taken with a Ponar grab. Samples were sieved through a 200 μm mesh sieve when possible, but when sandy substrate were encounted a 400 μm mesh sieve was used. Sample residues were preserved in 4–10% formalin and the animals subsequently sorted out under the low power of a dissecting microscope. Further details on sampling procedures, station locations, depths etc. are described in Flannagan (1979). For ease of comparison with other studies, and to delineate more clearly changes in benthic faunal composition, the lake is divided in this paper, as ih Flannagan (1979), into its three principal physical areas – North Basin, Narrows and South Basin (Fig. 3).

RESULTS

Eight of the eleven taxa of Trichoptera larvae collected during this survey were rare, occurring only sporadically or in very low densities. However, since they are mainly taxa considered as being generally lotic (MacKay and

[1] D.M. Cauvin, Freshwater Institute, personal communication.

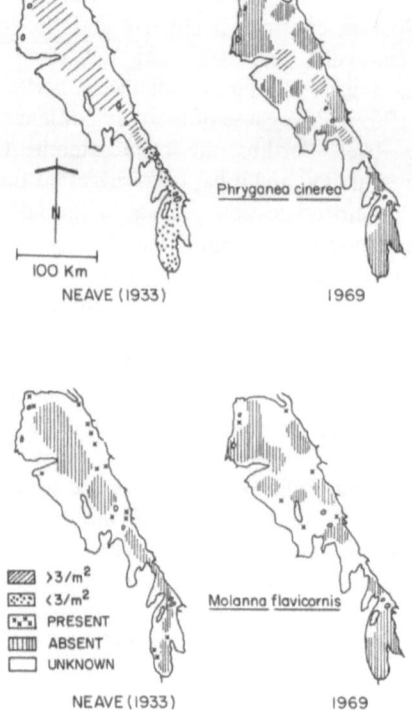

Fig. 2. A comparison of the density and distribution of the two species of profundal caddisflies in 1969 with those given by Neave (1933 June and July).

Wiggins 1979, and others) it seems important to list their presence and depth or range of occurrence in the lake: *Neureclipsis sp.* (16 m), *Polycentropus cinereus* Hagen (6–8 m), *Phylocentropus* sp. (4 m), *Cheumatopsyche* sp. (6 m), *Hydropsyche* sp. (4–10 m), *H. recurvata* Banks (12–15 m), *H. slossonae* Banks (5 m) and *Ceraclea* sp. (6 m).

Molanna flavicornis Banks, *Phryganea cinerea* Walker and *Oecetis* (*inconspicua* McLachlan?) were abundant in at least part of the Lake, *M. flavicornis* and *P. cinerea* were present in the North Basin and Narrows, but absent in the South Basin (Fig. 2). *O. inconspicua* was very common in the South Basin, but limited in distribution in the North Basin (Fig. 3).

The two profundal species, *M. flavicornis* and *P. cinerea*, have disappeared completely from the South Basin and at least *P. cinerea* has a more restricted distribution in the North Basin than previously when compared to the results of Neave (1933) (Fig. 2). The maximum densities of these two species, where they still occur, are similar to those recorded by Neave (1933) but the mean densities (using only samples collected at equivalent times of the year) (Table II) appear to be reduced throughout the whole lake.

Examination of the depth distribution of *O. inconspicua* throughout the

70

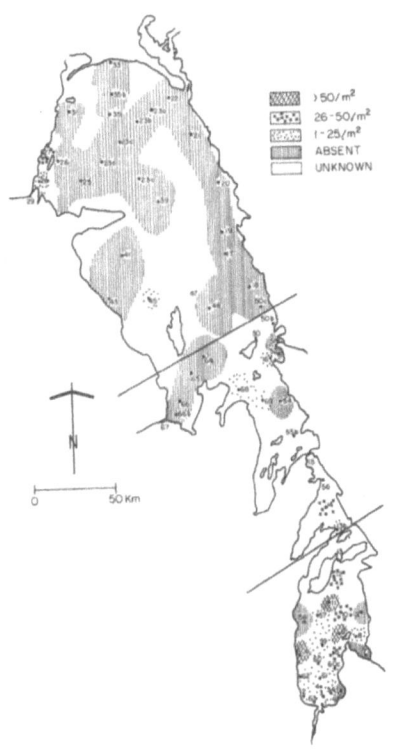

Fig. 3. Mean density distribution (all samples) of *Oecetis inconspicua*, summer 1969. Station locations for 1969 survey.

Table II. Comparison of profundal Trichoptera densities; Neave (1933) and present survey.

	Neave (1933) June and July 1928–32 #/m²	Present survey[1] June and July 1969 #/m²	Present survey[2] June and July 1969 #/m²
Phryganea cinerea			
North Basin	4.7	0.6	5.9
Narrows	not listed	0.6	4.9
South Basin	2.4	0	0
Molanna flavicornis			
North Basin	10–15	1.7	8.2
Narrows	10–15	0.8	4.2
South Basin	10–15	0	0

[1] Mean of all grabs taken during this period.
[2] Mean of grabs only where Trichoptera were present during this period.

71

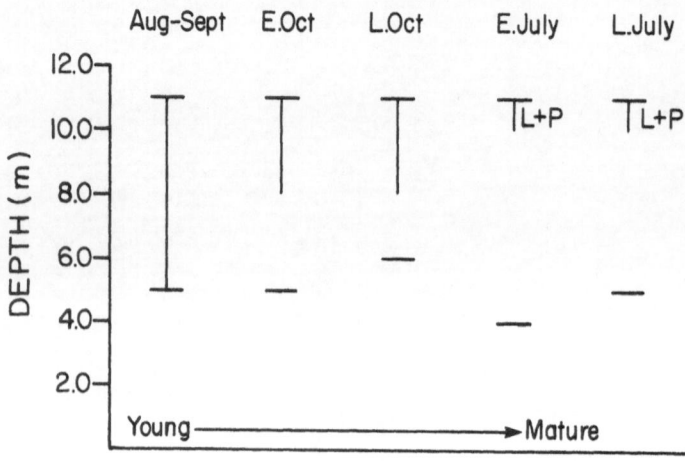

Fig. 4. Depth distribution of *Oecetis inconspicua* throughout its life cycle. (Horizontal bars indicate depth sampled on each sampling cruise, vertical bars indicate depth range over which specimens were collected.)

open-water period (Fig. 4), indicates that the smaller animals occur at the widest depth ranges while the mature larvae and pupae are restricted to the deeper water. Since in all samples the densities appeared to be relatively constant, throughout the summer, at the deep water stations, and since pupae were never found at the shallow stations, it appears that the depth restriction is brought about by differential survival rather than migration from, or earlier hatching at, the shallow stations.

DISCUSSION

Two other *Oecetis* species are mentioned, as adults, from the Lake Winnipeg area by Neave (1934) in collections from 1927–32, but larvae were apparently not collected in his Lake Winnipeg samples in 1928–32 (Neave 1933). In our study *O. inconspicua* was absent from most of the North Basin of the lake, but was locally common in the north-west and west parts of this basin, was common in the Narrows ($1-25/m^2$), and in the South Basin it was widespread and occurred up to $180/m^2$. Thus it seems unlikely that this species occurred in or near the Lake in Neave's time.

Direct comparisons of the density of caddisflies in our samples with those recorded by Neave (1933) is quite difficult, since Neave did not give exact locations for his samples. However, from his density/distribution maps, it is obvious that the two species of profundal caddisflies in Neave (1933) have become considerably more restricted in their distribution in the Lake, and may well be considerably reduced in density where they are still present, even if we only consider the maximum possible densities in column 3, Table II. Both *M. flavicornis* and *P. cinerea* are now completely absent from the South Basin of the Lake and at least *P. cinerea* and probably *M. flavicornis* appear to be more restricted in density and distribution in the North Basin. Since the major pollution sources to the Lake enter the South Basin (runoff from agricultural land to the south, industrial and domestic sewage effluents via the Red R., and mercury and other forestry and wood processing by-products via the Winnipeg R.) and since the changes in the original two caddis species are greatest in the South Basin and decrease, with the drainage direction of the Lake to the north, it appears that the faunal changes from 1928–32 to 1969 may well be linked to an increase in pollution originating in the populated southern part of the province. This, of course, would suggest that *O. inconspicua* is less sensitive to the pollutants than *P. cinerea* and *M. flavicornis*. Alternately, since *P. cinerea* and *M. flavicornis* are both omnivores (Neave 1933) while *O. inconspicua* appears to be largely carnivorous (Winterbourne 1971) it is possible that the increase in turbidity in the South Basin has reduced the amount of algal and organic detrital material available as food for the former two species, thus allowing the incursion of the more carnivorous *O. inconspicua*. However, this alone would not explain the invasion of *O. inconspicua* into the now more transparent North Basin of the lake. The increased densities of oligochaetes, recorded recently by Provincial Government biologists (Jo-Anne Crowe, personal communication), in the South Basin of the Lake may favour the more strongly carnivorous *O. inconspicua*.

Although it is not possible, because of the lack of published pollution tolerance data on these three species of caddisflies, and because of the extent and variety of physical and chemical changes in the lake over the forty-year period between the surveys, to positively determine the cause(s) of the changes in the Trichoptera, these changes are consistent with changes in the other benthic animals in the lake (Flannagan 1979, several papers on *Pontoporeia hoyi, Hexagenia* sp., in prep.).

ACKNOWLEDGEMENTS
We would like to thank Mr S.S. Chang, who did all of the sampling, Ms G. Decterow and G. Porth for typing the manuscripts, Ms D. Taite for preparing the figures, Dr D.M. Rosenberg and Mr A.P. Wiens for providing useful criticisms of the manuscript, Ms Jo-Anne Crowe for making manuscripts and other unpublished data available to us and Dr Glenn Wiggins for identifying specimens of *Oecetis inconspicua*.

REFERENCES
Bajkov, A., Contr. Can. Biol. Fish. 5: 381–422, 1930.
Brunskill, G.J., Verh. Int. Ver. Limnol. 18: 1755–1759, 1973.
———., Campbell, P. and Elliott, S.E.M., Can. Fish. Mar. Serv. MS Rep. 1526, 1979a.
———., Schindler, D.W., Elliott, S.E.M. and Campbell, P., Can. Fish. Mar. Serv. MS Rep. 1522, 1979b.
———., Elliott, S.E.M. and Campbell, P., Can Fish. Mar. Serv. MS Rep. 1556, 1980.
Flannagan, J.F., *In* Proc. 2nd Int. Conf. on Ephemeroptera, eds K. Pasternak and R. Sowa, pp. 103–114, Krakov, Poland, 1979.
MacKay, R.J. and Wiggins, G.B., Ann. Rev. Entomol. 24: 185–208, 1979.

Neave, F., Contr. Can. Biol. Fish. 7: 117–201, 1932.
————., Int. Rev. ges. Hydrobiol. Hydrograph. 29: 17–28, 1933.
————., Int. Rev. ges. Hydrobiol. Hydrograph. 31: 157–170, 1934.
Winterbourn, M.J., Can. J. Zool. 49: 623–635, 1971.

DISCUSSION

Badcock: You mentioned finding various hydropsychid species in Lake Winnipeg. In Britain, I have not found hydropsychid larvae in lakes, in Britain they are essentially running water species with an inherent current demand. In Lake Winnipeg, are the larvae found around the river mouths or are there appreciable currents in the lake or are the species there adapted to still water?

Flannagan: Hydropsychids occur, especially along the shorelines, of large and some of the smaller, northerly lakes in Canada. In the Lake Winnipeg study they were not particularly associated with river mouths or with the shoreline. Their distribution in the profundal zones of Lake Winnipeg is probably due to the strong currents throughout the lake.

STUDIES OF NEOTROPICAL CADDISFLIES, XXVII: ANOMALOPSYCHIDAE, A NEW FAMILY OF TRICHOPTERA

O.S. FLINT, JR

SUMMARY

The male and female, larva, and pupa of *Anomalopsyche minuta* (Schmid) are described and figured. The new family, Anomalopsychidae, is proposed for this species and, tentatively, *Contulma cranifer* Flint. Additionally, undescribed and unassociated larvae and adults from Brazil, Colombia and Ecuador belong to this family. The family is the first undoubted member of the leptocerid branch of the Limnephiloidea found to possess ocelli. Although relationships are not absolutely certain, this neotropical taxon appears to be related to the Beraeidae and Helicophidae.

INTRODUCTION

Recent collecting activities by entomologists of the National Museum of Natural History in Chile and adjacent Argentina have resulted in the discovery of many new or poorly known forms (Flint, 1967, 1969, 1974, 1979). One of the first discoveries was a small caddisfly, originally placed in the Sericostomatidae, but subsequently discovered to possess ocelli in the adult stage, a condition totally out of place in the Sericostomatidae and related families. Larvae and pupae, including many male and female metamorphotypes, were soon discovered inhabiting moss on the upper surfaces of stones in cold, rapidly flowing Andean brooks. Study of all life stages has convinced me that this species represents a distinct new family.

Family: Anomalopsychidae, fam. nov.

Type genus: *Anomalopsyche* Flint, 1967

Anomalopsyche minuta (Schmid)

Figs. 1–15, 19 –35

Myotrichia minuta Schmid, 1957, p. 392.

Anomalopsyche ocellata Flint, 1967, p. 66.

Anomalopsyche minuta (Schmid): Flint, 1974, pp. 84, 91.

During the preparation of this paper, I reinvestigated the adults and give the following information to supplement, or correct, the original descriptions. The larvae and pupae are described for the first time.

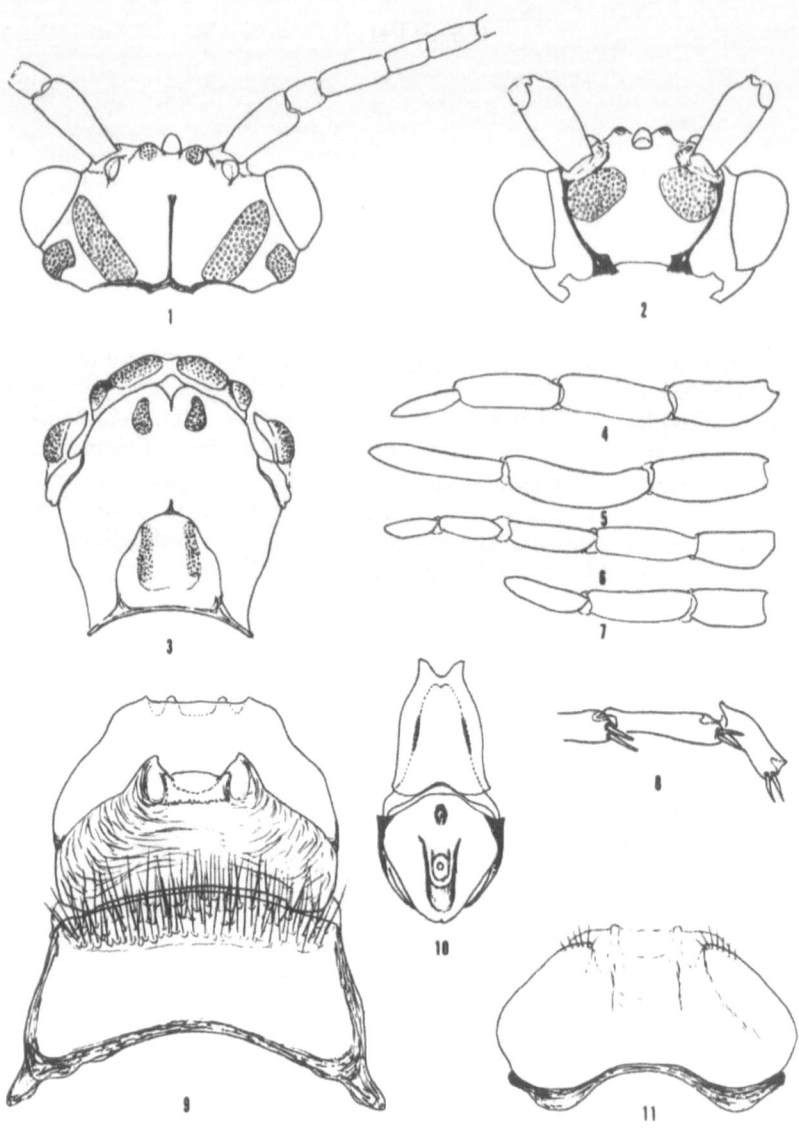

Figs 1–11. *Anomalopsyche minuta* (Schmid). 1, head, dorsal; 2, head, anterior; 3, pro- and mesonotum, dorsal; 4, ♂ maxillary palpus; 5, ♂ labial palpus; 6, ♀ maxillary palpus; 7, ♀ labial palpus; 8, tarsal segments lateral; 9, ♀ genitalia, ventral; 10, vaginal sclerites, ventral; 11, ♀ genitalia, dorsal.

76

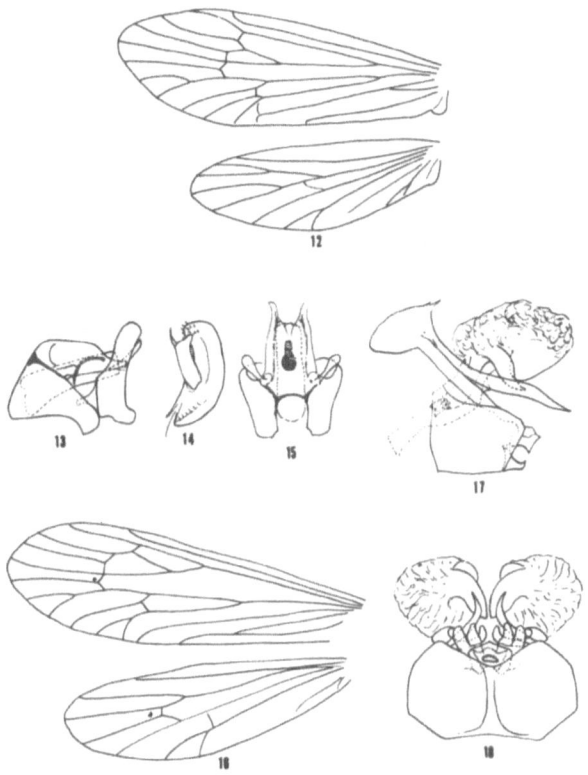

Figs 12–18. A. minuta. 12, wings; 13, ♂ genitalia, lateral; 14, clasper, posterior; 15, ♂ genitalia, dorsal. *Contulma cranifer* Flint. 16, wings; 17, ♂ genitalia, lateral; 18, ♂ genitalia, ventral.

Adult

Maxillary palpus 4-segmented, ♂; 5-segmented, ♀; basal 3 segments very lightly sclerotized, apical 1 or 2 segments strongly sclerotized. Labial palpus of ♂ and ♀ 3-segmented; 2 basal segments very lightly sclerotized, apical segment strongly sclerotized; palpus of male as long as maxillary palpus, of female distinctly shorter. Front of face with a large wart below each antenna; dorsally with a pair of small anteromesal warts and a pair of elongate posterolateral warts; with an erect, long wart on posterior of head. With a dorsomesal suture. With 3 ocelli. Pronotum with 2 pairs of dorsal warts. Mesonotum a pair of small anteromesal warts; mesoscutellum with a single large wart, but with setae borne laterally. Tibia of midleg with a row of

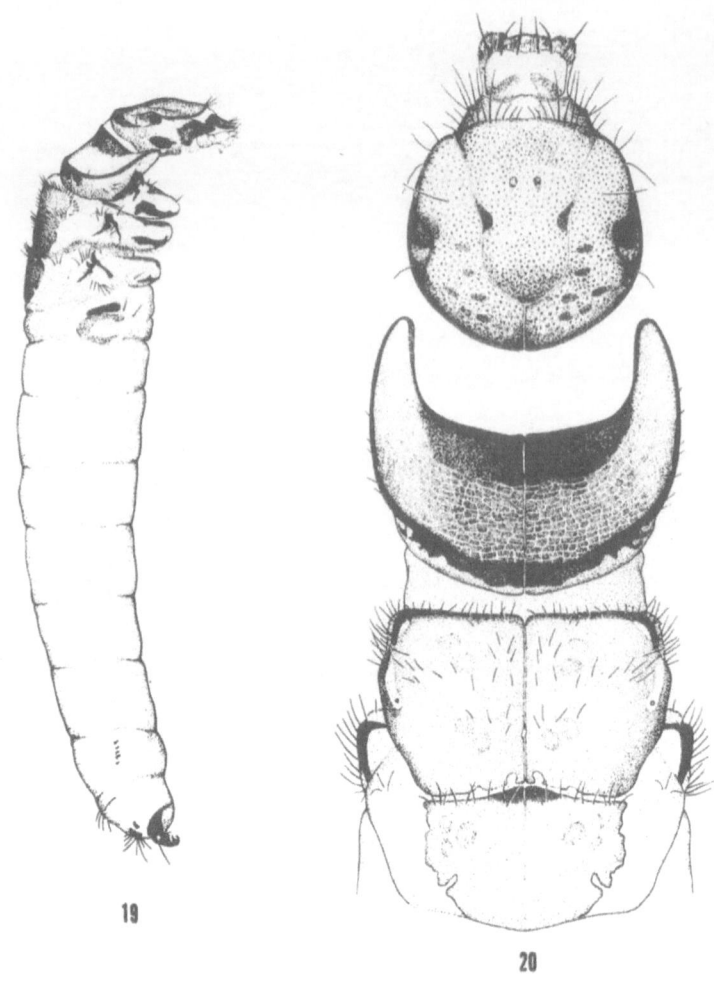

Figs 19–20. A. minuta. 19, larva, lateral; 20, larval head and thorax, dorsal.

5–6 short spines; all legs with basal tarsomere bearing 1 or 2 pairs of small spines, 4 basal tarsomeres with 2 pairs of apical spines. Female genitalia: Eighth sternum a large, rectangular plate with posterior third lightly sclerotized and bearing a dense brush of setae. A large membranous connection to ninth sternum (probably gaping to hold an egg ball). Ninth sternum with posterior margin produced in 2 submesal angles; surface striate. Ninth and tenth terga fused, no indication of cerci; posterior margin truncate, semimembranous.

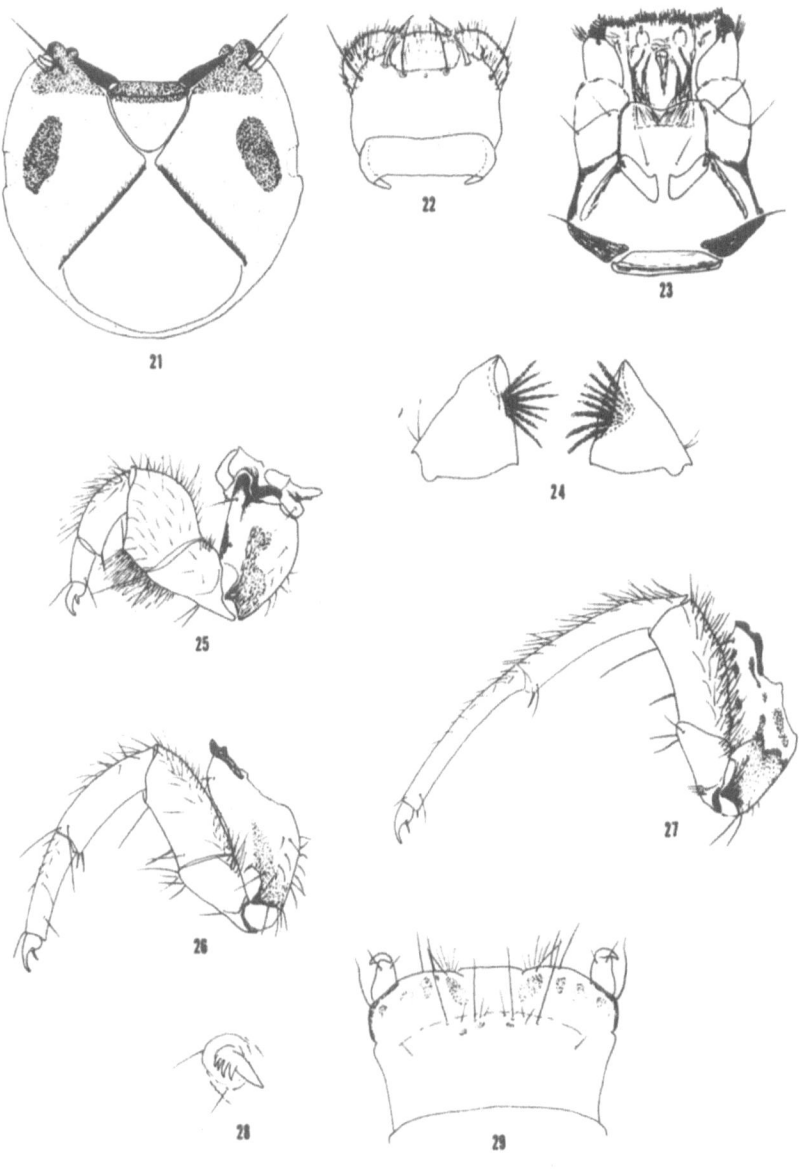

Figs 21—29. A. minuta. 21, larval head, ventral; 22, larval labrum, dorsal; 23, larval maxillolabium, ventral; 24, larval mandibles, ventral; 25, larval foreleg, lateral; 26, larval midleg, lateral; 27, larval hindleg, lateral; 28, larval anal claw, posterior; 29, apex of larval abdomen, dorsal.

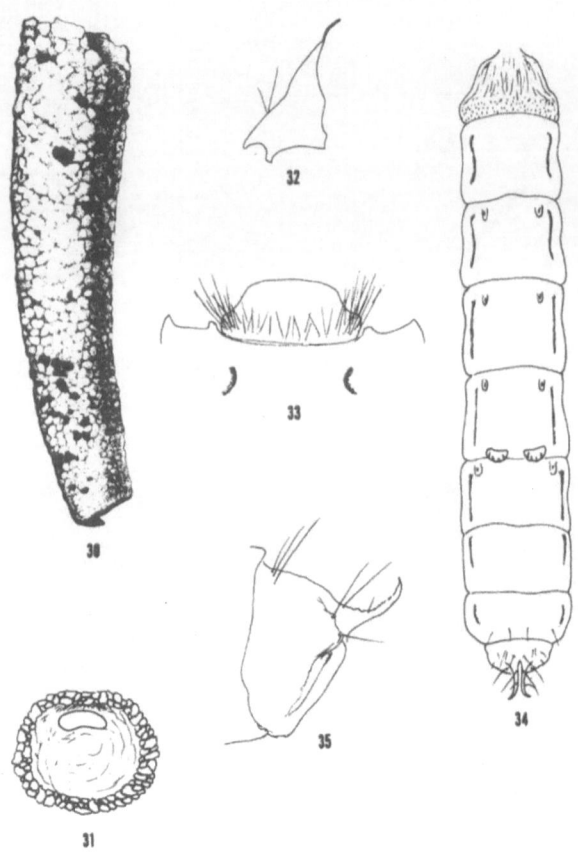

Figs 30–35. *A. minuta.* 30, larval case, lateral; 31, posterior closure of pupal case; 32, pupal mandible, dorsal; 33, pupal labrum, dorsal; 34, pupal abdomen, dorsal; 35, apex of pupal abdomen, lateral.

Larva

Length to 5 mm. Color of sclerites reddish-brown. Labrum with anterolateral and anteromesal membranous lobes bearing dense brushes of short setae; dorsum with 2 pairs of erect dark setae and 1 pair of decumbent pale setae. Mandibles triangular, apex scoop-like; both with brushes of feathered setae mesally, left mandible with brush borne from a deep pocket. Maxillolabium narrow, elongate; maxillary palpus short, segments almost completely lost, galea broad, membranous, with a fringe of short setae; labium ending in a truncate, membranous, lobe fringed with short setae. Head capsule heavily sclerotized dorsally, surface rugose; ventrally mostly pale except for a large ovoid spot near eye, surface covered with low rugosities. A well-marked carina from eye anteriad and across anterior margin of frontoclypeus, whose anterolateral margin bears a fringe of secondary setae; another carina starting beneath eye and continuing across posterior of head capsule. Gular sclerite broadly triangular, separating genal halves.

Pronotum with a transverse carina posteriorly, continuing anterolaterally and forming a large anterolateral rounded lobe; dorsal area heavily sclerotized, with polygonal markings; area beneath carina laterally pale, and bearing scattered setae. Trochantin small and mostly fused to pleuron. No prosternal horn or plate. Mesonotum covered with a dorsal plate, divided mesally, narrowed posteriad; bearing scattered short setae mostly on anterior half. Metanotum with dorsal plate only partially covering dorsum, divided mesally; with short setae along anterior margin. Foreleg short, with femur very broad, ventral and dorsal margins hairy. Midleg elongate, dorsal margin hairy, ventral margin with few setae. Hindleg elongate, tibia and tarsus especially elongate; dorsal margin hairy, ventral with few setae. Tarsal claws all short, with a small basoventral seta. Meso- and metasterna each with a pair of sublateral setae.

Abdomen without lateral line or gills. First segment unmodified dorsally, laterally with low humps bearing lightly sclerotized areas; venter produced in a large transverse hump, with a pair of submesal setae. Eighth segment with a row of 6–8 bifid tubercles laterally. Ninth tergum without a sclerotized shield, but with a few lightly sclerotized muscle attachments; posterior margin with a few setae. Anal proleg with lateral sclerite lightly sclerotized, except strongly so where in contact with ventral sole plate, muscle attachments noticeable, with a few setae posteriorly; with a narrow extension across venter, apex of which is darkened. Ventral sole plate strongly sclerotized. Anal claw with a dorsomesal comb of accessory teeth.

Pupa

Length to 5 mm. Mandibles broad basally, inner margin slightly sinuate, apex with a thread-like appendage; 2 basolateral setae. Labrum semicircular with a transverse brush of setae basally. Anterior tentorial pits well marked. A small cluster of setae on face at inner margin of eye. Mesonotum with a pair of anterior and 2 pairs of dorsal setae; metanotum with 2 pairs of anterodorsal setae. Legs lacking hair fringes. Legs and wing pads appressed to body; antennae, wing pads and hindleg sheaths reaching to segments 7 or 8 ventrally. Abdomen lacking gills and hair fringe. Segments 2–8 with dorsal and ventral sclerotized bars. Dorsal abdominal surface from posterior third of segment 1 through segment 6 with minute spicules (progressively smaller on posterior segments). Hook plates very weak; anteriorly on segments 3–6, each with a very small hook and spicules; posteriorly on segment 5, with several hooks. Apical appendages, short rod-like; tapering to an upturned point, dorsal surface rugose; apical abdominal segment and appendages with a few setae.

Case

Large enough only to house larva or pupa. Firmly constructed of sand grains and silk; tapered and slightly curved. Posterior of case almost closed by a silken sheet, with an ovoid dorsal opening, above which the silken sheet is slightly projecting in profile.

Pupal case differs only in possessing anteroventral attachments to substrate, and being closed by a silken sheet with radially arranged ridges; several small openings left ventrally between case and anterior closure.

In addition to the genus *Anomalopsyche*, I am provisionally placing *Contulma* Flint in the same family. Although the only synapomorphic state I find is the same simple, reduced type of female genitalia, they have several symplesiomorphic characteristics: presence of ocelli, the same placement of warts on the head dorsally, and venation in the forewing (except a more complete anal system). Unfortunately the larvae and pupae of *Contulma* are still unknown. In any case the continued placement of *Contulma* in the Sericostomatidae is totally untenable, and its placement in the Anomalopsychidae is much more logical.

I also have in the collection in Washington a few adults of undescribed species from Brazil and Ecuador, and larvae from Colombia, Ecuador, and Brazil that definitely belong to the same family unit. Considering the foregoing material, I propose the following family definition.

Adult

Ocelli 3. Maxillary palpus of male, 4 or 5 segmented; of female, 5 segmented; labial palpus, 3 segmented. Antenna shorter than forewing, scape cylindrical, about 3 times as long as broad; pedicel as long as broad; flagellar segments cylindrical, longer than broad. Head with a pair of warts beneath antennae; dorsally with a pair of warts anteromesally, a pair of elongate warts posterolaterally, and a pair on posterolateral face; with a dorsomesal suture.

Pronotum with 1 or 2 pairs of dorsal warts. Mesonotum with anterior margin bilobed; scutum with warts variable. Mesoscutellum less than half length of scutum, with a large wart. Spurs 2, 2, 4. Midleg with a row of short spines on tibia; basal tarsomere with 1 or 2 pairs of spines ventrally, all tarsomeres with 2 pairs of apical spines. Forewing with discoidal and thyridial cells; R_s with 4 apical branches, R_5 tending toward capture by M; M with 3 apical branches; Cu_1 forked apically; Cu_2 simple. Hindwing with R_s with 3 or 4 apical branches; M unbranched; Cu_1 forked apically.

Larva

Labrum with membranous lobes and with short hairs anteriorly. Mandibles trianguloid, edentate, with mesal brushes. Maxillolabium elongate, nearly truncate, membranous and hairy apically; maxillary palpi greatly reduced. Head heavily sclerotized, nearly encircled by carina dorsally. Pronotum with a transverse carina posteriorly, developed into a large anterolateral lobe. Foreleg short, femus very broad; midleg longer; hindleg longest with tibia and tarsus especially lengthened. Meso- and metanota with sclerites divided mesally; metanotal sclerites smaller than mesonotal. Abdomen lacking gills and lateral line. Spacing humps on first abdominal segment reduced. Eighth segment with bifid tubercles laterally. Ninth tergite and lateral plates of anal proleg not sclerotized, with a few setae. Anal claw with a dorsomesal comb of accessory teeth.

Pupa

Mandibles elongate, inner margin sinuate, with an apical filament. Labrum semicircular, with a basal brush of setae. Lacking lateral fringe on legs. Hook plates weakly developed: anteriorly on segments 3–6, posteriorly on segment 5. Apical appendages cylindrical, short, tips upturned.

Case

Cylindrical, tapered and slightly curved. Composed of sand grains cemented by silk. Posterior with a silken sheet, pierced dorsally by an ovoid opening.

DISCUSSION

The Anomalopsychidae clearly belongs to the superfamily Limnephiloidea (Ross 1967, 1978). This superfamily is primitively ocellate, and the trait is continued into the limnephilid branch where the ocelli are lost independently in several families. The ocelli are either lacking in all the leptocerid branch (Ross 1967) or retained in the Plectrotarsidae, a family transferred to the most primitive position in the branch by Ross (1978). On the basis of the published figures of the adults (the immature stages are undescribed) of this family, I have the feeling that its placement near the Limnephilidae is more likely correct than its placement in the leptocerid branch.

The leptocerid branch was characterized by Ross (1967) as follows: ocelli lost, supratentorium reduced, ♂ front wing with M_4 retained, and larval pronotum without crease. *Anomalopsyche* possesses ocelli. If his supratentorium is equivalent to the dorsal arms of the tentorium, then this structure is lacking. The forewing appears to lack M_4, but Ross modifies this characteristic in the text to 'present in several primitive genera only', so its lack is hardly significant. The larval pronotal crease is partially explained in the text as the posterior suture of the pronotum close to the posterior margin as opposed to being anteriad in the limnephilid branch. I am unable to see any consistant differences in the two branches in any characteristic of the pronotum. The larvae of the families of the leptocerid branch, in so far as known, all agree on the following characteristics: lack of prosternal horn (plesiomorphic), bifid tubercles on the eighth segment only (apomorphic), and lack of a single strongly sclerotized dorsal plate on the ninth segment (plesiomorphic). The larvae of the Anomalopsychidae are in complete agreement. It seems that the balance of the characteristics weigh strongly in favor of the leptocerid branch for the Anomalopsychidae.

It seems unrealistic to believe that ocelli could secondarily redevelop from a non-ocellate ancestor. Therefore, I must place the origin of the family in a very basal position in the leptocerid branch before the ocelli were permanently lost. Ocelli are apparently easily lost, as shown by their frequent disappearance in various taxa throughout the order. It would, therefore, hardly be surprising if their closest relatives are non-ocellate. A number of other plesiomorphic characteristic are found in *Anomalopsyche* adults: full complement of warts on head and thorax, only a slight reduction in venation of forewing (especially in the anal field), and simple genitalia with 2-segmented clasper.

Within the leptocerid branch what are the relationships? I believe that the closest relatives are the Beraeidae and the Helicophidae (and possibly the Antipodoecidae). They share the following derived character states as adults: spurs 2, 2, 4, tarsal spines reduced to 2 apical pairs on all but the basal segment, female genitalia with a large plate-like eighth sternum, the ninth sternum all or partially divided mesally, without apical processes or with only a short pair of cerci. As larvae they share the following derived states: lateral carina on head, loss of lateral line and gills (generally), and loss or reduction of humps on first abdominal segment. The pupae agree on: mandible with tip attenuate, hook plates on segments 3–6 anteriorly most of which are small and lightly sclerotized, posteriorly on segment 5, apical processes short, terete, with tip pointed and upturned.

The differences between the Anomalopsychidae and Beraeidae-Helicophidae are profound, however. The presence of ocelli, a more complete venation, reduction of palpal segments in the male, and form of male genitalia in adult Anomalopsychidae are distinctive. The larvae with pad-like anterior margins of the labrum and maxillolabium, non-dentate mandibles, unmodified tarsal claws, hairs on the lateral plate of the anal proleg, and comb on the anal claw are also easily recognized. The whip-like tip of the pupal mandible and cluster of hairs on the labrum are unique to *Anomalopsyche.*

REFERENCES

Flint, O.S., Jr, Beitr. Neotrop. Fauna 5: 45–68, 1967.
———., Proc. Ent. Soc. Wash. 71: 497–514, 1969.
———., Rev. Chilena Ent. 8: 83–93, 1974.
———., Proc. Biol. Soc. Wash. 92: 640–649, 1979.
Ross, H.H., Ann. Rev. Ent. 12: 169–206, 1967.
———., *In* Proc. 2nd Int. Symp. Trich., ed. I. Crichton, pp. 1–6, Junk, The Hague, 1978.
Schmid, F., Beitr. z. Ent. 7: 379–398, 1957.

DESCRIPTION OF *HYDROPSYCHE* LARVAE FOUND IN THE IBERIAN PENINSULA

D. GARCIA DE JALON

SUMMARY

The larvae of *Hydropsyche exocellata* Dufour and *Hydropsyche tibialis* McL. are described and a key to the seven Iberian *Hydropsyche* is given.

INTRODUCTION

There are known to be 15 species of the genus *Hydropsyche* in the Iberian Peninsula (Botosaneanu and Malicky 1978; González and Malicky 1980): *H. ambigua* Schmid, *H. brevis* Mosely, *H. bulvifera* McL., *H. dissimulata* Kum and Bots., *H. exocellata* Dufour, *H. hibera* Schmid, *H. infernale* Schmid, *H. instabilis* Curtis, *H. lobata* McL., *H. pellucidula* Curtis, *H. pictetorum* Bots. and Schmid, *H. siltalai* Döhler, *H. tibialis* McL., *H. urgorrii* González and Malicky and *H. angustipennis* that recently has been found in Catalonia (Puig, personal comunication).

To the date, only the larvae of *H. angustipennis, H. dissimulata, H. exocellata, H. instabilis, H. pellucidula* and *H. siltalai* have been described (Sedlak 1971; Hildrew and Morgan 1974; Szczesny 1974; Verneaux and Faessel 1976; Statzner 1976; Boon 1977, Moretti and Spinelli (1979).

During the studies on fauna in the Lozoya, Jarama and Henares rivers, of Central of Spain, some species of *Hydropsyche* were collected (García de Jalon 1978; García de Jalon and González del Tánago 1980). One, *H. tibialis,* is described for the first time and another, *H. exocellata* is redescribed.

RESULTS

Hydropsyche exocellata Dufour, *1841*

Although this species lives in Central Europe, the British Isles and the Iberian Peninsula (Botosaneanu and Malicky 1978), the only mention of its larval taxonomy are Verneaux and Faessel's (1976) key and Szczesmy (1974).

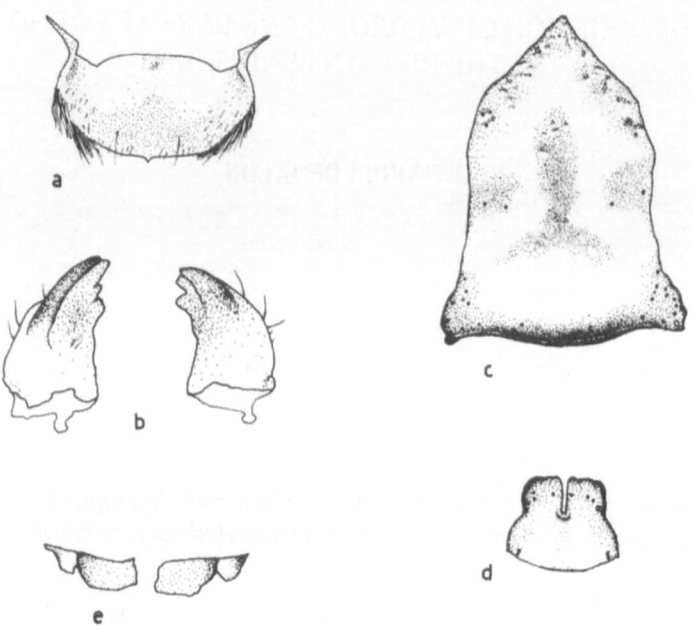

Fig. 1. *Hydropsyche exocellata*: (a) labrum; (b) mandibles; (c) frontoclypeus; (d) mentum; (e) posterior prosternites.

Description of the larvae. Length of full-grown larva: 17-20 mm. General colour of the body buff.

Head. Dorsal view rectangular almost square (Fig. 2b); light coloured. Fronto-clypeal apotome tapers sharply at the posterior, maximum width at the extreme anterior, slight narrowing at eye level (as in *H. pellucidula* and *H. angustipennis*). The frontoclypeus has a pattern of yellow-spots (Fig. 1c) that distinguishes *H. exocellata* from others hydropsychid larvae; the aboral, oral and both lateral flecks are continuous. Around and inside the frontoclypeus there is a brownish area, with a granular appearance, especially intensive on the posterior part. Frontoclypeus mediotransversal folding absent or litted marked (as in *H. ornatula* and *H. Contubernalis*).

Labrum. (Fig. 1a) ovoid shaped, with an anterior dark spot.

Mandibles (Fig. 1b) with a massive teeth. Ventral side of head (Fig. 2a) whitish with two longitudinal brownish stripes. Submentum darker than the rest of the head, elongated with concave margins. In mentum (Fig. 1d) the margins are usually darker than the centre, but in other specimens all the mentum is dark, or there is a dark central longitudinal stripe. The channel between the lobes of the mentum is straight-sided.

88

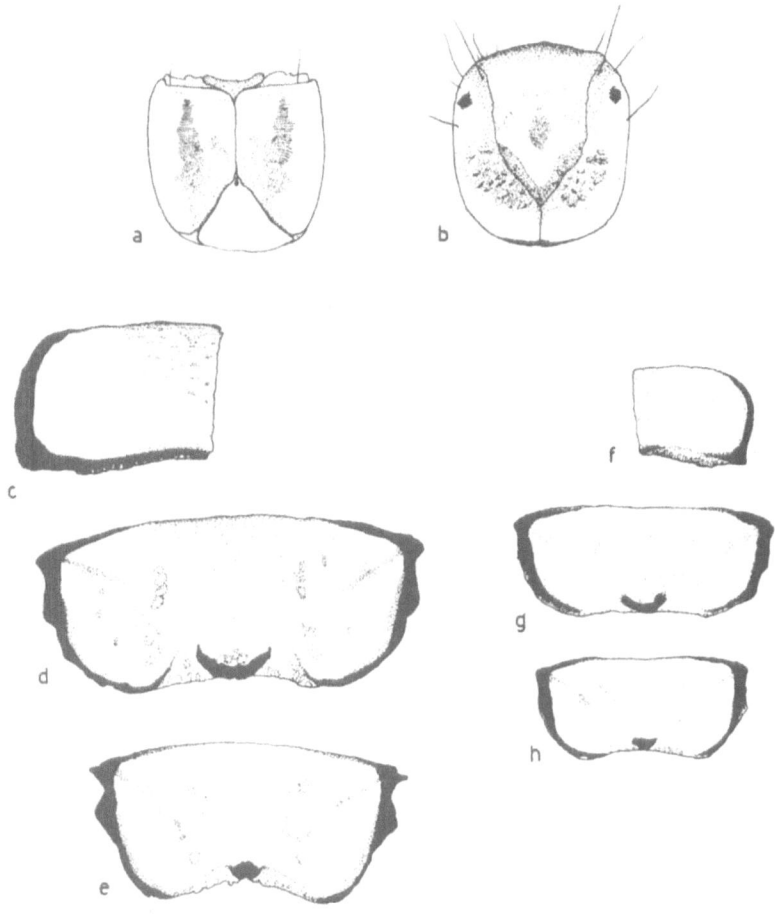

Fig. 2. Hydropsyche exocellata: (a) head, ventral view; (b) head, dorsal view; (c) hemi-pronotum; (d) mesonotum; (e) metanotum; *Hydropsyche tibialis*: (f) hemi-pronotum; (g) mesonotum; (h) metanotum.

Thorax. Pronotum (Fig. 2c) composed of two hemisclerites with an anterior triangular brownish spot, anteriorly. Meso and metanotum (Fig. 2d, e) have no special characteristics. In the posterior prosternites (rhomboidal sclerites) the medial regions are somewhat darker than the lateral ones (Fig. 1e), each region being at its darkest towards the exterior edges.

Abdomen. Ventral gills present on first seven segments. Anal legs have no special features.

Material and methods. H. exocellata was collected abundantly in the middle and lower reaches of the Jarama and Henares rivers. The material used for this

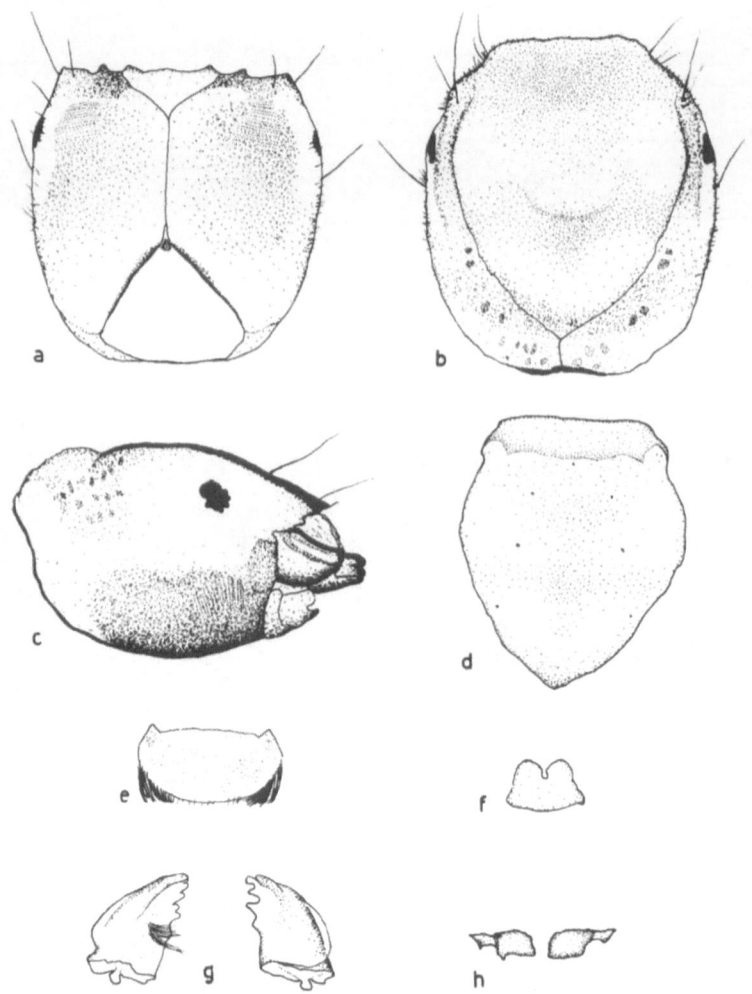

Fig. 3. Hydropsyche tibialis: (a) head, ventral view; (b) head, dorsal view; (c) head, lateral view; (d) frontoclypeus; (e) labrum; (f) mentum; (g) mandibles; (h) posterior prosternites.

description includes 30 specimens taken from these rivers. The adult-larva correlation was derived from mature male pupae, with fully developed genitalia and cocoons which include the larval exuvia and sclerites.

Hydropsyche tibialis McLachlan, *1884*

Species endemic to the Iberian Peninsula.

Description of the larvae. Size of full-grown larva small (length: 12–15 mm). General colour of the body yellowish brown.

90

Head. Distinctive shape of cephalic capsule which differentiates *H. tibialis* from other hydropsychids (Fig. 3a, b, c). The transversal section of head at eye level is semicircular ending at the frontoclypeus sutures. Frontoclypeus completely flat, occupying most of the dorsal part of the head. Characteristic shape of frontoclypeus (Fig. 3d) with maximum width at eye level. The pattern of yellow spots and the medio-transversal folding of the fronto-clypeus are very poorly defined.

Labrum. (Fig. 3e) ovoid, uniformly brownish coloured. Mandibles (Fig. 3g) with clearly prominent teeth. The ventral side of head (Fig. 3a) is brownish with two longitudinal sets of darker transversal striations. The submentum is triangular and has two anterior indentations. The two lobes of the mentum (Fig. 3f) are rounded so that the channel between them is curved, as in *H. instabilis* and *H. fulvipes.*

Thorax. Pro, meso and metanotum (Fig. 2f, g, h) without special character-istics. Posterior prosternites (rhomboidal sclerites) as in Fig. 3h.

Abdomen. There is no special characteristic which differentiates this from the typical abdomen of the genus *Hydropsyche.* Ventral gills present on first seven segments.

Material and methods. The present description is based on 18 specimens of *H. tibialis* taken from the upper reaches (Hipocrenon and Epirhithron) of the Lozoya and Jarama rivers (Central Spain). The adult larva correlation was derived from mature male pupae.

Key to the larvae of the Iberian Hydropsyche

1 Maximum width of frontoclypeus at eye level (Fig. 3b). Dorsal side of head markedly flattened. Small size . . . . *H. tibialis*
  Maximum width of frontoclypeus never at eye level. 2
2 Gills absent on 7th abdominal segment. Medium size . . . . *H. siltalai*
  Gills present on 7th abdominal segment. 3
3 Anterior lobes of mentum rounded, channel between them not straight-sided. Margins of submentum not markedly concave. Medium size. . . . .*H. instabilis*
  Anterior lobes of mentum more rectilinear (Fig. 1d), channel between them straightsided. Margins of submentum markedly concave 4
4 Lateral, aboral and oral flecks on the frontoclypeus joined (Fig. 1c). Posterior part of head with a brownish area of granular appearance. Medium size. . . . .*H. exocellata*
  Lateral, aboreal and oral flecks on the frontoclypeus separated 5
5 Frontoclypeus pattern of yellow spots composed only of two longitudinal

and short flecks. Posterior margin of mesonotum with a continuos black line.. *H. dissimulata*
Without these characteristics 6
6 Frontoclypeal anterior margin rough and ridged. Posterior part of head without a dorso-transversal dark stripe. Posterior prosternites rhomboidal shaped.. *H. angustipennis*
Frontoclypeal anterior margin smooth. Posterior part of head with a dorso-transversal dark stripe. Posterior prosternites trapezoidal shaped... *H. pellucidula*

REFERENCES

Boon, P.J., *In* Proc. 2nd. Int. Symp. Trich., ed. M.I. Crichton, pp. 165–173, Junk, The Hague, 1977.
Botosaneanu, L. and Malicky, H., *In* Limnofauna Europaea, ed. Illies, G. Fisher Verlag, Stuttgart, 1978.
Garcia de Jalon, D., Doct. Th. Univ. Politecnica de Madrid, pp. 231, 1978.
————., and Gonzalez del Tanago, M., Estimacion de la contaminacion de las aguas mediante indicadores biologicos: aplicacion al rio Jarama. Icona, Madrid, 1980 (in press).
Gonzalez, M. and Malicky, H., Entom. Zeits. 90: 28–32, 1980.
Hildrew, A.G. and Morgan, J.C., J. Ent. (B) 43: 217–229, 1974.
Moretti, G.P. and Spinelli, G., Riv. Idrobiol. 18: 133–171, 1979.
Sedlak, E., Acta Ent. Bohemoslav. 68: 185–187, 1971.
Statzner, B., Ent. Germ. 3: 265–268, 1976.
Szczesny, B., Pol. Arch. Hydrobiol. 21: 387–390, 1974.
Verneaux, J. and Faessel, B., Annls. Limnol. 12: 7–16, 1976.

DISCUSSION

Moretti: Have you found *Hydropsyche pictetorum* and, if so, was it abundant?
De Jalon: *Hydropsyche pictetorum* was first collected by Pictet on the Sierra de Guadarra and, so far as I know, this is the only record to date.

PREDATORY EXCLUSION OF *APATANIA ZONELLA* (ZETT.) BY *POTAMOPHYLAX CINGULATUS* (STEPH.) (TRICHOPTERA: LIMNEPHILIDAE) IN ICELAND

G.M. GÍSLASON

SUMMARY

Potamophylax cingulatus (Steph.) was recently recorded for the first time in Iceland. *Apatania zonella* (Zett.) was found to be common all over Iceland in studies made in the 1920s and 1930s; in 1974–78 it was no longer found in the East, and was scarce in the North. *P. cingulatus* now occupied most streams in both areas. Field studies and laboratory experiments suggest that *P. cingulatus* is excluding *A. zonella* by predation from these areas.

INTRODUCTION

In a comprehensive study on the distribution of Icelandic Trichoptera it was found that a recent addition to the Icelandic fauna, *Potamophylax cingulatus* (Steph.) limited the distribution of *Apatania zonella* (Zett.) (Gíslason 1974, 1977).

Since Gause (1934) published his hypothesis, that the competition of two species for the same resource would lead to the extinction of one of them, several workers have been occupied on the competitive exclusion principle (Hardin 1960). Several studies on predator exclusion from intertidal areas (e.g. Kitching and Ebling 1961; Reise 1977, 1978; Peterson 1979) have shown that predators can affect the spatial distribution patterns of prey species. However, only a few studies have shown that geographical distribution of organisms can be limited by predation. Kitching and Ebling (1967) studied the distribution of the common mussel (*Mytilus edulis* L.) in Lough Ine, S-Ireland. The distribution of the mussel was restricted to exposed areas or steep rock faces, where its predators, three species of crabs and one species of starfish, could not eliminate it. Lindsey (1964) has shown that the lake trout (*Salvelinus namaycush* (Walbaum) is distributed in the Canadian high arctic, in areas that seem to be beyond the range of marine lampreys (*Entosphenus* spp., *Petromyzon marinus* L.).

MATERIALS AND METHODS

n 1974–1978 a search for *A. zonella* and *P. cingulatus* larvae was made hroughout Iceland. At each sampling site, from one hour to a few days were spent searching thoroughly over extensive areas of lakes and river beds. One or both species were found at each sampling site. All larvae were collected and preserved in 70% alcohol for further examination.

In laboratory studies on the cohabitation of *A. zonella* and *P. cingulatus* stirring bowls designed by Philipson (1953) were used. Three series of experiments were designed. As controls, ten larvae of each species were placed in separate bowls. In experiments on cohabitation ten larvae of each species were placed together in each bowl. Each bowl was $0.03145 \, m^2$, and the densities of the larvae were thus $318/m^2$ in the control, and $636/m^2$ in the cohabitation experiments.

A. zonella larvae were collected from the stream Hafnarfjardarlaekur, SW-Iceland ($64°04'$ N, $21°56'$ W), and *P. cingulatus* larvae from Pond Burn, Co. Durham, England ($54°54'$ N, $1°46'$ W). Newcastle tap water was diluted four times with distilled water to reduce the conductivity to $80–100 \, \mu S/cm$ at $25°$ C, and pH 7–8, conditions similar to that of Icelandic freshwater. A substrate of a mixture of sand, gravel and stones was provided. The current velocity was about $9 \, cm/sec$ (O.T.T. small current flow meter), and the temperature $10°$ C $\pm 2°$ C. The living larvae were counted daily and dead larvae removed. Three sets of these experiments were made, one with the bare experimental substrate, one with the moss *Fontinalis antipyretica* in addition to the bare substrate (10–20% cover) and one with decomposing leaves (mainly beech (*Fagus silvatica*) and some oak (*Quercus robur*) to produce a cover of about 40–50%. Three or four replicates were made of each experiment.

INTERACTION BETWEEN A. ZONELLA AND P. CINGULATUS

Icelandic running waters are rather poor in species. Dominating invertebrates are chironomids (usually 90% of numbers) and, in some lake outlets, simuliids.

In Iceland, two species, *A. zonella* and *P. cingulatus*, are the dominating caddis larvae in running waters (Gíslason 1981). *P. cingulatus* is only found in the East and the North, both in the lowlands and in the highlands (Fig. 1). The first specimen from Iceland is an adult female, found in the East in July 1959 (Gíslason 1977). However, old people in eastern Borgarfjordur state that they have seen these larvae (which are easily recognizable) there for decades, and have a vernacular name for them ('hýdormur' – case worm). Fully grown, it is about 16–20 mm long, and about 80 mg wet weight, the largest insect in running waters in Iceland. *P. cingulatus* is mainly found in running waters (Gíslason 1974, 1977). Presumably, it colonized Iceland recently, probably windborne from the Faroe Islands, Scandinavia or Britain, and is spreading westward in Iceland.

A. zonella larvae are common in running waters in Iceland, except in the North and East, and in addition, it is also common on stony shores of lakes (Gíslason 1974, 1977, 1981) (Fig. 1). In studies by Lindroth (1931) and Fristrup (1942), *A. zonella* larvae were as common in the North and East as they were in other parts of Iceland. Fully grown larvae are about

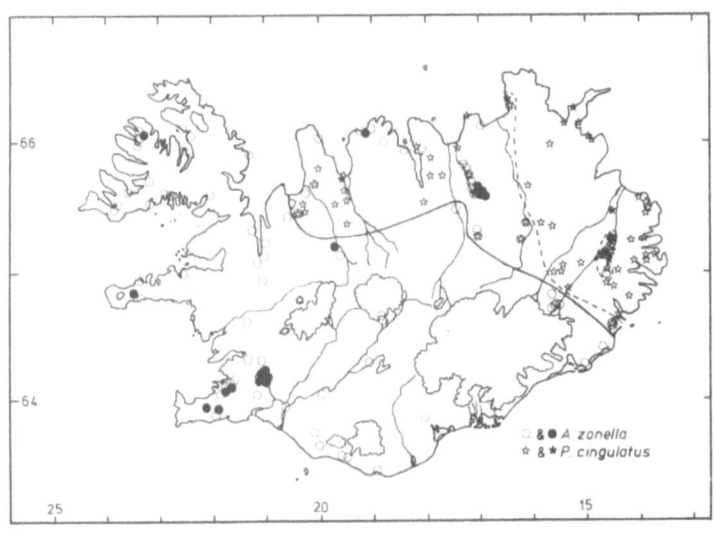

Fig. 1. Recorded occurrences of *Apatania zonella* (Zett.) and *Potamophylax cingulatus* (Steph.) larvae in 1974–78. Solid line: western limit of the distribution of *P. cingulatus*. Broken line: eastern limit of the distribution of *A. zonella*. Open symbols: larvae in lotic waters, solid symbols: larvae in lentic waters.

8–10 mm long and 15–20 mg wet weight, next in size to *P. cingulatus* of stream insects.

The distribution of these two species overlaps in a narrow zone in the eastern highlands and in the North. They occur together in three run-off rivers and seven lake outlets or spring-fed rivers (which generally have a very stable discharge and ionic composition and have mosses and algae growing in greater amounts than in run-off streams). A previous study (Gíslason 1974) suggested that chemical composition did not limit their distribution.

A. zonella larvae feed mainly on diatoms (Nielsen 1943). *P. cingulatus* larvae feed on tree leaves that fall into the streams and to some extent on Chironomidae larvae (Scott 1958; Elliott 1970; Otto 1974). A few larvae from streams in Iceland were examined, their gut contents were mainly decaying grass leaves. It is therefore unlikely that these species compete for food.

Woodland is rare in Iceland, forests covering about 1% of the country (Steindórsson 1964). Most streams run through unvegetated land, heaths or bogs. Field observations indicated that little allochtonous material is brought into the streams apart from some litter brought by surface water in the spring thaw. If *P. cingulatus* is an omnivorous species it would feed on the most available food at any time. If *A. zonella* are present they would be easy prey since they occupy the same habitat in the streams, i.e. under stones on the bottom. In the streams where both species occurred together, empty cases of *A. zonella* were often found ripped along the ventral side.

95

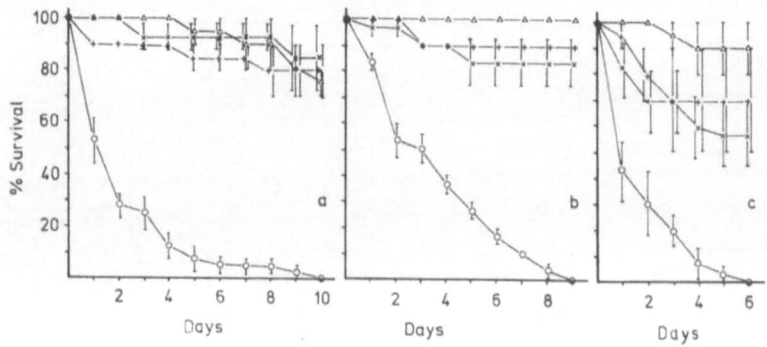

Fig. 2. Larvae of *A. zonella* and *P. cingulatus* cohabiting in artificial streams. (a) Substrate made of stones, gravel and sand only. Control 2 replicates and cohabitation 4 replicates. (b) Substrate made of stones, gravel, sand and the moss *Fontinalis anti-pyretica* (cover 10–20%). 1 control for each species and 3 replicates for cohabitation. (c) Substrate made of stones, gravel, sand and decomposing leaves (mainly beech (*Fagus silvatica*) and also oak (*Quercus robur*)). 40–50% cover of leaves. Control 2 replicates and cohabitation 3 replicates. Means shown with standard error. △ *A. zonella* control, ○ *A. zonella* cohabiting, + *P. cingulatus* control, × *P. cingulatus* cohabiting.

The results of the experiments on cohabitation of the two species are shown in Fig. 2. The number of *A. zonella* cohabiting, fell very sharply, about 50% of them died during the first day and they were all dead after 10 days in the bowls with bare substrate (Fig. 2a), after 9 days with moss (Fig. 2b) and after 6 days where leaf cover was provided (Fig. 2c). The larvae of both species in the control experiments and of *P. cingulatus* in the cohabition experiments showed very similar mortality rates, 10–20% of them dying during the duration of the experiments with the exception of the experiment employing leaf cover (Fig. 2c), where *P. cingulatus* had mortality of ca. 30–40%.

The dead *A. zonella* larvae had been eaten in nearly all cases. *P. cingulatus* was observed on several occasions eating them. They attacked them and ripped the cases along the ventral side, sometimes along the dorsal one, eating the larvae as they ripped, sclerotized parts being left uneaten. Most of the cases of *A. zonella,* some with half eaten larvae, had been opened in this way when they were removed from the experiments. Similarly damaged cases of *A. zonella* were also observed in material collected from streams in Iceland where *P. cingulatus* larvae were also found.

P. cingulatus larvae in cohabitation and control experiments, were mostly killed by cannibalism, and their cases were also ripped.

These studies suggest that in the process of spreading towards the West and South of Iceland, *P. cingulatus* limited the distribution of *A. zonella* by predation. At present *P. cingulatus* has not crossed the central highlands of Iceland. *A. zonella* forms an easy prey, where it occupies similar micro-habitats as *P. cingulatus* and generally pupates about 1–2 months earlier than *P. cingulatus* (Gíslason 1977). After pupation, they are unable to move and could easily be attacked.

As observed in the experiments, *P. cingulatus* larvae ate only the soft parts of *A. zonella* larvae and this could explain the absence of animal remains in the guts examined.

Where *P. cingulatus* larvae have eliminated all *A. zonella* larvae, they can turn to other food items, such as other invertebrates or detritus brought into the rivers, especially the litter brought there in the spring thawing. *P. cingulatus* larvae seem to be opportunistic in their food selection, eating what is most available or easiest to obtain at any time.

A possible disadvantage to *A. zonella* could be that it reproduces parthenogenetically, and might therefore be unable to respond genetically to changes in the environment, such as the introduction of a predator.

ACKNOWLEDGEMENTS

I wish to thank Dr G.N. Philipson, Newcastle University, for his advice and encouragement during the tenure of this study, the staff of the Institute of Biology, University of Iceland, especially Prof. Agnar Ingólfsson, for assisting in the field work. Prof. Arnthór Gardarsson, Mr M. Brazil, and my wife Kristín read the manuscript of this article. I thank Mrs Dóra Jakobsdóttir for typing the manuscript. Financial support was received from Icelandic Science Foundation, Icelandic Government Students Loan Fund and NATO Science Fellowship.

REFERENCES

Elliott, J.M., J. Zool. 160 (43–44): 279–290, 1970.

Fristrup, B., Zool. Icel. 3: 1–23, 1942.

Gause, G.F., The Struggle for Existence. Hafner, New York. (Reprinted 1964), 1934.

Gíslason, G.M., Natturufr. 44: 129–139, 1974.

———., Ph.D. Thesis, Univ. of Newcastle upon Tyne, p. 412, 1977.

———., *In* Proc., 3rd Int. Symp. Trichoptera, ed. G.P. Moretti, pp. 99–109, Junk, The Hague, 1981.

Hardin, G., Science 131: 1292–1297, 1960.

Kitching, J.A. and Ebling, F.J., J. Anim. Ecol. 30: 373–383, 1961.

———.,   ———., Adv. Ecol. Res. 4: 197–291, 1967.

Lindroth, C.H., Zool. Bidr. 13: 105–599, 1931.

Lindsey, C.C., J. Fish. Res. Bd. Can. 21: 977–994, 1964.

Nielsen, A., Ent. Medd. 23: 18–30, 1943.

Otto, C., J. Am. Ecol. 43: 339–361, 1974.

Peterson, C.H., Oecologia. 39: 1–24, 1979.

Philipson, G.N., Proc. R. ent. Soc. London (A) 28: 15–16, 1953.

Reise, K., Helgoländer wiss. Meeresunters. 30: 263–271, 1977.

———., Helgoländer wiss. Meeresunters. 31: 55–101, 1978.

Scott, D., Arch. Hydrobiol. 54: 340–392, 1958.

Steindorsson, S., Grodur a Islandi. A.B. Reykjavik, p. 186, 1964.

DISCUSSION

Flint: Did you observe *P. cingulatus* feeding on any other caddis or aquatic insect species?
Gislason: No.

DISTRIBUTION AND HABITAT PREFERENCES OF ICELANDIC TRICHOPTERA

G.M. GÍSLASON

SUMMARY

Eleven species of Trichoptera are found in Iceland. Four species occur both in running and stagnant waters, six species prefer very stagnant waters, shallow ponds and swamps and one species occurs mainly in streams. In general each species occupies a much greater range of habitats in Iceland than they do elsewhere in Europe, where the number of species is 20 to 30 times greater. Distributional patterns within Iceland fall into 4 categories: (a) found all over Iceland (5 species). (b) found where summer temperatures are high (4 species), (c) found in the East and North (1 species) and (d) found all over Iceland, except in the East (1 species). Species in the first category are those that occupy the widest range of habitats. In the second category are species that are found only in shallow stagnant waters, which freeze solid during the winter, and all larval growth takes place in the summer. The single species in the third category is a recent immigrant which is profoundly affecting the distribution of the single species in the fourth category.

INTRODUCTION

The first comprehensive survey of the geographical distribution of the Icelandic Trichoptera in the lowlands was made by Lindroth (1931) and of the whole country by Fristrup (1942). Other records have referred to isolated areas only (Tjeder 1964; Garside and Leak 1974; Sigurjónsdóttir 1974; Gíslason 1977a, 1980, Ólafsson pers. comm.), except for the distribution of *Apatania zonella* (Zett.) and *Potamophylax cingulatus* (Steph.) described by Gíslason (1974). Several new localities have been included in the present survey and a much more complete picture of the distribution of Trichoptera in Iceland is now available.

Most of Iceland is a plateau with a narrow zone (average of 1-2 km) of lowlands along the coast with the exception of the South and South-West (where lowlands extend 20–50 km from the coast), and valleys in the North and East. According to Poulsen (1924), in the only biological classification of Icelandic water bodies, based on entomostracan Crustacea, lakes between 0–100 m a.s.l. are covered with ice on average 207 days per year, lakes between 101–300 m a.s.l. 228 days and lakes above 300 m a.s.l. 250 days.

In the South, ice usually forms on large lakes in January and has completely melted by April. Whereas in the North and in the central highlands lakes usually freeze in October or November and the ice thaws in May (Rist 1969; Eythórsson and Sigtryggsson 1971). The ice thickness is usually 50–60 cm in the lowlands and about 90 cm in the highlands (Sigurjón Rist pers. comm.).

DEFINITION OF HABITATS

Definition of habitats is based on a subjective assessment of what is thought to be ecologically important.

Rivers in Iceland are classified into three groups: glacial rivers, direct run-off rivers and spring-fed rivers (Kjartansson 1945, 1965). Glacial rivers which originate in glaciers support no Trichoptera.

Run-off rivers dominate the basalt rock formation in West and East Iceland due to low permeability of the base rock. In general their discharge is greatest in the spring (snow melt) and in the autumn (rainfall) with a marked minimum in the summer and winter. Their flows are very variable and they can flood in early winter and spring. The water temperature is greatly affected by the air temperature. Anchor ice forms on their beds soon after the air temperature falls below freezing, greatly reducing the flow (Rist 1956). Drifting snow can also affect the flow. The pH of run-off rivers is usually between 6 and 7 and conductivity 40–60 μS/cm at 25°C.

Spring-fed rivers are found in the palagonite (móberg) region, which stretches from the South-West to the North-East across the country. The palagonite is very permeable, and precipitation which seeps into it may emerge in springs far away from the place where it fell. These rivers are very stable, the discharge is even and the temperature normally 3–5° C near the source all the year round. Spring-fed rivers never freeze near the source. Their pH is between 7.5–9 and conductivity 80–100 μS/cm at 25°C.

Lake outlets behave similarly to spring-fed streams, but have greater fluctuations of temperature. They are classified as spring-fed streams in Table 1.

Estuaries. The area defined as an estuary is where the river water mixes with the sea, so that animals living in that stretch of the stream are affected by salinity fluctuations during the tidal cycle.

Saltmarsh pools are small pools in saltmarshes, usually about 100 m² or less and are inundated by sea water during high spring tides. The salinity in these pools fluctuates greatly.

Lagoons are similar to lakes, except they have slightly saline water. Some lagoons are formed as lakes near the shore with only a narrow gravel bar parting them from the sea. Saline water is brought in by windspray or wave action during high spring tides. Some lagoons are formed at river estuaries where there are sand and gravel banks at the mouth of the rivers. There is always some running water in the latter. The lagoons are large enough to have exposed shores.

Lakes are large water bodies (often many square kilometers) and have exposed shores. They are of various depths, most are shallow ($<$ 5 m) but some are over 100 m deep. The shores are usually stony, but vegetated shores can be found in some places in the lakes.

Table I. Habitats of Icelandic Trichoptera larvae.
Occurrence of Icelandic Trichoptera larvae in different habitats expressed in percentage of collecting sites with each species and the percentage of collecting sites with 1, 2, 3 and 4 species.

Species \ Habitats	Spring-fed streams %	Run-off streams %	Lakes %	Ponds and pools %	Swamps and marshes %	Lagoons %	Estuaries %	Salt marsh pools %	Total %
A. zonella	84	47	62	26	10	22	20	100	45
L. affinis	25	4	60		10	22	40		22
L. decipiens					30				<1
L. elegans									1
L. fenestratus				2					<1
L. griseus	44	33	14	68	50	22			38
L. picturatus		6		32	30				7
L. sparsus	2		3	6		22	20		4
G. nigropunctatus			5	6					2
P. cingulatus	18	49				22	40		24
A. picta					20				2
Sites with 1 sp.	39	68	60	62	60	89	80	100	60
Sites with 2 sp.	51	26	38	36	30	11	20		34
Sites with 3 sp.	10	5	3	2	10				5
Sites with 4 spp.		1							<1
Mean number of species	1.7	1.4	1.4	1.4	1.5	1.1	1.2	1.0	1.5
Number of collecting sites	57	96	37	47	10	9	5	2	264

101

Ponds and pools are much smaller water bodies, with area less than 0.1 km². Pools are smaller than ponds, less than 500 m². They are very shallow, less than 1 m deep. They do not have exposed shores.

Swamps and marshes are overgrown wetlands, the water depth is variable but is usually about 20 cm. They have a dense cover of emergent sedges (mostly *Carex* spp.). They can cover large areas.

Pools and ponds are much more affected by the air temperature than lakes due to their small water volume. They freeze and thaw earlier than lakes, an ice cover usually forming on the first day of frost. Most of these small water bodies are shallow, often less than 0.5 m. Many of them freeze solid during the winter. In the South they can freeze and thaw a few times during the period October to April.

The chemical composition of the water in lakes and ponds is variable, and depends on the origin of the water. Lakes and ponds usually have a pH above 7, and often up to pH 9. Pools in bogs and marshes have a pH between 6 and 7.

The conductivity of pools, ponds and lakes is very variable, from about 20 to over 100 μS/cm at 25°C.

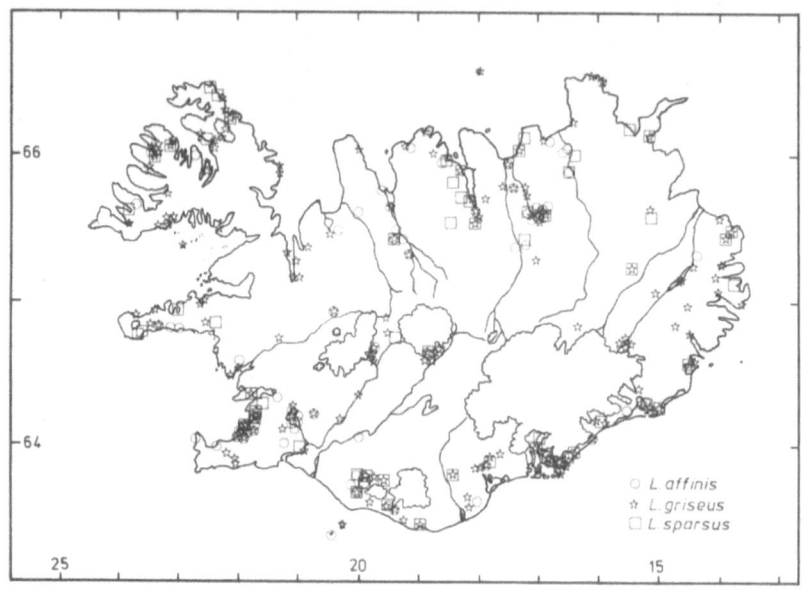

Fig. 1. Distribution of adults and larvae of *L. affinis*, *L. griseus* and *L. sparsus*.

Fig. 2 shows the July isotherms (after Eythórsson and Sigtryggsson 1971; Einarsson 1976 and per. comm.). The South is the warmest part of Iceland during the summer, but lowland areas in the North and East are also warm, due to their continental character. The coasts have an oceanic climate. The highlands are much colder than the lowlands, but weather stations in the highlands are few and data are less complete. The January temperatures

102

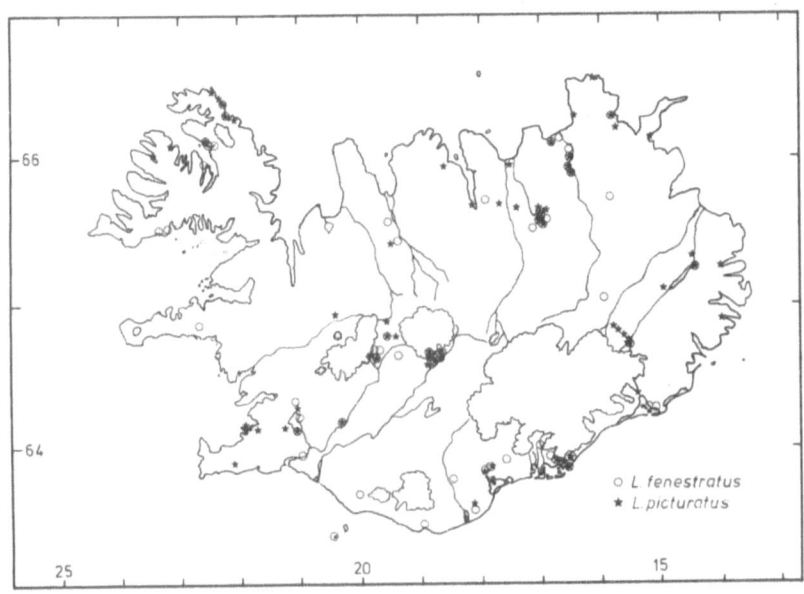

Fig. 2. Distribution of adults and larvae of *L. fenestratus* and *L. picturatus*.

relate more to height above seal level than the July temperatures. The south coast has an average temperature of about $0-1°$ C during January, whereas the north coast has about $-1°$ C. The average temperature falls gradually further inland and is about -4 to $-8°$C in the central plateau (Eyhórsson and Sigtryggsson 1971; Einarsson 1976).

GEOGRAPHICAL DISTRIBUTION AND HABITAT SELECTION

The geographical distribution shows distinct patterns and falls into 4 categories: (a) found all over Iceland (5 species), (b) found where summer temperatures are high (4 species), (c) found in the East and North (1 species) and (d) found all over Iceland, except in the East (1 species).

In the first category are *Limnephilus affinis* Curt. *L. fenestratus* (Zett.), *L. griseus* (L.), *L. picturatus* McL. and *L. sparsus* Curt. (Figs. 1–2). *L. affinis, L. griseus* and *L. sparsus* were mainly found in waters that do not freeze solid during winters (Table 1). *L. affinis* and *L. griseus* had the widest choice of habitats, mainly in lakes and rivers but were also found in ponds and swamps and in saline waters, *L. affinis* up to $2°/_{oo}$ and *L. griseus* up to $17°/_{oo}$. *L. sparsus* was never common in any habitat, but was found in streams, mainly run-off streams and in lagoons and estuaries saline to $1°/_{oo}$. Larval growth can take place in all seasons but is reduced in winter (Gislason 1977b, 1978a). *L. fenestratus* and *L. picturatus* were only found in stagnant waters. *L. picturatus* was common in ponds, pools and swamps in the highlands,

103

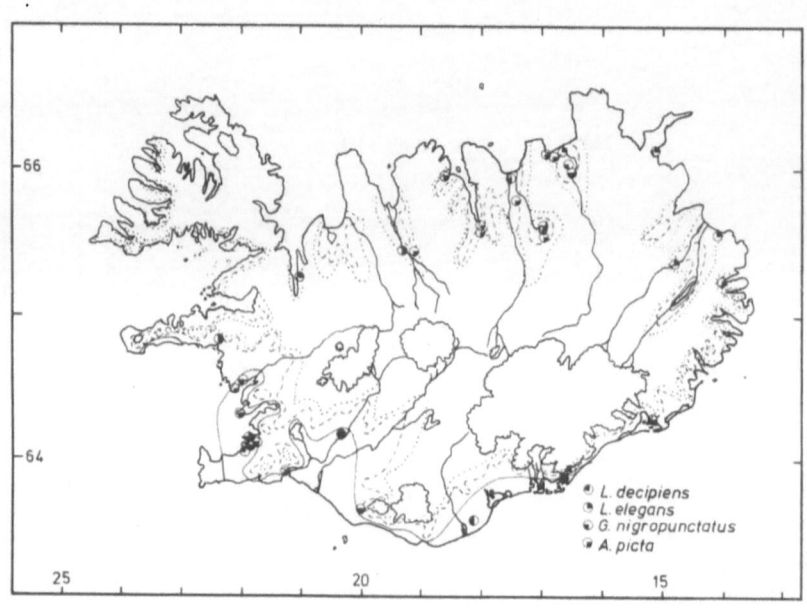

Fig. 3. Distribution of adults and larvae of *L. decipiens, L. elegans, G. nigropunctatus* and *A. picta.* ——— 11° C, – – – – 10° C, and . . . . 9° C July isotherm lines (based on Eythórsson and Sigtryggsson 1971; Einarsson 1976 and pers. comm.).

L. fenestratus larvae were only found in one pool in the South, but the adults were very common among sedges and cotton grass (*Eriophorum* spp.) in swamps. These two species have a northern distribution, *L. fenestratus* and *L. picturatus* have an arctic and a subarctic distribution respectively in the Holarctic (Fristrup 1942; Nimmo 1978).

Species found only where summer temperatures are high are *L. decipiens* (Kol.), *L. elegans* Curt., *Grammotaulius nigropunctatus* (Retz.) and *Agrypnia picta* Kol. (Fig. 3). *L. decipiens* was only found in the South and Southwest where the mean temperature for July is over 11° C, but the others were also found in other areas of Iceland, with July temperatures higher than 10° C (*G. nigropunctatus*) and 9° C (*L. elegans* and *A.picta*). None of these species are common. The larvae are only found in shallow stagnant waters, either in small ponds, pools and swamps (Table 1) which freeze solid during the winter. These species have only the summer for larval growth, the winter is spent in ice as eggs or early instars, with flight periods in late summer (*L. decipiens* and *L. elegans*) or as 3rd–5th instar larvae, with flight periods in early summer (*G. nigropunctatus* and *A. picta*) (Gíslason 1977b, 1978b). These species also have a southerly distribution in England (Hickin 1967; Crichton 1971) and Scandinavia (Fristrup 1942). Temperature could affect their flight activities, since most species require a minimum temperature to be able to fly (Solem 1969, 1976; Gíslason 1977b).

The species in the 3rd and 4th categories are *Apatania zonella* (Zett.) and *Potamophylax cingulatus* (Steph.) respectively (Fig. 4). *P. cingulatus*

104

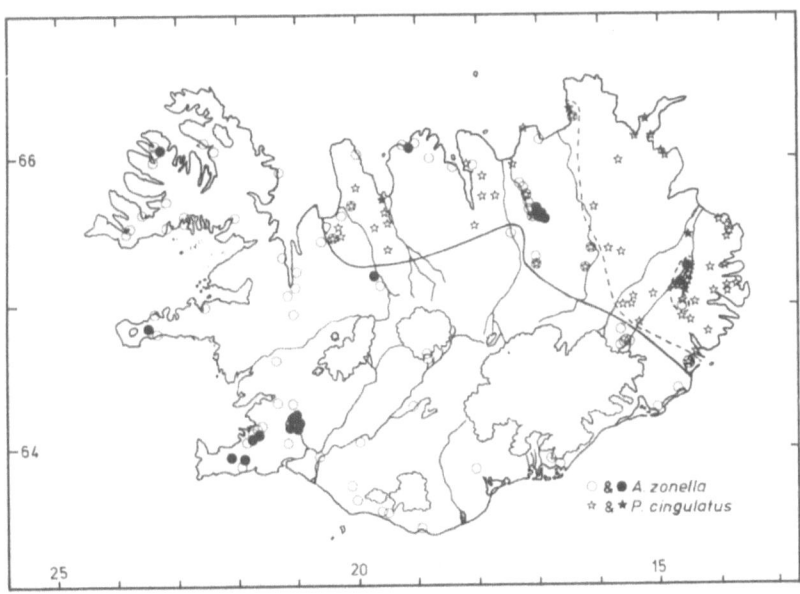

Fig. 4. Distribution of larvae of *A. zonella* and *P. cingulatus*, based on material collected in 1974–78. Solid line: western limit of the distribution of *P. cingulatus*. Broken line: eastern limit of the distribution of *A. zonella*. Open symbols: larvae in lotic waters, solid symbols: larvae in lentic waters.

presumably a recent immigrant to Iceland (Gíslason 1975, 1981), was probably windborne to the East from the Faroe Islands, Scandinavia or Britain and is spreading to other parts of Iceland, and at the same time excluding *A. zonella* from streams it colonizes (Gíslason 1981). Hitherto *P. cingulatus* has not crossed the central highlands to the South. Its southern limit is near 500 m a.s.l. although it is found at high altitudes in the East. It is mainly found in running waters, in the East it is the commonest species in run-off streams and it is also common in spring-fed streams. It rarely occurs in lakes, but has been found in lagoons and estuaries with salinity up to $1.2°/_{oo}$. *A. zonella* is very frequent in streams and lakes. It was found in 80% of spring-fed streams searched, 47% of run-off streams and 62% of lakes. It is abundant in lakes in the East and North-East. The larvae have been found in saline waters from 1 to $17°/_{oo}$. Until recently, *A. zonella* was found all over Iceland (Fristrup 1944; Gíslason 1974, 1981). The interactions of these species are discussed elsewhere (Gíslason 1981).

RANGE OF HABITAT SELECTION IN ICELAND AND ELSEWHERE

Comparing the range of habitats selected by Icelandic species with those selected by the same species from other countries shows that the Icelandic larvae select a much wider range of habitats.

105

The following species select the same habitats in Iceland as in other regions (Lepneva 1966; Hiley 1973): *L. affinis, L. decipiens, L. elegans, G. nigropunctatus* and *A. picta. L. affinis* is found in ponds, pools and salt-marsh pools in Britain (Sutcliffe 1960; Hiley 1973), and in lakes and streams in the U.S.S.R. (Lepneva 1966). It is also recorded from the Gulf of Finland in salinity up to $6^{\circ}/_{\circ\circ}$ (Siltala 1906). *L. decipiens, L. elegans, G. nigropunctatus* and *A. picta* are found in swamps and in grassy pools in Iceland and elsewhere (Lepneva 1966; Hiley 1973).

A. zonella is not recorded from lakes in the U.S.S.R. nor from saline water (Lepneva 1966).

L. sparsus is only recorded from stagnant water bodies in U.S.S.R. and Britain. In Iceland it is only recorded from running water, or water with wave action, some of them slightly saline.

L. griseus is usually only recorded from pools outside Iceland, but it is very common in streams and lakes in Iceland, where it is also found in slightly saline water; Siltala (1906) recorded it from the Gulf of Finland in $5-6^{\circ}/_{\circ\circ}$ salinity.

P. cingulatus in Britain is only found in streams and lakes (Hiley 1973), but in Iceland it is also found in estuaries and lagoons.

The reason for this wider range of habitats of Icelandic larvae may be that fewer species are found in Iceland compared with other European countries. This may enable species from stagnant water to occupy streams because of lack of competition with other species (cf. McArthur and Wilson 1967).

DISTRIBUTION OF ICELANDIC TRICHOPTERA OUTSIDE ICELAND

Table 2 gives a synopsis of the distribution of Icelandic Trichoptera outside Iceland. Most of the species found in Iceland also occur in Eurasia, eight of them in Britain and Denmark, and all in Fennoscandia. Only three of the Icelandic species are found in Greenland and three in North America. Two, *A. zonella* and *L. fenestratus*, are arctic and are found in northern Scandinavia and northern America. *A. zonella* is also found in Spitzbergen, where it is the only Trichoptera species recorded (Fristrup 1942). *L. picturatus* is subarctic and is only found in northern parts of Eurasia and America outside Iceland (Fristrup 1942; Nimmo 1978).

Iceland forms the northern limit of the distribution of *L. affinis, L. decipiens, L. elegans, L. sparsus, G. nigropunctatus* and *A. picta.* Four of these species, *L. decipiens, L. elegans, G. nigropunctatus* and *A. picta* have a southerly distribution and are not found in northern Norway. All the boreal species are found in Britain.

The Faroe Islands are a cluster of islands between the Shetlands and Iceland, extending in south to north direction over 130 km. The largest island is 47 km long and 12 km wide. Seventeen Trichoptera species are known from the Faroes (Henriksen 1929), but only four of the eleven species found in Iceland. Six of those not found in the Faroes are species mainly found in swamps. Swamps are rare in the Faroes, which are basalt mountains with steep slopes rising from the sea, with little lowland. The seventh species

106

Table II. Synopsis of the distribution of Icelandic Trichoptera outside Iceland. Based on Fristrup (1942), Ross (1944), Svensson and Tjeder (1975). In calculating the number of families present, the old family Sericostomatidae is regarded as three families (Brachycentridae, Lepidostomatidae and Sericostomatidae) according to Ross (1944).

	North America	Greenland	Iceland	Spitzbergen	The Faroes	Norway	Sweden	East Fennoscandia	Denmark	Britain
Apatania zonella	+	+	+	+		+	+	+		
Limnephilus affinis			+		+	+	+	+	+	+
L. decipiens			+			+	+	+	+	+
L. elegans			+			+	+	+		+
L. fenestratus	+	+	+			+	+	+	+	
L. griseus		+	+		+	+	+	+	+	+
L. picturatus	+		+			+	+	+		
L. sparsus			+		+	+	+	+	+	+
Grammotaulius nigropunctatus			+			+	+	+	+	+
Potamophylax cingulatus			+		+	+	+	+	+	+
Agrypnia picta			+			+	+	+	+	+
Total	3	3	11	1	4	11	11	11	8	8
Number of species in the regions ca.	1200	7	11	1	17	167	215	207	152	193
Number of families in the regions	15	1	2	1	6	14	15	14	15	15

Number of known species in the world ca. 7000

Number of known families in the the world ca. 40

not found in the Faroes is *A. zonella*, which is an arctic species. The southern *A. muliebris* McL. is however represented in the Faroes by a local subspecies, *A. muliebris kolteriana* Nielsen (Nielsen 1969). The thirteen Trichoptera species from the Faroes, which do not occur in Iceland are boreal species either confined to rivers or lakes.

Several factors could contribute to the difference between the caddis fauna of Iceland and of the Faroes. Iceland is many times larger than the Faroes, but lies two times further away from the mainlands of Britain and Scandinavia. Species have a better chance to reach the Faroes than Iceland, but less chance to survive there, because of the small size of the islands (cf. McArthur and Wilson 1967). This could possibly explain the casual character of the caddis fauna in the two regions. Also, the mild climate in the Faroes could favour a boreal caddis fauna.

ACKNOWLEDGEMENTS

I wish to thank Dr G.N. Philipson, Newcastle University, for his advice and encouragement during the tenure of this study. Thanks for assistance in field work are due to the staff at the Institute of Biology, Univ. of Iceland, especially to Prof. Agnar Ingólfsson, and the Messrs Erling Olafsson, Nat. Hist. Museum. Reykjavík and Hálfdán Björnsson, Kvísker, Öraefi. Prof. Arnthór Gardarsson, Univ. of Iceland, Mr M. Brazil, Univ. of Stirling and my wife Kristín read the manuscript of this paper. I am indebted to Mrs Dóra Jakobsdóttir for typing the manuscript.

Financial support was received from Icelandic Science Foundation, Icelandic Governments Students Loan Fund and NATO Science Fellowship.

REFERENCES

Crichton, M.I., J. Zool. Lond. 163: 533–563, 1971.

Einarsson, M.E., Vedurfar a Islandi. Idunn, Reykjavik, 1976.

Eythórsson, J. and Sigtryggsson, H., Zool. Icel. 1 (3): 1–62, 1971.

Fristrup, B., Zool. Icel. 3 (43–44): 1–23, 1942.

Garside, A. and Leak, S. Newcastle Univ. Exploration Soc. Mimeogr. rep.: 16–20, 1974.

Gíslason, G.M., Natturufr. 44: 129–139, 1974.

———., *In* Eyjabakkar, Landkonnun ogrannsóknir á gródri og dýralifi, ed. H. Guttormsson and G.M. Gíslason. Orkustofnun, Reykjavik, 1977a.

———., Ph.D. Thesis, Univ. of Newcastle upon Tyne, 1977b.

———., Verh. int. Verein. Limnol. 20: 2622–2629, 1978a.

———., *In* Proc. 2nd Int. Symp. Trichoptera, ed. M.I. Crichton, pp. 135–146, Junk, The Hague, 1978b.

———., Natturufr. 50: 35–45, 1980.

———., *In* Proc. 3rd Int. Symp. Trichoptera, ed. G.P. Moretti, pp. 93–97, Junk, The Hague, 1981.

Henrikson, K.L., Zool. Faroes. 2 (1, 38): 1–11, 1929.

Hickin, N.E., Caddis Larvae. Larvae of the British Trichoptera, Hutchinson, London, 1967.

Hiley, P.D., Ph.D. thesis, Newcastle, Univ. 1973.

Kjartansson, G., Natturufr. 15: 113–145, 1945.

————., Geogr. Tidsskr. 64: 174–187, 1965.
Lepneva, S.G., Fauna of the U.S.S.R. Trichoptera Vol. 2 No. 2, 1966. Israel Program for Scientific Translations. Jerusalem, 1971.
Lindroth, C.H., Zool. Bidr. 13: 105–599, 1931.
————., et al., Ent. Scand. Suppl. 5: 1–280, 1973.
MacArthur, R. and Wilson, E.O., The Theory of Island Biogeography. Princeton Univ. Press, Princeton, 1967.
Nielsen, A., Ent. Medd. 37: 313–318, 1969.
Nimmo, A.P., Syesis 11: 117–133, 1978.
Philipson, G.N., Verh. int. Verein. Limnol. 18: 312–319, 1972.
Poulsen, E.M., Vidensk. Medd. Dansk Naturh. Foren. 78: 81–124, 1924.
Rist, S., Icelandic fresh waters 1. Raforkumalastjori – Vatnamaelingar, Reykjavik, 1956.
————., Jökull 19: 121–127, 1969.
Ross, H.H., Bull. Ill. St. Nat. Hist. Surv. 23: 1–326, 1944.
Sigurjónsdóttir, H., Manuscript, Univ. Icel. 1974.
Siltala (Silfvenius), A.J., Acta Soc. Fauna Flora fenn. 28 (6): 1–21, 1906.
Solem, J.O., Cand. Real. thesis, Trondheim Univ., 1969.
————., Norw. J. Ent. 23: 23–28, 1976.
Sutcliffe, D.W., Proc. R. ent. Soc. Lond. (A) 35: 156–162, 1960.
Svensson, B.W. and Tjeder, B., Ent. Scand. 6: 261–274, 1975.
Tjeder, B., Opusc. ent. 29: 143–151, 1964.

DISCUSSION

Moretti: What is the maximum water temperature in which Trichoptera have been found?

Gíslason: The larvae of *A. zonella, L. affinis* and *L. griseus* have been found in thermal streams with temperatures of up to 20° C, and the larvae of *L. picturatus, L. decipiens* and *A. picta* have been found in swamps and pools which can be as warm as 25° C in the summer.

Moretti: Does *Grammotaulius nigropunctatus* have a diapause in Iceland?

Gíslason: No, none of the Icelandic Trichoptera undergo an imaginal diapause.

Crichton: Do any Icelandic species have a two-year life cycle?

Gíslason: No, they are all univoltine.

Nielson: I think that *A. zonella* may have a life cycle of two or more years. It is found in lake Gjuuvatn at about 2,000 metres above sea level, which is not ice-free every year.

Gíslason: In Iceland it is univoltine in all types of habitats.

Nielsen: In Central Norway, I have found *A. zonella* only in melt water lakes above the snow line. There I have found males to be quite common.

EFFECTS OF ACCLIMATION TO TEMPERATURE ON THE RESPIRATORY RATE OF SOME LIMNEPHILID LARVAE

J.D. HARRISON AND R.M. BADCOCK

SUMMARY

A micro-Winkler technique was used to study the effects of temperature acclimation on respiratory rates of two limnephilid larvae, *Chaetopteryx villosa* (Fabr.) and *Potamophylax cingulatus* (Stephens).

Larvae from two sites with different temperature regimes were compared; the results suggest that the respiratory abilities of the larvae were affected by their thermal history. The effects of acclimation of larvae to a constant temperature in the laboratory confirms the field results.

INTRODUCTION

It is well known that respiratory rates of caddis fly larvae vary with temperature (Collardeaux 1961; Collardeaux-Roux 1966; Phillipson and Moorhouse 1976; Roux 1979).

Phillipson and Moorhouse (1976) measured the respiratory rates of *Polycentropus flavomaculatus* (Pictet) and *Plectrocnemia conspersa* (Curtis), two species of stream dwelling polycentropodid, over a range of temperatures, and found differences between the metabolism-temperature curves of the two species, which they associated with the differences in stream temperature commonly experienced by the larvae. They concluded from their results that temperature may be an important factor affecting the downstream succession found in this species pair. Similar conclusions were drawn for three species of *Hydropsyche*, although oxygen consumption was not measured directly in these experiments (Phillipson and Moorhouse 1974).

However, a number of authors have shown that respiratory rates in invertebrate poikilotherms are affected by a variety of factors, including acclimation to temperature (e.g. Bullock 1955; Vernberg and Vernberg 1972; Meyer-Bornsen 1976). This would suggest that the results obtained by Phillipson and Moorhouse (1974, 1976) could be a result of distribution, rather than a demonstration of a factor affecting distribution.

The aim of the study described here was to discover to what extent caddis larvae may be acclimatised to conditions in the field, and to demonstrate acclimation to temperature in the laboratory. *Acclimation* can be defined as the compensatory response of an organism to an alteration of one environmental factor, while *acclimatisation* refers to a response to fluctuation in a number of factors (Vernberg and Vernberg 1972).

Materials and methods

Fourth or fifth instar larvae of two species of limnephilid were used in this investigation, *Chaetopteryx villosa* (Fabricius) and *Potamophylax cingulatus* (Stephens).

Oxygen consumption was measured using the closed bottle method and the oxygen content of the water was determined by micro-Winkler analysis (Fox and Wingfield 1938), using the azide modification. This method was slightly modified however. Instead of micro-titration with sodium thio-sulphate, the iodine level was measured with a spectrophotometer (Rees and Hilton 1977). Many trials were carried out using standard solutions and this method was proved to be quite accurate.

The larvae used in these experiments had cases composed mainly of mineral particles, so it was assumed that any errors due to respiration of the case components would be low and affect all larvae to the same extent. It was similarly assumed that epiflora and epifauna would have little effect on the experimental result. No larvae had a heavy infestation of rotifers. Larvae were weighed after oven drying at 60°C for 48 hours.

Field acclimatisation of metabolism

Two stream sites were chosen with differing temperature regimes (Fig. 1). Coombes Brook (C) is a typical stony troutbeck in Staffordshire, England, while Tumbling Brook (T) is a headstream of Coombes Brook. The graphs show (Fig. 1) that the main stream experienced more extreme temperature conditions during the spring and summer of 1980, so it was postulated that the thermal differences shown between the streams might be reflected in the respiratory metabolism of larvae from those streams.

Larvae of *C. villosa* were collected from both streams, then kept at 5° C in aerated pond water, without food, for at least 48 hours. Oxygen consumption was measured over several hours with three larvae per bottle. Absolute respiration rate was calculated in milligrams of oxygen per hour and plotted against the dry weight of the larvae (Fig. 2). The data were then subjected to an analysis of covariance. The regression coefficients do not differ significantly and it is reasonable to assume that the data were drawn from normal populations with common variance. Oxygen consumption of larvae from Coombes Brook was found to be significantly higher at any given weight ($F = 17.34$, $P < 0.005$).

Acclimation to constant temperature

Large numbers of *C. villosa* larvae were collected from one site, and kept in aerated pond water at 5° C. Larvae were removed from the collection after

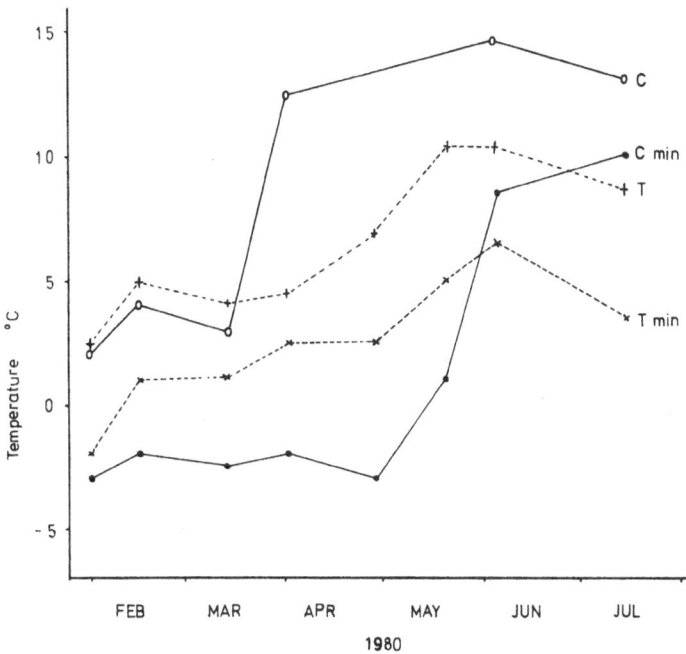

Fig. 1. Temperature for Coombes Brook (C ———) and Tumbling Brook (T ————) during 1980. Lower lines show minimum temperature, upper lines show means. Measurement using a maximum-minimum thermometer.

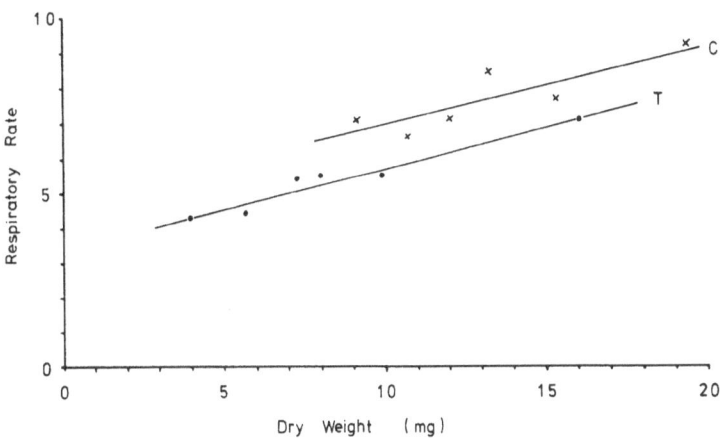

Fig. 2. Comparison of respiratory rates of larvae of *Chaetopteryx villosa* from Coombes Brook (C) and Tumbling Brook (T), calculated in mg O_2/hr ($\times 10^{-3}$) and determined at 5°C.

known periods of time, and rates of respiration determined in the manner described above. Rate of respiration per unit weight was then calculated using dry weight, and this was plotted against acclimation time in hours. Increase in rate between 25 hours and 96 hours was apparent, but

113

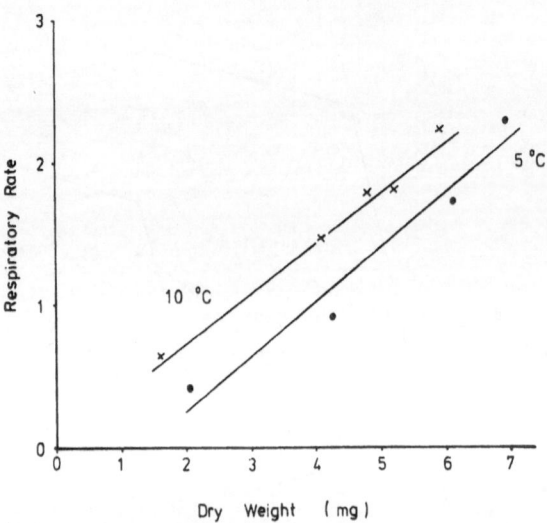

Fig. 3. Comparison of respiratory rates of larvae of *Potamophylax cingulatus* maintained at 10°C and 5°C, calculated in mg O_2/hour ($\times 10^{-3}$), and determined at 5°C.

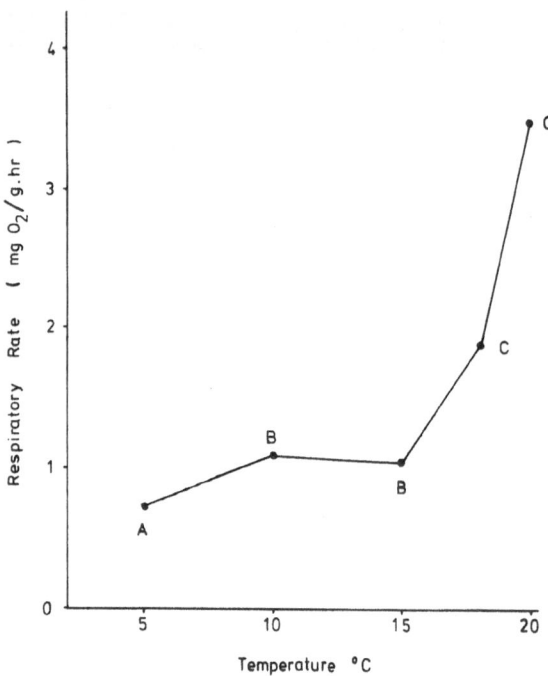

Fig. 4. Temperature-metabolism curve for *Chaetopteryx villosa* collected from streams at 13.5°C, determinations made after at least 48 hours at the experimental temperature. Differences A/B and B/C are significant ($P < 0.05$).

114

the regression coefficient did not differ significantly from zero ($t = 1.99$, $P > 0.05$).

Larvae collected at the same time but kept at a temperature of $13.5°$ C, similar to that experienced in the field, showed a slight decrease of rate over 50 hours, but the differences were not significant.

Larvae of *Potamophylax cingulatus* were collected from a local stream and maintained in two water baths at $5°C$ and $10°C$. Larvae were fed on partly decayed leaves.

After several months acclimation, larvae were removed to another water bath at $5°C$ without food for several days. The oxygen consumption of individual larvae was then measured using the method described above, and plotted against dry weight (Fig. 3). The data were again subjected to an analysis of covariance. The regression coefficients do not differ significantly, and it can again be assumed that the data were drawn from normal populations with common variance. Oxygen consumption of larvae kept at $10°C$ was significantly higher at any given weight ($F = 13.38, P < 0.01$).

DISCUSSION

Chaetopteryx villosa is a common British caddis fly, larvae of which are abundant in many small streams. Respiratory rates of the larvae vary with temperature in the same way as those of other caddis larvae (Fig. 4), metabolic rates increasing with temperature increase, but being insensitive over a range of temperatures. It has been suggested (Precht et al. 1973) that this plateau in the curve shows the temperature range to which an animal is adapted. Experimental results presented here were only for respiration rates at $5°C$, but it can be seen from these results that larvae of *C. villosa* possibly have the ability to acclimate to temperature, a process which, in effect, alters the position of the temperature-metabolism curve. The experiments on acclimation to constant temperature in *C. villosa* and *Potomophylax cingulatus* suggest that this occurs over a matter of months rather than days.

It is important to appreciate that temperature is one of many factors affecting respiratory rates. The rate can vary diurnally, with feeding, oxygen availability or activity (Phillipson and Moorhouse 1976; Roux 1979), or any combination of these and other factors. Feldmeth (1970) showed that acclimation to current velocity affected respiratory rates significantly in larvae of two species of *Pycnopsyche* (Limnephilidae). Adaptation may also be behavioural rather than physiological (Bournaud 1972), but this does not prevent interpretation of the results presented here.

It must also be pointed out that this is a preliminary study. The data base is low and could be very much affected by a few bad results, nevertheless the results of the experiments with *Chaetopteryx villosa* and *Potamophylax cingulatus* suggest that acclimation of metabolism to temperature does occur.

From this we can conclude that more experiments must be carried out on larval acclimation before it can be assumed that the effect of temperature on metabolic rate is an important factor in distribution, and not that differences in distribution account for differences in metabolic rate. In addition it may be that some species have a greater ability to acclimate to environmental factors than others; this could well affect distribution and needs further detailed investigation.

ACKNOWLEDGEMENTS

We thank the Royal Society for the Protection of Birds, the Staffordshire Trust for Nature Conservation and Mr M. Waterhouse for permission to collect in Coombes Valley Nature Reserve, also Mr R. Genn for assistance with field work. We are grateful to the Department of Biological Sciences, University of Keele for the research studentship which supported J.D.H. for three years during this and other studies on Trichoptera.

REFERENCES

Bournaud, M., Ann. Limnol. 8: 141–216, 1972.
Bullock, T.H., Biol. Rev. 30: 311–342, 1955.
Collardeaux, C., Hydrobiologia, 18: 252–264, 1961.
Collardeaux-Roux, C., Hydrobiologia 27: 385–394, 1966.
Feldmeth, C.R., Physiol. Zool. 43: 185–193, 1970.
Fox, H.M. and Wingfield, C.A., J. exp. Biol. 15: 437–445, 1938.
Meyer-Bornsen, E., Arch. Hydrobiol. 77: 176–204, 1976.
Phillipson, G.N. and Moorhouse, B.H.S., Freshwat. Biol. 4: 525–533, 1974.
————., ————., Freshwat. Biol. 6: 347–353, 1976.
Precht, H., Christophersen, J., Hensel, H. and Larcher, W., Temperature and Life, Springer Verlag, Berlin, Heidelberg, New York, 1973.
Rees, T.D. and Hilton, J., Lab. Pract. 26: 91–93, 1977.
Roux, C., Freshwat. Biol. 9: 111–117, 1979.
Vernberg, W.B. and Vernberg, F.J., Environmental Physiology of Marine Animals, Springer-Verlag, Berlin, Heidelberg, New York, 1972.

CADDIS FLY SYSTEMATICS UP TO 1960 AND A REVIEW
OF THE GENERA (INSECTA: TRICHOPTERA)

L.W.G. HIGLER

SUMMARY

A review of the families and genera of Trichoptera found in the
Trichopterorum Catalogus is presented. The total number of families, genera
and species increase through the years.

INTRODUCTION

While preparing volume XVI of the *Trichopterorum Catalogus,* I became
aware of the need for a complete review of world Trichoptera families and
genera. With the help of volumes I–XV and the index (Fischer 1973), I
compiled this and counted the number of species in each genus. I fully
realize that many new species and genera have been described in the past
twenty years; however, the 1960 catalogue is still the most comprehensive
review available today. In due time I hope to up-date this to 1970. In the
meantime, this review, which gives an insight into caddis fly systematics,
should prove useful.

The families, with the number of genera and species in each, are listed
in Table I; extinct families, genera and species are, also, included.
Hydropsychidae is the largest family; Limnephilidae and Leptoceridae follow.
The greatest number of genera are found in Limnephilidae (114),
Hydroptilidae (69) and Hydropsychidae (55). The successful evolutionary
development of Limnephilidae postulated by Ross in 1956 is reflected in
its high number of genera. A comparison of our results and the figures given
by Fischer (1965) for 1938 and 1951 shows that the number of families
increased from 20 in 1930 to 26 in 1951 and then to 35 in 1960; the number
of genera from 517 to 598 and then to 630. There is a parallel rise in the
number of species: 3702, 4537 and 5546, with a probable logarithmic
increase during the last two decades. Malicky (1973) estimates the total to
be about 10,000 species and with the ever rising number of trichopterologists,
we can expect our knowledge on this interesting order to expand.

The genera are listed in alphabetical order in Table II, a cross precedes the
names of extinct genera; the figure in brackets is the number of the family

Table I. Trichoptera described up to 1961. The first six families are extinct. The genera in the second column are extinct, those in the third column contain living as well as extinct species. The last column lists the total of extinct species in each family.

Family		Genera	+		Species	+
1. Necrotauliidae	+	15	15		43	43
2. Prosepididontidae	+	1	1		1	1
3. Microptysmatidae	+	2	2		2	2
4. Kalophryganeidae	+	1	1		1	1
5. Cladochoristidae	+	2	2		2	2
6. Prorhyacophilidae	+	1	1		1	1
7. Glossosomatidae		12		1	232	1
8. Rhyacophilidae		39	1	1	494	3
9. Philopotamidae		12	2	2	318	7
10. Hydroptilidae		69	1	5	428	6
11. Stenopsychidae		4		1	60	1
12. Polycentropodidae		37	7	6	396	81
13. Psychomyiidae		22	1	2	265	17
14. Xiphocentronidae		1			4	
15. Hydropsychidae		55		3	779	6
16. Arctopsychidae		3			26	
17. Phryganeidae		17	3	2	129	38
18. Plectrotarsidae		2		1	4	1
19. Phryganopsychidae		1			5	
20. Limnocentropodidae		2			10	
21. Molannidae		4		1	27	2
22. Helicophidae		2			3	
23. Calamoceratidae		11		2	97	2
24. Philorheithridae		6			8	
25. Odontoceridae		18	4	1	73	6
26. Leptoceridae		46		3	729	7
27. Goeridae		10		3	82	3
28. Limnephilidae		114	5	1	750	17
29. Philanisidae		2			2	
30. Lepidostomatidae		42	5		236	10
31. Brachycentridae		9		2	81	2
32. Bereidae		9		1	31	3
33. Sericostomatidae		44	4		130	4
34. Uenoidae (Thremmatidae)		3			12	
35. Helicopsychidae		13	8	1	85	10
Totals		630	63	39	5546	277

to which the genus belongs and the last figure shows the number of species in each genus. The largest genus is *Rhyacophila* with 322 species, the work of Schmid (1970) in India and Tibet must have now brought the number to about 500, followed by *Hydropsyche* (212), *Limnephilus* (180), *Chimarra* (164), *Athripsodes* (148) and *Oecetis* (126). The genus *Athripsodes* has now been divided into *Ceraclea* and *Athripsodes* (Morse and Wallace 1976).

Fig. 1 shows the distribution of the species over genera graphically and reveals that 261 (41.4%) are monospecific, that 97 (15.4%) have two species and that when we come to the genera with seven species (10), 80% have already been accounted for. The last 3%, starting with genera of 60 or more species, has by far the greatest value for ecological research.

Table II. Genera of Trichoptera, described up to 1961. Between brackets the figure representing the family (table 1); followed by the number of species. + extinct genera.

Abacaria (15) 3
Abaria (13) 3
Abtrichia (10) 2
Acostatrichia (10) 2
Acrocentropus (12) 1
Acrophylax (28) 4
+ Adelomyia (35) 1
Adicella (26) 24
Adinarthrella (30) 4
Adinarthrum (30) 4
Aethaloptera (15) 3
Agapetus (7) 113
Agoerodella (30) 1
Agoerodes (30) 8
Agraylea (10) 9
Agrypnia (17) 10
Allobiosis (8) 1
Allochorema (8) 1
Allocosmoecus (28) 1
Alloecella (32) 2
Alloecentrella (32) 1
Allogamus (28) 11
Allomyia (28) 9
Allosetodes (26) 2
Allotrichia (10) 9
+ Amagupsyche (17) 1
+ Amechanites (35) 1
Amphicosmoecus (28) 1
Amphipsyche (15) 21
Amphipsychella (15) 1
Anabolia (28) 24
Anachorema (8) 4
Anacrunoecia (30) 4
Anagapetus (7) 4
Anisocentropus (23) 38
Anisogamodes (28) 1
Anisogamus (28) 3
Annitella (28) 10
Anomalocosmoecus (28) 1
Anomalopterygella (28) 2
Antarctoecia (28) 1
Antarctopsyche (15) 2
Antillopsyche (12) 1
Antillotrichia (10) 2
Antipodoecia (33) 1
Antoptila (7) 4
Apatania (28) 54
Apataniana (28) 4
Apatanodes (28) 1
Apatelina (28) 4
Apatidelia (28) 1
Aphiloreithrus (24) 1
Aphropsyche (15) 1
Aplatyphylax (28) 4
Apsilochorema (8) 17

+ Archaeocrunoecia (30) 3
+ Archaeoneureclipsis (12) 2
Archaeophylax (28) 1
+ Archaeotinodes (13) 13
+ Archiptilia (1) 1
Archithremma (34) 1
+ Archotaulius (35) 1
Arctopora (28) 2
Arctopsyche (16) 17
Arctopsychodes (16) 1
Argyrobothrus (10) 1
Ascalaphomerus (23) 2
Aselas (33) 1
Asmicridea (15) 2
Astenophylina (28) 1
Astratodina (28) 2
Asynarchus (28) 20
Atanatolica (26) 1
Athripsodes (26) 148
Atomyiodes (30) 1
Atopsyche (8) 33
Atriplectides (25) 1
+ Aulacomyia (33) 1
Australobiosis (8) 1
Australochorema (8) 1
Austrheithrus (24) 2
Austrochorema (8) 1
Austrocosmoecus (28) 1
Austropsyche (15) 1
Austrotinodes (13) 6
Axiocerina (26) 1
Ayabeopsyche (30) 1

Bachorema (8) 2
Baliotrichia (10) 5
Banksiola (17) 5
Banyallarga (15) 2
Barbarochton (31) 1
Barynema (25) 2
Barypenthus (25) 6
Beaumontia (28) 1
Beraea (32) 12
Beraeamyia (32) 3
Beraeodina (32) 1
Beraeoptera (33) 1
Bereodes (32) 4
Betrichia (10) 1
Blepharopus (15) 1
Brachycentriella (31) 1
Brachycentrus (31) 26
Brachypsyche (28) 2
Brachysetodes (26) 6
Brethesella (26) 1

Table II. (continued)

Caenota (33) 2
Cailloma (8) 4
Calamoceras (23) 2
Caldra (15) 1
Caledopsyche (15) 1
Caloca (25) 4
Canoptila (7) 1
Catagapetus (7) 1
Catoxyethira (10) 3
Celaenotrichia (10) 1
Centromacronema (15) 8
Cerasma (33) 1
Cernotina (13) 14
Ceylanopsyche (35) 6
Chaetopterna (28) 1
Chaetopterygella (28) 2
Chaetopterygopsis (28) 2
Chaetopteryx (28) 15
Chaimacheramus (33) 1
Chathamia (29) 1
Cheumatopsyche (15) 91
Chiasmoda (15) 2
Chilocentropus (12) 1
Chiloecia (28) 1
Chilostigma (28) 3
Chilostigmodes (28) 2
Chimarra (9) 164
Chimarrhafra (9) 2
Chionophylax (28) 3
Chloropsyche (15) 2
Chrysotrichia (10) 4
Chyranda (28) 1
+ Cladochorista (5) 1
+ Cladochoristella (5) 1
Clavichorema (8) 5
Clistoronia (28) 4
Clostoeca (28) 2
Cochliopsyche (35) 1
Coenoria (33) 1
Conoesucus (33) 3
Consorophylax (28) 2
Conuxia (33) 1
Cordillopsyche (12) 1
Costachorema (8) 4
Costatrichia (10) 1
Costora (33) 3
Crunoecia (30) 4
Crunoeciella (30) 4
Cryptochia (28) 3
Cryptothrix (28) 1
Culoptila (7) 4
Cyrnellus (12) 3
Cyrnodes (12) 1
Cyrnopsis (12) 1
Cyrnus (12) 12

Dampfitrichia (10) 1
Dasystegia (17) 15
+ Derobrochus (12) 8
Desmona (28) 1
Dhatrichia (10) 1
Dibusa (10) 1
Dicaminus (10) 1
Dicosmoecus (28) 10
Dinarthrella (30) 2
Dinarthrena (30) 2
Dinarthrodes (30) 6
Dinarthropsis (30) 2
Dinarthrum (30) 29
Dinomyia (30) 1
Diplectrona (15) 67
Diplectronella (15) 5
Diplex (15) 1
Dipseudopsis (12) 82
Dolichocentrus (31) 1
Dolochorema (8) 1
Dolotrichia (10) 3
Drusus (28) 39
Dyschimus (30) 5

Ecclisomyia (28) 7
Ecclisopteryx (28) 2
Ecnomina (13) 4
Ecnomodellina (13) 1
Ecnomus (13) 71
Ecnopsyche (15) 1
Edpercivalia (8) 7
+ Electracanthinus (9) 1
Electragapetus (7) 3
+ Electraulax (30) 2
+ Electrocerum (25) 1
+ Electrocrunoecia (30) 1
+ Electrodiplectrona (15) 1
+ Electrohelicopsyche (35) 1
+ Electropsilotes (25) 1
+ Electrotrichia (10) 1
Enoicyla (28) 3
Enoicylopsis (28) 1
Eodinarthrum (30) 2
Eodipseudopsis (12) 1
Eoneureclipsis (12) 1
Eosericostoma (33) 2
+ Epididontus (1) 1
Episetodes (26) 1
Eremopsyche (30) 1
Ernodes (32) 6
Erotesis (26) 4
Eubasilissa (17) 9
Eutonella (15) 1
Evanophanes (28) 1
Exitrichia (10) 11

Table II. (continued)

Fabria (17) 2
Farula (28) 4
+ Folindusia (28) 2
Frenesia (28) 2

Ganonema (23) 25
Gastrocentrella (27) 1
Gastrocentrides (27) 2
Georgium (23) 2
Glossosoma (7) 68
Glyphopsyche (28) 2
Glyphotaelius (28) 3
Gnathotrichia (10) 1
Goera (27) 53
Goerita (27) 2
Goerodella (30) 3
Goerodes (30) 54
Goerodina (30) 2
Grammotaulius (28) 12
Grensia (28) 1
Grumicha (33) 1
Grumichella (26) 1
Guerrotrichia (10) 2
Gumaga (33) 1
Gunungiella (9) 6

Hagenella (17) 6
Halesochila (28) 1
Halesus (28) 13
Hampa (33) 1
Helicopha (22) 3
Helicopsyche (35) 68
Hemileptocerus (26) 2
Herbertorossia (15) 1
Hesperophylax (28) 7
Heteroplectron (23) 2
Himalopsyche (8) 23
Holocentropus (12) 29
Homilia (26) 13
Homophylax (28) 4
Homoplectra (15) 5
Hudsonema (26) 6
Hughscottiella (25) 1
Hummeliella (33) 1
Hyalopsyche (12) 11
Hyalopsychella (12) 1
Hydatomanicus (15) 3
Hydatophylax (28) 12
Hydatopsyche (15) 4
Hydrobiosis (8) 18
Hydrochorema (8) 2
Hydromanicus (15) 20
Hydronema (15) 1
Hydropneuma (10) 2

Hydropsyche (15) 212
Hydropsychodes (15) 43
Hydroptila (10) 118
Hydrosalpinx (35) 1

Iguazu (8) 3
Indocrunoecia (30) 1
+ Indusia (28) 6
Ironoquia (28) 5
Isogamus (28) 2
Ithytrichia (10) 4

Javanotrichia (10) 2

+ Kalophryganea (4) 1
Khandalina (13) 1
Kibuneopsychomyia (13) 2
Kizakia (23) 1
Kodala (30) 1
Kosrheithrus (24) 1
Kyopsyche (12) 1

Lamonganotrichia (10) 1
Larcasia (27) 2
Lasiocephala (30) 1
Lectrides (26) 1
Lenarchus (28) 13
Lepania (28) 1
Lepidostoma (30) 56
Leptecho (26) 3
+ Leptobrochus (12) 1
Leptocella (26) 48
Leptocellodes (26) 1
Leptocerina (26) 11
Leptocerodes (26) 1
Leptocerus (26) 18
Leptodermatopteryx (25) 1
Leptodrusus (28) 1
Leptonema (15) 53
Leptophylax (28) 1
Leptopsyche (15) 1
Leptorussa (26) 2
Leptotaulius (28) 1
Leucotrichia (10) 5
+ Liadoptilia (1) 1
+ Liadotaulius (1) 1
Liapota (18) 1
Limnephilus (28) 180
Limnocentropus (20) 9
Limnoecetis (12) 1
+ Limnopsyche (17) 1
Lingora (33) 3
Lithax (27) 6
+ Litobrochus (12) 1

Table II. (continued)

Lorotrichia (10) 1
Loxinum (23) 1
Loxotrichia (10) 4
Lype (13) 10

Macronema (15) 110
Macrostactobia (10) 1
Madioxyethira (10) 1
Magellomyia (28) 22
Maniconeura (30) 2
+ Maniconeurodes (30) 1
Marilia (25) 26
Martynomyia (30) 1
Matasia (33) 1
Matrioptila (7) 1
Mayatrichia (10) 4
Mecynostomella (33) 1
Melampophylax (28) 5
Melanotrichia (13) 3
Mellomyia (30) 2
+ Mesobrochus (12) 2
Mesopaduniella (13) 2
Mesophylax (28) 5
+ Mesotaulius (35) 1
+ Mesotrichopteridium (1) 2
Metachorema (8) 2
Metacosmoecus (28) 1
Metalype (13) 1
Metanoea (28) 3
+ Metarchitaulius (1) 1
+ Metatrichopteridium (1) 2
Metrichia (10) 5
Mexitrichia (7) 6
Micrasema (31) 45
Micrasemodes (31) 1
Microchorema (8) 2
Micropterna (28) 17
Microptila (10) 6
+ Microptysma (3) 1
+ Microptysmodes (3) 1
Microthremma (33) 3
+ Miopsyche (28) 1
Molanna (21) 19
Molanniella (21) 1
Molannodes (21) 5
Monocentra (28) 1
Monocosmoecus (28) 8
Moropsyche (28) 2
Mortoniella (7) 4
Moselyana (28) 1
Moselyella (10) 3
Myotrichia (33) 2
Mystacides (26) 17
Mystacopsyche (24) 1

Namamyia (25) 1
+ Necrotaulius (1) 20
Nectopsyche (26) 2
Nemotaulius (28) 7
Neoatopsyche (8) 3
Neobiosella (9) 1
Neochorema (8) 1
Neolepidostoma (30) 2
Neoleptonema (15) 1
Neophylax (28) 23
Neopsilochorema (8) 1
Neoseverinia (30) 1
Neothremma (28) 3
Neotrichia (10) 14
Nerophilus (25) 1
Neucentropus (12) 1
Neureclipsis (12) 11
Neurocentropus (12) 1
Neurochorema (8) 2
Neurocyta (17) 1
Nolganema (15) 1
Nostrafilla (28) 3
Notalina (26) 6
Notania (28) 2
Nothopsyche (28) 10
Notidobia (33) 8
Notidobiella (33) 2
Notiomyia (23) 4
+ Nyctiophylacodes (12) 1
Nyctiophylax (12) 40

Ochrotrichia (10) 26
+ Ocnerites (17) 1
Odontocerum (25) 1
Oeceotrichia (10) 1
Oecetinella (26) 3
Oecetis (26) 126
Oecetodella (26) 3
Oecismus (33) 3
Oeconesus (33) 4
Oestropsyche (15) 2
+ Ogmomyia (35) 1
Oligophlebodes (28) 5
Oligoplectrodes (31) 2
Oligoplectrum (31) 3
Oligostomis (17) 5
Oligotricha (17) 10
Olinga (33) 2
Onocosmoecus (28) 8
Oopterygia (17) 3
Oropsyche (15) 1
Orphninotrichia (10) 1
Orthotrichia (10) 20
Orthotrichiella (10) 1

Table II. (continued)

Oxydroptila (10) 2
Oxyethira (10) 50
Oxytrichia (10) 2

Padunia (10) 4
Paduniella (13) 18
Pahamunaya (12) 1
+ Paladicella (12) 1
Palaeagapetus (10) 3
+ Palaeocrunoecia (30) 3
+ Palaeohelicopsyche (35) 1
+ Palaeolepidostoma (30) 1
+ Palaeotaulius (1) 1
Pangullia (28) 1
Parachiona (28) 1
Parachorema (8) 1
Paraethaloptera (15) 2
Paranyctiophylax (12) 1
Paraphlegopteryx (30) 5
Parapsyche (16) 8
+ Pararchitaulius (1) 2
Parasericostoma (33) 3
Parasetodes (26) 8
Parastactobia (10) 1
Parastenopsyche (11) 6
+ Parataulius (1) 1
+ Paratrichopteridium (1) 6
Parecnomina (13) 1
Paroecetis (26) 1
Paroxyethira (10) 3
Parthina (25) 1
Pasirotrichia (10) 1
Paulianodes (9) 1
Pectinariopsis (28) no species
                described
Pedomoecus (28) 1
Peltopsyche (10) 2
+ Perissomyia (35) 1
Perissoneura (25) 2
Petroplax (33) 5
Petrothrincus (21) 2
Petrotrichia (10) 1
Phanocelia (28) 1
Phanostoma (15) 2
+ Phenacopsyche (25) 2
Philanisus (29) 1
Philarctus (28) 7
Philocasca (28) 4
Philocrena (8) 1
Philopotamus (9) 24
Philorheithrus (24) 2
Phryganea (17) 49
Phryganopsyche (19) 5
Phylloicus (23) 19

Phylocentropus (12) 10
Phylostenax (28) 1
Pielus (28) 1
Pisulia (30) 3
Placocentropus (12) 1
Platycentropus (28) 5
Platyphylax (28) 3
+ Plecophlebus (25) 1
Plectrocnemia (12) 62
Plectrocnemiella (12) 2
Plectromacronema (15) 1
Plectropsyche (15) 1
Plectrotarsus (18) 3
Plesiopsyche (15) 1
Plethotrichia (10) 1
Plethus (10) 8
Podomacronema (15) 1
Polycentropus (12) 60
Polymorphanisus (15) 20
Polyplectropus (12) 27
Potamophylax (28) 13
Potamyia (15) 4
Propaduniella (13) 1
+ Prorhyacophila (6) 1
+ Prosepididontus (2) 1
Protarra (9) 4
Protodipseudopsis (12) 4
Protomacronema (15) 6
Protoptila (7) 23
+ Pseudoberaeodes (33) 1
Pseudoeconesus (33) 7
Pseudogoera (27) 1
Pseudoleptocerus (26) 8
Pseudoleptonema (15) 3
Pseudomacronema (15) 3
Pseudoneureclipsis (12) 11
Pseudopotamorites (28) 1
Pseudopsilopteryx (28) 1
Pseudoradema (8) 1
+ Pseudorthophlebia (1) 2
Pseudosericostoma (33) 1
Pseudosetodes (26) 1
Pseudostenophylax (28) 40
Pseudostenopsyche (11) 1
Pseudoxyethira (10) 1
Psilochorema (8) 9
Psilopsyche (25) 5
Psilopterna (28) 4
Psilopteryx (28) 5
Psilotreta (25) 16
Psychoglypha (28) 7
Psychomyia (13) 17
Psychomyiella (13) 8
Psychomyiellina (13) 2

Table II. (continued)

Psychomyiellodes (13) 4
Psychoronia (28) 2
Psyllobetina (8) 1
Ptilocolepus (10) 6
Ptilostomis (17) 6
Ptochoecetis (26) 2
Pycnocentrella (32) 1
Pycnocentria (33) 5
Pycnocentrodes (33) 6
Pycnopsyche (28) 18

Quarelia (32) 1

Radema (28) 8
Rhadicoleptus (28) 3
Rheochorema (8) 2
Rhoizema (33) 4
Rhyacophilia (8) 322
+ Rhyacophilites (8) 1
Rhyacophylax (15) 9
Rhyacopsyche (10) 1
Rhynchopsyche (31) 1
Rossiana (28) 1

Saetotrichia (35) 1
Saranchanotrichia (10) 1
Scelotrichia (10) 1
Schizoplex (33) 5
Sciadorus (15) 3
Sciops (15) 5
Semblis (17) 4
Sericostoma (33) 32
Setodellina (26) 14
Setodes (26) 73
Setodina (26) 1
Silo (27) 12
Silonella (27) 2
Silvatares (23) 1
Sinion (27) 1
Smicridea (15) 27
Sortosa (9) 52
+ Sphaleropalpus (33) 1
Stactobia (10) 35
Stactobiella (10) 6
Stenochorema (8) 1
Stenophylax (28) 22
Stenopsyche (11) 47
Stenopsychodes (11) 6
+ Stenoptilomyia (33) 1
Stenoxyethira (10) 2
Sumatranotrichia (10) 1
Symphitoneuria (26) 4
Symphitoneurina (26) 1
Symphitopsyche (15) 2
Synagotrichia (10) 1
Synatopsyche (15) 6

Synchorema (8) 2
Synoestropsis (15) 7

Tagalopsyche (26) 2
Tamasia (33) 1
Tarapsyche (33) 1
Taschorema (8) 6
Tasimia (33) 3
Tasmanthrus (24) 1
Thamastes (28) 2
Theliopsyche (30) 5
Thremma (34) 4
Tinodes (13) 82
Tiphobiosis (8) 5
Tismana (33) 1
Tobikera (26) 1
Triaenodes (26) 93
+ Tricheopteryx (28) 1
Tricholeiochiton (10) 2
Trichophila (8) 1
+ Trichopterella (1) 1
+ Trichopteridium (1) 1
Trichosetodes (26) 11
Trichostegia (17) 1
Trichovespula (30) 1
Triplectides (26) 42
Triplectidina (26) 2
Triplexa (26) 1
Triplexina (26) 1

Uenoa (34) 7
Uenotrichia (10) 1
Ugandatrichia (10) 3
Ulmerochorema (8) 1
Ulmerodes (30) 3
+ Ulmerodina (9) 1

Vigarrha (9) 1

Wormaldia (9) 61

Xanthochorema (8) 1
Xiphocentron (14) 4
Xuthotrichia (10) 4

Ylodes (26) 2
Yphria (20) 1

Zelandopsyche (33) 1
Zelandoptila (10) 1
Zelolessica (22) 1
Zelomyia (13) 1
Zepsyche (33) 1
Zumatrichia (10) 2

124

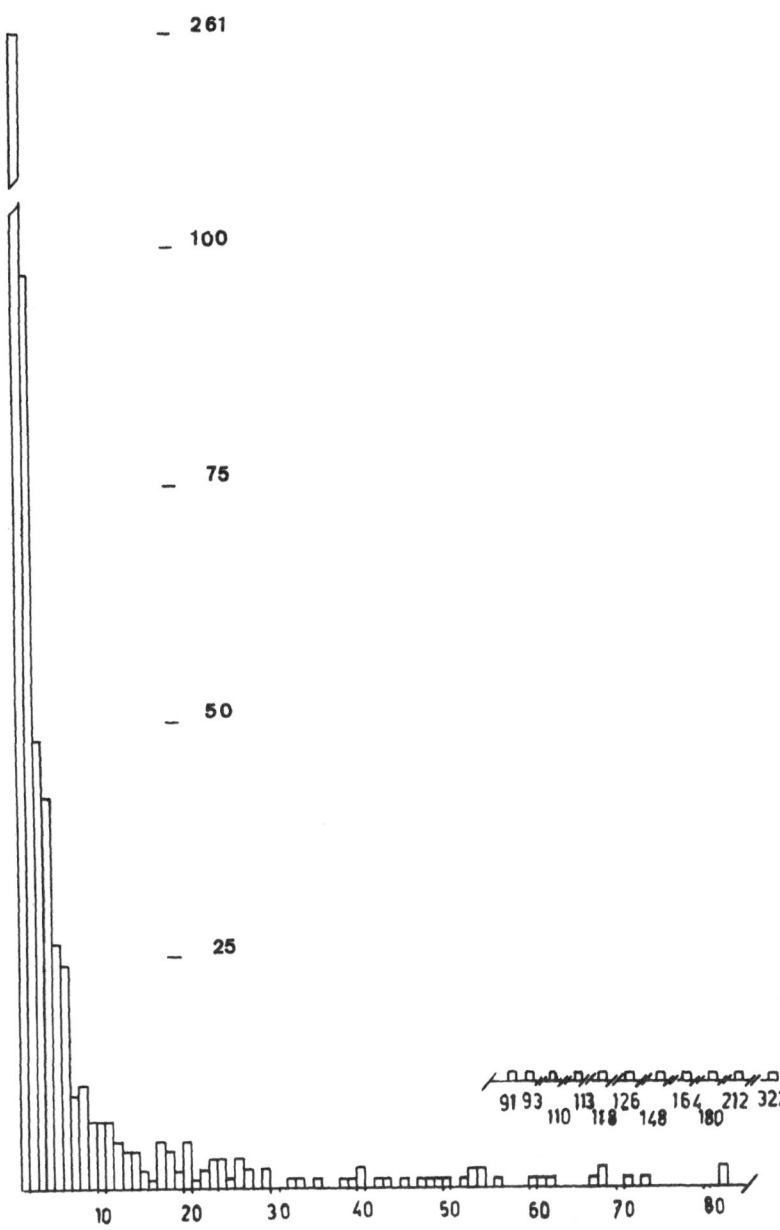

Fig. 1. Distribution of species of Trichoptera over genera. Horizontal: number of species per genus. Vertical: number of genera.

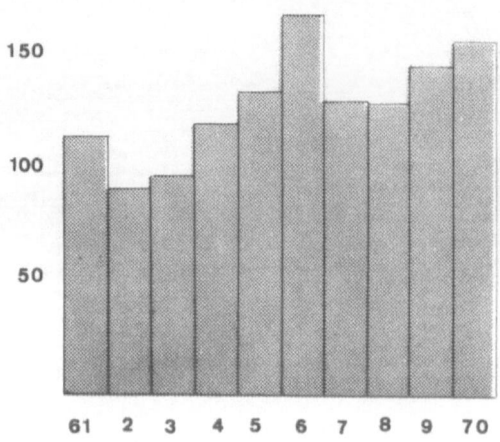

Fig. 2. Number of publications in which caddis flies are mentioned in the years 1961–1970.

Finally, the number of publications in which Trichoptera were mentioned during the ten-year period 1961–1970 are reported in Fig. 2. Short papers, as well as huge tomes, are included and the material is now being worked on for volume XVI of the *Trichopterorum Catalogus*. As the number of publications started to increase gradually in the 'seventies, the supplements will become bigger until a total revision will be required – one wonders if printing will be obsolete before then.

DISCUSSION

Morse: Comparison with bibliographies being prepared by Dr Unzicker (U.S.A.) for the North American Benthological Society and, of course, by Dr Malicky (Austria) for the Trichoptera Newsletter would, certainly, be worthwhile. Also, co-operation with someone with access to computer literature search facilities would, almost certainly, improve your list of references.
Higler: Experience with computer literature search is rather poor; but it can be a help.

REFERENCES

Fischer, F.C.J., *In* Proc. XII Int. Congr. Ent. London 1964, pp 124–125, 1965.
———, Trichopterorum Catalogus. Index volumes I–XI. NEV A'dam, 1973.
Malicky, H., Handb. Zool. 4, 1973.
Morse, J.C., *In* Proc. First Int. Symp. Trichoptera, ed. H. Malicky, pp 33–40, Junk, The Hague, 1976.
Ross, H.H., Evolution and Classification of the Mountain Caddisflies. Urbana: Univ. Ill. Press, pp 213, 1956.
Schmid, F., Mem. Soc. Ent. Canada 66: 230, 1970.

CADDIS LARVAE IN A DUTCH LOWLAND STREAM

L.W.G. HIGLER

SUMMARY

Lowland streams are found in the pleistocene part of the Netherlands. Most of them are polluted or canalized, but a few examples of relatively undisturbed ones can be found. These streams are characterized by the following features: fed by surface water and seapage water from superficial groundwater; an average current velocity below 50 cm/sec; great changes in discharge and current velocity in dependance of rainfall and groundwater storage and sometimes stagnation or drying up of stream stretches.

The fauna is adapted to these conditions and consists of species from stagnant water, species from running water and those, that are adapted to the changing conditions. Species of the first order streams that dry up in summer are *Stenophylax permistus*, *Micropterna lateralis*, *Ironoquia dubia* and *Limnephilus bipunctatus*. In the predominantly stagnant ditches we find *Limnephilus sparsus*, *L. auricula*, *L. extricatus* and *L. affinis*. *Bereodes minutus* and *Plectrocnemia conspersa* can be found as well, if the current velocity is not too low and the period of drying up is shorter than three months.

The stream under study is the 'Hierden Beek', a lowland stream of 20 km with a fall of 1:100. The drainage area is about 140 km^2, but the effective drainage area is about 50 km^2, because of the presence of a clay layer of that dimension at a depth of about eight meters in otherwise sandy soils. Before the mid sixties the stream was poor in nutrients and acid, especially in the upper course. Pollution by fattening farms caused an increase in the nutrient content of both the stream water and the ground water over the clay layer. The pH rose from 5.5 to 7. In 1977 a heavy pollution by beet pulp added to the deterioration of the stream fauna.

Caddis larvae were studied incidentally from 1965 up to 1970 and regularly in 1972–1974 and 1977–1979. Thirty species have been recorded. We studied the qualitative and quantitative distribution in the stream, the effects of pollution and the role of caddis larvae in the food webs of the stream.

The distribution of species shows a certain zonation over three zones, the last of which is hardly of importance for Trichoptera. In Table I the main characteristics of these zones are given.

Table I. Some characteristics of the zonation

	I	II	III
Width (m)	1.5–3	3–4	4–5
Depth (cm)	10–50	30–90	30–100
Current velocity (cm/s)	5–25	10–50	20–50
Bottom	sand + detritus	sand + gravel and leaves	sand + gravel sand + detritus
Species	*Limnephilus lunatus* *Plectrocnemia conspersa*	*Halesus radiatus interpunctatus* *Micropterna sequax*	

The distribution is influenced by the input of organisms from the acid seapage area in the upper course and by the presence of faster running stretches with boulders or stones as a result of human acitivities in the past (water mills). These stretches are narrower and in some cases small cemented cascades are present, so that the current velocity and the oxygen content are higher than could be expected in the natural course. Species from the acid seapage area can be restricted to it (*Beraea pullata, Adicella reducta*) or they add to the fauna of the upper course (*Plectrocnemia conspersa, Lype reducta, Bereodes minutus*). Species on the cemented boulders or the scattered stones in the narrow stretches are only to be found there (*Silo nigricornis*) or they add to the fauna of the upper course (*Hydropsyche angustipennis, Potamophylax rotundipennis, Anabolia nervosa*), or the middle course (*Hydropsyche pellucidula, Lype phaepa*).

The increased pollution of the last years caused the disappearance or diminishing of a number of species, among which the caddis species *Silo nigricornis* and *Hydropsyche angustipennis*. They belong to the most threatened species by their restricted distribution and their specialized diet (obligate scraper and obligate filtering collector). Most Limnephilids are generalists that are shredders in their last instars, but can be predators or collectors as well. They are not confined to the stony substratum and are to be found through most parts of the stream. The most numerous caddis larvae are *Limnephilus lunatus* in the upper course and *Halesus radiatus interpunctatus* in the middle course. The pollution in the upper course has caused a shift in the zonation as could be found before. The numbers of species from the upper course have diminished considerably.

TRICHOPTERA IN THE ILONA STREAM OF THE MÁTRA MOUNTAINS, NORTH HUNGARY

O. KISS

SUMMARY

A two-year study was carried out on Trichoptera in the Ilona Stream of the Mátra Mountains, which are volcanic in origin. Twenty species were collected; three are new to the Ilona Valley and five are only distributed on the volcanic substrate. The number of species is determined by the geological differences in the substrates, as well as by other ecological factors. The lack of pools seems to cause a slight increase in the number of species downstream and the spectacular increase in the number of individuals in the same area may be attributed to side waters which swell the stream, and to several other ecological factors.

INTRODUCTION

The Trichoptera of the Ilona Valley in the Mátra Mts have been studied in order to complete the data published earlier by Sátori (1939b) and Ujhelyi (1974) and to compare the number of Trichoptera species and individuals from two distinctly different geographical regions; one the limestone Bükk Mts, the other the volcanic andesite and andesite tuff Mátra Mts. During this research, which forms part of a survey on Hungarian fauna, the quantitative and qualitative species composition was analyzed according to the longitudinal division of the stream, the mosaic pattern theory and certain ecological factors.

MATERIALS AND METHODS

The Mátra Mts are the most homogeneous mountain range of volcanic origin in Hungary, and have the highest peak (Kékes, 1015 m). The Eastern part of the Mátra Mts, in which the Ilona Valley lies south of Paráfürdö, is a highly abraded mountain range consisting almost entirely of andesite and andersite tuff. Collapsed blocks of stone are typical of the higher region of the Ilona Valley at 270–420 m (Fig. 1); the mean annual temperature is below 8° C

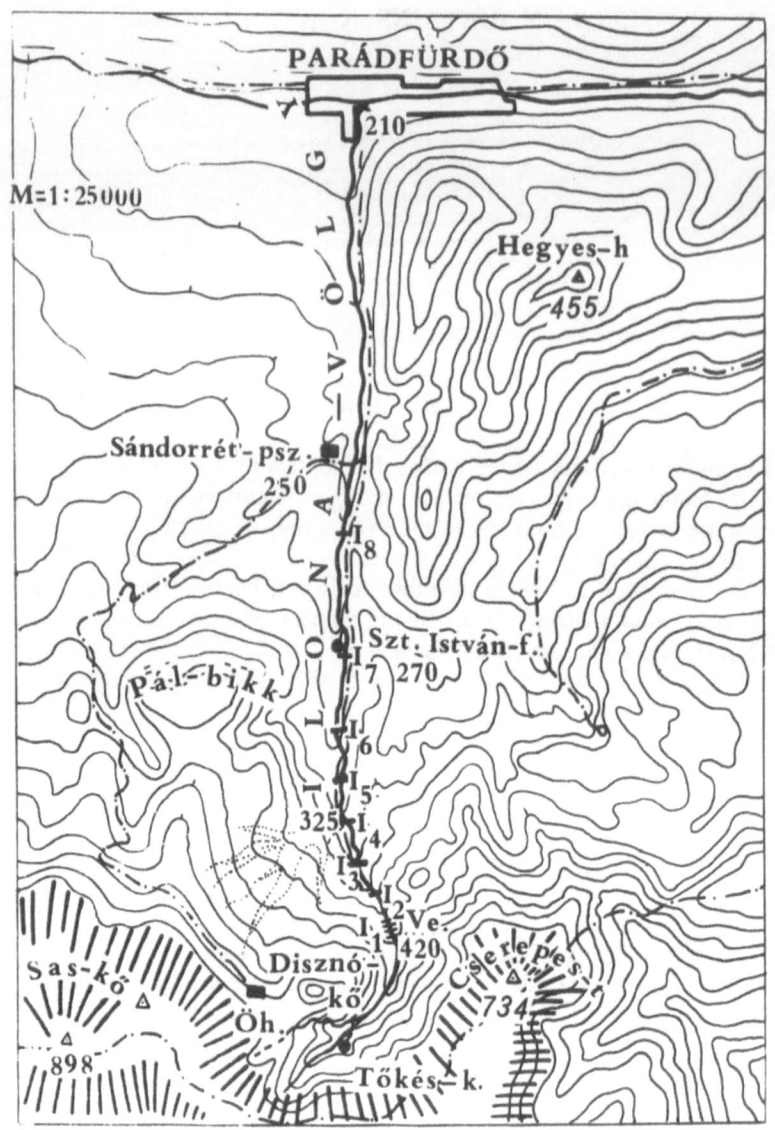

Fig. 1. The area under investigation in the Mátra Mts.

and the annual rainfall 700–800 mm. There are numerous rills, but no pools, in the lower reach of the Ilona Stream. The water output ranges from 25.5 to 100 1/sec.

Trichoptera larvae, pupae and imagos were sampled by the mosaic technique at intervals from July, 1977 to October, 1979. The methods

130

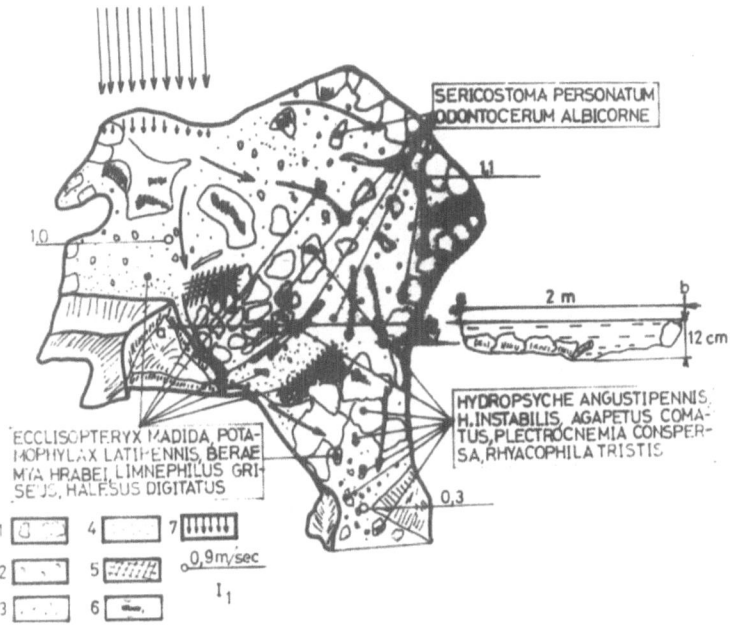

Fig. 2. At the waterfall (I₁). Substrate mosaics: 1. large stones, 2. small stones and gravel, 3. sand, 4. slime, 5. detritus, 6. moss, 7. waterfall.

of Kamler and Riedel (1960) and Macan (1958) were used. The various substrate mosaics were taken into consideration when the seven sampling sites were chosen. The profile diagrams of the sampling stations indicate stream bed characteristics, water depth, water velocity and the species which inhabit the different substrate mosaics. The quantitative and qualitative species composition is, also, given.

SOME TYPICAL TRICHOPTERA HABITATS, MOSAIC PATTERNS AND COMMUNITIES

The rhythmic water volume and velocity fluctuation cause a repeat design in the substrate mosaics which is reflected in their consequences and even the substrate mosaics, themselves are subjected to continuous appearance/space modifications. Thus, the complicated and disintegrated stream biocoenosis depends on the various mosaic pattern substrates and the communities inhabiting them. Trichoptera communities find the micro-substrate most adapted to their environmental needs when they are arranged according to the mosaic pattern.

Five of the seven Ilona Stream sampling stations will be described:

Station No. 1. At the waterfall (I₁). The water of the stream falls over a 5 m high vertical andesite rock wall covered with a thick 4 m wide carpet

131

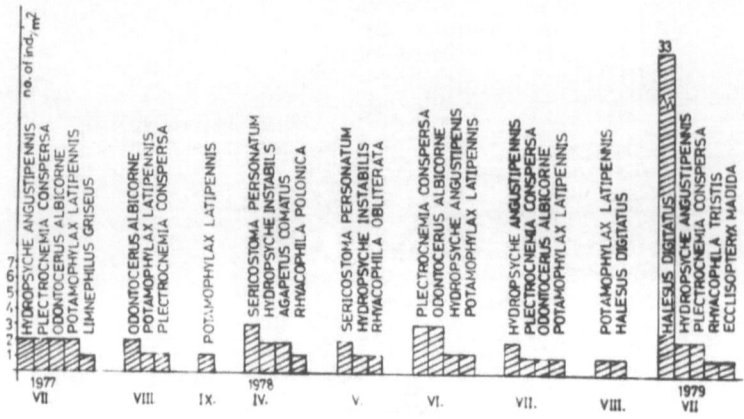

Fig. 3. The species found monthly and the number of individuals per m² at the station I₁.

of several moss species (*Rhynchostegium ripariodes* (Hedw.) Card and *Conocephalum conicum* (Hedw.) (B.S.C.). As the water falls it spreads out into a small round basin where the temperature is 7.6–13.0° C and where the dominant species are *Hydropsyche angustipennis, Plectronemia conspersa* and *Odontocerum albicorne. Rhyacophila tristis* is a very rare species and this is the first time it has been collected there (Figs. 2, 3).

Station No. 3. 150 m away from the waterfall (I₃). The area is shaded by *Fagus silvatica* L., the water turns into rapids and the temperature is in the 6.0–12.9° C range. The dominant species here are *Rhacophila obliterata, Agapetus comatus* and *Hydropsyche instabilis. Crunoecia irrorata* was found here for the first time (Figs. 4, 5).

Station No. 4. 250 m away form the waterfall (I₄). The valley widens and the rapids run between rocky walls. This reach of the stream is sunlit, the vegetation on the banks is rich in *Petasitetum hybridi* Dost. and *Euphorbia cyparissias* L. The water temperature ranges between 7.4–13.5° C. The dominant species are *Rhyacophila obliterata, Sericostoma personatum* and *Agapetus comatus* (Figs. 6, 7).

Station No. 5. Near the plot used for making fires (I₅). The stream bed is shaded by rich vegetation (*Petasitetum hybridi* Dost., *Euphorbia cyparissias* L.,) and water temperature ranges from 9.4–13.0° C. *Hydropsyche instabilis, Potamophylax latipennis* and *Silo pallipes* are among the predominating species (Figs. 8, 9).

Station No. 6. At the bridge (I₆). The stream widens to 3 m and the water flow slows down. The luxuriant vegetation on the banks half shades the

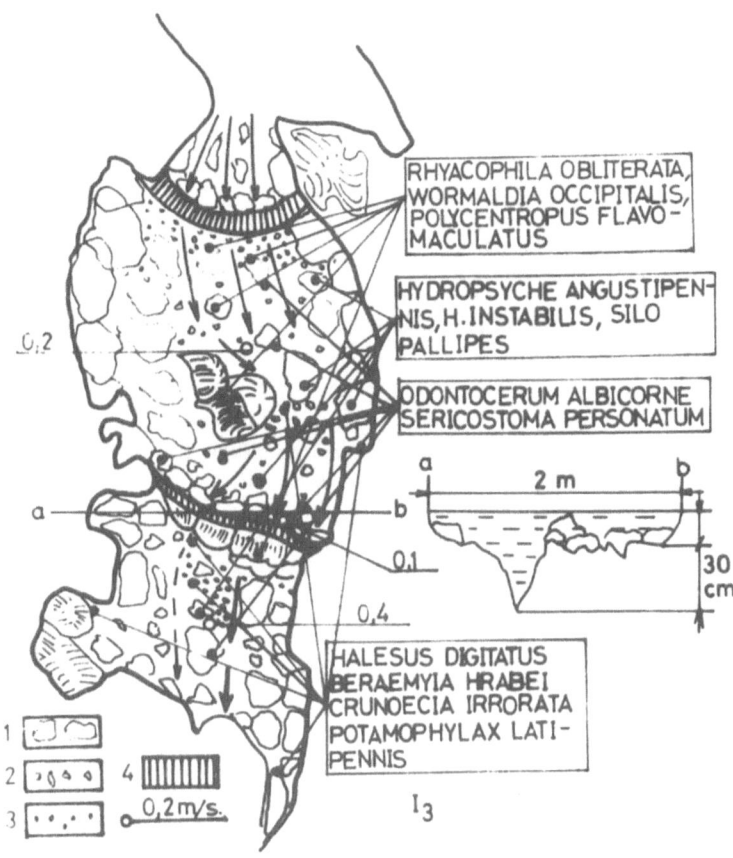

Fig. 4. 150 m away from the waterfall (I_3). Substrate mosaics: 1. large stones, 2. small stones and gravel, 3. sand, 4. waterfall.

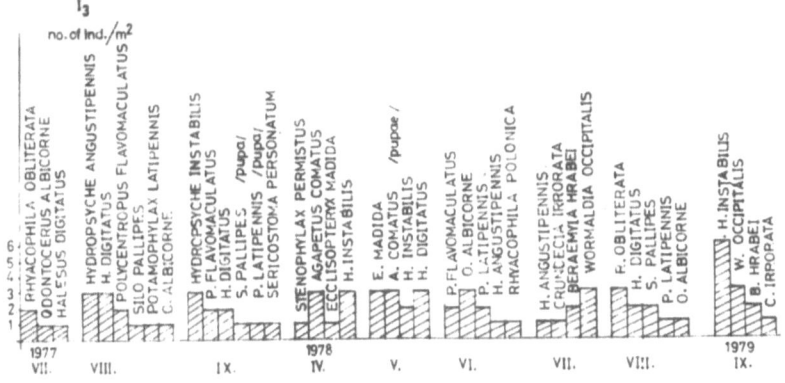

Fig. 5. The species found monthly and the number of individuals per m^2 at the station I_3.

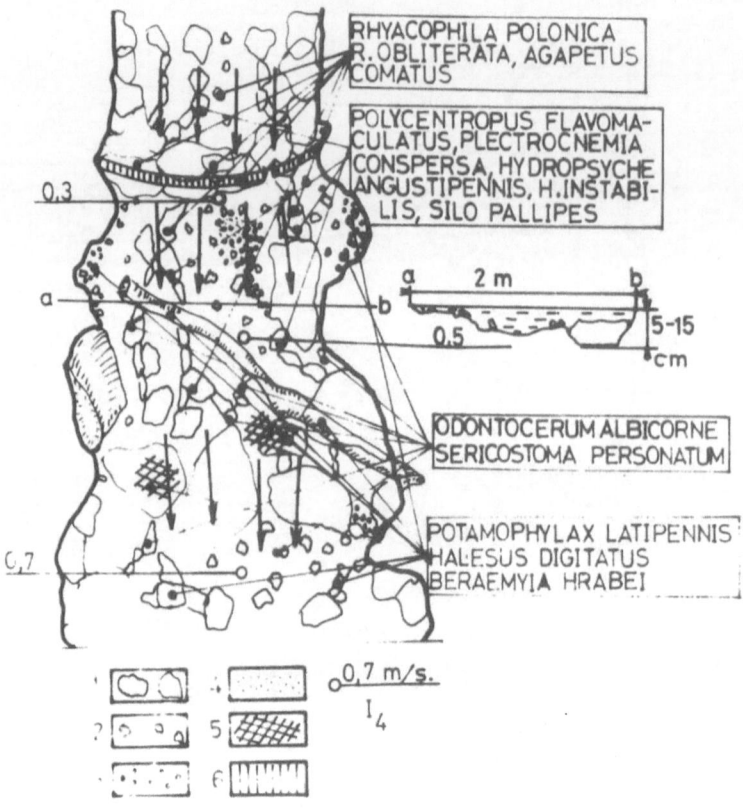

Fig. 6. 250 m away from the waterfall (I_4). Substrate mosaics: 1. large stones, 2. small stones and gravel, 3. sand, 4. slime, 5. detritus, 6. waterfall.

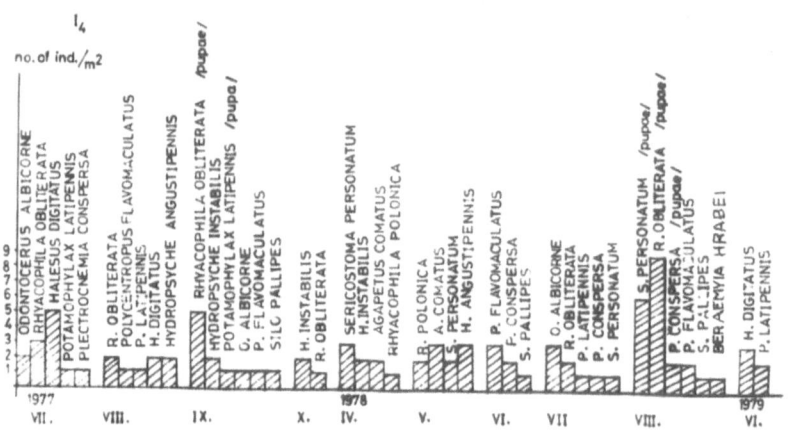

Fig. 7. The species found monthly and the number of individuals per m² at the station I_4.

134

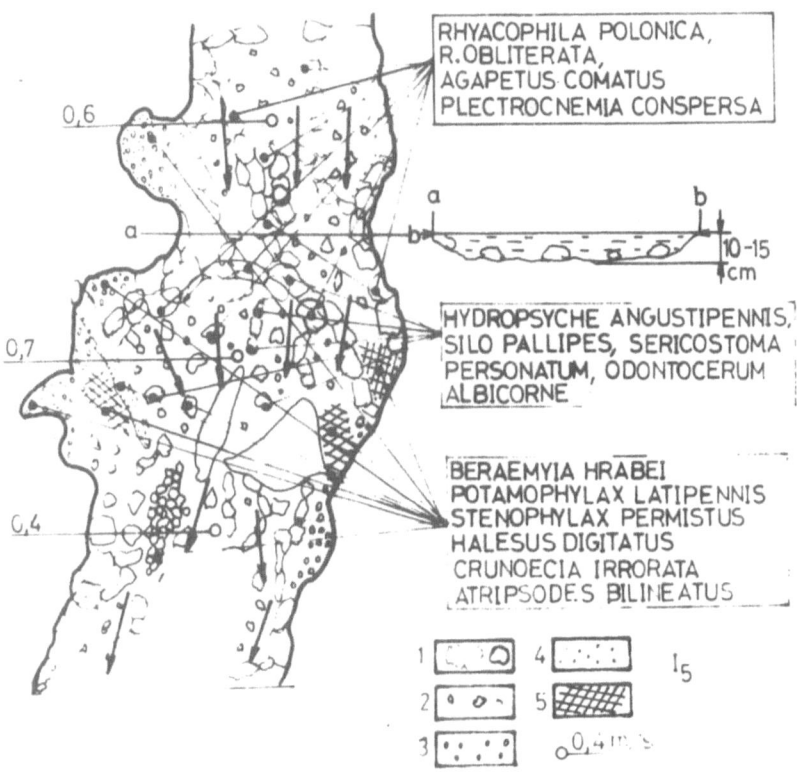

RHYACOPHILA POLONICA,
R. OBLITERATA,
AGAPETUS COMATUS
PLECTROCNEMIA CONSPERSA

HYDROPSYCHE ANGUSTIPENNIS,
SILO PALLIPES, SERICOSTOMA
PERSONATUM, ODONTOCERUM
ALBICORNE

BERAEMYIA HRABEI
POTAMOPHYLAX LATIPENNIS
STENOPHYLAX PERMISTUS
HALESUS DIGITATUS
CRUNOECIA IRRORATA
ATRIPSODES BILINEATUS

I_5

Fig. 8. Near the plot for making a fire (I_5). Substrate mosaics: 1. large stones, 2. small stones and gravel, 3. sand, 4. slime, 5. detritus.

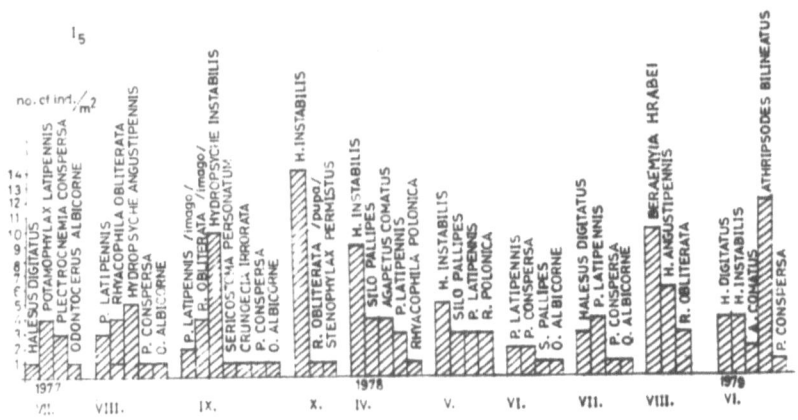

Fig. 9. The species found monthly and the number of individuals per m^2 at the station I_5.

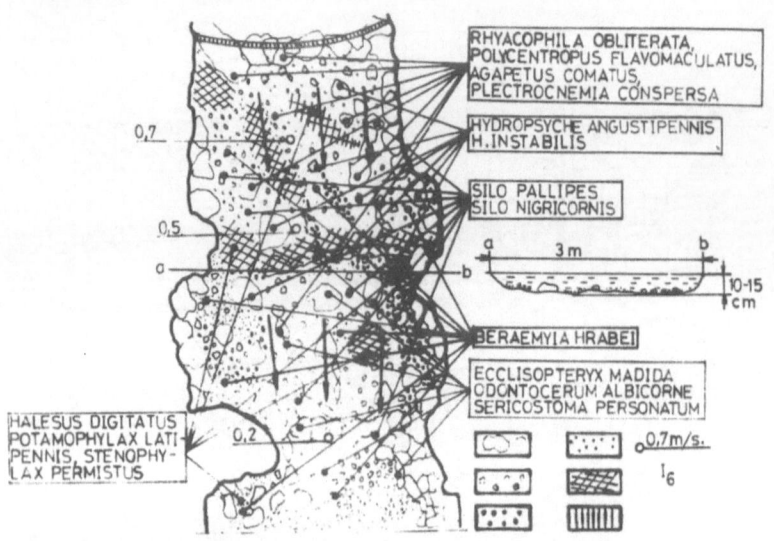

Fig. 10. At the bridge (I_6). Substrate mosaics: 1. large stones, 2. small stones and gravel, 3. sand, 4. slime, 5. detritus, 6. waterfall.

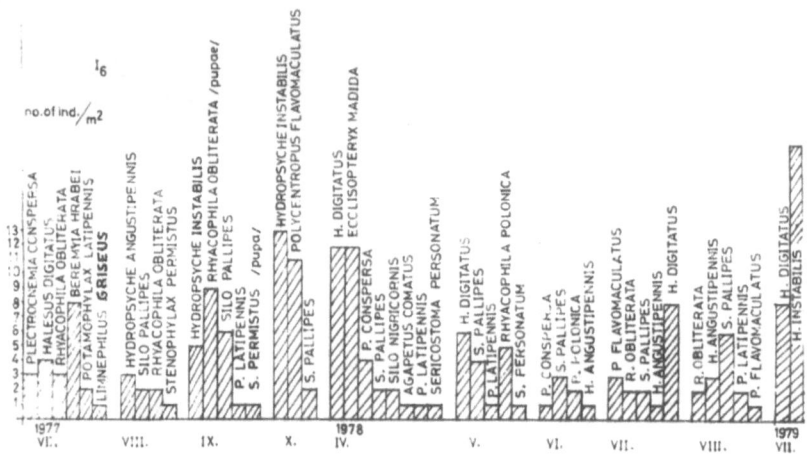

Fig. 11. The species found monthly and the number of individuals per m² at the stations I_6.

stream. Fifteen species are found here, *Hydropsyche instabilis, Silo pallipes, Halesus digitatus* and *Ecclisopteryx madida. Silo nigricornis*, which is seldom seen in Hungary, was collected here for the first time (Figs. 10, 11).

RESULTS AND CONCLUSIONS

Three of the 20 species collected in the Ilona Valley — *Rhyacophila tristis, Crunoecia irrorata* and *Sericostoma personatum* — are new to its fauna; 15 inhabit both andesite and limstone and five occur only on andesite or

136

Table I. List of species in the predominantly andesite Ilona Valley

Species collected	Species in all reaches	New species to the fauna of the Ilona Valley	Species found on both andesite and limestone	Species only found in the Ilona Valley
1. Rhyacophila obliterata (McL.)			+	
2. Rhyacophila tristis (Pict.)		+	+	
3. Rhyacophila polonica (McL.)				+
4. Agapetus comatus (Pict.)			+	
5. Wormaldia occipitalis (Pict.)			+	
6. Hydropsyche instabilis (Curt.)			+	
7. Hydropsyche angustipennis (Curt.)	+		+	
8. Plectrocnemia conspersa (Curt.)			+	
9. Polycentropus flavomaculatus (Pict.)			+	
10. Ecclisopteryx madida (McL.)				+
11. Limnephilus griseus (L.)				+
12. Potamophylax latipennis (Curt.)	+		+	
13. Halesus digitatus (Schrk.)			+	
14. Stenophylax permistus (McL.)			+	
15. Silo pallipes (Fbr.)			+	
16. Silo nigricornis (Pict.)				+
17. Crunoecia irrorata (Curt.)		+	+	
18. Sericostoma personatum (Spence)	+	+	+	
19. Beraemyia hrabei (Mayer)				+
20. Odontocerum albicorne (Scop.)			+	

andesite tuff (Table I). The finding of *Rhacophila polonica, Ecclisopteryx madida, Limnephilus incisus, Silo nigricornis* and *Beraeamyia hrabei* only on a substrate of volcanic origin demonstrates their preference for this limited biosphere.

Table II shows that the pools which precede the lower reaches of the stream cause a significant increase in the number of species, but have less influence on the number of individuals. However, the number of species fluctuates and the increase is only slight when there are no pools. It is noteworthy that the number of species found on limestone (6–7) was

Table II. Comparison of the number of species and individuals from the two different geological areas investigated.

	Stream Szalajka		Limestone		Stream Sebesuíz		Andesite and andesite tuff	
	The upper reach of the stream	The lower reach preceded by pools	The upper reach	The lower reach without any pools or rills	The upper reach		The upper reach	The lower reach with rills
No of species	5	17	6	7	14			15
No of individuals/m^2	107	206	232	216	50			201

considerably smaller than that found on andesite (14—15). This may be due to the geological origin of the substrate, as well as to a number of other ecological factors.

The spectacular increase in the number of individual Trichoptera larvae in the Ilona Stream (50—201) is attributable to the rills which run into the stream and swell it, the presence of plant-origin substrates, moss and algae, the variety of food and the increasing temperature of the downstream waters.

REFERENCES

Bournaud, M. and Keck, G., Acta Oecologica/Oecologia Generalis 1: 131—150, 1980.
Crichton, M.I.; Fisher, D. and Woiwod, I.P., Holoarctic ecology 1: 31—45, 1978.
Hickin, N.E., Caddis Larvae; Larvae of the British Trichoptera, London, Hutchinson, 1967.
Higler, L.W.G., Chandesse 9: 219—223, 1974—1975.
Kamler, E. and Riedel, W., Polskie Archwm Hydrobiol. 8: 87—94, 1960.
Kiss, O., Fol. Hist. nat. Mus. Matr. 4: 63—69, 1976—1977.
————., In Proc. 2nd Int. Symp. on Trichoptera, ed. M.I. Crichton, pp. 89—101, Junk, The Hague, 1978.
Lepneva, S.G., Fauna USSR/Akad. Nauk. USSR, Moszkva/Tom. I—II, 1966.
Macan, T.T., Methods of Sampling the Botton Fauna in Stony Streams, Int. Ass. Theor. and Appl. Limnol., Comm. 8, 1958.
McLachlan, R., A monographic revision and synopsis of the Trichoptera of the European Fauna (Reprint) Hampton: Classey, 1974—1980.
Malicky, V.H., Jahrgang 30: 170—183, 1978.
Moretti, G.P. and Mearelli, M., Riv. Idrobiol. 17: 137—187, 1978.
Statzner, B., In Proc. 2nd. Int. Symp. Trichoptera, ed. M.I. Crichton, pp. 121—134, Junk, The Hague, 1978.
Steinmann, H., Fauna Hungariae 15: 1—351, 1970.
Ujhelyi, S., Fol. Hist. Mus. Matr. 2: 99—115, 1974.
Wichard, W., Gewasser und Abwasser, pp 53)54; pp 85—90, 1974.

FAUNISTIC INVESTIGATIONS ON BULGARIAN
TRICHOPTERA TO JUNE, 1980 – WITH A
REVISED CHECK-LIST

K. KUMANSKI

SUMMARY

The first list of Bulgarian Trichoptera, which includes 218 species in 76 genera is presented. A detailed picture of the present position of faunistic studies in the different regions is given.

INTRODUCTION

Studies on Trichoptera, as well as on many other groups of invertebrates in Bulgaria, began about ninety years ago. *Phryganea varia* Fabr. was the first name recorded (Christovic, 1892). Its classification was evidently wrong and the report now has historical rather than scientific value.

Klapálek with his series of four papers at the turn of the century made the first important contribution on Bulgarian caddis flies and there were but a few, isolated publications in the following half century. Botosaneanu added considerably to our knowledge on Bulgarian Trichoptera with several papers published between 1956 and 1967; while Sykora (1960) and Novak (1971) were responsible for longer papers dealing with Bulgarian material.

Regular and systematic investigations were started by the writer in 1966. Up to that time, 39 publications with information on 130 species had been published. The number of publications now stands at 75 and they contain faunistic data on 226 valid species. As a result of critical analysis, accompanied in most cases by re-examination of the material, 15 species have been found to be wrongly classified; several species, either new or unknown to this country, have been added and the taxonomic status of a few others clarified.

RESULTS

The list now presented includes 218 species belonging to 76 genera and 20 families. The synonymic information contains all other names under which a species has been recorded – both junior synonyms and corrections in determinations published after re-examination of the material.

Anabolia furcata Brau (Locality: River Struma, railway station Skakavitsa, 21.10.1978, 1 ♂, J. Ganev leg. at light) and *Parachiona picicornis* (Pict.) (Locality: Ryla Mts., streamlet in the resort Borovez, 1400 m a.s.l., 6 ♂♂ and 2 ♀♀, A. Popov leg.) are genera and species both new to Bulgaria. While the former was to be expected in this country, *Parachiona* is a new genus for the

CHECK-LIST OF TRICHOPTERA IN BULGARIA

Rhyacophilidae
Rhyacophila Pictet 1834

c 1. *armeniaca* Guérin-Meneville 1843
vicaria Mart. – Botosaneanu & Sykora, 1963

v 2. *fasciata* Hagen 1859
septentrionis McL. – Sykora, 1960

+ 3. *joosti* Mey 1979

v 4. *loxias* Schmid 1970
Rh. sp. "larva monilibranchia" (nov.) – Botosaneanu, 1956

c 5. *mocsaryi* Klapálek 1898

v 6. *nubila* (Zetterstedt 1840)

v 7. *obliterata* McLachlan 1863

v 8. *polonica* Mclachlan 1879
hageni McL. - Klapálek, 1895

c 9. *fischeri* Botosaneanu 1957
philopotamoides McL. – Kumanski, 1969

c 10. *furcifera* Klapálek 1904

+ 11. *kownackiana* Szcsesny 1970

r 12. *braaschi* Malicky & Kumanski 1975

v 13. *obtusa* Klapálek 1894

+ 14. *pirinica* n. sp. (Kumanski, in press)

v 15. *tristis* Pictet 1834

Glossosomatidae
Glossosoma Curtis 1834

r 16. *boltoni* Curtis 1834

+ 17. *conformis* Neboiss 1963

c 18. *discophorum* Klapálek 1902

r 19. *intermedium* (Klapálek 1892)
Synagapetus McLachlan 1879

c 20. *iridipennis* McLachlan 1879
Pseudagapetus insons McL. – Klapálek, 1895
Synagapetus moselyi (Ulm.) – Novak, 1971 (part.)

+ 21. *moselyi* (Ulmer 1938)

r 22. *ater* Kalapálek 1904 (?); n. sp. (?)
iridipennis McL. – Kumanski, 1975a (part.)
moselyi (Ulm.) – Novak, 1971 (part.)
Agapetus Curtis 1834

r 23. *belareca* Botosaneanu 1957

c 24. *delicatulus* McLachlan 1884

c 25. *laniger* (Pictet 1834)

r 26. *ochripes* Curtis 1834

r 27. *rectigonopoda* Botosaneanu 1957

r 28. *slavorum* Botosaneanu 1960

Hydroptilidae
Stactobia McLachlan 1880

r 29. *caspersi* Ulmer 1950

+ 30. *mclachlani* Kimmins 1949
Stactobiella Martynov 1924

+ 31. *risi* (Felber 1908)
ulmeri (Silt.) – Kumanski, 1979a
Orthotrichia Eaton 1873

+ 32. *costalis* (Curtis 1834)
Ithytrichia Eaton 1873

r 33. *lamellaris* Eaton 1873
Oxyethira Eaton 1873

c 34. *falcata* Morton 1893
Hydroptila Dalman 1819

c 35. *forcipata* (Eaton 1873)

+ 36. *ivisa* Malicky 1972

r 37. *kalonichtis* Malicky 1972
taurica Mart. – Novak, 1971
bureschi (n. sp.) – Kumanski, 1972

r 38. *lotensis* Mosely 1930

r 39. *occulta* (Eaton 1873)

c 40. *simulans* Mosely 1920

r 41. *sparsa* Curtis 1834

+ 42. *taurica* Martynov 1934

r 43. *tineoides* Dalman 1819
femoralis (Eat.) – Sykora, 1960

+ 44. *uncinata* Morton 1893
angulifera (n. sp.) – Kumanski, 1974 nov. syn.

c 45. *vectis* Curtis 1834

+ 46. *vichtaspa* Schmid 1959
Agraylea Curtis 1834

c 47. *sexmaculata* Curtis 1834
pallidula McL. – Botosaneanu, 1956
Allotrichia McLachlan 1880

r 48. *pallicornis* (Eaton 1873)
Microptila Ris 1897

r 49. *minutissima* Ris 1897

Philopotamidae
Philopotamus Stephens 1829

v 50. *montanus* (Donovan 1913)

v 51. *variegatus* (Scopoli 1763)
amphilectus McL. – Klapálek, 1913
Wormaldia McLachlan 1865

c 52. *bulgarica* Novak 1971
khourmai bulgarica (n. ssp.) – Novak, 1971 (part.)

r 53. *juliani* Kumanski 1979

+ 54. *khourmai balcanica* Kumanski 1979

140

c 55. *occipitalis* (Pictet 1834)
    triangulifera McL. – Klapálek, 1894
    khourmai bulgarica (n. ssp.) – Novak, 1971 (part.)
r 56. *pulla* (McLachlan 1878)
r 57. *subnigra* McLachlan 1865
c 58. *triangulifera asterusia* Malicky 1972
    triangulifera McL. – Kumanski, 1969

Hydropsychidae
Diplectrona Westwood 1840
r 59. *felix* McLachlan 1878
    Hydropsyche Pictet 1834
v 60. *bulbifera* McLachlan 1878
    subguttata Mart. – Botosaneanu & Sykora, 1963
    guttata Pict. – Novak, 1971
    fallaciosa (n. sp.) – Kumanski & Botosaneanu, 1974
v 61. *bulgaromanorum* Malicky 1977
    guttata Pict. – Klapálek, 1894
    ornatula McL. – Russev, 1962
v 62. *contubernalis* McLachlan 1865
v 63. *dissimulata* Kumanski & Botosaneanu 1974
+ 64. *dentata* Kumanski 1974
c 65. *fulvipes* (Curtis 1834)
    mahrkusha Schm. – Novak, 1971
v 66. *instabilis* (Curtis 1834)
c 67. *tjederi* Botosaneanu & Marinković 1966
r 68. *valkanovi* Kumanski 1974
    tjederi Bots. & Marink. – Novak, 1971 (part.)
v 69. *angustipennis* (Curtis 1834)
v 70. *pellucidula* (Curtis 1834)
c 71. *tabacarui* Botosaneanu 1960
    pellucidula (Curt.) – Sykora, 1960
    Cheumatopsyche Wallengren 1834
c 72. *lepida* (Pictet 1834)

Polycentropodidae
Neureclipsis McLachlan 1864
c 73. *bimaculata* (Linnaeus 1758)
    Plectrocnemia Stephens 1836
c 74. *brevis* McLachlan 1871
v 75. *conspersa* (Curtis 1834)

    Polycentropus Curtis 1835
c 76. *excisus* Klapálek 1894
v. 77. *flavomaculatus* (Pictet 1834)
+ 78. *ierapetra* Malicky 1972
r 79. *irroratus* Curtis 1835
    Holocentropus McLachlan 1878
r 80. *stagnalis* (Albarda 1874)
    Cyrnus Stephens 1836
v 81. *trimaculatus* (Curtis 1834)

Psychomyidae
Psychomyia Latreille 1829
v 82. *pusilla* (Fabricius 1781)
    Lype McLachlan 1878
c 83. *reducta* (Hagen 1868)
    Tinodes Curtis 1834
+ 84. *janssensi* Jaquemart 1957
c 85. *kimminsi* Sykora 1960
c 86. *pallidulus* McLachlan 1878
    manni McL. – Novak, 1971 (part.)
c 87. *popovi* Kumanski 1975
    manni McL. – Novak, 1971 (part.)
+ 88. *raina* Botosaneanu 1960
r 89. *rostocki* McLachlan 1878
r 90. *unicolor* (Pictet 1834)
v 91. *unidentata* Klapálek 1894
r 92. *polifurculatus* Botosaneanu 1956
    valvata Mart. – Kumanski, 1975b

Ecnomidae
Ecnomus Rambur 1842
c 93. *tenellus* Rambur 1842

Phryganeidae
Trichostegia Kolenati 1848
+ 94. *minor* (Curtis 1834)
    Agrypnia Curtis 1835
+ 95. *pagetana* Curtis 1835
c 96. *varia* (Fabricius 1793)
    Phryganea Linnaeus 1758
+ 97. *ochrida* Malicky 1975
    grandis L. – Kumanski, 1975b
    Oligotricha Rambur 1842
r 98. *striata* (Linnaeus 1758)
    Neuronia ruficrus (Scop.) – Botosaneanu & Sykora, 1963

Brachycentridae
Brachycentrus Curtis 1834
c 99. *montanus* Klapálek 1892
    Oligoplectrum McLachlan 1868

r 100. *maculatum* (Fourcroy 1785)
      Micrasema McLachlan 1876
v 101. *minimum* McLachlan 1876

      Limnephilidae
      Drusus Stephens 1837
v 102. *discolor* (Rambur 1842)
      annulatus (Steph.) – Navas,
      1929 (part.)
c 103. *romanicus meridionalis*
      Kumanski 1973
      romanicus Murg. & Bots. –
      Kumanski, 1969
c 104. *biguttatus* (Pictet 1834)
v 105. *botosaneanui* Kumanski 1968
      annulatus (Steph.) – Navas,
      1929 (part.)
      tenellus (Klap.) – Klapálek,
      1913
      Ecclisopteryx madida
      (McL.) – Navas, 1929 (part.)
r 106. *balcanicus* Kumanski 1973 stat.
      nov.
      discophorus balcanicus
      (n. ssp.) – Kumanski, 1973b
r 107. *bureschi* Kumanski 1973
+ 108. *discophoroides* Kumanski 1979
c 109. *discophorus* Radovanović 1942
      rectus McL. – Klapálek,
      1913
      discophorus discophorus
      Rad. – Kumanski, 1973b
+ 110. *osogovicus* n. sp. (Kumanski,
      in press)
+ 111. *popovi* n. sp. (Kumanski, in
      press)
      Ecclisopteryx Kolenati 1848
v 112. *dalecarlica* Kolenati 1848
      guttulata dalecarlica Kol. –
      Botosaneanu & Sykora,
      1963
      Limnephilus Leach 1815
v 113. *rhombicus* (Linnaeus 1758)
c 114. *flavicornis* (Fabricius 1787)
+ 115. *subcentralis* Brauer 1857
c 116. *stigma* Curtis 1834
c 117. *flavospinosus* (Stein 1874)
c 118. *decipiens* (Kolenati 1848)
v 119. *lunatus* Curtis 1834
c 120. *griseus* (Linnaeus 1758)
v 121. *bipunctatus* Curtis 1834
v 122. *affinis* Curtis 1834
+ 123. *incisus* Curtis 1834
c 124. *hirsutus* (Pictet 1834)
r 125. *tauricus* Schmid 1964
c 126. *centralis* Curtis 1834
v 127. *sparsus* Curtis 1834
c 128. *auricula* Curtis 1834

v 129. *vittatus* (Fabricius 1798)
c 130. *extricatus* McLachlan 1865
+ 131. *fuscicornis* Rambur 1842
c 132. *coenosus* Curtis 1834
      Grammotaulius Kolenati
      1848
c 133. *nigropunctatus* (Retzius 1783)
      atomarius (Fabr.) –
      Kumanski, 1968
r 134. *nitidus* (Müller 1764)
      Glyphotaelius Stephens 1837
r 135. *pellucidus* (Retzius 1783)
      Anabolia Stephens 1837
+ 136. *furcata* Brauer 1857
      Asynarchus McLachlan 1880
r 137. *lapponicus* (Zetterstedt 1840)
      fusorius (McL.) – Klapálek,
      1913
      Rhadicoleptus Wallengren 1891
c 138. *alpestris macedonicus*
      Botosaneanu & Riedel 1965
      alpestris (Kol.) – Kumanski,
      1969
      Potamophylax Wallengren
      1891
r 139. *borislavi* Kumanski 1975
v 140. *cingulatus* (Stephens 1837)
      Stenophylax latipennis
      (Curt.) – Botosaneanu, 1956
v 141. *latipennis* (Curtis 1834)
      Stenophylax stellatus
      (Curt.) – Botosaneanu, 1956
c 142. *luctuosus* (Piller 1783)
      excisus (Mart.) – Kumanski,
      1969
r14143. *nigricornis* (Pictet 1834)
c 144. *pallidus* (Klapálek 1900)
      Chionophylax Schmid 1951
+ 145. *mindszentyi bulgaricus*
      Kumanski 1973
      czarnohoricus (Dzied.) –
      Kumanski, 1971
+ 146. *monteryla* Botosaneanu 1957
      czarnohoricus monteryla
      (n. ssp.) – Botosaneanu, 1957

      Halesus Stephens 1836
c 147. *digitatus* (Schrank 1781)
r 148. *tesselatus* (Rambur 1842)
      Parachiona Thomson 1891
+ 149. *picicornis* (Pictet 1834)
      Stenophylax Kolenati 1848
c 150. *mitis* McLachlan 1875
v 151. *permistus* McLachlan 1875
v 152. *speluncarum* McLachlan 1875
      vibex (Curt.) – Guéorguiev &
      Beron, 1962
      vibex speluncarum McL. –

Botosaneanu, 1965
Micropterna Stein 1874
c 153. *caesareica* Schmid 1959
r 154. *fissa* (McLachlan 1875)
r 155. *malaspina* Schmid 1955
v 156. *nycterobia* McLachlan 1875
v 157. *sequax* McLachlan 1875
c 158. *testacea* (Gmelin 1788)
Stenophylax mucronatus
McL. – Kumanski, 1971
Mesophylax McLachlan
1882
c 159. *aspersus* (Rambur 1842)
Allogamus Schmid 1955
c 160. *uncatus* (Brauer 1857)
Chaetopteryx Stephens 1837
+ 161. *bosniaca* Marinković 1959
c 162. *cissylvanica* Botosaneanu 1960
v 163. *stankovici* Marinković 1966
r 164. *bulgarica* Kumanski 1969
r 165. *maxima* Kumanski 1968
Psilopteryx Stein 1874
c 166. *montana* Kumanski 1968
c 167. *schmidi* Kumanski 1970
Chaetopterygopsis Stein
1874
r 168. *maclachlani* Stein 1874
r 169. *sisestii* Botosaneanu 1961
Annitella Klapálek 1907
c 170. *triloba* Marinković 1955

Goeridae
Goera Stephens 1829
c 171. *pilosa* (Fabricius 1775)
Lithax McLachlan 1876
+ 172. *musaca* Malicky 1972
r 173. *obscurus* (Hagen 1859)
Silo Curtis 1830
r 174. *graellsi* Ed. Pictet 1865
c 175. *pallipes* (Fabricius 1781)
c 176. *piceus* (Brauer 1857)

Thremmatidae
Thremma McLachlan 1876
c 177. *anomalum* McLachlan 1876

Lepidostomatidae
Lepidostoma Rambur 1842
r 178. *hirtum* (Fabricius 1775)
Lasiocephala Costa 1857
c 179. *basalis* (Kolenati 1848)

Leptoceridae
Arthripsodes Billberg 1820
r 180. *aterrimus* (Stephens 1836)
c 181. *bilineatus* (Linnaeus 1758)
albifrons (L.) – Kumanski,
1979

r 182. *cinereus* (Curtis 1834)
Leptocerus cinereus (Curt.)
– Botosaneanu, 1956
r 183. *longispinosus* (Martynov 1909)
Ceraclea Stephens 1829
r 184. *alboguttata* (Hagen 1860)
r 185. *annulicornis* (Stephens 1836)
Leptocerus annulicornis
(Steph.) – Botosaneanu &
Sykora, 1963
r 186. *aurea* (Pictet 1834)
Leptocerus aureus (Pict.) –
Botosaneanu & Sykora,
1963
c 187. *dissimilis* (Stephens 1836)
r 188. *riparia* (Albarda 1874)
Athripsodes riparius (Albda.)
– Kumanski & Malicky,
1976
r 189. *senilis* (Burmeister 1839)
Leptocerus senilis (Burm.) –
Botosaneanu, 1956
Mystacides Berthold 1827
c 190. *azurea* (Linnaeus 1761)
r 191. *nigra* (Linnaeus 1758)
Triaenodes McLachlan 1865
r 192. *bicolor* (Curtis 1834)
+ 193. *conspersus* (Rambur 1842)
r 194. *kawraiskii* Martynov 1909
r 195. *simulans* Tjeder 1929
conspersus (Ramb.) –
Novak, 1971
Oecetis McLachlan 1877
r 196. *furva* (Rambur 1842)
r 197. *notata* (Rambur 1842)
c 198. *ochracea* (Curtis 1825)
r 199. *testacea* (Curtis 1834)
r 200. *tripunctata* (Fabricius 1793)
alexanderi (n. sp.) –
Kumanski, 1976 nov. syn.
Setodes Rambur 1842
r 201. *punctatus* (Fabricius 1793)
r 202. *viridis bulgaricus* Kumanski
1976
Leptocerus Leach 1815
c 203. *interruptus* (Fabricius 1775)
Setodes interrupta (Fabr.) –
Klapálek, 1895
r 204. *tineiformis* Curtis 1834
Ymymia tineiformis (Curt.)
– Botosaneanu & Sykora,
1963
Setodes tineiformis (Curt.) –
Soffner, 1973
Adicella McLachlan 1877
c 205. *altandroconia* Botosaneanu &
Novak, 1965
r 206. *balcanica* Botosaneanu

Novak, 1965

c 207. *filicornis* (Pictet 1834)
+ 208. *reducta* (McLachlan 1865)
c 209. *syriaca* Ulmer 1907

Sericostomatidae
Oecismus McLachlan 1876
v 210. *monedula* (Hagen 1859)
Sericostoma Latreille 1825
c 211. *flavicorne* Schneider 1845
timidum Hag. – Buresch,
1939
Beraeidae
Beraea Stephens 1833
r 212. *maura* (Curtis 1834)
c 213. *pullata* (Curtis 1834)
Ernodes Wallengren 1891
c 214. *articularis* (Pictet 1834)

Beraea articularis (Pict.) –
Klapálek, 1895
Beraemyia Mosely 1930
r 215. *hrabei* Mayer 1937

Helicopsychidae
Helicopsyche von Siebold
1856
c 216. *bacescui* Orghidan &
Botosaneanu 1953

Odontoceridae
Odontocerum Leach 1815
v 217. *hellenicum* Malicky 1972
albicorne (Scop.) – Navas,
1929

Calamoceratidae
Calamoceras Brauer 1865
r 218. *illiesi* Malicky & Kumanski 1974

Note: + one locality only: 34 spp. (15%); r 2–5 locs. (rare): 70 spp. (32%); c 5–20 locs. (common): 74 spp. (34%); v more than 20 locs. (very common): 40 spp. (19%).

Balkan peninsula. Another species unpublished up till now is one of the genus *Synagapetus*. It is closely related to *S. iridipennis* McL. and has been confused both with it and *S. moselyi* (Ulm.). As it is not clear whether Kalpálek's *S. ater* is a close relative of *iridipennis* (Klapálek, 1904) or a synonym of it (auct.), the identification of this species must be delayed. *A priori*, I am inclined to think it is a new species and not *ater*. The material originates from two high localities: Ryla Mts., ca. 1600 m a.s.l. (Kumanski, 1975a), and Pirin Mts., ca. 2100 m a.s.l. (author's unpublished data).

Another four newly discovered species are included in the list. Their distribution is, most probably, restricted to particular mountains: *Rhyacophila joosti* Mey – Stara-planina, *Rhyacophila pirinica* n. sp. – Pirin, *Drusus popovi* n. sp. – Stara-planina, and *Drusus osogovicus* n. sp. – Mt. Osogovska.

Recent data on varieties of genus *Drusus* species in the Balkan peninsula confirm Kumanski and Malicky's (1976) supposition on the specific independence of *Drusus discophorus balcanicus* Kum. Here I give it specific status.

Stenophylax mucronatus McL. was known from one female only. However, it turned out to be *Micropterna testacea* (Gmel.) – a common species in this country. *S. mucronatus* is therefore withdrawn from the list. Although I have not studied the material, I exclude several other species previously recorded: *Rhyacophila aquitanica* McL., *Rhyacophila valkanovi* Bots. and *Rhyacophila vulgaris* Pict. (explanation in Kumanski, 1973a), *Philopotamus ludificatus* McL. (given only by Buresch, 1939), *Lype phaeopa* Steph. (Botosaneanu, 1956, after a larva), *Limnephilus politus* McL. (only by Joakimov, 1899) and *Notidobia* sp. (*ciliaris* L.), only from larval stages (Kumanski, 1968).

Two species are to be synonymized. Surprisingly, *Hydroptila angulifera* Kum. was found to be identical with *H. uncinata* Mort. The latter, considered

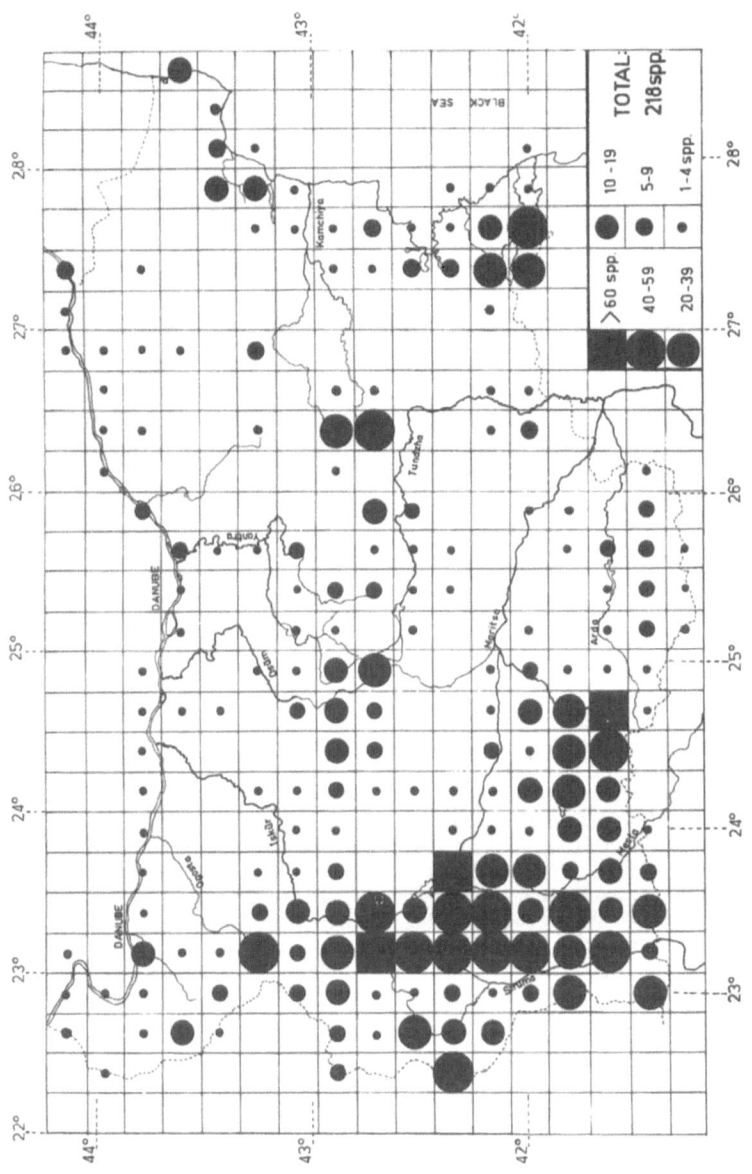

as endemic to Corsica, has been discovered in the continent (Malicky, in litt.). It was also Malicky who compared material of *Oecetis alexanderi* Kum. with typical specimens of *O. tripunctata* (Fabr.) and established their synonymy.

Five other species, not found personally or re-examined, are left in the list. These are: *Rhyacophila joosti* Mey, *Hydroptila taurica* Mart., *Oligotricha striata* (L.), *Triaenodes conspersus* (Ramb.) and *Adicella reducta* (McL.). Among them only *T. conspersus* must be accepted with some reserve. It is probable, that in recording it from Bulgaria, Klapálek (1894) could have had material either of *T. kawraiskii* Mart. or *T. simulans* Tjed. instead.

The present position of faunistic studies on Bulgarian Trichoptera is shown in Fig. 1. Each square is equal to a 20 km by 20 km area of Bulgaria. Much of the land is agricultural and the influence on the fauna there is strong. This added to the pollution from industry and farms causes a sharp reduction in caddis flies in such vast regions as the Danubian plain, Dobrudja and the Upper Thracian lowland. However, as no special investigations have been carried out in such regions, it can be presumed that a number of euribiotic representatives of *Hydropsyche, Limnephilus* etc. will be discovered in each of the squares at present empty.

The mountain regions are, of course, the most interesting. Up to June, 1980, 97 species had been registered from Mt Vitosha, 91 from the Ryla Mts, 73 from the Pirin Mts, 111 from the Rhodops, 113 from the Stara-planina Mts and 72 from the Strandzha Mts.

It could be hypothesized that Hydroptilidae, Leptoceridae, Rhyacophilidae and, naturally, Limnephilidae will be the families most responsible for extending the list. The last two are, also, the most promising for new species.

REFERENCES

Botosaneanu, L., Beitr. Ent. 6, 3/4: 354–402, 1956.
———, Beitr. Ent. 7, 5/6: 598–603, 1957.
———, Latv. entomol. 10: 53–60, 1965.
———, and Sykora J., Acta faun. ent. Mus. nat. Pragae 9, 77: 121–142, 1963.
Buresch, I., Mitt. bulg. ent. Ges. 10:153, 1939, (in Bulg.).
Christović, G., Sborn. nar. umotv. nauka i knizhn. 8: 337–346, 1892, (in Bulg).
Guéorguiev, V. and Beron P., Ann. Spéléol. 17: 285–441, 1962.
Joakimov, D., Period. spis. 60: 858–884, 1899, (in Bulg.).
Klapálek, F. Věstn. Ces. kral. Acad. 3: 308–310, 894.
———, Sborn. nar. umotv. nauka i knizhn. 11: 458–471, 1895, (in Bulg.).
———, Věstn. Čes. Akad. Cis. Fr. Jos. Praze 13, 8: 729–730, 1904, (in Czech).
———, Acta Soc. ent. Bohemiae 10, 1: 15–17, 1913.
Kumanski, K., Faun. Abh., Dresden 2: 109–115, 1968.
———, Bull. Inst. zool. and Musée, Sofia 29: 25–38, 1969.
———. Bull. Inst. zool. and Musée, Sofia 33: 99–109, 1971.
———, C. r. Acad. bulg. Sci. 25: 1261–1263, 1972.
———, Bull. Inst zool. and Musée, Sofia 38: 25–38, 1973a.
———, Tijdskr. Ent. 116: 107–121, 1973b.
———, Reichenbachia 15: 71–75, 1974.
———, Acta zool. bulg. 3: 48–58, 1975a.
———, Acta zool. bulg. 2: 58–69, 1975b.
———, Acta zool. bulg. 13: 3–20, 1979a.
———, Acta zool. bulg. 13: 71–75, 1979b.

————, Three New Caddis Species from Bulgaria (Trichoptera). (in press).
————, and Botosaneanu L., Acta Mus. Maced. Sci. Nat. 14: 25–43, 1974.
————, and Malicky H., Bull. ent. Pol. 46: 95–126, 1976.
Navas, L., Mitt. kgl. Nat.-Wiss. Inst. Sofia 2: 140–142, 1929.
Novak, K., Acta faun. ent. Mus. Nat. Pragae 14: 101–114, 1971.
Russev, B., Bull. Experiment. Stat. Freshwater Fish. Plovdiv 1: 115–128, 1962.
Soffner, J., Mitt. DEG 32: 12–14, 1973.
Sykora, J., Acta faun. ent. Mus. Nat. Pragae 6: 131–136, 1960.

DISCUSSION

Moretti: Is *Parachiona* found in the Bulgarian Mountains, as it is in the Alps?
Kumanski: Yes, in the Ryla Mts, 1400 m above sea level.

DIVERSITY OF ADULT TRICHOPTERA IN A 'NON-SEASONAL' TROPICAL ENVIRONMENT

E.P. McELRAVY, V.H. RESH, H. WOLDA, AND O.S. FLINT, Jr.

SUMMARY

The caddisfly fauna of a wet tropical area (Fortuna, Chiriqui Province, Panama, 8°44′ N., 82°16′ W., elev. 1050 m) that is 'non-seasonal' in terms of temperature, photoperiod, and rainfall was examined from nightly light trap collections, November 1976 to December 1977. At least four undescribed genera and 32 undescribed species were among the 67 taxa (plus 24 taxa of unassociated females) present in the total sample of 7,264 individuals. A comparison between the number of species/number of individuals-relationship of the Fortuna site with selected Nearctic sites indicated that Fortuna was more diverse than most Nearctic sites examined, although the faunas of a South Carolina and a Pennsylvania stream were approximately as diverse. Comparison with selected Palearctic sites showed a similar trend.

INTRODUCTION

The proposed construction of a hydroelectric facility in western Panama provided an opportunity to examine the fauna of a very unique wet tropical area — one that may be described as 'non-seasonal' with respect to temperature, photoperiod, and rainfall. Although temperature and photoperiod are often quite constant in tropical environments, most wet tropical areas are characterized by a marked seasonality in the annual distribution of precipitation. In this paper the diversity of the adult caddisfly fauna in a 'non-seasonal' environment will be described and compared with that reported from studies in seasonal temperate environments.

It has generally been observed that tropical areas of the world exhibit greater faunal and floral diversity than that seen in corresponding temperate habitats (MacArthur 1972). This apparent latitudinal gradient in the diversity of biotic communities has lead to considerable discussion about the reasons for these diversity differences (e.g. Connell and Orias 1964; Pianka 1966; Pielou 1975). Faunal comparisons of Nearctic-Neotropical diversity have been

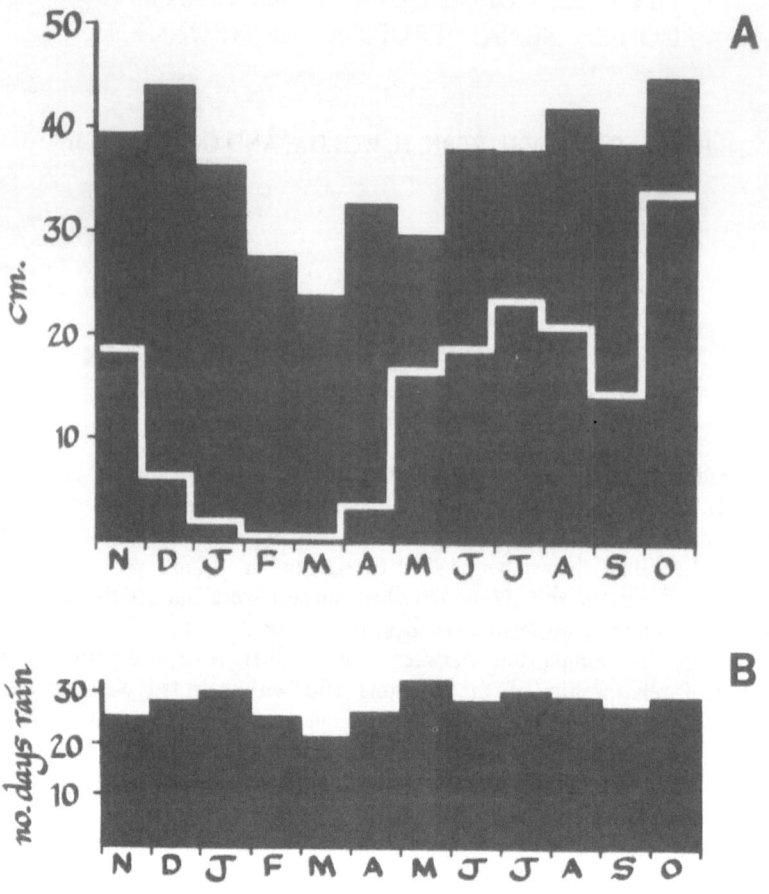

Fig. 1. A. Mean monthly rainfall for two sites in Panama: upper histogram represents the non-seasonal study site (Fortuna, 15-yr means); lower histogram represents a seasonal tropical area (Barro Colorado Island, located within the Panama Canal, 3-yr means). B. Number of days with rain per month at the Fortuna site in 1974.

done with selected terrestrial groups, but meaningful comparisons of aquatic insects are limited to the studies of Patrick et al. (1966) and Stout and Vandermeer (1975), both of which considered only immature stages of aquatic insects. Patrick et al. (1966) reported that diversity of Peruvian Amazon headwater streams was not significantly different from temperate-zone streams. In contrast, Stout and Vandermeer (1975) observed that most Costa Rican and Colombian streams, especially those at mid-altitudes, showed higher diversities than a series of Nearctic streams. However, a few of their tropical streams demonstrated diversity similar to, or lower than, some temperate ones.

150

METHODS

Nightly (dusk to approximately 2200 hours) collections of adult Trichoptera were made with a Pennsylvania-type (Frost 1957) black light trap from November, 1976 until July, 1979 at Fortuna (near Hornitos), Chiriqui Province, Panama (8° 44' N., 82° 16' W., elevation 1050 m). This area, in the upper portion of the Rio Chiriqui drainage basin, is in a region that averages over four meters of rain annually. The distinct 'wet' and 'dry' seasons that are typical of many tropical environments are lacking (Fig. 1A) and some rainfall occurs nearly every day of the year (Fig. 1B). The trap was located near a small tributary some 200 m from the Rio Chiriqui, which at Fortuna is a 4th-order stream. Initially suspended in the canopy of the rain forest (28 m above the ground), the trap was lowered to a height of two meters in July, 1977 and kept at this height for the remainder of the study. Although the trap was operated at two levels during the period under consideration here (November 1976–December 1977) no pronounced faunal distinctions were evident. Advantages and limitations of light trap data for estimating species richness and diversity are summarized by Wolda (1977, 1978).

RESULTS AND DISCUSSION

The Fauna. Adult Trichoptera were identified to described species when possible, but when this could not be done they were grouped into species-level Operational Taxonomic Units (OTU's). At least 67 species of caddisflies occurred in light trap collections made from November 1976 to December 1977; an additional 24 OTU's were determined from unassociated females (Table 1). Since we do not know how many of these females correspond to males included in the original total of 67 species, as many as 91 species may be present in the total sample of 7,264 individuals. Thus, the actual number of species collected lies somewhere between a minimum of 67 and an upper total of 91.

Taxonomic problems in the collections from Fortuna, typical of studies in tropical environments, include at least four undescribed genera and 32 undescribed species. Eleven families of Trichoptera have been collected at Fortuna. Although one-half of the species present probably have net-spinning larvae, these species make up over three-fourths of all individuals collected. Adults of species with case-making larvae are quite rare (less than one percent of the individuals collected) but comprise about one-third of all species.

Diversity. The relative diversity of the Fortuna fauna, compared with selected temperate sites, can be examined by using the number of species/number of individuals-relationship (i.e. by plotting the number of species vs. the number of individuals collected per site). If we compare the results from the Fortuna light trap samples with several Nearctic light trap studies of adult caddisflies (Fig. 2), it is evident that the Panamanian site is more diverse than most, but not all, of the Nearctic sites. For example, sites that are as diverse or nearly

Table 1. Species of Trichoptera obtained in light trap collections from Fortuna, Panama, November 1976–December 1977 (no asterisk – species identified from both males and females; * – species identified from males only; ** – species identified from females only; MS – Manuscript name; 0.0 – less than 0.05%).

Species	Number of Individuals Collected	Percent Samples Present	Percent Total Catch
Rhyacophilidae			
* *Atopsyche callosa* (Navas)	23	22.4	0.3
* *Atopsyche dampfi* Ross and King	16	19.0	0.2
Atopsyche implexa (Navas)	3	5.2	0.0
* *Atopsyche majada* Ross	856	96.6	11.8
* *Atopsyche* undescribed sp. "A" (nr. *talamanca* Flint)	11	13.8	0.2
* *Atopsyche* undescribed sp. "B"	9	10.3	0.1
Unassociated *Atopsyche* Females:			
Atopsyche sp. #1	26	22.4	0.4
Atopsyche sp. #2	2	3.4	0.0
Atopsyche sp. #3	468	86.2	6.5
Atopsyche sp. #4	3	5.2	0.0
Atopsyche sp. #5	1	1.7	0.0
Glossosomatidae			
Protoptila laterospina Flint	32	24.1	0.4
Mexitrichia undescribed sp. "A"	245	50.0	3.4
* *Mexitrichia* undescribed sp. "B"	6	8.6	0.1
Philopotamidae			
* *Wormaldia* undescribed sp. "A"	98	43.1	1.4
* *Wormaldia* undescribed sp. "B"	21	17.2	0.3
* *Wormaldia* undescribed sp. "C"	1	1.7	0.0
Unassociated *Wormaldia* Female:			
Wormaldia sp. #1	23	24.1	0.3
Chimarrhodella ulmeri Ross	163	82.8	2.2
* *Chimarra centralis* Ross	7	6.9	0.1
* *Chimarra emima* Ross	1	1.7	0.0
* *Chimarra mexicana* (Ulmer)	31	25.9	0.4
* *Chimarra spatulata* Ross	66	31.0	0.9
* *Chimarra wilsoni* Flint	32	32.8	0.4
* *Chimarra* undescribed sp. "A" (nr. *platyrhina* Flint Ms)	263	87.9	3.5
* *Chimarra* undescribed sp. "B" (nr. *poolei* Flint Ms)	56	50.0	0.8
* *Chimarra* undescribed sp. "C" (nr. *spangleri* Trivette Ms)	13	15.5	0.2
Chimarra undescribed sp. "D"	859	100.0	11.9
* *Chimarra* undescribed sp. "E"	189	27.6	2.6
Unassociated *Chimarra* Females:			
Chimarra sp. #1	107	53.4	1.5
Chimarra sp. #2	152	60.3	2.1
Chimarra sp. #3	68	51.7	0.9
Psychomyiidae (S.L.)			
* *Polycentropus dentoides* Yamamoto	2	1.7	0.0
* *Polycentropus digitus* Yamamoto	5	8.6	0.1
* *Polycentropus* undescribed sp. "A"	38	46.6	0.5
* *Polycentropus* undescribed sp. "B"	43	43.1	0.6
* *Polycentropus* undescribed sp. "C"	6	10.0	0.1
Unassociated *Polycentropus* Females:			
Polycentropus sp. #1	1	1.7	0.0
Polycentropus sp. #2	9	13.8	0.1
Polycentropus sp. #3	12	12.1	0.2
Polycentropus sp. #4	17	15.5	0.2
Polycentropus sp. #5	1	1.7	0.0
Polyplectropus deltoides (Yamamoto)	1	1.7	0.0
Polyplectropus undescribed sp. "A"	3	5.2	0.0
Unassociated *Polyplectropus* Female:			
Polyplectropus sp. #1	1	1.7	0.0
* *Xiphocentron* probably *aureum* Flint	1	1.7	0.0
* *Xiphocentron* undescribed sp. "A"	2	3.4	0.0
* *Xiphocentron* uncertain species	3	5.2	0.0
Austrotinodes undescribed sp. "A"	5	8.6	0.1
* *Austrotinodes* undescribed sp. "B"	4	6.9	0.1

Species	Number of Individuals Collected	Percent Samples Present	Percent Total Catch
Hydropsychidae			
* *Leptonema cheesmanae* Mosely	14	20.7	0.2
Leptonema intermedium Mosely	939	84.5	13.0
Leptonema simulans Mosely	211	72.4	2.9
Leptonema undescribed sp. "A" (nr. *salvini* Mosely)	874	98.3	12.1
* *Leptonema* undescribed sp. "B"	11	17.2	0.2
Unassociated *Leptonema* Females:			
Leptonema sp. #1	1	1.7	0.0
Leptonema sp. #2	1	1.7	0.0
Leptonema sp. #3	1	1.7	0.0
Leptonema sp. #4	1	1.7	0.0
* *Centromacronema auripenne* (S.L.) (Rambur)	1	1.7	0.0
* *Smicridea matagalpa* Flint	33	24.1	0.5
Smicridea talamanca Flint	30	32.8	0.4
* *Smicridea* undescribed sp. "A"	1	1.7	0.0
* *Smicridea* undescribed sp. "B"	2	3.4	0.0
Unassociated *Smicridea* Female:			
Smicridea sp. #1	5	6.9	0.1
Undescribed genus "A" (nr. *Calosopsyche*) undescribed sp. "A"	963	94.9	13.3
Undescribed genus "B" (nr. *Calosopsyche*) undescribed sp. "A"	37	41.4	0.5
Calamoceratidae			
Phylloicus aeneus (Hagen)	29	27.6	0.4
**Phylloicus* uncertain species	1	1.7	0.0
Undescribed genus "A" undescribed sp. "A"	2	3.4	0.0
Hydroptilidae			
Zumatrichia angulata Flint	16	10.3	0.2
* *Ochrotrichia panamensis* Flint	3	3.4	0.0
* *Ochrotrichia* undescribed sp. "A"	6	8.6	0.1
**Oxyethira uncertain species	3	5.2	0.0
**Oxyethira uncertain species	1	1.7	0.0
Unassociated *Hydroptilidae* Female:			
Hydroptilidae sp. #1	1	1.7	0.0
Leptoceridae			
* *Nectopsyche* uncertain species	1	1.7	0.0
* *Nectopsyche* undescribed sp. "A"	3	5.2	0.0
Unassociated *Nectopsyche* Females:			
Nectopsyche sp. #1	1	1.7	0.0
Nectopsyche sp. #2	1	1.7	0.0
**Oecetis probably *avara* (Banks)	15	19.0	0.2
**Oecetis uncertain species	1	1.7	0.0
***Triplectides gracilis* (Burmeister)	1	1.7	0.0
* *Triaenodes* undescribed sp. "A"	4	6.9	0.1
Odontoceridae			
* *Marilia* undescribed sp. "A"	1	1.7	0.0
Lepidostomatidae			
**Lepidostoma uncertain species	2	3.4	0.0
Helicopsychidae			
* *Helicopsyche incisa* Ross	1	1.7	0.0
* *Helicopsyche* probably *truncata* Ross	2	3.4	0.0
* *Helicopsyche* undescribed sp. "A"	18	25.9	0.2
Unassociated *Helicopsyche* Female:			
Helicopsyche sp. #1	20	25.9	0.3
Family uncertain			
* Undescribed genus, undescribed sp. "A"	1	1.7	0.0

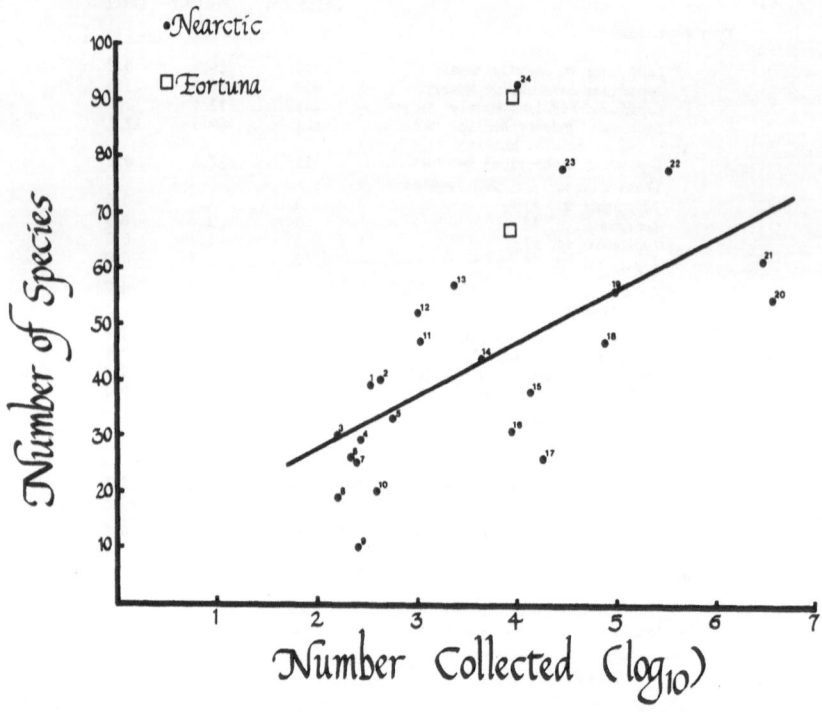

Fig. 2. Nearctic-Fortuna Trichoptera diversity comparison. Squares represent Fortuna collections having 67 or 91 taxa (with the actual number somewhere in between, see text); circles represent temperate sites. Regression line (p = 0.001) derived from temperate sites only. Locations and sources of data: South Carolina (Morse 1970: points 1–10; Morse et al. 1980: point 24); Pennsylvania (Masteller and Flint 1979: points 11–13; B.G. Swegman, unpublished data: point 23); Ohio (McElravy and Foote 1978: points 14–15; Marshall 1939: point 18); Illinois (Harris 1971: point 16); Texas-Oklahoma (Resh et al. 1978: point 17); Kentucky (Resh et al. 1975: point 19); Quebec, Canada (Corbet et al. 1966: points 20–21; Nimmo, 1966: point 22).

as diverse as Fortuna are located in South Carolina (Fig. 2, point 24) and Pennsylvania (Fig. 2, point 23). Comparison with results of several Palearctic studies (Fig. 3), mainly from Scandinavia, indicate a trend similar to that observed with the Nearctic sites.

Attempts at broad zoogeographical comparisons (i.e. Nearctic with non-seasonal Neotropical) must be tentative because of the limited data available from this type of tropical environment. However, a comparison of the Fortuna caddisfly diversity with that of seasonal Neotropical areas, which are by far the more usual type of tropical environment, could provide information to further clarify the 'climatic stability' hypothesis of latitudinal diversity gradients (Pianka 1966). This theory attributes the higher diversity of tropical systems to a presumed stability of physical conditions over the course of a year. Such stability is far less in the seasonal tropics where wet and dry seasons prevail than at the non-seasonal Fortuna site. Although appropriate information on Trichoptera that could be used to test this

154

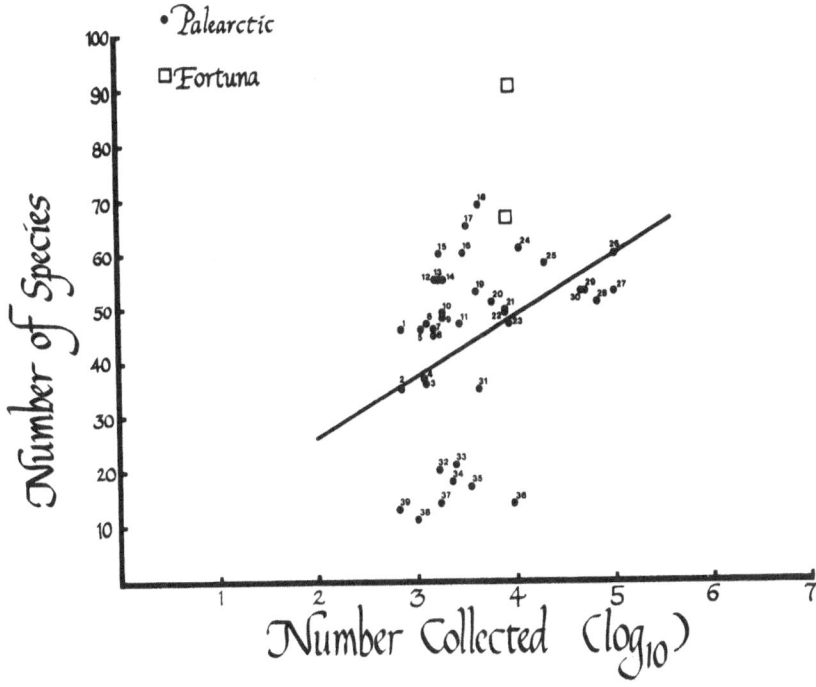

Fig. 3. Palearctic-Fortuna Trichoptera diversity comparison. Squares represent Fortuna collections having 67 or 91 taxa (with the actual number somewhere in between, see text); circles represent temperate sites. Regression line (p = 0.008) derived from temperate sites only. Locations and sources of data: Sweden (Svensson 1974: points 1–7, 9–18; J.O. Solem, unpublished data: points 8, 19, 27, 31; Solem 1979: point 23; Ulfstrand 1970: point 16); Norway (J.O. Solem, unpublished data: points 20, 22, 25, 29, 33–39; Solem 1979: points 21, 24, 32); England (Crichton 1960: points 28, 30).

hypothesis is not currently available for seasonal sites in the Neotropics, we hope that the information provided here for a non-seasonal area will permit such a comparison to be made in the future.

ACKNOWLEDGMENTS

We wish to thank Dr J.O. Solem, Royal Norwegian Society of Sciences and Letters, University of Trondheim, Norway, and Mr B.G. Swegman, University of Pittsburgh, Pennsylvania for the use of unpublished data, and Ms K.L. Sorg for figure preparation. We are grateful to the Panamanian Electricity Company (IRHE) for permitting the use of its facilities at Fortuna. We also thank Cecilio Estribi of the RENARE in the province of Chiriqui for his cooperation in the light trap program.

REFERENCES

Connell, J.H. and Orias, E., Am. Nat. 98: 399–414, 1964.

Corbet, P.S.; Schmid, F. and Augustin, C.I., Can. Ent. 98: 1284–1298, 1966.

Crichton, M.I., Trans. Roy. Ent. Soc. 112: 319–344, 1960.

Frost, S.W., J. Econ. Ent. 50: 287–292, 1957.

Harris, T.L., J. Kansas Ent. Soc. 44: 295–301, 1971.

MacArthur, R.A., Geographical Ecology., Harper and Row, Inc. New York, 1972.

Marshall, A.C., Ann. Ent. Soc. Am. 32: 665–688, 1939.

Masteller, E.C. and Flint, O.S. Jr, Great Lakes Ent. 12: 165–177, 1979.

McElravy, E.P. and Foote, B.A., Great Lakes Ent. 11: 143–154, 1978.

Morse, J.C., Unpublished thesis, Clemson University, South Carolina, U.S.A., 1970.

Morse, J.C.; Chapin, J.W.; Herlong, D.D. and Harvey, R.S., J. Ga. Ent. Soc. 15: 73–101, 1980.

Nimmo, A.P., Quaest. Ent. 2: 217–242, 1966.

Patrick, R.; Aldrich, F.A.; Cairns, J.; Drouet, F.; Hohn, M.H.; Roback, S.S.; Skuja, H.; Spangler, P.J.; Swabey, Y.H. and Whitford, L.A., Monogr. Acad. Nat. Sci. Phila. 14: 1–495, 1966.

Pianka, E.R., Am. Nat. 100: 33–46, 1966.

Pielov, E.C., Ecological Diversity. New York: Wiley and Sons, 1975.

Resh, V.H., Haag, K.H. and Neff, S.E., Envir. Ent. 4: 241–253, 1975.

————., White, D.S. and White, S.J., Southwestern Nat. 23: 381–388, 1978.

Solem, J., In Ecological Diversity in Theory and Practice, eds J.F., Grassle; G.P., Patil; W., Smith and C. Taillie, pp 255–267, Int. Coop. Publish. House, Fairland, Maryland, 1979.

Stout, J. and Vandermeer, J., Am. Nat. 109: 263–280, 1975.

Svensson, B.W., Oikos 25: 157–175, 1974.

Ulfstrand, S., Ent. Tidskr. 91: 46–63, 1970.

Wolda, H., Geo-Eco-Trop. 3: 229–257, 1977.

————., Am. Nat. 112: 1017–1045, 1978.

THE PHENOLOGY OF DISPERSAL OF SEVERAL CADDISFLY (TRICHOPTERA) SPECIES IN THE ISLAND OF CRETE*

H. MALICKY

SUMMARY

Light trap catches show regional and specific differences in the phenology of several species of Stenophylacini which are adapted to intermittent streams. The lack of a dispersal flight in springtime in Crete is striking; it explains the absence of the adults in caves in summer. A correlation between the specific vagility and the size of the distribution area is found in several species.

INTRODUCTION

Adults of the caddisfly genera *Stenophylax*, *Mesophylax* and *Micropterna* are often and abundantly found in caves. This presence in caves is an aestivation and is a part of the adaptations to intermittent streams. The Mediterranean Region, where intermittent streams are most common, is the distribution center of this group. Many papers deal with the ecology of this group. Those by Bouvet (1977) and Malicky (1977) contain additional references.

A number of these caddisflies are abundant in Crete, but in my field investigations it was striking that, despite intensive research, not one specimen could be found in caves in summer. Therefore, from May 1977 to May 1979, one light-trap was operated in each of the villages of Sises and Kastellakia in Crete. The traps were of JERMY type, with mercury tungsten lamps (Osram HWL 160 W). The traps were placed at distances of more than 3 km from any streams, because it was intended to find out the period of dispersal flights, and not to determine emergence periods which were already known. The vagility is another adaptation to life in intermittent streams, because it ensures a fast re-population of isolated streams in which the population might have been destroyed occasionally by unpredictable events. In both traps caddisflies, other than Stenophylacini, which live in permanent streams, were also caught.

*Mit Förderung durch den Fonds zur Förderung der wissenschaftlichen Forschung in Österreich, Projekte Nr. 1796 und 2986.

RESULTS AND DISCUSSION

In both traps the following species of Stenophylacini (Limnephilidae) were caught: *Mesophylax aspersus* Rambur, *Stenophylax minoicus* Malicky, *Micropterna caesareica* Schmid, *Micropterna fissa* McL., *Micropterna sequax* McL., *Micropterna taurica* Martynov. – *Micropterna malaspina* Schmid also lives in Crete but was not caught in the traps. At Kastellakia which is exposed to the wind, most insects were caught in lower numbers than at Sises. The species were caught in similar proportions in both traps. Distribution over the seasons is shown in Fig. 1, where the results from both places and both years are added together. *M. caesareica*, *M. fissa* and *M. sequax* were similar in their phenology and are also added together.

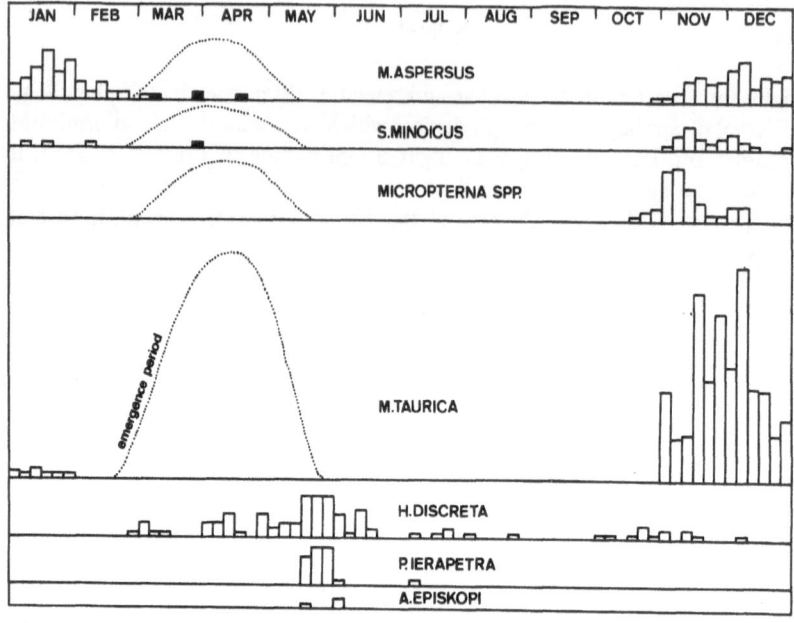

Fig. 1. Caddisfly records in light traps in Crete. Black columns: Newly emerged specimens of *M. aspersus* and *S. minoicus*. The dotted lines indicate the emergence period which was not represented in the light trap catches.

1. Stenophylacini: It is striking that these species in Crete had only one dispersal flight, which was in the autumn. The dispersal flight in spring was lacking. Under comparable conditions it is present in Continental Europe. The phenological curve is therefore unimodal and not bimodal as expected. The lack of the dispersal flight in spring explains the fact that these animals were never found in caves in summer, because the caves are usually some distance away from streams. It may be suggested that the adults hide in other places, such as soil fissures, in the vicinity of their emergence sites. It is however a mystery why this is so in Crete and not on the Continent.

It is difficult to compare the results from Crete with literature data from other regions because most species are different. In the literature, there is much information about the phenology of such species, but mainly from inside caves. Data from light traps outside caves were given by Crichton (1971) and Svensson (1972). In my light traps in Austria such caddisflies were never abundant. In Fig. 2 I have added all the scattered records from Austria known to me. I have also done the same for *M. aspersus* from Sardinia where I have seen plenty of material.

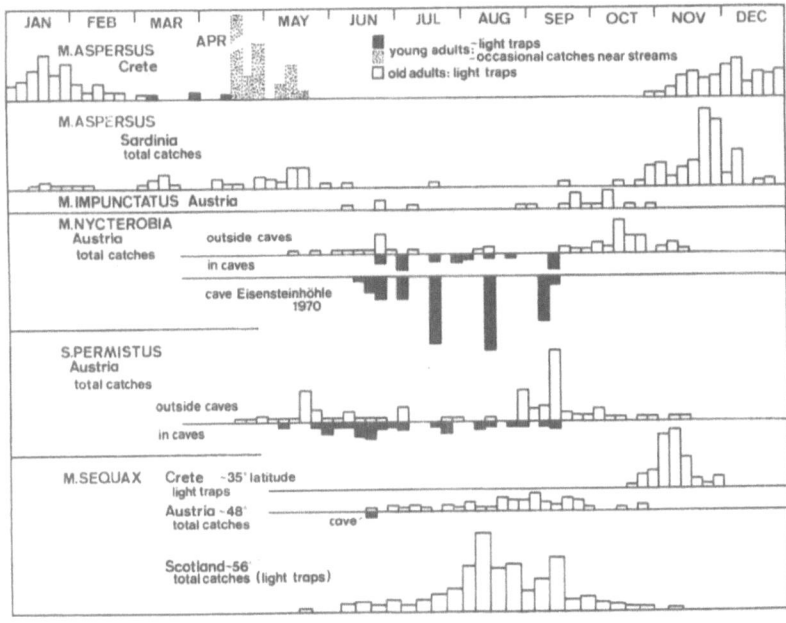

Fig. 2. Comparison of caddisfly phenologies from various regions. For the data from Scotland I express my gratitude to M.I. Crichton and the Rothamsted Insect Survey.

The histograms of *Micropterna nycterobia* McL. and *Stenophylax permistus* McL. from Austria show flight periods from April and May to November. The records from outside caves have two maxima: for *M. nycterobia* in June and October, and for *S. permistus* in May and September. On the other hand, the records from caves of *M. nycterobia* are from June to September, and of *S. permistus* from May to September, which are the periods in which the records outside caves show minima. The early summer maximum represents the newly emerged adults on their dispersal flight looking for hiding places in caves, where the majority stay for some weeks in summer. The autumn maximum represents the adults which have left caves and are dispersing in their search for oviposition sites. In Fig. 2 the total catches in one cave over one year (Eisensteinhöhle, 1970) coincide well with the sum of the scattered records.

The scattered records of *M. aspersus* from Sardinia showed fairly constant presence from October to June, with a maximum in November. The catches were from various sites where distances from streams were not recorded, but many of these sites were far away from any streams. In contrast to Crete, there was a clear presence in April and May which coincides well with a springtime dispersal which was lacking in Crete. In Fig. 2 I have added also the records from light catches in Crete in the immediate vicinity of streams; they agree well with the corresponding catches in Sardinia.

The age of the females may be determined by examination of the contents of the abdomen. Old females have either mature eggs or the abdomen is empty and shrunken. Young females which have not yet undergone aestivation have no eggs, and the abdomen is filled with fat body. Such young specimens are shown in black in Figs 1 and 2. They were scarce in the traps in Crete. However, light trapping close to a stream may in Crete yield large numbers of freshly emerged specimens even in April and May.

It could be argued that the adults do not fly to the light trap in spring in Crete because of cool weather. However, in Sardinia the weather at this time is certainly as cool as in Crete, if not cooler. But even in Crete they fly to the light in cool weather from nearby streams.

Fig. 1 demonstrates that *M. aspersus* and the *Micropterna* species have different phenologies. *S. minoicus* is perhaps intermediate. In Crete *M. aspersus* was trapped in fairly constant numbers from the end of October, which is the onset of the rainy period, to mid-February, without significant maxima. *Micropterna* species were first caught at the same time, but they had a maximum in November/December and then disappeared. In *M. taurica* a few specimens were caught in January, perhaps because this species is more abundant than the others. No freshly emerged specimens were caught in the traps in March, but they were found near streams. It is evidently necessary to distinguish between species in studies of their development. This was not sufficiently done by Bouvet (1977).

The two common Central European species, *M. nycterobia* and *S. permistus*, are absent from Crete and so cannot be compared. In the light traps in Crete, *M. sequax* was caught only in October and November, and I have also records of newly emerged specimens in April and May from springs and streams, but it was never found in the summer months. The phenology of this species is quite different in Central Europe where adults are found from June to October without a minimum in summer. *M. sequax* is rarely found in caves. The records by Crichton (1971) from Scotland and Svensson (1972) from Southern Sweden show light trap maxima in August. Denis (1974) found that in freshly emerged females of *M. sequax* kept at room temperature, the gonads developed faster than in control animals kept at $8°C$. This suggests a connection between phenology and temperature, which is also supported by the records from different latitudes. The results of Denis might indicate that this species passes through a diapause in cooler regions and higher latitudes, while in warmer regions such as Crete they mature directly. In fact, we have the contrary condition. It would be interesting to discover if there are genetic differences in the induction of dormancy between populations of different latitudes.

The records of *Mesophylax impunctatus* McL. in mountainous Austria are from the summer months only, when *M. aspersus* aestivates in South Europe. I cannot distinguish these two species morphologically, and there is considerable variation in the genital structures of the adults from various regions. Bouvet (1977) says that she has found *M. impunctatus* in the Hoggar Mts. in the Central Sahara. This may be true, but it illustrates the difficult taxonomy of these two species. The striking phenological similarity of *M. sequax* and *M. aspersus – impunctatus* at different latitudes suggests that *M. aspersus* and *M. impunctatus* are two ecological strains of the same species. Two more observations may be added for illustration. In cold springs in high mountains in Crete where the water temperature never exceeds 7°C, young larvae of *M. aspersus* may be found as early as September, while at lower altitudes oviposition starts in November. – Dr J. Margalit (Jerusalem) tells me that in the high mountains of the Sinai Peninsula the larvae are found in the warm season. It should also be remembered that *Mesophylax aethiopicus* Malicky, which is a descendant of this otherwise purely Mediterranean genus, is adapted to stagnant water at high altitudes in Ethiopia.

Micropterna lateralis Stephens is the only species of the group absent from the Mediterranean area but is widely distributed in Europe. According to Crichton (1971), Svensson (1972) and my own data it has a short flight period in June and July in Britain, Sweden and Austria. It was never found in caves in Austria, and cave records from other countries are scarce. This species may not aestivate.

2. The Limnephilus species: Only one or two specimens of *Limnephilus decipiens* Kolenati and *L. affinis* Curtis were caught. They were not known previously from Crete. Their larvae live in temporary pools or ponds, which are lacking or extremely rare in Crete, so it is not clear where these specimens had developed. Perhaps they came from the Continent. The phenology of these species (Novák & Sehnal, 1963) is similar to that of *Mesophylax* etc., but they do not aestivate normally in caves but rather in conifer forests on mountains where they feed on honeydew produced by homopterans. *L. affinis* and *L. decipiens* have a wide distribution in temperate Europe and Asia.

3. Other caddisflies: Several species which live in permanent streams were caught in small numbers. Adaptation to intermittent conditions is not known in these species. The histograms in Fig. 1 may therefore show the true emergence periods. It is known that at least some of these species may be found in Crete in any stage at any season, which means that they are acyclic, but they are not present in equal numbers all the time. In *Hydropsyche discreta* Tjeder maxima of adults may be found in May and September/ October, and of young larvae in July. In this species, and perhaps in others, it is suggested that individual development takes about half a year. *H. discreta* is widespread in the Eastern Mediterranean region, and will certainly be found on every island with suitable conditions for hydropsychids. *Polycentropus ierapetra* Malicky and *Agapetus episkopi* Malicky are widespread in the

Aegean Islands and live also in Peninsular Greece. *Polycentropus flavomaculatus* Pictet, *Wormaldia subnigra* McL. and *Hydroptila vectis* Curtis are widespread in Europe. *Tinodes rethimnon* Malicky is the only of these species endemic to Crete. The above mentioned Stenophylacini have also a wide distribution in the Mediterranean, except *S. minoicus* which is known only from Crete. Thus, the light traps caught mainly species which have large areas; 27 of the Cretean species are known to have large areas, and 13 of them were in the traps. On the other hand, of the twelve species known to be endemic to Crete, only two were in the traps. A correlation between the flight range and the size of the area is therefore evident. Both are expressions of a specific vagility (Malicky, in press). It is possible that individuals of vagile species may occasionally fly over the sea. The distances between the islands in the Aegean are often short, and thus colonies may be founded on other islands. It is not known how frequently this may happen. Newcomers must try to establish in an existing and usually very stable biocoenosis. Nevertheless, the correlation between the vagility and the size of the area demonstrates whether a species may be trustworthy for zoogeographical conclusions or not. An analysis of the subspecific structure of species such as *Polycentropus ierapetra* may answer these questions.

ACKNOWLEDGEMENTS

I express gratitude to my friends Charalambos Tsikalas and Jannis Delibasis who helped me with the operation of the traps. My sincere thanks are due to Ian Crichton for correcting the English text.

REFERENCES

Bouvet, Y., Thèse, Univ. Claude Bernard, Lyon, 1977.
Crichton, M.I., J. Zool. Lond. 163: 533–563, 1971.
Denis, C., Bull. Soc. Sci. Bretagne 49: 125–129, 1974.
Malicky, H., Biol. Gallo-Hellen 6: 171–238, 1977.
———, Acta Biol. Debrecina (in press).
Novak, K. and Sehnal, F., Acta Soc. Ent. Čsl. 60: 68–80, 1963.
Svensson, B.W., Oikos 23: 370–383, 1972.

DISCUSSION

Kumanski: *Micropterna sequax,* which is common in Crete as well as in Austria and the Balkans, shows great differences in behaviour; it does not estivate in the caves of Crete and is very rarely found in Austrian caves. This species is amongst the most abundant of the Stenphylacini in the Bulgarian caves, although it is not necessarily found only there. Have you any explanation of these facts?
Malicky: At present one can only speculate. A collaborative laboratory study

on the genetic differences between the populations will be carried out with Dr Denis.

Denis: Do you know when the larval development of the Crete species occurs?

Malicky: The larval development of the Stenophylacini species takes place in the rainy season between November and April.

Botosaneanu: 1) Are there true caves in the vicinity of the light traps operated in Crete. 2) I believe that *Mesophylax aspersus* and *impunctatus* are distinct species — consider their larvae. Of course there is much taxonomic work still to be done.

Malicky: 1) There are plenty of caves in Crete and they have been searched for caddis flies intensively during the appropriate season. Although many other insects, including moths, were found, there were no caddisflies.

Giudicelli: Quelques informations sur une population de *Mesophylax aspersus* vivant au Moyen Atlas (Maroc) dans une rivière soumise à un sévère étiage en été. La station se situe à 1900 mètres d'altitude, sur un plateau aride recouvert de lave basaltique (volcanisme quaternaire). Dans un tel environnement, les imagos de *M. aspersus*, dont les stades aquatiques vivent dans la rivière, n'ont pas la possibilité de trouver des refuges de diapause dans les grottes ni d'émigrer vers des zones boisées (absence de ripisylve et de forêt dans le voisinage). En fait, les imagos ont été trouvées en abondance (juillet et septembre) dans les zones rivulaires de la rivière, sur la face inférieure de blocs de basalte presque totalement émergés; elles se tenaient immobiles dans les nombreuses alvéoles du basalte. Il est probable que ces imagos trouvent, dans ce microhabitat, les conditions d'obscurité, de température et d'hygrométrie assez voisines de celles des grottes.

NEW TRICHOPTERA SPECIES AND SUBSPECIES
FOUND IN ITALY

G.P. MORETTI

SUMMARY

A number of new species and new subspecies — *Rhyacophila foliacea,
R. italica, R. italica ilvana, Oxyethira hartigi, O. pirisinui, Hydroptila ruffoi,
Wormaldia copiosa botosaneanui, W. mediana nielseni, W. pulla marlieri,
W. variegata denisi, Polycentropus malickyi, Tinodes apuanorum, T. bruttius,
Halesus radiatus vaillanti, Beraea botosaneanui, B. crichtoni, B. ilvae* — are
described and illustrated by a series of drawings. Others — *Drusus aprutiensis,
D. camerinus, Mesophylax aspersus sardous, Sericostoma cianficconii, S.
italicum* — are further described and the females of these species are reported
for the first time. The previously unknown *Halesus nurag* Mal. and
Micropterna wageneri Mal. females are also described. The *Polycentropus*
listed as *Polycentropus* sp. Moretti (1941) in Limnofauna Europaea (1978) is
now considered a good species and has been named *P. sardous*.

INTRODUCTION

A revision of my collection, undertaken with the intention of compiling a
first list of Italian Trichoptera, made it evident that some of the new species
and subspecies, as well as previously unknown ♀♀ required description.

Some of the species belonging to the *Polycentropus*, *Drusus* and
Sericostoma genus, already reported (Moretti 1941, Moretti and Cianficconi
1976, 1978) and figured, were awaiting the choice of a specific name and
holotype and allotype citation. This information is now presented.

Owing to the fact that the distinguishing characteristics of some species
and subspecies are taxonomically more complex they have been described
and discussed in greater detail, while others, being more easily distinguishable,
are dealt with only briefly.

Seven of the 22 new taxa are from the Italian islands, nine from the
Central Italy, three from the Central-Southern and three from Southern
Italy. The paratypes are very numerous because my investigation has been
carried out for almost 40 years and, in consequence, I have accumulated a
great number of specimens from different regions of Italy. The exact locality,

the date and the name of the collector have been omitted for reasons of space. They will, however, be reported in a future detailed catalogue of Italian Trichoptera.

DESCRIPTION OF NEW SPECIES

Rhyacophila foliacea n. sp. (Fig. 1)

Antennae and legs greyish-brown, anterior wings brownish with small darker markings in two transversal lines and a pale triangular field delimited by a dark line on the middle of the posterior margin. Wing spread: ♂ 24–27 mm, ♀ 25–30 mm.

Clearly differentiated from *R. vulgaris* (B), to which it is related, by the following ♂ genitalia characters: dorsal appendage of aedeagus narrower and more elongated with convex upper margin (a and a'), aedeagus with a ventral, proximal protruding triangle and pointed apex (b and b'), ventral lobe of aedeagus shaped like an oval leaf not like a hockey-stick (c and c'), dorsal apical lobe of 9th segment is not dilated half way along (d and d'), the preanal appendages are regularly curved, not sinuous on the outer margin (e and e').

R. vulgaris is diffused in the Alps, Prealps and North Apennines, while *R. foliacea* inhabits a well-defined area of the Central Apennines. The related *R. hartigi* is found in Calabria and Sicilia. All these species are reophyls of fast-running streams.

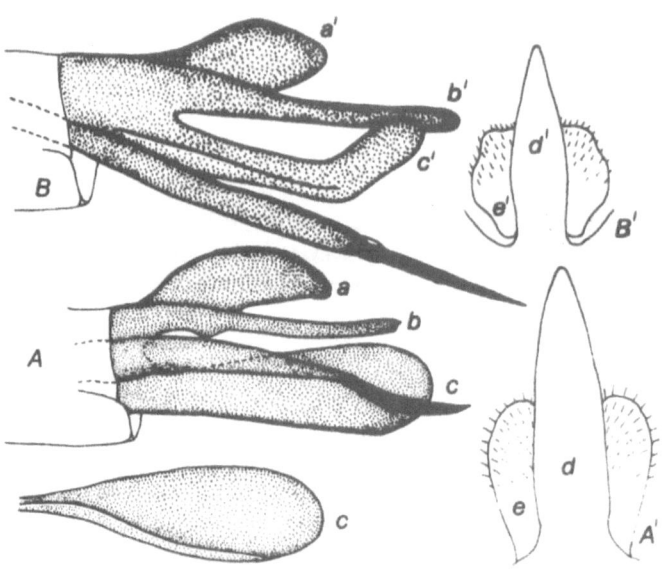

Fig. 1. A and A' = *Rhyacophila foliacea* n. sp.; B and B' = *Rhyacophila vulgaris* Pictet, ♂ genitalia: a and a' = dorsal appendage of aedeagus, b and b' = aedeagus, c and c' = ventral lobe of aedeagus, d and d' = dorsal apical lobe of the 9th segment, e and e' = preanal appendages.

Holotype ♂ and allotype ♀: Marche, Torrente Scarzito, 600 m, Camerino, Macerata, 22.7.1953, leg. Moretti; paratypes ♂♂ ♀♀ collected between 1953 and 1979: Toscana 1♂ 3♀, Umbria (the Menotre and Sordo rivers) 24♂ 6♀, Marche (the Potenza, Esino and Nera rivers) 46♂ 8♀, Lazio (River Velino) 13♂ 4♀, Abruzzi (River Sangro, several streams) 44♂ 1♀, Molise (River Biferno and tributaries) 45♂ 10♀. In Moretti's collection, Perugia. Marche and Umbria 7♂ 4♀, in Malicky's collection, Lunz am See, Austria.

Derivatio nominis. Named for the characteristics leaf-like ventral aedeagus lobe.

Rhyacophila italica n. sp. (Fig. 2)

Antennae and legs greyish-brown. Anterior wings yellow-brown with clear visible greyish-brown markings, also at the apex, pale triangular area edged by a brown line at posterior margin, pterostigmatic region more visible in ♂ than ♀. Wing spread: ♂ 21–30 mm, ♀ 26–34 mm. Related to *R. rougemonti* McL., *R. pallida* Giudicelli, *R. trifasciata* Mosely, *R. tarda* Giudicelli, but differs from them in its dorsally turned parameres. In *R. rougemonti* and *R. tarda* they have the regular down curve, while in *R. pallida* they swell before the apex and in *R. trifasciata* they form a down-turned right-angled hook. In

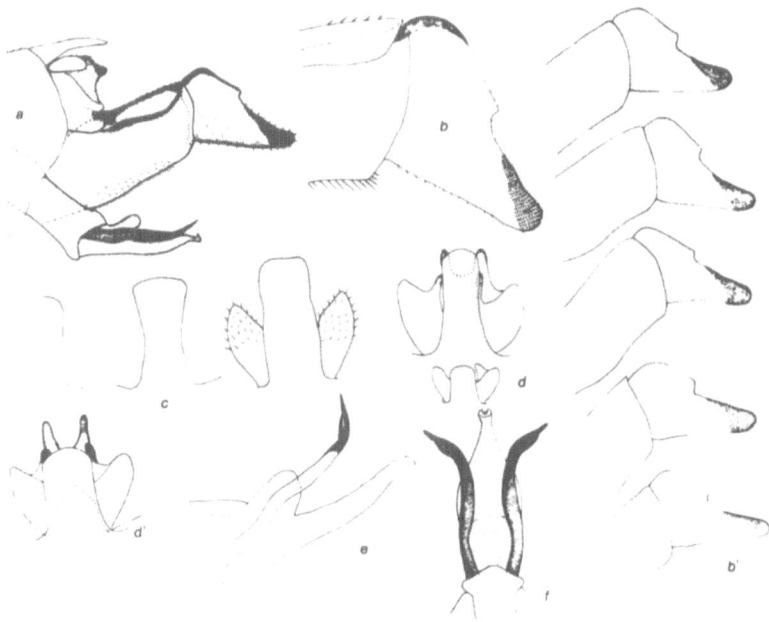

Fig. 2. Rhyacophila italica n. sp. ♂ genitalia: a = side view, b and b' = second joint of the inferior appendages and its variability, c = apical lobe of the 9th segment and its variability, d and d' = idem and preanal appendages, e = aedeagus and parameres side view, f = the same from above.

the Sardinian and Corsican species the dorsal branch of aedeagus is not very pronounced, while in *R. italica* it is markedly lobed as in *R. rougemonti*. The parameres of *R. italica* have a forked apex with branches of unequal length that slightly overlap (e, f). *R. italica* is easily distinguishable from the South Italian and Sicilian *R. rougemonti* as the indentation on the second joint of the inferior appendages is less marked and shows a certain variability (b and b'). The 9th segment dorsal apical lobe is subrectangular with the lateral margins sometimes turned in and a more or less convex anal margin (c). Variability has also been noted in the preanal appendages (d and d'). *R. rougemonti* is a reophyl which prefers clear waters, whereas *R. italica* is also found in sluggish, torbid, stagnant waters. The larvae and eggs of this species have recently been described (Moretti et al. 1978).

Holotype ♂ and allotype ♀: Marche, Fonti di Selvazzano, 655 m, Camerino, Macerata, 2.7.1953, leg. Moretti; paratypes ♂♂ ♀♀ collected between 1953 and 1980: Emilia Romagna (River Tiber) 24♂ 5♀, Toscana (River Tiber and tributaries) 13♂ 6♀, Umbria (tributaries of the River Tiber) 78♂ 90♀, Marche (several springs) 13♂ 5♀. In Moretti's collection. Umbria 1♂ 1♀ in Malicky's collection. 1♂ 1♀ was presented to all trichopterologists who attended to the 3rd Symposium on Trichoptera.

The name indicates its Central Italian distribution.

Rhyacophila italica ilvana n. ssp. (Fig. 3)

This *Rhyacophila* qualifies as a subspecies because the dorsal apical lobe of the 9th segment of the ♂ (d) forms a longer narrower cone than it is in *R. italica*

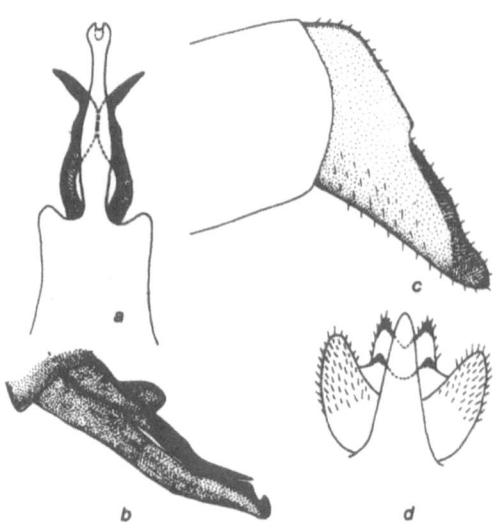

Fig. 3. *Rhyacophila italica ilvana* n. ssp. ♂ genitalia: a and b = aedeagus and parameres, c = second joint of the inferior appendages, d = dorsal apical lobe of the 9th segment and preanal appendages.

and the aedeagus (a, b) is more slender. The second joint of the inferior appendages (c) and preanal appendages (d) as in *R. italica*.

Holotype ♂: Isola Elba, Fosso Pedalta, 31.8.1957, 3 paratypes ♂, 2-6.10.1958, leg. Viganò, Gianotti, in Moretti's collection; 1♂ in Malicky's collection.

REFERENCES

Giudicelli, J., Vie et Milieu, 19: 49–54, 1968.
MacLachlan, R., A monographic Revision and Synopsis of the Trichoptera of the European Fauna. Reprint 1968, Hampton, Classey, 1874–1880.
Moretti, G.P.; Corallini Sorcetti, C. and Vignaroli, F., Riv. Idrobiol. 17: 85–134, 1978.
Schmid, F., Mem. Soc. Ent. Canada 66: 230 pp., 1970.

Oxyethira hartigi n. sp. (Fig. 4)

Is related to *O. fischeri* Higler and *O. falcata* Morton. The 8th segment is larger than the 9th and the tergite is longer than the sternite. Three black spines on each side on the 8th tergite, anal margin flanked by a more slender, shorter dorsally placed black spine (b). There is a large black spine on the allongated lobe on the lateral margin of the 8th segment (a). The subgenital appendages lateral to the aedeagus are very sclerified, notched at the apex with a ventrally directed process with an overhanging strong apical seta (a). The under process has a long strong seta (a). The inferior sclerified appendage turns up like the beak of a bird (a). The aedeagus apex is characterized by a triangular edge, free of teeth (b). There is a spiney longitudinal endophallic process, whereas the pointed titillator is on the upper part of aedeagus.

Holotype ♂ and paratypes 4 pupae ♂: Sardegna, Sorgente Monti, 450 m, Sassari, 27.3.1964, leg. Pirisinu, in Moretti's collection.

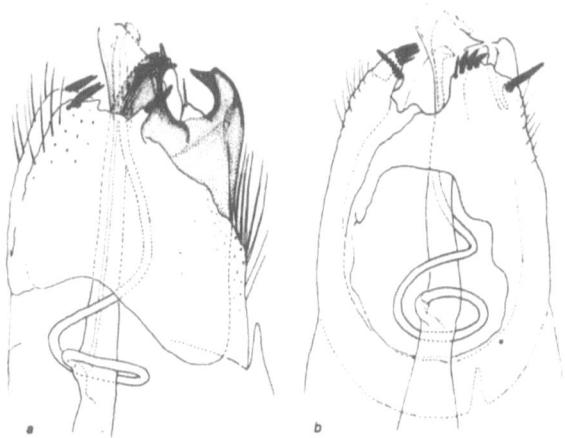

Fig. 4. Oxyethira hartigi n. sp. ♂ genitalia: a = side view, b = dorsal view.

169

I dedicate this species to the memory of my much missed friend F. Hartig who collected many trichoptera for me in various parts of Italy.

Oxyethira pirisinui n. sp. (Fig. 5)

Related to *O. simplex* Ris. The centre of the apex of 8th sternite seen laterally extends into a conical appendix (b). Dorsally the 9th segment is reduced to a narrow strip. Inferior appendages terminate in a pair of long pigmented claws, which seen ventrally curve outwards; seen laterally they curve downwards (b). The finger-like cylindrical-conical lobes of the 9th segment are equipped with two robust apical setae (a). Wide-based bilobed process with a robust seta on the lateral apical border. Seen laterally the subgenital appendages are spatula-like subtriangular with a ventrally turned vertex (b). Spiney scythe-shaped titillator. Two thorn-like sclerites run the full length of the fleshy apically-dilated aedeagus, which is hooked ventrally.

Holotype ♂, allotype ♀ and paratypes 6♂ 1♀ : Isola Capraia, Vado del Porto, 10 m, 9.6.1972, leg. Pirisinu; 1 paratype ♂: Sardegna, Siniscola, Sorgente S. Giuseppe 50 m, Nuoro, 22.10.1971, leg. Cacchiani, in Moretti's collection.

I dedicate this species to my student Q. Pirisinu who collected it on the edge of a small residual pool in the dry bed of a stream and in hygropetric surroundings on the Isle of Capraia.

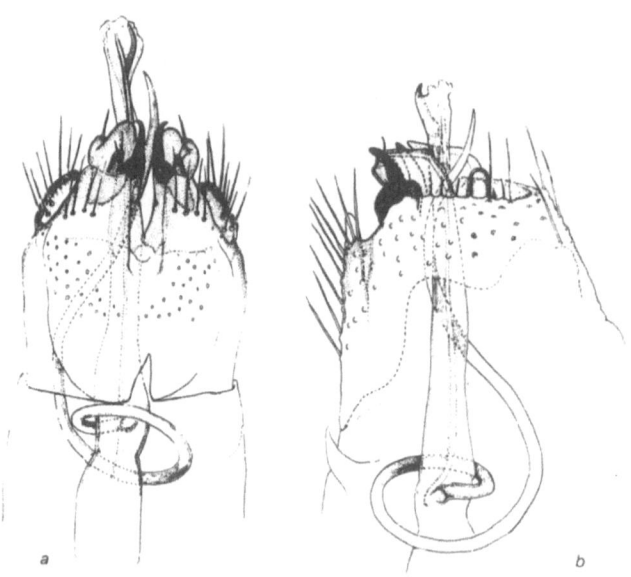

Fig. 5. Oxyethira pirisinui n. sp. ♂ genitalia; a = from below, b = side view.

REFERENCES

Higler, L.W., G. Entom. Berich. Deel 34: 62–63, 1974.
Kimmins, D.E., Ent. Gaz. 9: 7–17, 1958.
Ris, F., Mitt. schweiz. ent. Ges. 9: 420p, 1897.

Hydroptila ruffoi n. sp. (Fig. 6)

Similar to *H. fuentaldeala* Schmid. Body length: 2 mm (alcohol specimen). Prolongations of the 8th sternite have three robust, short, squamous comb-like spines (a, b). The 9th segment has sinuous posterior and convex anterior borders. The inferior appendages are not fused at the base, but form two long narrow processes, the finger-like outer one is furnished with 2 long setae, one preapical the other apical; the inner one has a cuspidate apex and a preapical seta (a, b, c). Ventral medial plate elongated with conical appendix hooked at apex (b, c). The 10th segment elongates into a dorsal plate which divides into two narrow lobes at the apex (a, b). The aedeagus is long, jointed into two partially interpenetrated terminal segments; the second is shaped like a beak (a, b). The paramere is twisted around the proximal aedeagus (b).

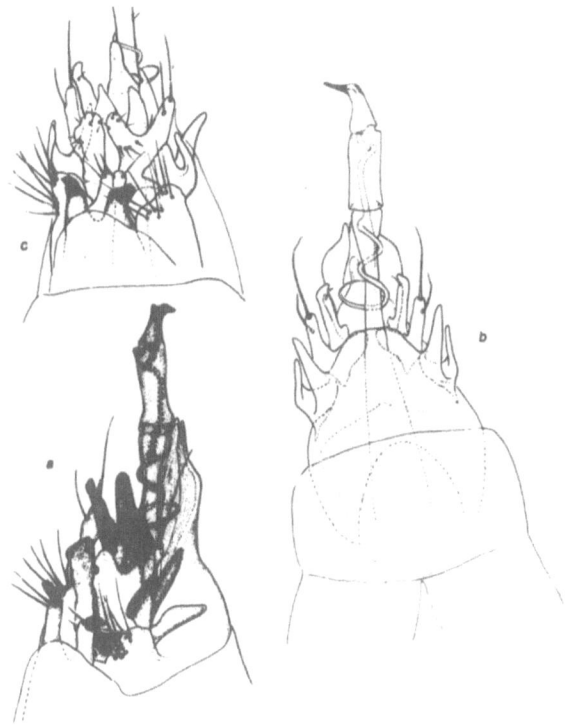

Fig. 6. Hydroptila ruffoi n. sp. ♂ genitalia: a = lateral view turned slightly to the left, b = ventral view, c = dorsal view turned 50° to the left.

171

Holotype ♂: Abruzzi, Monti della Laga, Rio Castellana, 1070 m, Teramo, 25.7.1978, leg. Moretti, in Moretti's collection.

This lovely but complicated species is dedicated to Prof. Sandro Ruffo, Director of the Natural Sciences Museum at Verona, in recognition of the interest he has always shown in my studies on Trichoptera.

REFERENCES

Kumanski, K.P., Acta Zool. Bulgar. Sofia: 1–20, 1979.
Morton, K.I., Trans. Ent. Soc. Lond.: 75–81, 1892.
Mosely, M.E., Eos, Rev. Espan. Entom. 6: 165–167, 1930.

Wormaldia copiosa botosaneanui n. ssp. (Fig. 7)

Size and colouring as for *W. copiosa*. ♂ genitalia: Wide-based superior appendages which taper abruptly after the outer medial angle (A: a). In *W. copiosa copiosa* these appendages are cylindrical-conical and do not form

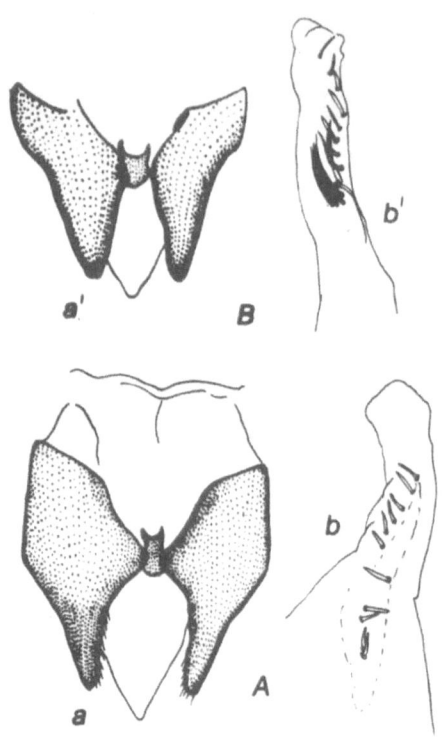

Fig. 7. A = *Wormaldia copiosa botosaneanui* n. ssp.; B = *Wormaldia copiosa*, ♂ genitalia: a and a' = superior appendages, b and b' = phallus endotheca.

172

a medial angle (B: a'). The phallus endotheca has a comb of a dozen short spines (A: b). In *W. copiosa copiosa* the proximal spines are longer, more supple and closer packed than the apical ones. Claspers as in *W. copiosa*.

Holotype ♂: Marche, Monti Sibillini, Sorgente del F. Tenna, 1180 m, Ascoli Piceno, 11.8.1955, leg. Tomasi; paratypes ♂♂ ♀♀ collected between 1955 and 1973: Emilia Romagna (Spring of the River Tiber) 1♂, Toscana (Pistoiese Apennine) 17♂ 9♀, Marche (Avellana spring) 7♂ 3♀, in Moretti's collection. Emilia Romagna 1♂ in Malicky's collection.

This new subspecies is dedicated to my friend and colleague L. Botosaneanu.

Wormaldia mediana nielseni n. ssp. (Fig. 8)

Size and colouring as for *W. mediana*. ♂ genitalia: Differs from *W. mediana* in its longer harpago which is less curved at the lower margin and more densely spiney on the inner apical surface (a). The phallus endotheca has ten apical spines arranged in the form of a hair-pin, the last four being longer and irregularly orientated (b). The other spines and the basal indented sclerites as in *W. mediana mediana* (c).

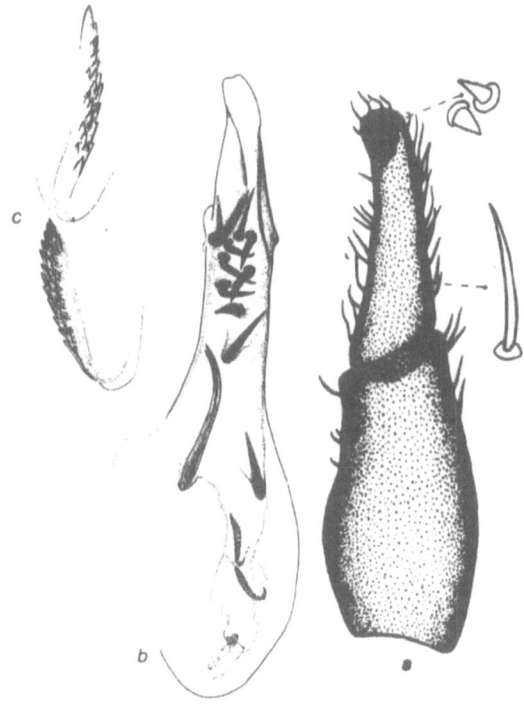

Fig. 8. *Wormaldia mediana nielseni* n. ssp., ♂ genitalia: a = harpago, b = phallus endotheca, c = basal indented sclerites.

173

Holotype ♂, allotype ♀, paratypes 24♂ 7♀: Calabria, Sila Grande, Fossiata, 1300 m, Cosenza, 8.8.1970, leg. Rotoloni. Several specimens collected between 1970 and 1976: Calabria (Sila Piccola, Aspromonte) 100♂ 44♀, Basilicata 2♂ 3♀, in Moretti's collection. Specimens collected from 1978 to 1979: Calabria (Sila Grande, Aspromonte) 37♂ 25♀, Sicilia (Peloritani Mounts) 17♂ 3♀, in Malicky's collection.

This subspecies is dedicated to the illustrious trichopterologist Prof. A. Nielsen.

Wormaldia pulla marlieri n. ssp. (Fig. 9)

Colouring and size as in *W. pulla.* Wide-based cerci with squat apical upturned hook, which is absent in *W. pulla pulla* (a). Phallus endotheca terminates in a large curved spine and a fine dorsal sclerite (d, e, f). Claspers have short wide coxopodite and the harpagones are narrowed at the base and rounded at the apex (b, c).

Holotype ♂, allotype ♀, paratypes 7♂: Toscana, Alpi Apuane, 150–800 m, Lucca and Massa, 19.6.1970, leg. Moretti and collaborators. In Moretti's collection.

I dedicate this new subspecies to my very capable colleague G. Marlier.

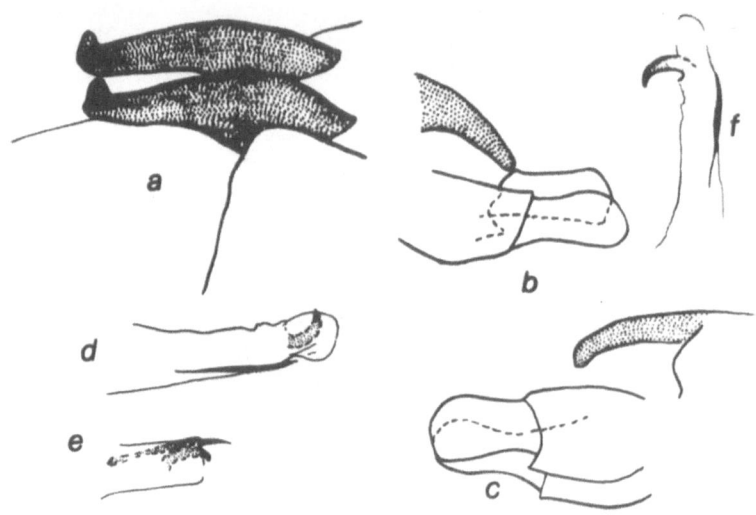

Fig. 9. *Wormaldia pulla marlieri* n. ssp., ♂ genitalia: a = cerci lateral view, b and c = claspers and cerci lateral view turned slightly to the left; d, e, f = apex of the phallus endotheca.

Wormaldia variegata denisi n. ssp. (Fig. 10)

This subspecies is similar to *W. variegata corsicana* Vaillant. (1974)

♂ genitalia: Eighth tergite with subtriangular notch at centre of anal margin (a). Tenth segment forms a long hook with rounded apex and has a preapical dorsal spicule which curves upwards and backwards (b). Slightly outcurving cerci with convex inner border (c). Phallus endotheca with 4 large spines (d), the proximal one a very curved grooved hook with double border (b), the 3 apical ones are squat, two are flattened and of different length, one, which flanks the previous 2, has a double border which looks as though it had been made up of two spines placed one upon the other (f). Seen laterally, the clasper coxopodites are almost as long as the harpagones, but when viewed from below they are narrower in the middle than at the apex (g).

Holotype ♂, allotype ♀: Isola Elba, Fosso della Leccia, M. Capanne, 22.8.1957; paratypes ♂♂ ♀♀, 10.1958, leg. Viganò, Gianotti. In Moretti's collection.

I dedicate this new subspecies to my friend C. Denis.

Fig. 10. Wormaldia variegata denisi n. ssp., ♂ genitalia: a = 8th tergite, b = 10th segment and cerci from above, b′ = apex of the 10th segment side view, c = cercus side view, d = phallus endotheca, e and e′ = proximal spine, f and f′ = distal spines, g = claspers.

REFERENCES

Botosaneanu, L., Deuts. Entom. Zeits. 7:261–293, 1960.
———, Acta Soc. Ent. Ceckosl. 57: 223–238, 1960.
Kimmins, D.E., Ann. Mag. Nat. Hist. 71: 801–808, 1953.
Vaillant, F., Ann. Soc. Ent. Fr. 10: 969–985, 1974.
Vigano', A., Boll. Mus. Zool. Univ. Torino 4: 25–32, 1974.

Polycentropus malickyi n. sp. (Fig. 11)

Related to *P. corsicus* Mos. and *P. flavomaculatus* Pictet. Head brown, cylindrical antennae with yellow and brown rings, meso and metanotum tergites dusky towards the wing joint. Brown anterior wings freckled with golden yellow. Posterior wings yellow-ochre. Abdomen brownish dorsally, paler ventrally. Legs testaceous yellow. Length of body: 9 mm, anterior wing length: 8 mm. ♂ genitalia: Middle lobe of 10th segment membranous, laminar seen from above, rectangular with straight border distally (c). Supple 10th segment processes bent like an S, its point is turned outwards (b, c) and it has a number of hairs. The superior appendages consist of one inner and one outer part. Seen from above the outer part is narrow at the base, then it widens and narrows again to form a cone at the apex (c). The inner part is wide and well sclerified from the base, it forms a robust elbow bent hook with the point turned ventrally (d, e). Seen from below the inferior appendages are long and rectangular, finish in a triangular point at the apex and have a proximal cusp (b). From the side there is a regular scope at the dorsal border after the cusp from where it tapers normally (a). The aedeagus is sinuous and equipped with 3 spines, one ventral preapical, two apical (a, b).

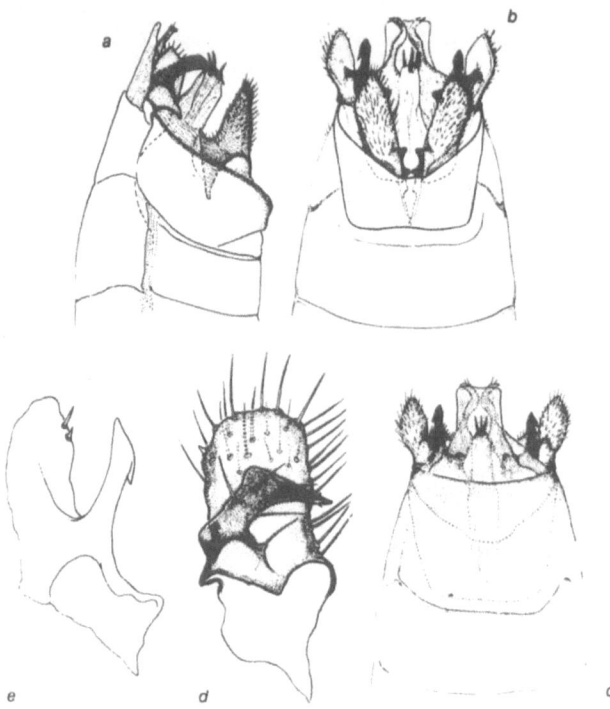

Fig. 11. *Polycentropus malickyi* n. sp., ♂ genitalia: a = side view, b = ventral view, c = dorsal view, d = superior appendage inner view, e = idem from above.

176

Holotype ♂, paratypes 2♂: Marche, F. Tronto, Ponte d'Arli, Ascoli Piceno, 30.8.1954, leg. Bellini; Toscana (River Tiber) 3♂, Campania pupa ♂. In Moretti's collection. Allotype ♀, paratypes 7♂ 3♀: Marche (River Tronto), Sicilia (Peloritani mounts) in Malicky's collection.

I dedicate this species to my colleague H. Malicky in friendship and in gratitude for his having named a beautiful Italian *Polycentropus* species after me.

Polycentropus sardous n. sp. (Fig. 12).

This species which is listed as *Polycentropus* sp. Moretti (1941) in Limnofauna Europaea (1978), is now redescribed and named *P. sardous* after the island where it is found. Similar to *P. mortoni* Mos. The ♂ mature pupa is completely sclerified and pigmented. The wings are black and scattered with tiny golden markings. Genitalia: Membranous middle lobe of 10th segment is rectangular with a straight distal border. The narrow and long superior appendages have a sinuous dorsal border and a slightly tapered turned-up apex (a). The inner part has a hooked point turned outwards and downwards which is less pointed than that of *P. mortoni* (b). Inferior appendages have a wide deep incision on the dorsal margin, so that they form 2 triangular lobes, the upper has a pointed apex, the lower has a sinuous dorsal margin with a spicule. Seen from the side these appendages have an angular notch which is much deeper than that of *P. mortoni*, seen from above they are lance-like with converging blades (b) and from below they are shaped like a pyramid with a pointed vertex (c).

Holotype pupa ♂: Sardegna, Alto Cedrino, demaniale Orgosolo, Nuoro, 7.1939, leg. Pomini.

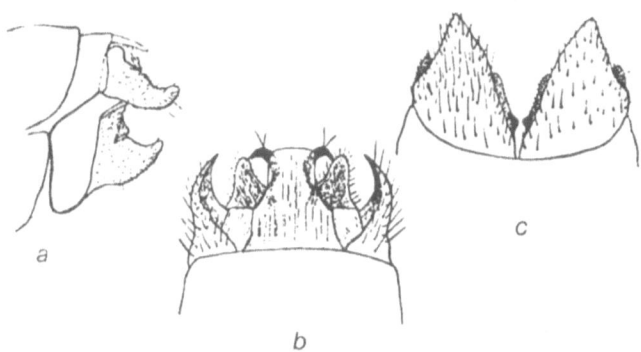

Fig. 12. Polycentropus sardous n. sp., ♂ genitalia: a = side view, b = from above, c = from below (Moretti, 1941).

REFERENCES

Botosaneanu, L. and Malicky, H. In Limnofauna Europaea, ed. J. Illies, p 345, Fischer, Stuttgart, 1978.

Malicky, H., Mitt. Ent. Ges. 25:81–100, 1975.

————, Sond. Nachr. Bayer. Entom. 4:65–77, 1977.

Moretti, G.P., Mem. Soc. Entom. Ital. 19: 259–291, 1941.

Mosely, M.E., Eos Rev. Espan. Entom. 6: 147–184, 1930.

————, Eos Rev. Espan. Entom. 8: 165–184, 1932.

Tinodes apuanorum n. sp. (Fig. 13)

Similar to *T. unicolor* Pictet and *T. bruttius* n. sp., although paler and smaller. Head, meso- and metathorax tawny with wide testaceous markings, prothorax pale. Antennae with rarely visible rings. Wings greyish-brown, slightly pubescent with large and evident pterostigma. Abdomen pale cream ventrally, brownish dorsally. Legs light yellow. Anterior wing length: 4 mm. (alcohol specimen). ♂ genitalia: Dorsal plate sclerified with sinuous lateral margin when view from above (a), arched and pointed when seen from the side (b). Fine, long and pubescent superior appendages angled at the base (a). Well developed coxopodite with straight anterior margin when viewed from the side and above (a, b). The harpagones are made up of two branches, one outer dorsal finger-like and scooped like a spoon on the inner surface (a), one ventral short and shaped like a curved thorn with a sinuous outer and convex inner margin (a, c). The basal plate apodeme is elongated into 2 sclerified triangular appendices, the right angle is rounded and points dorsally towards the aedeagus. There are two teeth at the apices and at the same point there are two dorsal setae (c). The aedeagus paraprocts

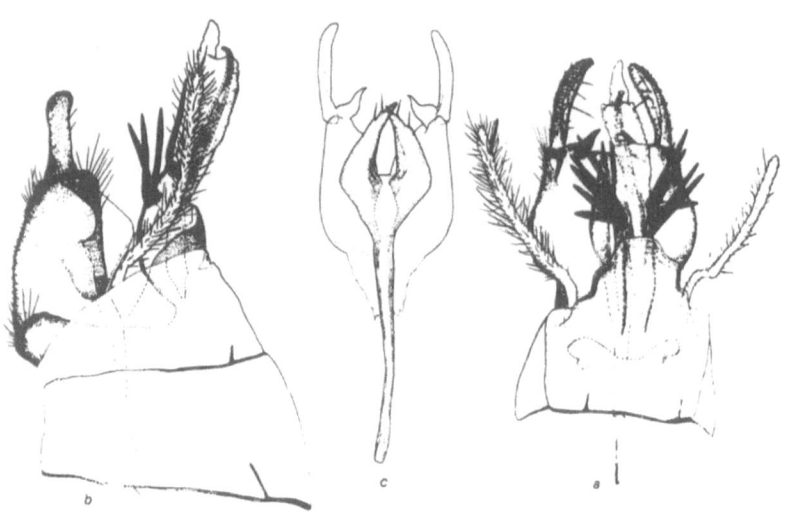

Fig. 13. Tinodes apuanorum n. sp., ♂ genitalia: a = from above, inclined to the front, b = side view, c = inferior appendages inner view.

are sclerified and have 5 strong spines on each side. The fleshy ventral extension at the extremity of the aedeagus. The ejaculatory duct turns down (b).

Holotype ♂, allotype ♀ and paratypes 59 ♂ 20♀: Toscana, Alpi Apuane, igropetrici, Passo del Vestito, 870 m., Massa, 19.6.1970; rhythron, Lucca, 6.8.1970, leg. Moretti and collaborators. In Moretti's collection; 1♂ 1♀ in Malicky's collection.

The name comes from the ancient Apuani people.

Tinodes bruttius n. sp. (Fig. 14)

Similar to *T. unicolor* Pictet and *T. apuanorum* n. sp. Head and thorax with a slightly paler zone. Antennae brown with lighter rings. Wings greyish-brown. Abdomen brownish dorsally, pale yellow ventrally. Legs light yellow. Anterior wing length: 5.5 mm. (alcohol specimen). ♂ genitalia: Chitinous dorsal plate, wide at base with pointed vertex if seen from the side (a). The superior appendages are fine, long pubescent and shaped like a twig (a, c). Well developed trapezoidal coxopodites with bilobed apical border if seen from the side (a). The harpago is short and shaped like a thumb, the apex is rounded when seen from the side (a), sinuous if viewed from above (b). Towards the inner side and at the base of the harpago there is a robust wide-based thorny appendix, pointed at the top and converging inwards (b, c). It is longer and narrower than in *T. apuanorum*. The intermediate appendages grow from behind the coxopodite, they are sickle-shaped, angled downwards, blade-like and terminate in a pointed apex. They have two black, robust preapical setae which point forward (a). The aedeagus is sclerified at the base, membranous

Fig. 14. Tinodes bruttius n. sp., ♂ genitalia a = side view, b = from above, c = inferior appendages from below.

distally, straight dorsally, ventrally concave and has six long, black, robust spines on the sclerified paraproctal appendices. The extremity of the aedeagus lengthens ventrally into a carnous protuberance. The ejaculatory duct curves down at the apex (a).

Holotype ♂: Calabria, Le Sezze, T. Macchinante, Brognaturo, 850 m, Catanzaro, 20.6.1972, leg. Iozzo; paratypes 12♂, allotype ♀ collected between 1972 and 1979: Calabria (Sila Piccola, Pollino Mount). In Moretti's collection.

Derivatio nominis. From the ancient latin name of the Calabrians people.

REFERENCES

Botosaneanu, L. and Taticchi Vigano', M., Boll. Mus. Zool. Univ. Torino 2: 9–14, 1974.
Marlier, G., Boll. Ann. Soc. Roy. Entom. Belgique 95: 197–204, 1959.
Vaillant, F., Ann. Mag. Nat. Hist. 7: 58–62, 1954.

Drusus genus

The *D. aprutiensis* and *D. camerinus* ♂ have already been described, but no species name has yet been given them. I shall now give them a species name and describe the female of each species, as Botosaneanu suggested there might be a difference in the philetic line based on the structure of the ♀ genitalia.

Drusus aprutiensis (= taxon 2, Moretti and Cianficconi, 1974) (Fig. 15)

Wing spread: ♂ 19–21, ♀ 24–25 mm; closed wing: ♂ 10.5–11.5 mm, ♀ 13–14 mm. ♂ genitalia: The 8th tergite has three distinct spiney lobes, the intermediate appendages are black, wide, squat, rounded at the apex and, seen dorsally, spread open (c). The inferior appendages are large, conical and angulate towards the middle of the upper edge (a) ♀ genitalia: The proximal and distal part of the 9th segment are distinct. The distal part and the 10th segment form a short cube-shaped tubular piece with a more or less rounded upper spicule, the lateral lobes of the segment 9 are slender oval (i, l, m). From above the segment 10 has two well-separated lobes (d, e). The central 9th segment lobe is long and arched ventrally like that of *D. monticola* (i, l, m). The lateral lobes of the vulvar squama are large and curve towards the shorter median lobe (f, g, h). *D. aprutiensis* is, therefore, similar to the *D. monticola* ♀ as Botosaneanu rightly supposed.

Holotype ♂, allotype ♀: Abruzzi, P.N.A., Jannanghera, 1200 m, L'Aquila, 25.6.1954, leg. Consiglio. Paratypes ♂♂ ♀♀ collected between 1955 and 1979: Abruzzi (several springs) 364♂ 83♀, Lazio 12♂ 4♀ + 3♂ 3♀ sent to us by Botosaneanu. In Moretti's collection. Abruzzi 1♂ 1♀ in Malicky's collection.

Derivatio nominis. The latin name of the Abruzzi.

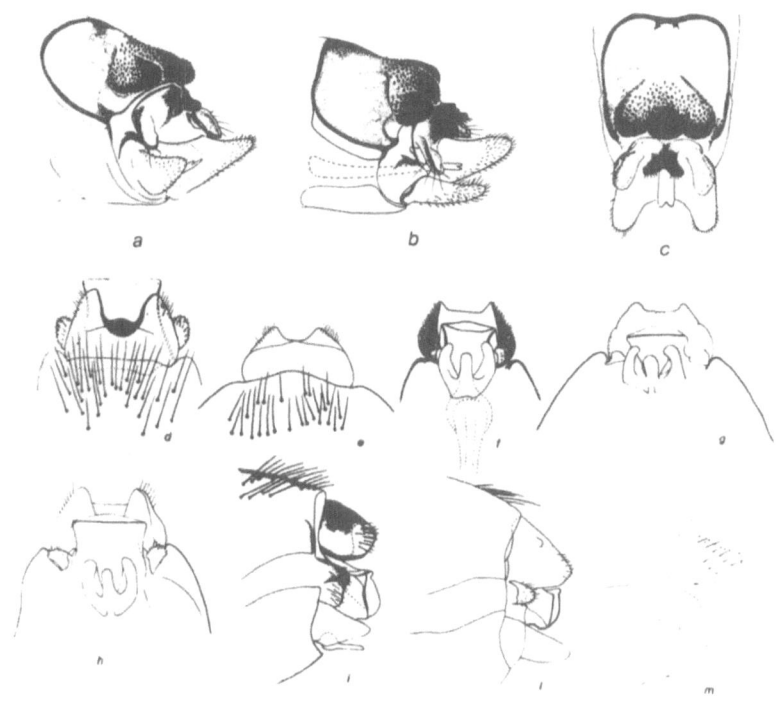

Fig. 15. Drusus aprutiensis n. sp., ♂ genitalia: a = front lateral view, b = lateral view, c = from above; ♀ genitalia: d, e = from above; f, g, h = from below; i, l, m = side view.

Drusus camerinus (= taxon 1, Moretti and Cianficconi, 1974) (Fig. 16)

Wing spread: ♂ 18–19 mm, ♀ 19–21.5 mm. ♂ genitalia: The 8th tergite is trilobed as in *D. aprutiensis,* but the middle lobe is larger than the lateral ones (a, c). The intermediate appendages are short, not hooked at the apex, and blackened at the apex only (b). The superior appendages are large and suboval, from the side they are hatchet-shaped (a). The inferior appendages are large, subcylindrical and rounded at the apex (b, c).

♀ genitalia are similar to those of *D. improvisus,* particularly the central lobe on the ventral area of the 9th segment and the tubular piece, although this is more supple at the anal margin than in the *D. improvisus* (f, g, h). The lateral appendages have a narrow connecting stalk (g). The lateral lobes of the vulvar squama, as well as the medial lobe which is narrow and long, are very similar to those of *D. improvisus,* for this reason *D. camerinus* has been assigned to the *improvisus* groupe.

Holotype ♂, Allotype ♀: Marche, Fonti di Selvazzano, 655 m., Camerino, Macerata, 10.5.1954, leg. Verdarelli; Paratypes ♂♂ ♀♀ collected between 1954 and 1976: Umbria 22♂, Marche (several springs) 361♂ 92♀, In Moretti's collection. Marche 1♂ 1♀ in Malicky's collection.

Fig. 16. Drusus camerinus n. sp., ♂ genitalia: a = front lateral view, b = lateral view, c = from above; ♀ genitalia: d = from above; e = from below, f, g, h = side view.

Derivatio nominis. From the latin name of a Marche's town in the springs of which it was found.

REFERENCES

Moretti, G.P. and Cianficconi, F., *In* First Int. Symp. on Trichoptera, ed. H. Malicky, pp. 93–104, Junk, The Hague, 1974.
Schmid, F., Mem. Inst. Roy. Sc. Nat. Belgiques 55:1–92, 1956.

Halesus nurag Mal. ♀ (Fig. 17)

The female has not previously been described. The colour and markings of the anterior wing are identical to the male, the size is slightly smaller. Wing spread: ♂ 37–40 mm., ♀ 36–39 mm. Genitalia: Tubular piece rather long,

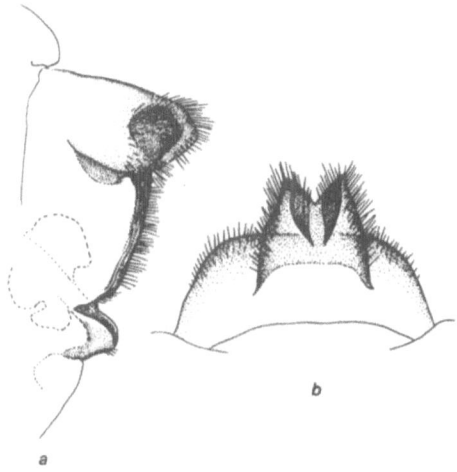

Fig. 17. Halesus nurag Mal., ♀ genitalia: a = side view, b = from above.

wide based and narrow tipped, with long golden hairs; in profile specimens cleared in KOH solution have a slightly bulging apical border (a). Lateral pieces of the 9th segment are large and straight. Seen from above 10th segment has a deep triangular notch in the anal border with long hairs and pigmented internal apical zones (b).

Allotype ♀ and paratypes several ♀: Sardegna, Tempio, 3.11.1964, leg. Prota. In the Moretti's collection and in Istituto di Entomologia, Università, Sassari.

Halesus radiatus vaillanti n. ssp. (Fig. 18)

Closely related to *H. nurag* Mal. and *H. radiatus* Curt. Head testaceous yellow antennae yellowish tawny, thorax brown. The yellowish anterior wings have elongated dusky markings edged with white arranged longitudinally between the nervures which form a distinct marbling between the apical forks. Hyaline yellowish-grey posterior wings. Upper abdomen brown, lighter underneath. Legs straw-yellow. Wing spread: ♂ 44–48 mm., ♀ 44–49 mm. ♂ genitalia: Large superior appendages, concave and bend to form a longitudinal edge on the inside. Intermediate appendages black at the apex, subrectangular or sub-elliptical and sclerified. Inferior appendages well-sclerified and blackened at the apices where they form a wide, flat, external branch, which is separated from a short inner proximal tooth by a pronounced notch. (a, b). Long aedeagus with 4 black teeth arranged like a butterfly's wings at the apex, these are often reduced to 2 (a). Parameres are narrow at the base, wide at the centre and tapered at the apex. There are about 10 black teeth which become progressively longer on the upper margin. The lower margin is markedly convex (d, c).

183

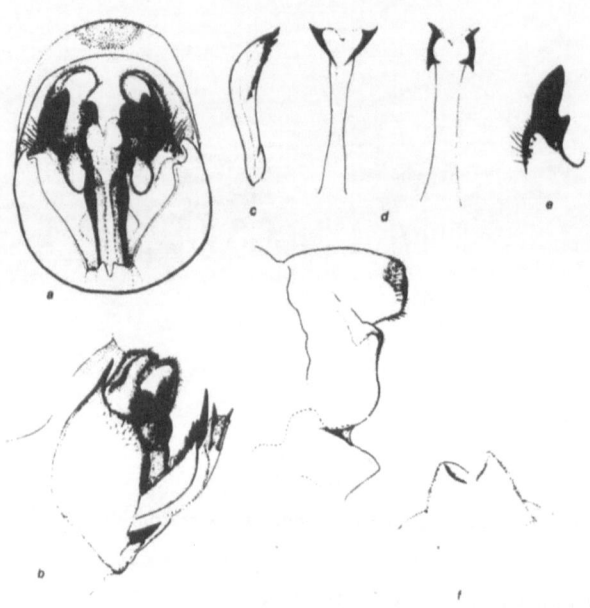

Fig. 18. Halesus radiatus vaillanti n. ssp., ♂ genitalia: a = front, b = side view, c = paramere, d = two types of aedeagus, e = inferior appendage; f = ♀ genitalia side view and from above.

♀ genitalia: Cylindrical-conical tubular piece with short golden hairs and an almost straight apical border when seen from the side (f), divided by a deep triangular incision when seen from above. Lateral pieces of 9th segment large and slightly concave. It inhabits the hill and mountain rhythron and hyporhithron.

Holotype ♂: Umbria, F. Sordo, Norcia, 500 m, 8.11.1967, leg. Venturi; allotype ♀: Campania, F. Lete, Matese, 1000 m, 22.9.1967, leg. Pangia; paratypes ♂♂ ♀♀ collected between 1967 and 1980: Emilia Romagna (River Tiber) 3♂ 4♀, Umbria (Nese stream) 33♂ 25♀, Lazio 1♂, Abruzzi 1♂, Calabria (Sila Piccola) 7♂. In Moretti's collection.

I dedicate this subspecies to my friend F. Vaillant with whom I have exchanged various taxonomic data.

The problem involved in the validity of this subspecies is complex owing to the variable characters of the ♂ genitalia on the superior, intermediate and inferior appendages, as well as of the aedeagus and parameres. In my opinion it could also be considered a subspecies of *H. nurag* which it resembles particularly in the shape of the inferior appendages that are identical to those of the Sardinian species. On the other hand, the inferior appendages differ from those of all other *Halesus* of the *digitatus* group including *H. radiatus*. The variability in the apical aedeagus sclerites reflects one of the main characteristics of *H. radiatus*, as Schmid (1951) demonstrated. However, no variations in the aedeagus apical teeth were seen in about 100 *Halesus nurag* specimens kindly sent for study by Prota; they were always 2 and their shape was always the same. In addition, the *H. radiatus vaillanti* parameres are convex and swollen as in *H. radiatus* and not like those of *H. nurag*. Furthermore, *H. nurag* is considerably smaller than *H. radiatus vaillanti* and has more distinct wing markings. The ♀♀ resemble the *H. radiatus* rather than the *H. nurag* ♀♀. For these reasons, provisionally, I consider that this is a new Italian peninsula subspecies of *H. radiatus* to which it is closely related, but, at the same time, not excluding that *H. nurag* and *H. radiatus vaillanti* could have a closer relationship and, therefore, a more direct philetic origin.

REFERENCES

Curtis, J., Philos. Mag. 125, 1834.
Malicky, H., Ent. Zeit. 84: 86–87, 1974.
Moretti, G.P. and Spinelli Batta, G., Boll. Zool. Suppl. 46: 156–157, 1979.
Schmid, F., Trab. Mus. Cienc. Nat. Barcelona 1: 4–27, 1951.

Micropterna wageneri Mal ♀ (Figg. 19)

Although Malicky (1971) described the male of this species, this is the first report on the female. The ♀ is much bigger than the male but the same yellow. Antennae and legs clear yellow. The anterior wings are strawy-rust, the posterior wings paler. It is similar to the *M. sequax* ♀ but much smaller. Wing spread: 27–28 mm, body length: 10 mm. Genitalia: The dorsal face of the segment 9 is long, very wide and form convex protruding lateral angles (a). The 10th segment dorsal scale is subtrapezoid and reaches half-way along the internal margin of the lateral lobes which are large-based short-tipped

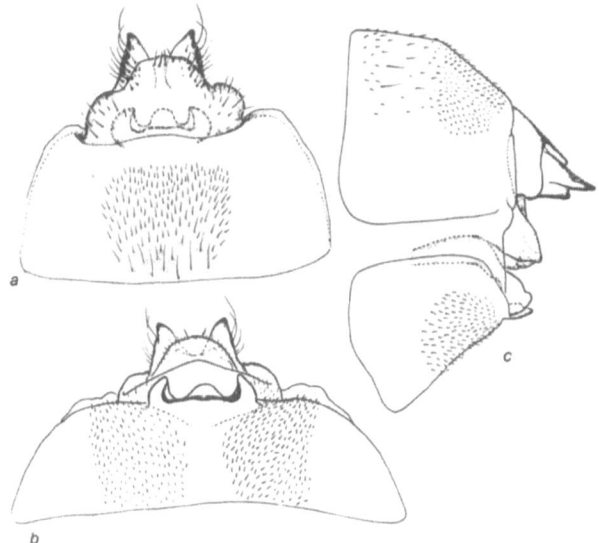

Fig. 19. Micropterna wageneri Mal., ♀ genitalia: a = from above, b = from below, c = side view.

185

triangle with slightly turned outward apices with inner margin that is more convex than that of *M. sequax*. When the lobes are seen from the side they are narrower, slenderer and less protruding than those of *M. sequax* (c). The ventral 10th segment scale forms a large dome (b). The ventral part of the 9th segment is prominent and convex, when seen from the side, unlike that of *M. sequax* it forms two large, only slightly convex, lobes which do not appear to be separated (b). The vaginal vestibule is elliptical. The wide convergent lateral lobes take up most of the two poles. The distal internal prolongation of the two lateral lobes are widened, rounded and convergent; the central lobe is large and semicircular, almost triangular in dry specimens. Like the male, the ♀ *wageneri* is therefore not unlike *M. sequax,* but is the smaller species of the *Micropterna* genus found in Italy.

Allotype ♀, paratypes 6♂ 7♀: Toscana, Pratomagno, 1000 m, 19.6.1980 Arezzo, leg. Moretti, in Moretti's collection.

REFERENCES

Malicky, H., Die Höhle 1: 15–20, 1971.

Mesophylax aspersus sardous Moret. and Gian. (Fig. 20)

The taxonomic characteristics are briefly described and illustrated; a more detailed explanation is at present in press (Moretti and Cianficconi). The lower appendages of the male genitalia differ from those of *M. aspersus* as they have a pronounced notch which terminates in a well-sclerified cusp on the interior of the apex; while on outside there is a poorly sclerified lobe covered with long golden hairs on the outer margin. The black pigmentation of the T 7 anterior wing venation is interrupted by paler areas. The insect is smaller than the apennine form.

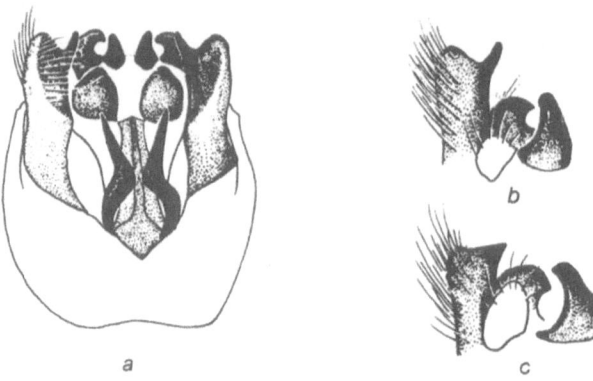

Fig. 20. Mesophylax aspersus sardous Moret. and Gian., ♂ genitalia: a = front, b = details of inferior appendage, c = the same in the Italian peninsula *M. aspersus.*

186

Holotype ♂: Sardegna, Grotta "Su Coloru", Laerru, Sassari, 11.10.1961, leg. Prota; paratypes 8♂ collected between 1967 and 1976, Sardegna in caves.

REFERENCES

Moretti, G.P. and Gianotti, F.S., Mem. Soc. Ent. It. 46: 73–125, 1967.
——, and Cianficconi, F., Lav. Soc. It. Biog. in press.

Sericostoma cianficconii Moretti (Fig. 21)

The male genitalia have already been described (Moretti and Cianficconi, 1978). The ♂ mask with conical apex is very prominent, twice the size of the eye (a). Antennae yellowish-brown. Pale legs. Wing spread: ♂ 20–27 mm, ♀ 22–26 mm.

♀ genitalia: The 9th tergite is triangular, convex and sclerified with an apex divided into a pair of small processes. Two oval anterior-dorsal branches and two narrower but longer conical and pubescent postero-ventral branches form laterally at the base of the tergite 9 (b). There is a wide triangular incision bordered by two tiny only slightly protruding teeth on the 10th segment, this gives the anal margin a sinuous, horizontal look when it is viewed either

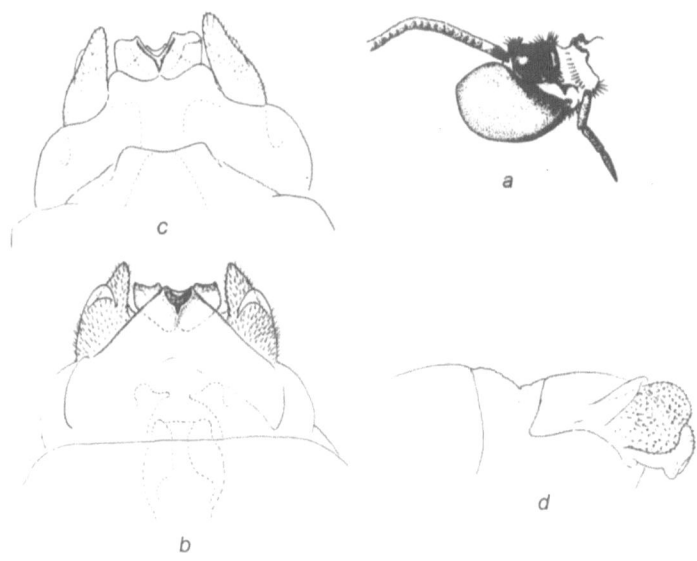

Fig. 21. Sericostoma cianficconii Moretti, a = ♂ maxillary palpus side view, ♀ genitalia: b = from above, c = from below, d = side view.

187

ventrally or dorsally (b, c). Side view, the 9th segment ventral branch which roofs the dorsal lobe, is sclerified and blade-like. The ventral lobe is very protruding and it forms a bristly protuberance (d). The segment 10 looks like a turned down sclerified eave (d).

Holotype ♂: Marche, Fiume Potenza, Torre del Parco, 330 m, Castelraimondo, Macerata, 1.6.1948, leg. Moretti; allotype ♀: Toscana, Camaldoli, acque sorgive, Arezzo, 26.6.1965, leg. Gianotti; paratypes ♂♂ ♀♀ collected between 1955 and 1970: Emilia Romagna 2♂, Toscana 1♂, Umbria 20♂ 17♀. In Moretti's collection.

I dedicate this to the most expert of my collaborators, F. Cianficconi, who is tenacious in her Trichoptera studies.

Sericostoma italicum Moretti (Fig. 22)

The general aspect of the ♂ and genitalia have already been illustrated (Moretti and Cianficconi, 1978). Here I have limited myself to adding certain valid distinguishing characteristics for better recognition of the species. Wing spread: ♂ 24–29 mm, ♀ 28–29 mm; body length: ♂ 8–15 mm, ♀ 11–16 mm; closed wing: ♂ 14–15 mm, ♀ 13–16 mm. Bronze insect with even dark brown antennae which are large and only slightly tapered at the apex. The protuberances at the point of insertion of the antennae are large and cylindrical. The blackish-brown head is covered with tufts of thick bronze hairs. Small not very prominent mask as wide as the diametre of the eye (a). Brown only slightly pubescent labial palpi (a). The membranous part of the

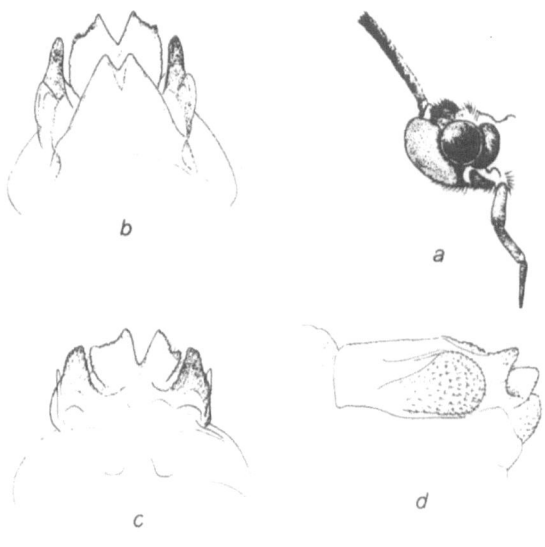

Fig. 22. Sericostoma italicum Moretti, a = side view of the ♂ maxillary palpus; ♀ genitalia: b = from above, c = from below, d = side view.

188

prothorax is almost white, the black sclerites are covered with tufts of thick hairs. Meso- and metathorax are smooth, shiney dark brown with lighter scutellum. Brown wings densely covered with bronze pubescence from which the venation stand out. Legs with almost black coxa, brown femur, pale tibia and tarsa. Abdominal tergal and sternal sclerites dark brown, lateral line wide and whitish, as is the intersegmental area of the urites. ♀ genitalia: 9th tergite as in *S. cianficconii*, but much more convex, narrower and pigmented (b). Laterally there are two oval lateral-dorsal branches, which are narrower than those of *S. cianficconii* and two narrower, conical pubescent posterior-ventral branches. The 10th tergite extends beyond the posterior-ventral branches into a squarish shape with saw-edged margin where there is a deep medial triangular notch bordered by two spicules (b). The two pubescent lobes of the 9th sternite are also serrated and do not reach the extremity of the 10th sternite, as do the same appendices in *S. cianficconii*. The 10th sternite is bilobed and toothed (d).

Holotype ♂: Lazio, Acilia, Roma, 7.8.1939, leg. Castellani; allotype ♀: Umbria, prati marcitoi, Norcia, 18.6.1976, leg. Moretti; paratypes ♂♂ ♀♀ collected between 1952 and 1979: Toscana 1♂ 1♀, Umbria 37♂ 7♀, Lazio 18♂ 5♀, Abruzzi 121♂ 79♀, Molise 5♂, Campania 1♂. In Moretti's collection.

Derivatio nominis. Named after Italy.

REFERENCES

Giudicelli, J., Entomops. 46: 201–212, 1978.
Malicky, H., Entomol. Z. 83: 249–251, 1973.
Moretti, G.P. and Cianficconi, F., *In* Proc. 2nd Int. Symp. on Trichoptera, ed. M.I. Crichton, pp 7–30, Junk, The Hague, 1978.
Nielsen, A., Det. Kong. Dans. Videns. Selsk. Biol. Skz. 23: 115–120, 1980.

Beraea botosaneanui n. sp. (Fig. 23).

Length with wings closed: 4.5–4.7 mm. Black elliptical androconia surrounded by a pale area. Similar to *B. crichtoni,* from which it differs in the

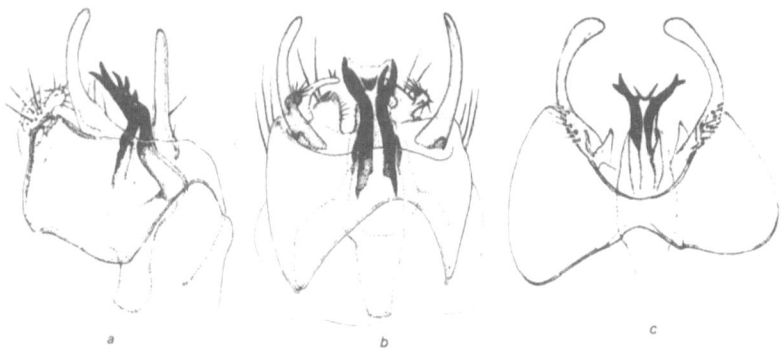

Fig. 23. Beraea botosaneanui n. sp., ♂ genitalia: a = lateral view, turned slightly to the left, b = from above, c = from below, balsam specimen.

pincer-like superior appendages that broaden from the centre to the tip when seen from above (c), but not when seen from below or from the side (a, b). The inferior appendages are bilobed as in *B. crichtoni;* however, the lower lobe is longer and slenderer (b). The ventral coxopodite branches are conical with a narrow notch before the external apical tooth and there are long thorn-like setae on the ventral face (b). The extremity of the aedeagus is bilobed and two large, robust, thorn-like sclerites run the full length, they are black distally, brown proximally and diverge at the apex where they are superimposed by two short black spines (a, b, c). On the inside of these there is a hemispherical sclerified plaque which is also present in *B. crichtoni.*

Holotype ♂ and paratypes 4♂: Sardegna, Guttusu Mannu, Cagliari, 18.4.1972, leg Hartig. In Moretti's collection.

This species is dedicated to the trichopterologist Botosaneanu in recognition of the suggestions he has frequently offered me.

Beraea crichtoni n. sp. (Fig. 24)

Similar to *B. alva* Mal. and to *B. terrai* Mal. Brown unringed antennae. The vertex of the head elongates into a conical very protruding cusp equipped

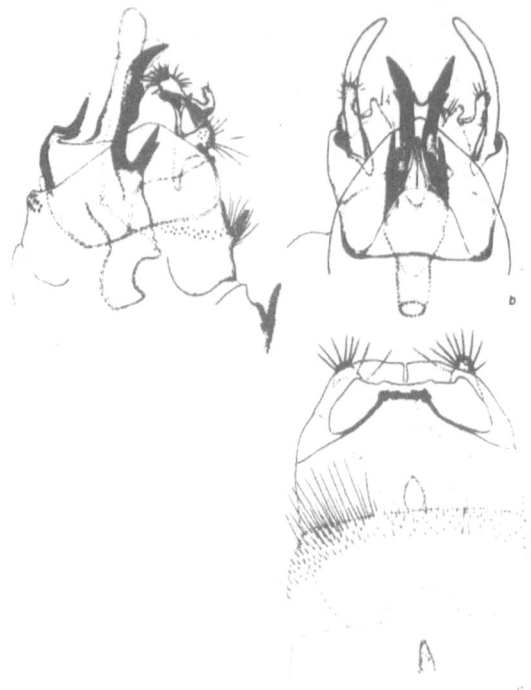

Fig. 24. Beraea crichtoni n. sp., ♂ genitalia: a = lateral view, b = from above; ♀ genitalia: c = from below.

with a tuft of long setae. Maxillary and labial palpi carpeted with thick hairs. Mesothorax of a uniform chestnut-brown with a pale circular wart on the anterior area of each scutum. Legs light brown. Fore wings bronze and pubescent with a long fringe of apical hairs. The elliptical spindle-like androconia at the base of the wing are clearly visible and surrounded by a pale zone. Length wings closed: 5 mm (alcohol specimen). ♂ genitalia: The 9th tergite terminates in a triangular, convex, swollen elongation of the anal margin (a). This lies above the sclerified laminar extension of the 10th segment. In the part where the convex margin of the 9th tergite lies over the 10th segment there is an evident step-like spicule (a). The 10th segment apex is also sharp and triangular. The dark, sclerified scissor-like superior appendages are large and have a lobular apex when viewed from the side (d). The inferior appendages are bilobed, the dorsal lobe being more swollen than in *B. botosaneanui* and bristling with rayed robust setae. The elongated ventral lobe is shorter and narrower and has 2—4 tiny thorn-like setae. The ventral coxopodite branches are oval with spiney setae on the ventral face (a). The apical border is furnished with inwardly turning teeth (a). The fleshy aedeagus is bilobed at the extremity armed with long spiney sclerites; 2 proximal lateral short ones with inner preapical tooth, 2 much longer and stronger on the interior which curve down and have two preapical teeth pointing outwards. A hemispherical, convex at the back, sclerified plaque is situated between the two larger spine as in *B. botosaneanui.*

♀ — There is a conical cusp at the vertex of the head as in the ♂ but it is shorter and squatter. Mesothorax brown with a tiny white spot on the anterior zone of each scutum; in addition there are two larger white oval marks on the anterior part of the scutellum, as in the *B. maura* ♀, that are not present in the male. The venation of the wings of the ♀ differs from that of the ♂, as is in *Beraea* genus. Length as for ♂. ♀ genitalia: The 7th sternite has a strong sagitally compressed black tooth on a pale field (c). The distal half of the 8th sternite is covered with a thick fringe of long golden setae that project well beyond the anal margin. Viewed dorsally the 9th tergite has two dark conical lateral processes joined to the base of the 9—10th segment by a membranous area. The 10th segment terminates in a trapezoidal sclerified protuberance at the anal border.

Holotype ♂: Calabria, La Sila, Camigliatello, 22.6.1960, leg. Ruffo; allotype ♀ and paratype 84♂ 9♀: Basilicata, Sorg. Lago Sirino, Lagonegro, 860 m, Potenza, 24.8.1976, leg. Petroni. In Moretti's collection.

I dedicate this species to my colleague M.I. Crichton in memory of his friendliness at the 2nd Symposium on Trichoptera.

Beraea ilvae n. sp. (Fig. 25)

Closely related to *B. maura* and with similar colouring. Clearly visible androconia at the base of the anterior wing as in *B. crichtoni.* ♂ genitalia: A broad triangular projection starts from the middle of the 9th segment, it projects slightly when viewed from the side. The underlying median part of the 10th segment is wide based and has a pointed apex. Superior appendages strongly chitinized fuscous curving inward with a pointed apex when viewed from above (b), obliquely rounded if seen from the side (a). The coxopodite dorsal branch is bilobed, the dorsal lobe is rounded at the apex and has erect, rayed setae, the ventral lobe is slender and longer, curved inward and is equipped with three short spiniform setae. Oval ventral coxopodite branch with several spine-like setae on the apical border and a well defined notch on the outer apical margin, the ventral face is equipped with long setae. Fleshy aedeagus armed with a line of 4 pairs of large black spines, three of which point to the sides. The apical pair point towards the estremity.

Holotype ♂, allotype ♀ and paratypes ♂ ♀: Isola Elba, Fonte della Chiava, M. Capanne, 22.8.1957, leg. Viganò, Gianotti. In Moretti's collection.

Derivatio nominis. From the latin name of the Isle of Elba.

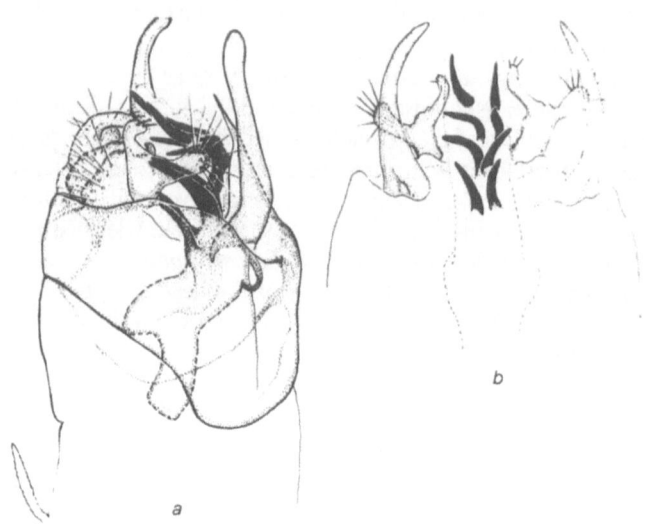

Fig. 25. Beraea ilvae n. sp., ♂ genitalia: a = lateral view turned slightly to the left, b = from above.

ACKNOWLEDGEMENTS

I should like to thank A. Sensidoni for having executed most of the drawings. The research was supported in part by the CNR programme: 'Promozione della qualitá dell'ambiente', linea di ricerca 'Zoocenosi delle acque interne' — Coordinatore: Prof. S. Ruffo.

REFERENCES

Kimmins, D.E., The Entomologist. 84: 19–22, 1951.
Malicky, H., Mitt. Ent. Ges. Basel. 25: 81–100, 1975.
Mosely, M.E., Eos. Rev. Espan. Entom. 6: 165–167, 1930.
Nielsen, A., Biol. Skr. Dan. Vid. Selsk. 8: 1–159, 1957.

COMPARATIVE SEM AND TEM STUDIES ON THE ANDROCONIAL STRUCTURES OF *LASIOCEPHALA BASALIS* KOL. AND OTHER TRICHOPTERA

G.P. MORETTI AND M.C. BICCHIERAI

SUMMARY

The results of histological as well as scanning electron microscope (SEM) and transmission electron microscope (TEM) ultrastructural studies carried out on the *Lasiocephala basalis* Kol. androconia were compared with other previously studied Trichoptera species. Although the androconial histological features were similar, marked differences were encountered in the morphology of the various scales and setae.

INTRODUCTION

It is well known that scent organs are usually present in the male Trichoptera, although it has been reported that the female *Enoycila pusilla* Burm. emits a scent that attracts males from far off (Kelner-Pillault 1975).

This is a continuing investigation and SEM and TEM ultrastructural studies, as well as histological studies, have previously been carried out on *Hydroptila aegyptia* Ulm. (Moretti and Bicchierai 1979a), *Drusus improvisus* McL. (Moretti, Cianficconi and Bicchierai 1979), *Silo nigricornis* Pict. (Moretti and Bicchierai 1979b) and *Sericostoma pedemontanum* McL. (Moretti and Bicchierai 1979c). *Beraea crichtoni* Moret. (Moretti and Bicchierai 1979d) and *Beraeamyia squamosa* Mos. (Moretti and Bicchierai 1979e) were studied by SEM only. The present research deals with the androconia of *Lasiocephala basalis* Kol. which, to the best of our knowledge, has not been reported before.

RESULTS AND DISCUSSION

The average length of the male *L. basalis* anterior wing is 9 mm. It has a fringe of long (aprox. 800 μ), thick bristles directed backwards and outwards on the anterior margin. The dense line of bristles begins at the point where the wing joins the body and terminates just before the anastomosis between

Fig. 1. Fringe of androconial bristles on the male *Lasiocephala basalis* Kol. anterior wing × 65

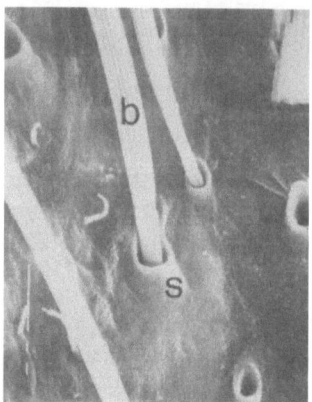

Fig. 2. Basal region of the androconial bristles × 900
b = scent bristle, s = crater-like socket

the costa and the subcosta. This anterior wing scent-organ differs from both that found in *Beraea crichtoni*, where there is a rounded callosity full of scales at the base of the wing behind which are seen seed-like scales, and that of *Beraeamyia squamosa*, where a long spatula like process bordered by a dense area of scales starts at the wing joint and lies along the proximal depressed zone between the radius and the beginning of the anal veins.

Total staining of the *L. basalis* wing shows that the bristles are inserted in a raised tapered border which contains the androconial cells. Transverse sections stained with methylene blue reveal a large cell under each bristle

194

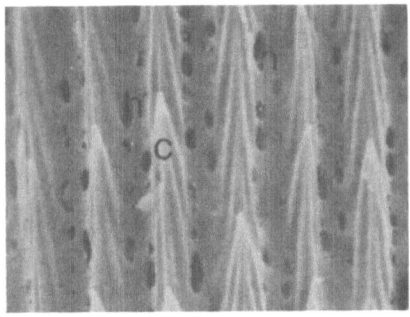

Fig. 3. Longitudinal ribs of the androconial bristle × 10,000
c = crest, h = hole

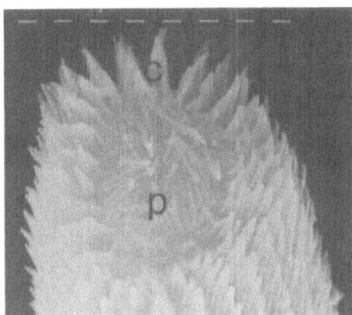

Fig. 4. Basket like arrangement of the androconial bristle apex × 3,000
c = crest, p = projections

socket with a nucleus about $14\,\mu$ in diameter, that is, five times the size of an epidermal cell nucleus. This nucleus is highly chromatophilic, rich in chromatin clumps and has well defined nucleoli. Contrast phase microscope observations had already revealed secretion droplets at the base of some of the bristles. In transversal sections the androconial region is seen to be an elongated, curved sac lined by epidermal cells on the inner surface. The large androconial cells are amassed on the concave face and there is a nerve fibre and few trachea on the convex face.

At SEM the bristles appear as cilindrical, thickened hairs of a reasonably even diameter (Fig. 1), except at the base where they enter the raised crater-like socket, at which point they taper very slightly (Fig. 2). The sockets of the other species manifest different forms; in *Drusus improvisus* it is the shape of a doughnut ring, in *Silo nigricornis* it is an irregular circle tilted and raised to one side and in *Beraeamyia squamosa* it is like a life-buoy.

The *L. basalis* bristles are longitudinally grooved, from the base to the apex (Fig. 3). The raised rib of each groove appears to be crested. The spire-like projections on the crests overlap, are pleated and point towards the

195

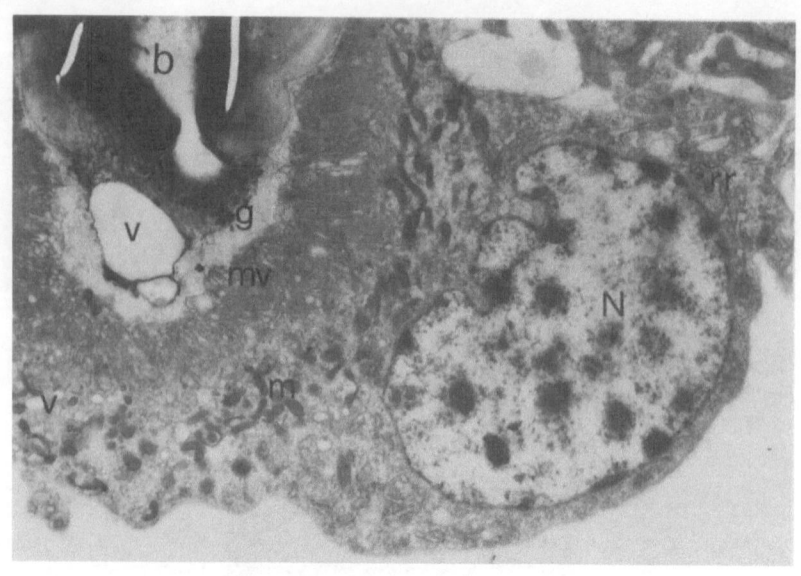

Fig. 5. Cell at the base of the androconial bristle × 8,000
b = bristle, g = electron dense granules, m = mitochondria, mv = microvilli, N = nucleus,
rr = rough endoplasmic reticulum, v = vacuoles

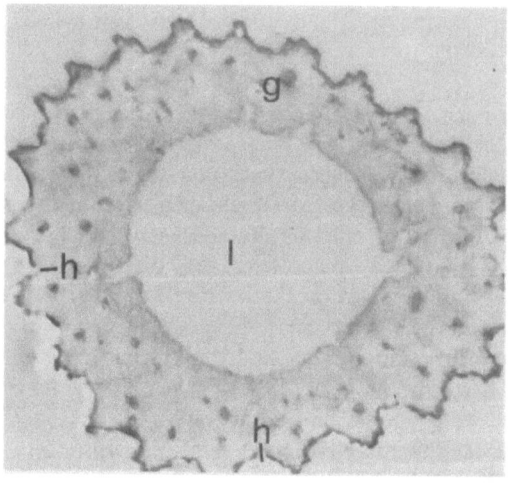

Fig. 6. Cross-section of the scent-bristle × 11,000
g = electron dense granules, h = holes, l = lumen

196

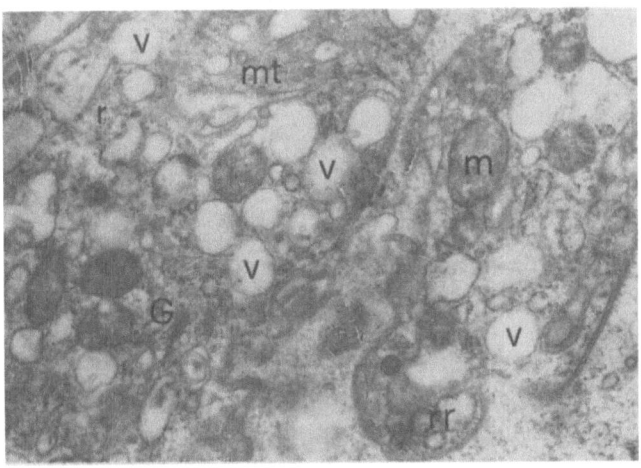

Fig. 7. Androconial cell cytoplasm ✕ 18,000
G = Golgi complex, m = mitochondrion, mt = microtubles, r = free ribosomes,
rr = rough endoplasmic reticulum, v = vacuoles

apex. Tiny holes that range between 0.1 and 0.5 μ in diameter are scattered in the base of the grooves. This type of rib shaped bristles is not unlike the scales that fill the callosity in *Beraea crichtoni* or the banana like scales in *Beraeamyia squamosa,* but are very different from the inverted flask-shaped scales of *Drusus improvisus* in which the longitudinal ribs are made up of dense, tiny, only slightly raised pleats or the club-like bristles of *Silo nigricornis* with their deep, grooved and feathered ribs. The end of the *L. basalis* bristle is blunt; the apex (Fig. 4) is depressed below the surrounding crests and covered with fine finger-like projections that are ocasionally arranged like the spokes of a wheel.

When transverse sections are examined in the TEM, the bristles appear as hollow cogs. A thick, ocasionally interrupted, loosely compacted, scalloped band encircles the lumen and the electron dense particles seem to move towards the holes (Fig. 6).

The voluminous cells (Fig. 5) with long, well defined apical microvilli, characteristic of all the scent-organs described, are seen at the base of the bristles. The large nuclei have dense chromatin clumps scattered throughout the karyoplasm; the perinuclear cisterna often broadens. The nuclei are lobed like those of *Drusus improvisus,* while in the other species examined, although the nuclei are large with respect to those of the surrounding epidermal cells, they are rounded or elliptical in *Silo nigricornis* and *Hydroptila aegyptia* and kidney shaped in *Sericostoma pedemontanum.* The cytoplasm of the *L. basalis* androconial cell is rich in organelles and there are abundant polymorphous mitochondria mainly amassed in the apical region, as they are in *Sericostoma pedemontanum.* The rough endoplasmic reticulum is well developed both in the cisternal and in the vescicular form.

197

Extensive rough endoplasmic reticulum is present in the other species studied, except *Hydroptila aegyptia* where the endoplasmic reticulum is mostly smooth. Microtubules are common in this type of cells. In *Drusus improvisus* they are often arranged in bundles, in *Silo nigricornis* they are sometimes directed towards the apex and in *Sericostoma pedemontanum* they are directed towards the hemidesmosomes forming characteristic attachment devices in the apical region of the cell. The microtubules are randomly distributed in the androconial cell cytoplasm of *L. basalis* (Fig. 7) and free ribosomes, sometimes joined to form polysomes, are observed. The Golgi apparatus is frequently evident and numerous various sized vacuoles with an electron transparent matrix are present. In some preparations they are very abundant. The vacuoles found at the base of the bristles, where there are granules of electron dense material (Fig. 5), often join together to form large vesicles. The vesicles scattered in the cytoplasm and the small, more or less, electron dense deposits at the base of the bristles and in their lumen were also encountered in the scent organs of the other species examined.

As these data suggest there is active secretion in the *Lasiocephala basalis* androconia; it is hypothesized that the secretion discharges into the bristle lumen and is then emitted and dispersed through the holes of the chetes.

ACKNOWLEDGEMENTS

TEM research was carried out with the technical assistance of the Electron Microscopy Centre of the Perugia University.

REFERENCES

Cummings, B.F., Proc. Zool. Soc. London, p 459, 1914.
Eltringham, H., Trans. R. ent. Soc. London, p 420–430, 1919.
Jacobson, M., Insect Sex Pheromones, p 57, Academic Press, New York, 1972.
Kelner-Pillault, S., Bull. Soc. Entomol. Fr. 80: 252–257, 1975.
Kimmins, D.E., Country Side, London, 17, 368–369, 1956.
Moretti, G.P. and Bicchierai, M.C., Riv. Idrobiol. 18: 173–195, 1979a.
————., and ————., Riv. Idrobiol. 18: 379–390, 1979b.
————., and ————., Riv. Idrobiol. 18: 197–213, 1979c.
————., and ————., Riv. Idrobiol. 18: 229–235, 1979d.
————., and ————., Riv. Idrobiol. 18: 329–336, 1979e.
————., and Cianficconi, F., Atti Acc. Naz. It. Ent. Rend. 11: 199–202, 1963.
————., and ————., Boll. Zool. 31: 1179–1189, 1964.
————., and ————., Boll. Zool. 32: 975–986, 1965.
————., ————., and Bicchierai, M.C., Riv. Idrobiol. 18: 337–351, 1979.
Mosely, M.E., Trans. R. ent. Soc. London, p 291–294, 1923a.
————., Proc. R. ent. Soc. London, p 103–106, 1923b.
Noirot, C. and Quennaday, A., Annu. Rev. Entomol. 19: 61–80, 1974.
Osmani, Z., Sci. and Cult. 36: 642–646, 1970.
Percy, J.E. and Weatherston, J., *In* Pheromones, ed. M.C. Birch, pp. 11–34, North-Holland Publ. Co., Amsterdam, 1974.

FIRST LIST OF ITALIAN TRICHOPTERA

G.P. MORETTI AND F. CIANFICCONI

SUMMARY

A first list of Trichoptera in the geographical territory of Italy including Corsica, Canton Ticino, Istria up to July, 1980, together with a map showing the species collected in each region, is presented. Three hundred and thirty-one species and 17 subspecies belonging to 90 genera and 19 families have been recorded; 27% of these are endemic to Italy, 15 species and 7 subspecies are new. Piemonte, Toscana and Lombardia have the greatest number of taxa. The classification problems are investigated for some of the taxa.

INTRODUCTION

This is the first list of the Trichoptera of the Italian region. It is mainly compiled on the basis of a research started by Moretti in 1933 and continued with the collaboration of his colleagues and students. The classification of some specimens was confirmed by L. Navas, G. Ulmer, W. Döhler, M.E. Mosely, D.E. Kimmins, F. Schmid, A. Nielsen as has been reported in preceding partial faunistic and ecological lists. Part of the Moretti collection has recently been reviewed with the collaboration of Malicky, who has added valuable information from his personal experience. Botosaneanu kindly agreed to check the list of Italian Trichoptera and also supplied us with notes and specimens of an Italian species sent to him by Consiglio. We believe that our inquiry into the Trichoptera of Italy, preserved in private collections or in the collections of Italian-foreign Museums, the review of the bibliographies and the exchange of opinions with Malicky and Botosaneanu have lead to a satisfactory, up to date, picture of the Italian Trichoptera fauna. We should, however, be grateful for information regarding any gaps or omission.

The species are presented by region in seven tables (I–VII).

LIST OF ITALIAN TRICHOPTERA

Table I.

REGIONS and ISLES / TAXA	CT	PIEMONTE	VALLE D'AOSTA	LIGURIA	LOMBARDIA	TRENTINO A.ADIGE	VENETO	FRIULI V.GIULIA	EMILIA ROMAGNA	TOSCANA	UMBRIA	MARCHE	LAZIO	ABRUZZI	MOLISE	CAMPANIA	PUGLIA	BASILICATA	CALABRIA	CORSICA	CAPRAIA	ELBA	SARDEGNA	SICILIA
PHYACOPHILIDAE																								
1 *Rhyacophila albardana* McL.		+	+		+					+									+					
2 " *appennina* McL.									+										+					
3 " *aquitanica* McL.				+	+	+																		
4 " *aurata* Brauer					+	+	+																	
5 " *casasi* Navas		+			+																			
6 " *dorsalis* Curtis	CT	+			+	+		+																
" *ssp.acutidens* McL.		+							+	+	+	+	+	+										
7 " *fasciata* Hagen							+																	
8 " *foliacea* Moretti									+	+	+	+	+	+										
9 " *giareosa* McL.				+			+																	
10 " *hartigi* Mal.																			+					
11 " *hirticornis* McL.					+		+																	
12 " *intermedia* McL.		+			+	+			+	+														
13 " *italica* Moretti									+	+	+	+												
" *ssp.ilvana* Moretti																						+		
14 " *kelnerae* Schmid		+								+														
15 " *laevis* Pictet										+														
16 " *meyeri* McL.	CT	+								+	+		+											
17 " *nubila* Zett.														+										
18 " *occidentalis* McL.										+	+					+	+							
19 " *pallida* Mosely																			+				+	
20 " *pascoei* McL.		+			+							+												
21 " *producta* McL.									+															
22 " *pubescens* Pictet				+					+	+	+	+	+						+	+				
23 " *rectispina* McL.	CT	+			+																			
24 " *rougemonti* McL.												+	+	+	+	+	+	+	+				+	+
25 " *simulatrix* McL.		+		+	+	+	+	+	+			+		+										
26 " *stigmatica* Kol.		+			+	+																		
27 " *tarda* Giudicelli																			+					+
28 " *torrentium* Pictet		+	+		+	+	+	+																
29 " *trifasciata* Mosely																			+				+	
30 " *tristis* Pictet	CT	+	+	+	+	+		+		+	+	+	+	+					+	+	+			+
31 " *vulgaris* Pictet	CT	+	+		+	+	+	+	+															
GLOSSOSOMATIDAE																								
32 *Glossosoma bifidum* McL.						+																		
33 " *boltoni* Curtis	CT	+	+		+	+																		
34 " *conforme* Neboiss							+	+		+									+					
35 " *spoliatum* McL.										+	+	+												
36 *Catagapetus nigrans* McL.		+						+	+	+	+	+	+			+	+			+				
37 *Synagapetus dubitans* McL.							+																	
38 *Agapetus cyrnensis* Mosely																				+			+	+
39 " *fuscipes* Curtis					+	+			+											+				
40 " *laniger* Pictet									+	+	+													
41 " *nimbulus* McL.	CT				+		+			+	+	+							+				+	
42 " *quadratus* Mosely		•														+								+
HYDROPTILIDAE																								
43 *Ptilocolepus granulatus* Pictet	CT	+	+	+	+	+			+	+	+	+		+					+					
44 *Stactobia beatensis* Mosely																							+	+
45 " *caspersi* Ulmer										+	+													+
46 " *eatoniella* McL.						+			+															
47 " *furcata* Mosely					+				+	+												+	+	
48 " *fuscicornis* Schneider	CT	+			+				+							+							+	+
49 " *maculata* Vaillanti									+														+	+
50 " *moselyi* Kimmins									+			+	+	+								+		
51 *Orthotrichia angustella* McL.					+																		+	+
52 " *costalis* Curtis	CT				+							+	+	+									+	
53 " *tragetti* Mosely					+								+					+						+

Table II.

TAXA (REGIONS and ISLES)	PIEMONTE	VALLE D'AOSTA	LIGURIA	LOMBARDIA	TRENTINO A.ADIGE	VENETO	FRIULI V.GIULIA	EMILIA ROMAGNA	TOSCANA	UMBRIA	MARCHE	LAZIO	ABRUZZI	MOLISE	CAMPANIA	PUGLIA	BASILICATA	CALABRIA	CORSICA	CAPRAIA	ELBA	SARDEGNA	SICILIA	
54 Ithitrichia sp.(lamellaris Eaton?)																						+		▶
55 Oxyethira falcata Morton									+		+	+		+	+								+	▲
56 " flavicornis Pictet	+			+					+														+	
57 " frici Klap.															+									
58 " hartigi Moretti																					+			N
59 " pirisinui Moretti																			+			+		N
60 " simplex Ris												+												O
61 " unidentata McL.																						+		●
62 Hydroptila acuta Mosely																		+						
63 " aegyptia Ulmer									+	+	+													
64 " angulata Mosely		+						+	+	+	+	+		+				+				+	+	●
65 " cognata Mosely						+																		
66 " cornuta Mosely		+							+		+													
67 " cortensis Mosely																		+				+		E
68 " forcipata Eaton	+		+						+	+														O
69 " insubrica Ris	CT			+					+		+		+											
70 " martini Marshall								+	+	+	+	+	+		L			+	+					O
71 " occulta Eaton	+																							
72 " ruffoi Moretti												+												N
73 " serrata Morton																		+			+			P
74 " simulans Mosely		+							+	+	+	+												
75 " sparsa Curtis		+		+					+	+		+												E
76 " stellifera Morton									+															
77 " tigurina Ris		+																+						
78 " tineoides Dalman	CT	+							+						+	+		+						E
79 " uncinata Morton								+	+		+		+									+		
80 " vectis Curtis	+					+		+	+	+	+	+		+				+			+	+	+	O
81 Agraylea multipunctata Curtis ?																						+		O
82 " sexmaculata Curtis	+				+		+		+		+													
83 Allotrichia pallicornis Eaton	+								+	+	+				+									
84 Tricholeiochiton fagesi Guinard									+	+														
PHILOPOTAMIDAE																								
85 Philopotamus corsicanus Mosely																			+					E
86 " ludificatus McL.	CT +	+	+	+	+	+	+	+	+	+	+		+											
87 " montanus Don.	CT															+		+					+	
88 " variegatus Scop.	CT +	+		+	+	+	+	+	+			+												
" " ssp.flavidus Hagen																		+						E
89 Wormaldia copiosa McL.	CT +			+	+	+	+																	
" " ssp.boiscnoarai Moretti								+	+	+														N
90 " mediana McL.		+						+	+	+	+	+	+	+	+									
" " ssp. nielseni Moretti																		+	+				+	N
91 " occipitalis Pictet		+	+	+	+	+	+	+	+	+	+	+	+					+	+		+			
" " ssp.morettii Viganò									+	+		+						+						E
92 " pulla McL.		+		+					+			+						+						
" " ssp.marlieri Moretti		+							+															N
93 " triangulifera McL.		+							+															●
94 " variegata Mosely																		+				+		E
" " ssp. maclachlani Kimmins	+																							E
" " ssp. denisi Moretti																					+			N
95 Chimarra marginata L.		+		+							+								+			+		
HYDROPSYCHIDAE																								
96 Diplectrona atra McL.	CT				+	+																		
97 " felix McL.	+								+						+	+								
98 " magna Mosely									+						+	+	+							E
99 " meridionalis Hagen																	+							F
100 Hydropsyche angustipennis Curtis	CT +	+		+	+	+		+	+		+		+		+	+	+						
101 " dissimulata Kum.Bots.	+			+		+		+	+	+	+	+	+	+	+	+		+				+		
102 " doehleri Tobias																		+				+		E

201

Table III.

REGIONS and ISLES / TAXA	CT	PIEMONTE	VALLE D'AOSTA	LIGURIA	LOMBARDIA	TRENTINO A.ADIGE	VENETO	FRIULI V.GIULIA	EMILIA ROMAGNA	TOSCANA	UMBRIA	MARCHE	LAZIO	ABRUZZI	MOLISE	CAMPANIA	PUGLIA	BASILICATA	CALABRIA	CORSICA	CAPRAIA	ELBA	SARDEGNA	SICILIA	
103 *Hydropsyche fumata* Tobias																			+						E
104 " *instabilis* Curtis	CT	+			+	+		+	+	+									+			+	+		
105 " *klefbecki* Tjeder											+			+	+	+	+					+	+	+	E
106 " *pellucidula* Curtis	CT	+	+		+	+		+	+	+	+	+	+	+	+		+	+	+			+	+	+	E
107 " *sattleri* Tobias		+				+			+	+												+	+		E
108 " *tenuis* Navas		+				+			+	+			+						+		+				O
109 *Hydropsyche* sp.									+	+				+					+		+				
110 *Cheumatopsyche lepida* Pictet		+			+			+	+	+	+	+	+	+	+	+		+				+	+		
POLYCENTROPODIDAE																									
111 *Neureclipsis bimaculata* L.				+	+																				
112 *Plectrocnemia appennina* McL.										+								+							E
113 " *conspersa* Curtis	CT	+			+	+	+	+		+	+		+				+	+				+	+		
114 " *geniculata* McL.					+				+	+	+	+	+	+			+	+		+			+		E
" " ssp. *calabrica* Mal.																	+	+					+		
" " ssp. *corsicana* Mosely																			+			+			
115 " *praestans* McL.		+																							E
116 *Polycentropus corsicus* Mosely																			+						E
117 " *divergens* Mosely																			+						E
118 " *flavomaculatus* Pictet	CT	+		+	+	+			+	+	+	+	+	+	+	+		+				+			E
119 " *ichnusa* Mal.																							+		E
120 " *irroratus* Curtis	CT			+				+	+	+															O
121 " *kingi* McL.				+							+											?			
122 " *malickyi* Moretti					+				+		+				+								+		N
123 " *morettii* Mal.		+			+				+														+		E
124 " *mortoni* Mosely																	+						+		E
125 " *radaukles* Mal.																							+		E
126 " *sardous* Moretti																							+		E
127 *Polycentropus* sp.n.										+															
128 *Holocentropus dubius* Ramb.							+																		O
129 " *picicornis* Steph.									+	+		+													O
130 " *stagnalis* Albarda												+													O
131 *Cyrnus crenaticornis* Kol.				+																					N
132 " *insolutus* McL.									+	+		+													
133 " *trimaculatus* Curtis	CT			+	+		+		+	+	+	+													O
PSYCHOMYIDAE																									
134 *Psychomyia pusilla* Fbr.		+			+	+	+		+	+	+	+	+	+	+			+			+				
135 *Lype phaeopa* Stephens					+				+	+	+	+	+	+			+	+				+			▶
136 " *reducta* Hagen	CT			+	+				+	+	+	+	+			+	+	+			+	+	+		O
137 *Metalype fragilis* Pictet				+																					
138 *Tinodes agaricinus* Mosely																			+			+			E
139 " *antonioi* Bots.Vig.Tat.				+				+	+	+	+	+	+	+	+		+				+				E
140 " *apuanorum* Moretti									+	+		+													N
141 " *bruttius* Moretti																	+								N
142 " (*yr.canariensis* McL.?)																			+			+			●
143 " *cortensis* Mos.																			+			+			E
144 " *dives* Pict.				+	+		+		+			+		+		+									
145 " " ssp. *consiglioi* Bots.								+	+	+	+	+	+	+	+		+								
146 " *locuples* McL.																	+						+		E
147 " *luscinia* Ris	CT			+	+																		+		E
148 " *maclachlani* Kimmins	CT	+		+	+		+		+	+	+	+	+	+		+	+		+	+	+	+	+	+	
149 " *maculicornis* Pictet										+			+												O
150 " *sylvia* Ris	CT									+			+												E
151 " *unicolor* Pictet				+		+			+	+															
152 " *waeneri* L.	CT			+	+		+		+	+			+				+				+	+	+		
153 " *zelleri* McL.				+																					
ECNOMIDAE																									
154 *Ecnomus tenellus* Ramb.	CT	+			+	+	+			+	+		+					+					+	+	

202

Table IV.

REGIONS and ISLES / TAXA	PIEMONTE	VALLE D'AOSTA	LIGURIA	LOMBARDIA	TRENTINO A.ADIGE	VENETO	FRIULI V.GIULIA	EMILIA ROMAGNA	TOSCANA	UMBRIA	MARCHE	LAZIO	ABRUZZI	MOLISE	CAMPANIA	PUGLIA	BASILICATA	CALABRIA	CORSICA	CAPRAIA	ELBA	SARDEGNA	SICILIA	
PHRYGANEIDAE																								
155 *Agrypnia obsoleta* Hagen						+																		
156 " *pagetana* Curtis					+																			
157 " *varia* Fabr.	+				+	+						+	+	+										
158 *Phryganea bipunctata* Retz.					+	+		+																
159 " *nattereri* Brauer	+				+	+	+		+														O	
160 *Oligotricha striata* L.					+	+	+		+															
161 *Oligostomis reticulata* L.					+	+																		
BRACHYCENTRIDAE																								
162 *Brachycentrus montanus* Klap.	+										+							+					O	
163 " *subnubilus* Curtis									+	+	r							+					O E	
164 *Micrasema cinereum* Mosely																							E	
165 " *minimum* McL.				+					+	+	+	+	+	+		+	+						O	
166 " *morosum* McL.				+					+	+		+	+					+					O	
167 " *setiferum* Pictet										+			+					+				+	E	
168 " *togatum* Hagen																			+		+	+	E	
LIMNEPHILIDAE																								
169 *Apatania fimbriata* Pictet	CT +	+		+																				
§ " *muliebris* ssp.*helvetica* Schmid		+																						
170 *Drusus aprutiensis* Moretti												+	+										N	
171 " *alpinus* Meyer-Dür	CT +				+																			
172 " *biguttatus* Pictet	+				+	+	+																	
173 " *camerinus* Moretti											+	+											N	
174 " *schmidi* McL.	CT																							
175 " *chrysotus* Ramb.	CT				+																			
176 " *discolor* Ramb.	CT +	+		+	+	+			+	+														
177 " *improvisus* McL.									+	+	+	+	+	+									E	
178 " *melanchaetes* McL.	CT +				+																			
179 " *monticola* McL.						+																		
180 " *muelleri* McL.	CT +																							
181 " *nigrescens* Meyer-Dür	CT +			+																				
§182 " *trifidus* McL.												+											●	
183 *Ecclisopteryx guttulata* Pictet		+			+	+					+	+	+	+			+							
184 *Cryptothrix nebulicola* McL.		+			+	+																		
185 *Metanoea flavipennis* Pictet	CT +																							
186 " *rhaetica* Schmid						+																		
187 *Leptodrusus budtzi* Ulmer																			+			+	E	
188 *Monocentra lepidoptera* Ramb.		+	+	+																				
189 *Limnephilus affinis* Curtis		+				+		+	+										+	+	+			
190 " *auricola* Curtis											+	+	+	+					+	+				
191 " *binotatus* Curtis						+																		
192 " *bipunctatus* Curtis	+					+		+	+	+	+	+	+					+	+	+		+	+	
193 " *borealis* Zett.?						+																		
194 " *centralis* Curtis	CT +					+																		
195 " *coenosus* Curtis	+					+																		
196 " *extricatus* McL.	CT +					+			+														O	
197 " *flavicornis* Fabr.					+	+	+	+	+	+	+	+	+	+		+	+	+				+		
198 " *flavospinosus* Stein.	CT +				+	+				+	+								+	+	+			
199 " *griseus* L.	+								+									+						
200 " *helveticus* Schmid					+						+		+	+									O	
201 " *hirsutus* Pictet	+					+			+	+				+					+	+		+	+	
202 " *ignavus* McL.	CT +					+							+						+					
203 " *italicus* McL.									+	+									+			+		
204 " *lunatus* Curtis	+				+	+	+	+	+	+	+	+	+		+				+	+	+			
205 " *marmoratus* Curtis									+	+									+	+				
206 " *rhombicus* L.	CT +	+			+	+	+	+	+	+	+	+	+		+				+					
207 " *sericeus* Say						+																		
208 " *sparsus* Curtis		+				+			+				+	+					+	+			+	

203

Table V.

TAXA	CT	PIEMONTE	VALLE D'AOSTA	LIGURIA	LOMBARDIA	TRENTINO A.ADIGE	VENETO	FRIULI V.GIULIA	EMILIA ROMAGNA	TOSCANA	UMBRIA	MARCHE	LAZIO	ABRUZZI	MOLISE	CAMPANIA	PUGLIA	BASILICATA	CALABRIA	CORSICA	CAPRAIA	ELBA	SARDEGNA	SICILIA	
209 *Limnophilus stigma* Curtis		+				+			+																O
210 " *subcentralis* Brauer						+								+											O
211 " *vittatus* Fabr.		+	+		+		+			+			+	+	+		+		+	+			+	+	
212 *Colpotaulius incisus* Curtis																			+						►
213 *Grammotaulius nigropunctatus* Retz.		+					+		+	+			+	+	+		+								
214 " *nitidus* Müller							+																		
215 *Glyphotaelius pellucidus* Retz.		+			+				+	+							+								
216 *Anabolia lombarda* Ris	CT	+			+																				
217 *Phacopteryx brevipennis* Curtis		+																							
218 *Potamophylax cingulatus* Steph.		+		+	+	+	+	+	+	+	+	+	+	+	+	+	+		+		+		+		
" " ssp.*gambaricus* Mal.																			+					+	E
219 " *latipennis* Curtis						+	+																		
220 " *nigricornis* Pictet		+					+			+	+			+											
221 *Acrophylax zerberus* Brauer					+																				
222 *Halesus digitatus* Schrank				+	+					+	+														
223 " *nurag* Mal.																							+		E
224 " *radiatus* Curtis					+																				
" " ssp.*vaillanti* Moretti										+	+	+	+	+					+					+	N
225 " *rubricollis* Pictet	CT	+	+	+		+																			
226 " *tessellatus* Ramb.					+																				
227 *Platyphylax frauenfeldi* Brauer														?											
228 *Melampophylax melampus* McL.	CT									+	+	+	+	+					+						O
229 *Anisogamus difformis* McL.		+			+																				
230 *Parachiona picicornis* Pictet		+	+			+	+																		
231 *Enoicyla costae* McL.												+	+					+							
232 " *pusilla* Burm.		+																							
233 " *reichenbachii* Kol.	CT	+								+		+													
234 *Stenophylax crossotus* McL.		+		+								+	+						+				+		
235 " *mitis* McL.		+	+	+	+		+			+	+	+	+	+	+	+				+			+		
236 " *mucronatus* McL.		+								+	+	+	+	+	+		+	+		+					
237 " *permistus* McL.	CT	+	+	+	+	+	+		+	+	+	+	+	+	+			+	+				+		
238 " *vibex* Curt.				+	+					+	+	+	+					+	+	+					
239 *Micropterna fissa* McL.	CT	+	+	+	+	+	+			+	+	+	+	+			+	+					+		
240 " *lateralis* Steph.		+		+		+								?				+	+						
241 " *malatesta* Schmid.																							+		
242 " *nycterobia* McL.	CT	+	+	+	+		+			+	+	+	+	+			+	+					+		
243 " *sequax* McL.	CT	+	+	+	+		+			+	+	+	+	+				+	+				+	+	
244 " *testacea* Gmelin		+	+	+	+		+			+	+	+	+	+	+	+	+		+				+		
245 " *wageneri* Mal.										+															
246 *Mesophylax aspersus* Ramb.		+		+	+	+	+	+	+	+	+	+	+	+	+	+	+	+	+	+		+	+	+	
" " ssp.*sardous* Moret.Gian.																							+		
247 " *impunctatus* McL.	CT			+	+	+	+																		
248 *Allogamus antennatus* McL.	CT				+					+	+	+	+	+			+		+						
249 " *auricollis* Pictet	CT	+	+		+	+		+						+	+			+	+						O
250 " *corsicus* Ris																		+		+					E
251 " *hilaris* McL.	CT	+			+					+	+	+	+				+	+		+					►
252 " *illiesorum* Bots.																							+		E
253 " *mendax* McL.	CT	+	+		+							+		+											O
254 " *uncatus* Brauer		+			+	+				+															
δ 255 *Consorophylax consors* McL.					+	+																			
256 " *piemontanus* Bots.		+																							E
257 *Chaetopteryx clara* McL.or sp.n.?					+		+																		
258 " *gessneri* McL.	CT	+		+	+					+	+			+											
259 " *vulture* Mal.	CT																	+	+						E
260 *Pseudopsilopteryx zimmeri* McL.	CT			+	+	+												+	+						O
GOERIDAE																									
261 *Goera pilosa* Fabr.					+									+											
262 *Lithax niger* Hagen	CT	+				+	+																		
263 *Silo mediterraneus* McL.										+	+			+	+	+		+	+				+		E

204

Table VI.

TAXA	PIEMONTE	VALLE D'AOSTA	LIGURIA	LOMBARDIA	TRENTINO A.ADIGE	VENETO	FRIULI V.GIULIA	EMILIA ROMAGNA	TOSCANA	UMBRIA	MARCHE	LAZIO	ABRUZZI	MOLISE	CAMPANIA	PUGLIA	BASILICATA	CALABRIA	CORSICA	CAPRAIA	ELBA	SARDEGNA	SICILIA	
264 *Silo nigricornis* Pictet	+			+	+	+	+	+	+	+	+	+		+	+			+				+		
265 " *pallipes* Fabr.	CT +		+	+	+		+											+						
266 " *piceus* Brauer	+					+												+						
267 " *rufescens* Ramb.											+							+				+		E
268 *Silonella aurata* Hagen																		+				+		E
THREMMATIDAE																								
269 *Thremma sardoum* Costa																						+		E
LEPIDOSTOMATIDAE																								
270 *Lepidostoma hirtum* Fbr.	CT			+				+	+									+						
271 *Lasiocephala basalis* Kol.				+				+	+	+	/	+		+				+						
272 *Crunoecia irrorata* Curt.	CT +			+		+		+	+	+	+	+	+			+	+			+	+	+	+	
LEPTOCERIDAE																								
273 *Athripsodes albifrons* L.	+							+																
274 " *aterrimus* Steph.		+		+		+		+	+			+	+					+						
275 " *bilineatus* L.								+	+	+	+	+	+											
276 " *cinereus* Curtis	CT +			+					+										'			+		
277 *Athripsodes* sp.n. ?																		+				+		E
278 " *genei* Ramb.																		+				+		
279 " *leucophaeus* Ramb.				+																				
280 *Ceraclea alboguttata* Hagen				+																				
281 " *aurea* Pictet					+																			
282 " *dissimilis* Steph.	+			+	+				+		+											+		
283 " *fulva* Ramb.				+					+		+	+												▶
284 " *senilis* Burm.					+			+			+											+		O
285 *Mystacides azurea* L.	+	+		+	+		+	+	+	+	+		+		+		+				+	+		
286 " *longicornis* L.	+			+						+														
287 " *nigra* L.	+			+																				
288 *Triaenodes bicolor* Curtis	+			+								+									+			
289 " *conspersus* Ramb.												+												●
" *ochreellus*.ssp. *lefkas* Mal.									+															
290 *Erotesis baltica* McL.	+			+					+		+	+						+						
291 *Oecetis furva* Ramb.	+			+					+		+	+												
292 " *lacustris* Pictet	+						+				+	+												▶
293 " *notata* Ramb.	+			+	+	+					+					+								
294 " *ochracea* Curtis	+								+															
295 " *testacea* Curtis				+					+															O
296 " *tripunctata* F.	+			+														+	+		+	+		
297 *Setodes argentipunctellus* McL.	+			+					+						+									▶
298 " *punctatus* Fabr.									+			+												
299 " *viridis* Fourcroy				+	+	+			+															O
300 *Leptocerus interruptus* Fabr.									+									' +						
301 " *lusitanicus* McL.				+					+		+	+									+	+		
302 " *tineiformis* Curt.	+			+	+				+															
303 *Adicella filicornis* Pictet	+							+																
304 " *reducta* McL.						+																		
SERICOSTOMATIDAE																								
305 *Notidobia ciliaris* L.				+		+			+															
306 *Sericostoma cianficconii* Moretti							+	+	+	+								+						E
307 " *clypeatum* Hagen							+	+				+			+							+		
308 " *galeatum* Ramb.	+			+				+														+		
309 " *italicum* Moretti							+	+	+	+	+	+	+						+					N
310 " *maclachlanianum* Costa								+														+		E
311 " *pedemontanum* McL.	CT +	+		+	+		+	+	+	+						+					+		
312 " *personatum* Spence	+		+	+	+			+							+									
? " *romanicum* Navas								+																
313 " *siculum* McL.															+	+	+					+		E
314 " *subaequale* McL.				+	+																			E

205

Table VII.

REGIONS and ISLES / TAXA		PIEMONTE	VALLE D'AOSTA	LIGURIA	LOMBARDIA	TRENTINO A.ADIGE	VENETO	FRIULI V.GIULIA	EMILIA ROMAGNA	TOSCANA	UMBRIA	MARCHE	LAZIO	ABRUZZI	MOLISE	CAMPANIA	PUGLIA	BASILICATA	CALABRIA	CORSICA	CAPRAIA	ELBA	SARDEGNA	SICILIA	
315 *Sericostoma timidum* Hag.						+																			
" ssp. *turbatum* McL. ?		+				+	+																		
316 " *vittatum* Ramb.													+									+		●	
BERAEIDAE																									
317 *Beraea aureomarginata* Mos.																			+			+			E
318 " *botosaneanui* Moretti																			+			+			N
319 " *arichtoni* Moretti																+	+								N
320 " *dira* McL.	Is						+	+																	▲
321 " *ilvae* Moretti																						+			N
322 " *maura* Curtis				+	+	+			+	+	+	+	+			+	+			+		+	+		
323 *Beraeodes minutus* L.										+	+														
324 *Ernodes articularis* Pict.		+			+	+			+	+															
325 " *nigroauratus* Mos.										+								+					+		E
§ 326 " *vicinus* McL.	CT	+																					+		
327 *Beraeodina palpalis* Mos.																			+						E
328 *Beraeamyia squamosa* Mos.							+	+	+				+						+						
HELICOPSYCHIDAE																									
329 *Helicopsyche revelieri* McL.																		+	+			+			E
330 " *sperata* McL.	CT				+				+				+	+	+	+	+	+	+			+		+	
ODONTOCERIDAE																									
331 *Odontocerum albicorne* Scop.		+	+			+	+	+		+	+	+	+	+	+	+	+	+	+	+		+	+	+	

Tables I–VII. Species and subspecies (no number) are listed alphabetically as in Limnofauna Europaea according to the regions from West to East and from North to South. The findings in foreign territories which geographically belong to Italy (Canton Ticino = CT, Istria = I, Corsica) have been included. The isles of Elba and Capraia are given separately because of their zoogeographical interest. E = species and subspecies endemic to Italy, N = new species or subspecies, ● = species and subspecies not previously recorded in Italy, o = species and subspecies not previously reported in peninsular and insular Italy, ▲ = species reported by Moretti et al. not listed in Limnofauna Europaea, ꝺ = found by Malicky, § = found by Botosaneanu.

OBSERVATIONS ON CERTAIN TAXA

1 Found in the Marche and Calabria by Malicky. *5* Found in Lombardia by Botosaneanu. *10* R. *vulgaris hartigi* is now raised to the rank of species in agreement with Malicky. *11* According to Malicky R. *philopotamoides* det. Ulmer (Trentino-Moretti 1937a) is R. *hirticornis*. *24* R. *rougemonti sicula* is regarded as a synonym of the species R. *rougemonti*. *54* The species identification is based on larvae and prepupae and is, therefore, uncertain. *Oxytrichia* sp. which would appear between *61* and *62* has not be included as only a single damaged specimen was available. It was found on the Isle of Elba and was characterized by a sudden tapering of the wings, ocelli and 0, 3, 4 spurs. *71* Listed as H. *occulta* in Umbria in previous papers (Moretti 1952; Moretti and Gianotti 1966, 1967). *73* Syn. *bifurcata* Mosely according to Botosaneanu and Malicky (in lit.). *81* Although only the cases were found, they had evident characteristics of *A. multipunctata*. *84* Encountered only in the larval stage. *87* Experts disagree about the validity of *Philopotamus montanus siculus*. Botosaneanu and Schmid (1973), Vaillant (1974), Gonzalez and Terra (1979) consider this a true geographical subspecies, however Malicky does not. Our examination of numerous specimens from Calabria, Basilicata and Sicilia revealed *a)* a single black elongated spine on the penis endotheca in most specimens, with the exception of an occasional sample from Basilicata where there was a supplementary plaque at the side of the elongated spine; *b)* well-developed, pointed internal ventral coxopodite; *c)* ribbon-like inferior branch of the gonopodes in the specimens from Basilicata, clearly club-like in those from Calabria and Sicilia, *d)* oval cerci with convex anal border and *e)* the stalk of apical fork no 4 of the anterior wing short. Therefore, the characteristics in both the Southern Italian and Sicilian specimens were mainly those of *Ph. montanus*. If, on the other hand, the penis endotheca is taken into account, *Ph. montanus siculus* is a good Southern Italian subspecies. *90 W. mediana* also includes specimens previously attributed to *W. subnigra* (Moretti 1944, 1945, Marcuzzi 1956). *94 W. variegata corsicana*, similar to the *denisi* ssp., indicated as dubious in Limnofauna Europaea, was reported from Corsica by Vaillant (1974). *109* Further studies are required on the *Hydropsyche* genus in Italy as elsewhere. Revision could establish whether this is a new species or not. We believe, therefore, that the previous mention of *Hydropsyche guttata* in Trentino (Moretti 1937a) should be reconsidered. *121* Only larvae were available for classification. *126* Listed in Limnofauna Europaea (1978) as *Polycentropus* sp. Moretti 1941. *127* Awaiting description. *136 Lype flavospinosa* found in Corsica is, according to Barnard who compared the *L. flavospinosa* holotype with *L. reducta* (in lit. Malicky), a synonym of *L. reducta*. *139* Mentioned as *Tinodes* sp. A indet. for the Apuanian Alps (Moretti and al. 1970). *140* Mentioned as *Tinodes* sp. B indet. for the Apuanian Alps. *144 T. dives* of Marche, Abruzzi and Campania classified by Malicky. *145* Subspecies founded by Botosaneanu on the basis of Lazio specimens. It was previously indicated as *T. divitisimilis* by Moretti (Moretti and Gianotti 1966, 1967, Moretti and Cianficconi 1975) who found it had a wide Italian distribution. *159* Classified as *Phr. grandis* in previous papers (Moretti 1937a, Moretti et al. 1978). *177* Reported by Botosaneanu for Lazio. *193* The ♀♀ specimens manifest diagnostic characteristics of *L. borealis*, the ♂♂ a certain difference at the level of the superior appendages (Moretti 1937a). For which reason, it is listed with reserve. *216 Anabolia nervosa* of Mendrisio turned out to be *A. lombarda* (revised by Barnard). *218* The number and length of the ♂ parameres are very variable. The markings and colouring of the anterior wings vary according to zone. Not numbered *H. radiatus vaillanti* mentioned as *H. radiatus interpunctatus* by Moretti, Cianficconi (1975). *234* The Stenophylacinae have been found to be widely distributed throughout Italy as a result of careful research into cave-dwelling. *235* Found, also, on the Isle of Montecristo (Fanfani and Groppali 1979). *238* The Calabrian *St. vibex* is very similar to *St. meridio-orientalis* Malicky (1980). *248* The Apennine population is smaller in size than that of the Alps. *257* Is waiting to be compared with *Ch. clara* specimens. *277* Presumably a new species, but, owing to the fact that it is represented by a single specimen, more extensive study is required. <u>Not numbered</u> *T. ochrellus lefkas* mentioned as ssp. of *T. unanimis* in agreement with Botosaneanu, by Moretti and Corallini Sorcetti (1978b). *312* The findings of *S. personatum* given to us by Malicky for Toscana and Calabria are included in this list, although a revision of all European *Sericostoma* is needed. For the time being, *S. pedemontanum* of the Apennines is differentiated from the *S. personatum* of the Alps. *? S. romanicum* is a doubtful species, as it was not possible to examine its holotype. *315* The validity of the distinguishing features of *S. timidum* and *S. turbatum* were discussed in a paper dedicated to the *Sericostoma* genus (Moretti and Cianficconi 1978). After exchanging opinions with Malicky and Barnard this species was convalidated. Moretti's accurate study of male specimens showed that they could be distinguished from the *maura-aureomarginata* group by the spiney sclerified process of the aedeagus.

CONCLUSIONS

The list presented includes 331 species and 17 subspecies belonging to 90 genera and 19 families. 15 are new species and 7 new subspecies; 13 species and 1 subspecies are registered for the first time in Italy. Forty-five of the species already mentioned for the Alps (Illies region 4) are now, also, reported for the peninsular and insular Italy (Illies region 3), information on 12 of these was given in previous papers (Moretti et al. 1967, 1968, 1969, 1972, 1974, 1976). The families most widely represented in Europe are also those with the highest species incidence in Italy: Rhyacophilidae (31 species), Hydroptilidae (42), Limnephilidae (95), Leptoceridae (33).

This list shows a significant difference between the composition of the apennine and alpine Trichoptera fauna. The Sardinian fauna is related to the fauna of Corsica and the Sicilian fauna to that of Calabria. The highest numbers of species encountered to date are those for Piemonte, Lombardia, Trentino-Alto Adige in North Italy; Toscana and Umbria in Central Italy; Calabria in South Italy and Sardegna for the islands (Fig. 1). The fact that there are extensive mountain waters in some regions and that certain regions have been more careful studied than others may account for this picture.

The available information on distribution shows that 81 species and 13 subspecies from those listed, or 27%, are endemic to Italy. Calculated from the total number of species recorded for each region, the highest proportion of Italian endemics are found in the islands: Sardegna (48.6%), Corsica (46.2%), Sicilia (23%). The highest values in the peninsula are, at the moment, in South Italy (Calabria, Basilicata, Molise).

It is obvious that this evaluation of Italian species is provisional, as it will, doubtlessly, be modified as investigations progress, particularly as there is still unclassified material in our possession. Our intention has been to list the species and indicate some of the problems still requiring clarification.

208

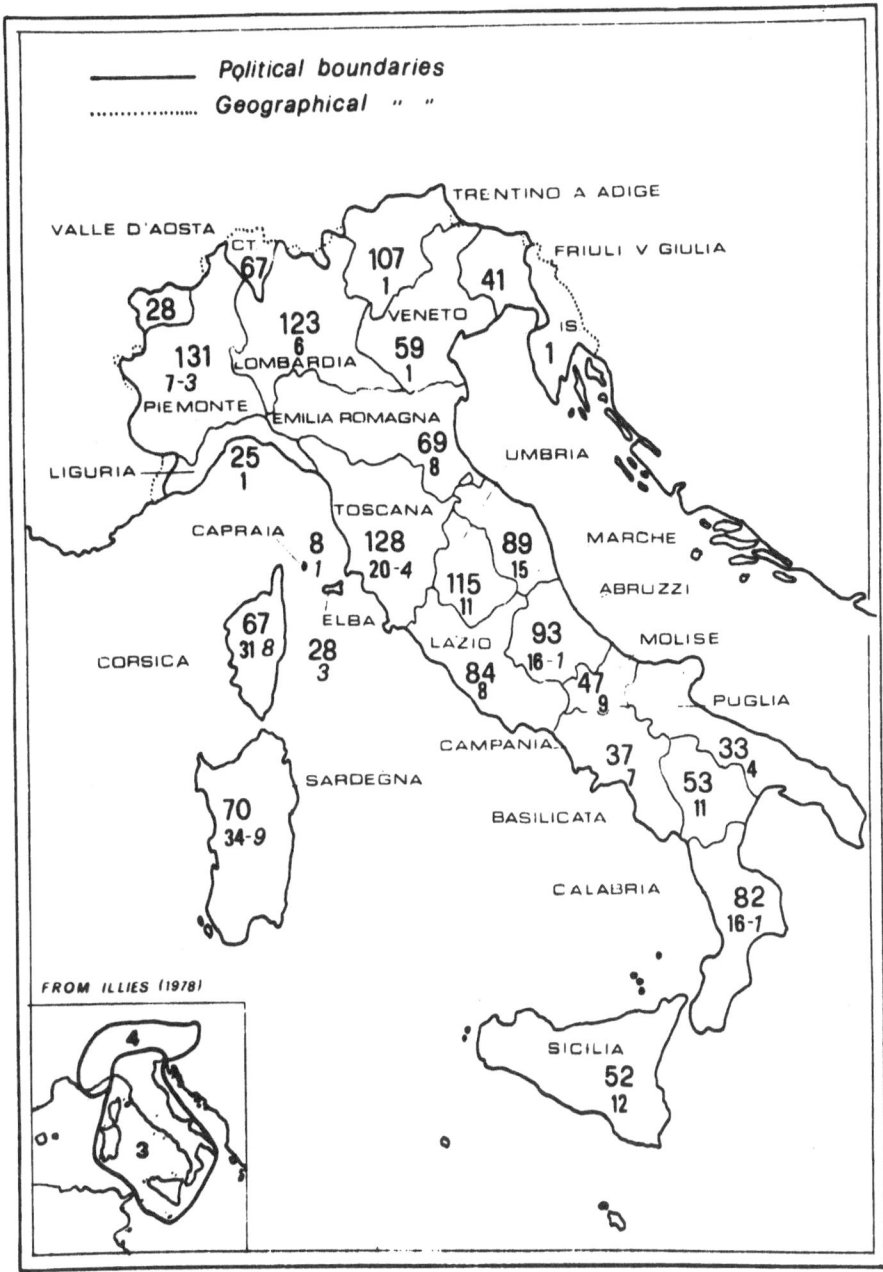

Fig. 1. No of species found in each region to July, 1980.
Top number = total
Bottom number = endemic to Italy. Italics show endemic regional species
Inset = Illies regions from Limnofauna Europaea for comparison.

NOTE

During the printing of this volume *Hydropsyche cyrnotica* sp. n. from Corsica was described by Botosaneanu and Guidicelli (Bull. Zool. Mus. Amsterdam 8: 14–19, 1981).

ACKNOWLEDGEMENTS

We would like to express our appreciation to Dr H. Malicky for having worked with us on the classification of certain species here in Perugia and to Dr L. Botosaneanu for his careful revision of this list.

REFERENCES

Botosaneanu, L., Bull. Zool. Mus. Univ. Amsterdam 7: 73–80, 1980.
———., and Malicky, H. In Limnofauna Europaea, ed. J. Illies, pp. 333–359, Fischer, Stuttgart, 1978.
———., and Schmid, F., Rev. Suisse Zool. 80: 221–256, 1973.
Cianficconi, F.; Moretti, G.P.; Pirisinu, Q. and Tucciarelli, F., Lav. Soc. Ital. Biog. 6: 479–523, 1979.
Fanfani, A. and Groppali, R., Publ. Ist. Entom. Univ. Pavia 9: 1–52, 1979.
Felber, J., Arch. Naturg. 74: 1–90, 1908.
Fischer, F.C.J., Trichopterorum Catalogus. Ned. Ent. Vereen. Amsterdam 1–15, 1960–1973.
Giudicelli, J., Ecologia Mediterranea 1: 133–147, 1975.
Gonzalez, M.A. and Terra, L.S.W., Boll. Assoc. Esp. Ent. 3: 163–172, 1979.
Kimmins, D.E. and Botosaneanu, L., Acta Zool. Acad. Sci. Hung. 13: 353–361, 1967.
MacLachlan, R., A monographic Revision and Synopsis of the Trichoptera of the European Fauna. Reprint 1968, Hampton, Classey, 1874–1880.
Malicky, H., Die Höhle 1: 15–20, 1971a.
———., Ent. Zeit. 81: 257–265, 1971b.
———., Nachr. Bayer. Ent. 26: 65–77, 1977.
———., Mitt. Abt. Zool. 8: 11–42, 1979.
———., Entomofauna, Zeit. Ent. 1: 95–102, 1980.
Marcuzzi, G., Mem. Cl. Sci. Mat. Nat. Venezia: 200–202, 1956.
———., Riv. Idrobiol. 15: 200–319, 1976.
——— and Lorenzoni, A.M., St. Trent. Sc. Nat. 45: 165–212, 1968.
——— and Nigro Faccipieri, L., Riv. Idrobiol. 17: 230–232, 1978.
Moretti, G.P., Atti Soc. Ital. Sc. Nat. 73: 93–145, 1934.
———., Studi Trent. Sc. Nat. 18: 33–73, 1937a.
———., Mem. Ist. Lomb. Sc. Lett. 23: 139–189, 1937b.
———., Mon. Comit. Scient. CAI Varallo 16: 49–72, 1938.
———., Boll., Ist. Ent. Univ. Bologna 11: 88–94, 1939.
———., Boll. Zool. Agr. Bachic. Univ. Milano 10: 1–15, 1940a.
———., Mem. Soc. Ent. Ital. 19: 259–291, 1940b.
———., Mem. Ist. Ital. Idrobiol. 20: 295–306, 1942.
———., Boll. Zool. Agr. Bachic. 12: 1–51, 1944.
———., Atti Soc. It. Sc. Nat. 84: 5–12, 1945.
———., Boll. Soc. Eustachiana 4: 203–208, 1950.
———., Boll. Zool. 19: 245–269, 1952.
———., Mem. Ist. Ital. Idrobiol. 8: 257–270, 1954a.
———., Boll. Soc. Eustachiana 47: 59–123, 1954b.
———., Boll. Zool. 21: 503–529, 1954c.

————., Boll. Zool. Agr. Bachic. 22: 189–214, 1956.

————., and Cianficconi, F., *In* Proc. 1st Int. Symp. Trichoptera, ed. H. Malicky, pp. 93–104. Junk The Hague, 1974.

Moretti, G.P. and Cianficconi, F., *In* Atti V Simp. Naz. Conserv. Nat. 2, ed. L. Scalera Liaci, pp. 69–83, Cacucci, Bari, 1975.

————., and Cianficconi, F., *In* U. Ferrarese and B. Sambugar, Riv. Idrobiol. 15: 98–105. 1976.

————., and Cianficconi, F., *In* Proc. 2nd Int. Symp. Trichoptera, ed. M.I. Crichton, pp. 7–30, Junk The Hague, 1978.

————., and Cianficconi, F., Boll. Zool. Suppl. 45: 35, 1978.

————., and Cianficconi, F., Boll. Zool. Suppl. 46: 150–151, 1979.

————., and Cianficconi, F., Lav. Soc. It. Biog., in press.

————., and Corallini Sorcetti, C., Boll. Zool. Suppl. 45: 36, 1978.

————., and Corallini Sorcetti C., Riv. Idrobiol. 19: 1–7, 1980.

————., and Gianotti, F.S., Riv. Idrobiol., 5: 51–67, 1966.

————., and Gianotti, F.S., Riv. Idrobiol. 6: 103–114, 1967a.

————., and Gianotti, F.S., Mem. Soc. Ent. It. 46: 73–125, 1967b.

————., and Mearelli, M., Riv. Idrobiol. 17: 137–186, 1978.

————., and Pirisinu, Q., Boll. Zool. 36: 393–394, 1969.

————., and Spinelli Batta G., Boll. Zool. Suppl. 46: 156–157, 1979.

————., and Taticchi, M.I., Riv. Idrobiol. 8: 89–104, 1969.

————., and Viganò, A., Boll. Zool. 26: 573–588, 1959.

————., and Viganò, A., Atti Acc. Naz. It. Ent. 8: 254–261, 1960.

————., and Viganò-Taticchi, M.I., *In* Proc. 1st Int. Symp. on Trichoptera ed. H. Malicky, pp. 87–92, Junk The Hague, 1974.

————.; Cianficconi, F., Pirisinu, Q., and Ponziani, G., Boll. Zool. 34: 145, 1967.

————.; Di Giovanni, M.V. and Viganò, A., Note App. Sper. Ent. 12: 35–60, 1967.

————.; Cianficconi, F.; Gianotti, F.S.; Pirisinu, Q. and Viganò, A., Lav. Soc. It. Biog. 1: 488–532, 1970.

————.; Pirisinu, Q.; Ravizza, C. and Fiorelli, M.A., Riv. Idrobiol. 11: 79–101, 1972.

————.; Cianficconi, F. and Tucciarelli, F., Lav. Soc. It. Biog. 6: 525–568, 1976.

————.; Cianficconi, F. and Tucciarelli, F., Boll. Zool. 45: 1p, 1978a.

————.; Corallini Sorcetti, C. and Vignaroli, F., Riv. Idrobiol. 17: 85–134, 1978b.

————.; Gattaponi, P. and Corallini Sorcetti, C., Boll. Zool. 45: 1p, 1978c.

————.; Tucciarelli, F. and Cruccolini, E., Riv. Idrobiol. 17: 27–82, 1978.

————.; Cianficconi, F.; Corallini Sorcetti, C.; Gattaponi, P. and Tucciarelli, F., Boll. Soc. It. Biol. Sper., 55: 1279–1294, 1979.

Mosely, M.E., Eos 8: 165–184, 1932.

Schmid, F., Mem. Soc. Ent. Canada 66: 230pp, 1970.

Touring Club Italiano, L'Italia fisica pp. 12–14, Milano, 1967.

Vaillant, F., Ann. Soc. Ent. Fr. 10: 969–985, 1974.

GREGARINES IN TRICHOPTERA LARVAE

G.P. MORETTI AND C. CORALLINI SORCETTI

SUMMARY

In Italy, 39 of the 71 Trichoptera species examined housed Gregarines. The 29 taxa recorded belong to the *Gregarina-Leidyana-Pileocephalus-Asterophora-Globulocephalus-Ancyrophora* genera.

There does not appear to be a strict specificity between the protozoa and the host. It would not seem that the Trichoptera larvae are harmed by the presence of the Gregarines and even when there is considerable infestation they regularly complete their cycle.

The cycle of the Gregarines terminates as the Trichoptera reaches the aquatic stage; they are no longer present in the pupa.

The type of food regimen and the biology of the host exert considerable influence on these protozoa.

INTRODUCTION

This paper is an up-to-date report on Gregarines in Italian Trichoptera. These protozoa are usually found in insect intestines and for some time their presence in Trichoptera has been reported from other countries (Baudoin-Frantzius-Geus-Hoside-Kölliker-Léger-Schneider-Stein-Zwetkow).

Seventy-one mainly Central and Southern Italian species belonging to 17 families were investigated.

Although the study was mostly carried out on larvae, the pupae, imagos and eggs were also examined despite the fact that Gregarines are not normally found in these stages.

Larvae were collected from a very wide variety of environments, such as springs, head waters, streams, fast-running brooks, rivers, lakes and trout-breeding beds.

In order to study the Gregarines, the digestive tube of the Trichoptera was removed, fixed in 4% Bouin formaline and some sections stained with aniline red fuchsin or Mayer's acid haemalum and examined at the light microscope.

Table 1. Trichoptera examined

° =Trichoptera with Gregarines

RHYACOPHILIDAE

1)°Rhyacophila dorsalis ssp.acutidens Mc L.

2)°   "   foliacea Moret.

3)°   "   hartigi Mal.

4)°   "   italica Moret.

5)°   "   rougemonti Mc L

GLOSSOSOMATIDAE

6) Glossosoma sp.

7)°Catagapetus nigrans Mc L.

HYDROPTILIDAE

8) Orthotrichia costalis Curt.

9) Hydroptila aegyptia Ulm.

PHILOPOTAMIDAE

10) Philopotamus ludificatus Mc L.

11)   "   montanus Don.

12) Wormaldia occipitalis Pict.

13) Wormaldia sp.

HYDROPSYCHIDAE

14)°Diplectrona magna Mos.

15) Hydropsyche angustipennis Curtis

16)°   "   dissimulata Kum. Bots.

17)°   "   instabilis Curt.

18)°   "   pellucidula Curtis

19)°   "   sattleri Tobias

20)°Cheumatopsyche lepida Pict.

POLYCENTROPODIDAE

21) Plectrocnemia geniculata Mc L.

22) Plectrocnemia sp.

23) Polycentropus flavomaculatus Pict.

PSYCHOMYIDAE

24) Tinodes antonioi Bots. Vig.Tat.

25)°   "   maclachlani Kimm.

26)°   "   waeneri L.

ECNOMIDAE

27)°Ecnomus tenellus Ramb.

PHRYGANEIDAE

28)°Agrypnia varia Fbr.

BRACHYCENTRIDAE

29) Brachycentrus montanus Klap.

30) Micrasema minimum Mc L.

31)   "   togatum Hagen

32) Micrasema sp.

LIMNEPHILIDAE

33)°Drusus aprutiensis Moret.

34)°   "   camerinus Moret.

35)°   "   improvisus Mc L.

36)°Leptodrusus budtzi Ulm.

37) Limnephilus flavicornis Fbr.

38)   "   flavospinosus Stein.

39)°   "   rhombicus L.

40)°Limnephilus sp.

41)°Grammotaulius nigropunctatus Retz.

42) Glyphotaelius pellucidus Retz.

43) Potamophylax cingulatus Steph.

44)°   "   nigricornis Pict.

45)°Halesus radiatus ssp.vaillanti Moret.

46)   "   rubricollis Pict.

47) Melampophylax melampus Mc L.

48)°Stenophylax sp.

49)°Micropterna wageneri Mal.

50)°Mesophylax aspersus Ramb.

51)°Mesophylax sp.

52)°Allogamus antennatus Mc L.

53)°   "   hilaris Mc L.

54)   "   uncatus Brauer ·

55) Allogamus sp.

56)°Chaetopteryx sp·

GOERIDAE

57) Silo nigricornis Pict.

LEPIDOSTOMATIDAE

58) Lepidostoma hirtum Fbr.

59) Lasiocephala basalis Koll.

60) Crunoecia irrorata Curt.

LEPTOCERIDAE

61)°Mystacides azurea L.

62)   "   longicornis L.

63) Triaenodes okrella ssp.lefkas Mal.

64) Oecetis furva Ramb.

65) Leptocerus tineiformis Curt.

SERICOSTOMATIDAE

66)°Sericostoma italicum Moret.

67)°   "   pedemontanum Mc L.

68)°   "   siculum Mc L.

BERAEIDAE

68) Beraea maura Curtis

HELICOPSYCHIDAE

70) Helicopsyche sperata Mc L.

ODONTOCERIDAE

71)°Odontocerum albicorne Scop.

RESULTS

So far sporozoa have been observed in 38 Trichoptera taxa (Table 1).

Rhyacophilidae-Hydropsychidae-Ecnomidae-Limnephilidae were the families with the highest number of infected species and the highest degree of single infestation, while in Hydroptilidae-Philopotamidae-Polycentropodidae-Brachicentridae-Goeridae-Lepidostomatidae-Beraeidae-Helicopsychidae there was no trace of infection.

Gregarines were found only in larvae, the eggs, images and pupae were all free of infestation. Sometimes a large number of protozoa was found in single insects (*Tinodes maclachlani-Ecnomus tenellus-Drusus improvisus-Potamophylax cingulatus-Allogamus antennatus*), at others, a case of incidental parasitism was observed (*Catagapetus nigrans*). A high degree of infestation was found in *Rhyacophila foliacea* (75%)-*Tinodes maclachlani* (80%)-*T. waeneri* (58%)-*Ecnomus tenellus* (63.33%)-*Drusus improvisus* (56%).

Twenty-nine sporozoa belonging to three Eugregarine families were revealed (Table II). As the stages required for definition were not always available, systematic classification was not always possible.

It is reasonably certain that two new forms were recorded, one in *D. improvisus* from the River Tiber, the other in *A. hilaris* from the Trocchi Springs (Mount Catria). The first, which was small (length $35\,\mu m$, width $10\,\mu m$) with an elongated narrow deutomerite and an unmarked septum can be attributed to the *Gregarina* genus. The second, which was large with a cephaline that measured $380\,\mu m$ and a sporadine which could reach a length of $1,130\,\mu m$ we named Eugregarinide indet. When the gamont is mature the protomerite disappears and the specimen assumes a trapezoid shape. Sometimes three specimens were seen together. Unfortunately, studies were halted due to alterations in the natural environment.

The most frequently observed Gregarine was *Gregarina* genus; the least frequent, *Ancyrophora*.

These protozoa are not usually host-specific; infact, the same Gregarine (*Gregarina lunata-G. mystacidarum-G. sericostomae-G. stenophylacis-Pileocephalus lanceatus-Asterophora heeri*) may be present in different host species, even those belonging to different families, and a single host species (*D. improvisus-E. tenellus-A. antennatus*) may house numerous sporozoan species.

When several species of Gregarines are found in a single host, usually one or two are present in greater numbers, while the others are incidental. We identified *Globulocephalus hydropsyches* and *Asterophora hydropsyches*, the last only in the larvae of the *Hydropsyche* genus, these findings agree with reports from other investigators. *Asterophora* was seen in *H. sattleri*, although it seemed to differ from other species recorded in the literature. These terms will be described elsewhere.

The only species seen to date in *Rhyacophila* is *A. moucronata.* Limnephilidae larvae house three genera, *Gregarina* and *Pileocephalus*, which generally co-exist and *Leidyana*, which is usually present in small numbers. The last can be seen moving rapidly when viewed through the intestinal wall. The host's feeding habits have a decisive influence on the Gregarine species

Table II. Gregarines found and host species

GREGARINIDAE	
1) Gregarina fontinalis Zwetkow	Drusus aprutiensis Moret.-D.camerinus Moret.-D.improvisus Mc L.
2) " limnophili Zwetkow	Leptodrusus budtzi Ulm.-Limnephilus rhombicus L.-Allogamus an= tennatus Mc L.-Chaetopteryx sp.
3) " lunata Rauchelles	Tinodes waeneri L.-Ecnomus tenellus Ramb.
4) " mystacidarum Frantzius	D.camerinus Moret.-D.improvisus Mc L.-Mystacides azurea
5) " pusilla Baudoin	Catagapetus nigrans Mc L.-T.maclachlani Kimm.-T.waeneri L.- D.improvisus Mc L.
6) " sericostomae Baudoin	Potamophylax cingulatus Steph.-Allogamus sp.-Sericostoma itali= cum Moret.-S.pedemontanum Mc L.-S.siculum Mc L.-Odontocerum albicorne Scop.
7) " stenophylacis Zwetkow	L.budtzi Ulm.-P.cingulatus Steph.-P.nigricornis Pict.-Stenophy= lax sp.-A.antennatus Mc L.-A.hilaris Mc L.-Allogamus sp.
8) Gregarina n.sp.	D.improvisus Mc L.
9) Gregarina sp.	Diplectrona magna Mos.
LEIDYANIDAE	
10) Leidyana vierlingi Geus	Halesus sp.-Stenophylax sp.
11) Leidyana sp.	Halesus radiatus ssp.vaillanti Moret.
ACTINOCEPHALIDAE	
12) Pileocephalus agilis Geus	P.cingulatus Steph.-Micropterna wageneri Mal.
13) " glyphotaeli Stein.	Grammotaulius nigropunctatus Retz.
14) " lanceatus Baudoin	D.improvisus Mc L.-Allogamus antennatus Mc L.-Allogamus hilaris Mc L.-Odontocerum albicorne Scop.
15) " schyphoides Baudoin	P.cingulatus Steph.
16) " sinensis Schneider	Mesophylax aspersus Ramb.-Mesophylax sp.
17) Pileocephalus sp.1	M.wageneri Mal.
18) Pileocephalus sp.2	P.cingulatus Steph.
19) Asterophora heeri Kölliker	E.tenellus Ramb.-Agrypnia varia Fbr.
20) " hydropsyches Baudoin	Hydropsyche pellucidula Curtis-H.sattleri Tobias
21) " moucronata Léger	Rhyacophila dorsalis acutidens Mc L.-R.foliacea Moret.-R.harti- gi Mal. - R.italica Moret. - R. rougemonti Mc L.
22) " tiaroides Baudoin	E.tenellus Ramb.
23) Asterophora sp.1	A.varia Fbr.
24) Asterophora sp.2	H.sattleri Tobias
25) Globulocephalus hydropsyches Baudoin	H.dissimulata Kum. Bots.-H.instabilis Curt.-H.pellucidula Curtis - H.sattleri Tobias
26) Globulocephalus sp.	Cheumatopsyche lepida Pict.
27) Ancyrophora sp.1	E.tenellus Ramb.
28) Ancyrophora sp.2	R.italica Moret.
29) Eugregarinide indet.	A.hilaris Mc L.

observed. The *Gregarina* predominates in omnivorous but mainly vegetarian larvae; whereas the *Pileocephalus* genus is highest when the diet contains numerous animal components.

Asterophora and *Ancyrophora* are mainly found in predatory larvae (*Rhyacophila-Ecnomus-Agrypnia*). Gregarines were not revealed in larvae that feed on detritius rich in lime nor in Hydroptilidae, probably because, as they feed on algae, they are unlikely to ingest gametocysts.

Gregarines found in the intestinal lumen are most often located in the

216

mesenteron between the intestinal wall and the peritrophic membrane, while the gametocysts are situated in the end tract of the mesenteron and the proctodeum.

The Gregarine and larva life cycles evolve together so the parasite is not found in pupa – we at no time observed celomic Gregarines. The protozoa may pass through more than one life cycle during the larva's life. However, although the protozoa can be completely eliminated with antibiotics, even when infestation was high, the larvae did not seem to suffer and it is interesting to note that the larval stage of the life cycle of these insects evolved naturally under these conditions. As Gregarines are ancient parasites their relationship with Trichoptera is, probably, highly specialized and they may be regarded as commensals rather than true parasites. Obviously, species whose life cycle evolves within a cell, or that are celomic, are an exception.

Krenal and rhitral environments are those which normally yield high numbers of Gregarines.

The trout-breeding beds at Visso (Umbria, Central Italy), favour the multiplication of the *A. antennatus* and their Gregarines and this may, in part, be due to the eutrophic conditions which result from the fish food.

This is a first report on the finding of Gregarines in: *Rhyacophila rougemonti-R. hartigi-Hydropsyche instabilis-H. sattleri-Leptodrusus budtzi-Micropterna wageneri-Allogamus antennatus-Chaetopteryx* sp.-*Sericostoma italicum.*

REFERENCES

Baudoin, J., Ann. Stat. Biol. 2: 15–160, 1967.
Frantzius, A., Arch. f. Naturgesch. 14: 188–196, 1848.
Geus, A., Die Tierwelt Deutschlands 57: 3–608, 1969.
Kölliker, A., Zeitsch. f. Wissenscheftl. Zool. 1: 1–37, 1848.
Léger, L., Tabletes Zool. 3, 1892.
Moretti, G.P. and Sorcetti Corallini, C., Boll. Zool. 43: 69–73, 1976.
Schneider, A., Ibid. 2: 199–207, 1892.
Stein, G.A., Journ Zool. 39: 1135–1144, 1960.
Zwetkow, W.N., Trav. Inst. Sc. Nat. Petzrhof 6: 191–198, 1929.

DISCUSSION

Tachet: Je suis étonné que vous n'ayez jamais trouvé de Grégarines chez les Polycentropodidae pourtant connus pour leurs moeurs carnivores.

Corallini Sorcetti: j'ai examiné des larves de *Polycentropus flavomaculatus* Pictet. du lac de Piediluco et je n'ai pas retrouvé jusqu'à présent de Grégarines. Botosaneanu: 1) Avez-vous remarqué des particularités de l'aspect extérieur des larves parasitées par les Grégarines? 2) Avez-vous trouvé des espèces de Grégarines trouvées aussi en Tchécoslovaquie chez les Trichoptères?

Corallini Sorcetti: 1) je n'ai pas remarqué de particularités de l'aspect extérieur. 2) j'ai trouvé *Gregarina mystacidarum* Frantzius, qu' on a re-trouvé en Slesia.

THE TRICHOPTERA COMPONENT IN THE HYDROPHYTON
OF LAKE CHIUSI (TUSCANY)

G.P. MORETTI AND M.V. DI GIOVANNI

SUMMARY

The larval stages of Trichoptera in the "hydrophyton" of Lake Chiusi, a small Tuscany Lake, and their relationship to various species of aquatic plants are reported. The list of the species found is confirmed by the adult Trichoptera collected along the banks of the lake during random follow-up studies.

INTRODUCTION

The animal component in the Lake Chiusi hydrophyton was investigated during the year July, 1979 to June, 1980. The findings on the Trichoptera in this ecotope are presented here.

The term "hydrophyton" was coined by us following a research on Lake Trasimene in 1968 and it is used to define an environment made up of floating aquatic and submerged aquatic plants, either alive or in a state of decomposition, on which living communities of micro- or macro-organisms settle. It, also, includes temporary components belonging to different communities. For this reason such an environment, even if always well defined, is very polymorphous. Therefore, the endophyton, epiphyton and periphyton together with the hydrophytes make up the hydrophyton.

Lake Chiusi is in Tuscany, in the Province of Siena and almost on its Umbrian border. It is the residue of a large lake that, during the Pliocene Age, covered the entire Valle di Chiana. It is linked to another small lake, Lake Montepulciano, by an outflow canal and Lake Montepulciano is, also, linked to a series of canals, the Chiana Canals, that drain the valley, once a vast marsh. Today the Lake Chiusi watershed has been reduced in size, as several streams on the South of the lake have been diverted through other canals, the Anguillara, in order to increase the capacity of nearby Lake Trasimene (Fig. 1a, 1b).

Lake Chiusi has a surface area of 4 square kilometres and lies in a modest basin at 251 m. above sea level. It is a regularly formed lake running North to South which at its deepest point, on the Eastern shores, is only about 6 m.

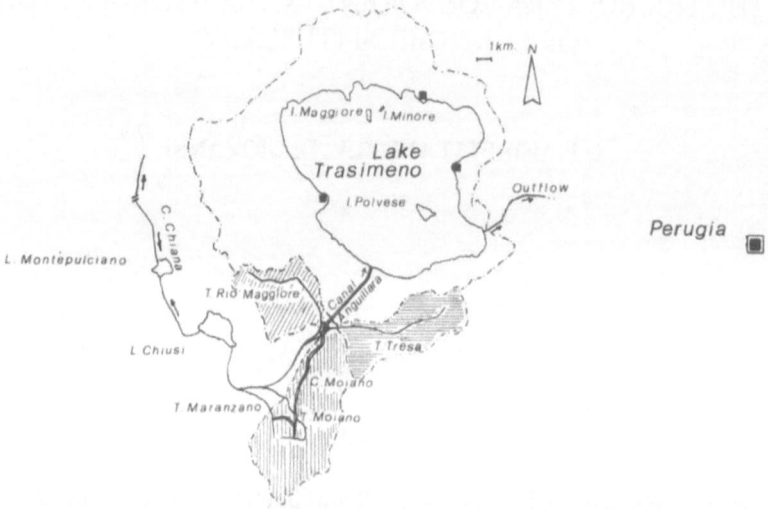

Fig. 1a. The position of Lake Chiusi and Montepulciano in relationship to Lake Trasimene and its water shed.

Fig. 1b. General panoramic view of Lake Chiusi. (June 1979)

Taking into account the morphology of the shores, the inflowing streams, the outflowing canal and the direction of the prevailing winds, 5 research stations were established in the neritic zone.

Samples of aquatic plants were taken in a pre-established line starting from the reeds that border the lake and working towards the centre of the lake at each of these 5 stations. Plastic bags were used for the collection of single plants, which were then carefully washed and gently brushed in the laboratory. All animals were analyzed, counted and the animal/vegetation per kg ratio recorded.

RESULTS

Trichoptera larvae had colonized 14 of the 27 plants examined. (Table I). The incidence varied from plant to plant and from station to station. The

220

Table I. Trichoptera larvae per kg for each plant, station and date.

PLANT	TRICHOPTERA	No.SPECIMENS per Kg.	STATION	DATE
Cladophora	– *Leptocerus tineiformis*	5.46	4	October, 1979
Nymphaea alba	– *Ecnomus tenellus*	5.08	3	Aprile, 1980
" "	– " "	13.16	4	" "
" "	– *Leptocerus tineiformis*	3.05	2	October, 1979
" "	– " "	18.98	1	Aprile, 1980
Nuphar lutea	– " "	10.60	1	June, "
Callitriche	– " "	17.02	1	Aprile, "
Ceratophyllum demersum	– *Ecnomus tenellus*	10.71	1	December 1979
" "	– " "	44.71	2	" "
" "	– *Limnephilus flavospinosus*	17.86	1	" "
" "	– *Leptocerus tineiformis*	9.56	1	September, "
" "	– " "	5.65	1	October, "
" "	– " "	126.58	4	" "
" "	– " "	15.77	5	" "
" "	– " "	160.71	1	December, "
" "	– " "	30.44	3	" "
" "	– " "	123.87	1	February, 1980
" "	– " "	6.31	1	Aprile, "
" "	– " "	51.30	1	June, "
Myriophyllum verticillatum	– *Ecnomus tenellus*	25.64	2	July, 1979
" "	– " "	12.67	4	February, 1980
" "	– *Agrypnia varia*	3.03	2	" "
" "	– *Leptocerus tineiformis*	21.19	4	October, 1979
" "	– " "	9.18	1	February, 1980
" *spicatum*	– " "	3.57	4	December, 1979
Potamogeton natans	– *Ecnomus tenellus*	25.90	4	July, "
" *perfoliatus*	– *Ceraclea fulva*	1.10	5	September, "
" "	– *Leptocerus tineiformis*	337.30	2	June, 1980
" *lucens*	– *Orthotrichia costalis*	7.35	2	July, 1979
" "	– *Ecnomus tenellus*	14.70	2	" "
Scirpus	– *Limnephilus flavospinosus*	7.44	1	December, "
Phragmites australis	– *Ecnomus tenellus*	25.72	2	April, 1980
" "	– *Limnephilus flavospinosus*	11.58	2	" "
" "	– *Leptocerus tineiformis*	8.24	2	October, 1979
Typha latifolia	– " "	11.17	4	September, "
" *angustifolia*	– *Ecnomus tenellus*	27.62	1	December, "
" "	– " "	15.59	2	" "
" "	– " "	23.77	2	February, 1980
" "	– " "	40.28	4	April, "
" "	– *Limnephilus flavospinosus*	36.52	1	December, 1979
" "	– *Leptocerus tineiformis*	2.96	1	October, "
" "	– " "	3.45	1	December, "
" "	– " "	9.38	2	February, 1980

plants most densely populated by Trichoptera larvae were *Ceratophyllum demersum* subspecies *demersum* and *Typha angustifolia*.

C. demersum which grows in extensive submerged masses reaches the water surface during the flowering period in June and July. It is the host of large numbers of *Leptocerus tineiformis*, which are found between the whorls of the twice-forked lance-like leaves, as well as small numbers of *Limnephilus flavospinosus* and *Ecnomus tenellus*.

At times *Typha angustifolia* forms a mixed plant population with the reeds furthest from the land and, at others, isolated small islands. The section of this plant immediately beneath the water is the home of *Leptocerus tineiformis*, *Limnephilus flavospinosus* and *Ecnomus tenellus* larva. On the other hand, *Leptocerus tineiformis* were only rarely observed on *Typha latifolia*.

221

Miriophyllum verticillatum is the only plant on which *Agrypnia varia* was seen, whereas *Leptocerus tineiformis* and *Ecnomus tenellus* were both commonly found on this plant.

Miriophyllum spicatum is less frequently the host of Trichoptera and only one specimen of *Leptocerus tineiformis* was collected here.

Both *Ecnomus tenellus* and *Leptocerus tineiformis* occassionally attach themselves to the under leave of *Nymphaea alba,* whereas *Leptocerus tineiformis* is the only Trichoptera which lives on *Nuphar lutea.*

The larvae of *Ecnomus tenellus* are found on the submerged stalk of *Phragmites australis.*

The stems and leaves of pondweed, *Potamogeton perfoliatus,* harbour the highest number of *Leptocerus tineiformis* and this was the only plant which yielded *Ceraclea fulva.*

Orthotrichia costalis inhabited the axis of *Potamogeton lucens* and no other plant, as did *Ecnomus tenellus.*

Only one Trichoptera species, *Ecnomus tenellus,* was found on *Potamogeton natans.*

There were rare specimens of *Leptocerus tineiformis* on *Callitriche* sp. and this species was, also, collected from *Scirpus* sp. as late as December.

The *Cladophora* sp., alga, wrapped around aquatic plants or floating on the surface, provided low numbers of *Leptocerus tineiformis.*

The other zoological exponents that made up the hydrophyton belonged to various taxa (Fig. 2). Insect life represented 22% of the total, of which $12\frac{1}{2}$ were Trichoptera. The incidence of the Trichoptera in aquatic stages per

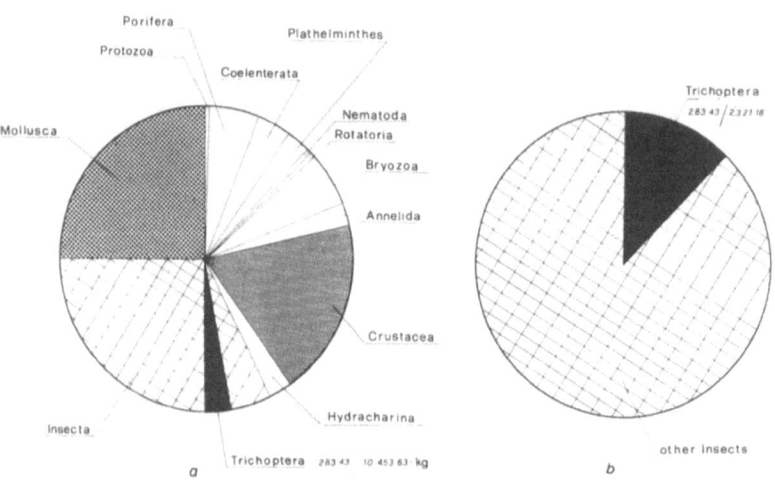

Fig. 2. Numerical density of Trichoptera larvae compared with a) other taxa and b) other insects.

222

kg hydrophytes was 283.43 specimens in a total of 2,321.18 insect specimens and 10,453.53 total zoological exponents.

A comparison of the populations at the various stations showed them to be more dense at Stations 1 and 2, in the South and South-East of the basin (Fig. 3).

Species list (larvae) :

Orthotrichia costalis Curt.
Ecnomus tenellus Ramb.
Agrypnia varia Fbr.
Limnephilus flavospinosus Stein.
Leptocerus tineiformis Curt.

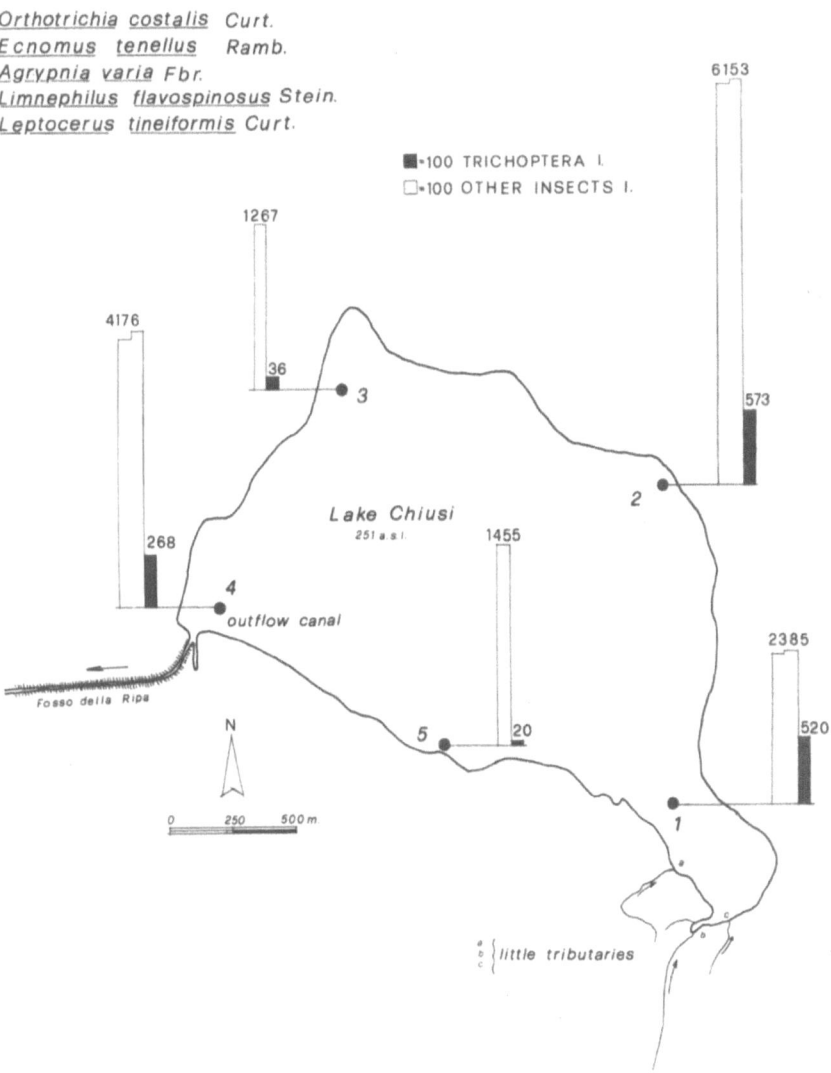

Fig. 3. Trichoptera distribution compared with that of other insects at the different stations expressed as N° specimens/kg.

Table II. Physical-Chemical characteristics for Lake Chiusi from July, 1979 to June, 1980

	Stn.1	Stn.2	Stn.4	Stn.5
Water depth cm.	1.50/ 2.00	2.20/ 3.20	1.80/ 2.60	0.50/ 1.20
Trasparency cm.	1.50/ 1.60	1.50/ 3	1.80/ 2.00	0.50/ 1.20
Water T°C	5.69/ 22.00	6.64/ 21.80	6.38/ 22.50	6.50/ 22.10
Oxygen %s.v.	55.04/106.65	81.04/118.60	83.70/113.45	85.38/124.80
Hardness F.degrees	19.07/ 28.20	20.30/ 24.10	19.40/ 25.80	20.60/ 29.70
Chloride mgr./1.	51.50/ 61.00	49.70/ 59.60	47.90/ 58.20	47.90/ 53.90
Air T°C	7.00/ 29.50	7.00/ 27.50	7.00/ 29.00	7.00/ 26.00

It is suggested that the factor which has the greatest influence on distribution is the wind, because the environmental factors, such as depth, transparency and, above all, the temperature and O_2 s.v. of the water are reasonably uniform throughout all the basin, in all season (Table II). In fact, Station 1 (Fig. 4) and 2 are more sheltered, while Station 3 is the most exposed to the prevailing North Tramontana and West Fagogno winds. The same conditions exist at Station 4 which, in addition, is in the precincts of the outflow canal and affected by the continuous though modest movement of the outflowing water. Station 5 is situated on the water side of a small artificial port and its physionomy is partially modified by man.

There was no significant distribution pattern of Trichoptera species with respect to the various stations. However, there was a relationship between the hydrophyte population, the water turbulence, and Trichoptera colonization density. The adults captured along the shore and the total species collected are reported in Table III. To date, the adults but no larvae of *Hydroptila aegyptia*, *Holocentropus picicornis*, *Ceraclea senilis*, *Oecetis furva* have been recorded, while *Agrypnia varia* has been found in the larval, but not the adult, stage. In all, five species typical of different environments, particularly ponds, were collected in the larval stage. *Orthotrichia costalis* alone presents a slightly different picture.

The presence of *Ceraclea fulva* in epiphyte sponges (*Spongilla lacustris* L.) is an interesting confirmation of the close food-chain link between this insect and Porifera. Only a short time earlier this species was found in Lake Trasimene.

224

Fig. 4. Station 1 in June, 1980.

Table III   *TRICHOPTERA: species in Lake Chiusi*

	larvae	adults	
1. Orthotrichia costalis Curt.	●	June , 1980	1♀
2. Hydroptila aegyptia Ulm.		July , 1979	49♀
3. Holocentropus picicornis Steph.		" "	1♂
4. Ecnomus Tenellus. Ramb.	●	" "	1♀
5. Agrypnia varia Fbr.	●		
6. Limnephilus flavospinosus Stein.	●	October, 1979	1♀
7 Ceraclea fulva Ramb.	●	June ,1980	1♂
8. Ceraclea senilis Burm.		" "	2♂
9. Oecetis furva Ramb.		" "	1♀
10. Leptocerus tineiformis Curt.	●	July , 1979 June 1980	23♂ 27♀ 16 ♂ 16♀

The adult of *Ceraclea senilis* was first reported in Tuscany (Stn. 5; 26 June, 1980). In fact, in 1980 *Ceraclea* was found to be widely distributed in the Central Italian lakes and this was probably due to the heavy spring rains. These findings form the first report on the Trichoptera component of Lake Chiusi and they demonstrate a physionomy which is almost superimposable on that of Lake Trasimene. The only exception being *Limnephilus flavospinosus* which is lacking in Lake Trasimene.

225

The major role played by the hydrophytic component in determining the ecology of Lake Chiusi is confirmed by this research, as it previously was in Lake Trasimene (Moretti, 1954, 1958) and, once again, it has been demonstrated that Trichoptera are sensitive indicators of the environmental phytotrophic resources.

REFERENCES

DI Giovanni, M.V., Boll. Zool. 26: 615–636, 1959.
——, Riv. Idrobiol. 1: 189–234, 1962.
——, Boll. Zool. 31: 1371–1385, 1964.
Gianotti, F.S. and Di Giovanni, M.V., Lav. Soc. Ital. Biogeogr. 2: 575–581, 1971.
Granetti, B. and Bencivenga, M., Riv. Idrobiol. 19, 1980. (in press)
Moretti, G.P., Verh. Int. Ver. Theor. Angew. Limnol. 10: 344–352, 1949.
——, Boll. Zool. 21: 503–529, 1954.
——, Quaderni 21, Soc. It. Biol. Sper., pp. 153–185, 1958.
——, and Corallini Sorcetti, C. Riv. Idrobiol. 19, 1980.
——, and Taticchi, M.I., Riv. Idrobiol. 8: 89–104, 1969.
Riccardi, R., Boll. Reale Soc. Geogr. Ital. 4: 143–164, 1939.
Tiberi, O.; Taticchi Vigano', M.I. and Di Giovanni, M.V., Riv. Idrobiol. 10: 37–233, 1971.

DISCUSSION

Tachet: Do you find many *Ecnomus tenellus* larvae in Lake Chiusi, and how many?
Di Giovanni: Yes, I do, and it has the highest density of any Trichoptera in the lake.

ECOLOGICAL PROFILES IN THREE
RHYACOPHILA SPECIES

G.P. MORETTI AND M. MEARELLI

SUMMARY

Ecological profiles calculated by 27 parameters in a small Umbrian sub-tributary of the River Tiber and their effect on the *Rhyacophila* species, *Rhyacophila dorsalis-acutidens* McL., *Rhyacophila foliacea* Moret. and *Rhyacophila* gr. *tristis* Pictet, were investigated. The research was carried out over a period of one year. This methodology clearly revealed the needs and the ecological tolerability of the three species.

INTRODUCTION

A research was carried on the chemical and physical characteristics of the River Topino in order to determine the ecological background of three Trichoptera species-*Rhyacophila dorsalis-acutidens*, *Rhyacophila foliacea*, *Rhyacophila* gr. *tristis* that inhabit the mountain and the foot-hill tracts of this river.

RESULTS

The variation in population density and stability were compared with 27 physico-chemical factors divided into three profiles to see which of these exerted the greatest influence on the populations. The results of the analysis were used to establish the behaviour of these three taxa in relationship to each single environmental factor and to calculate the possible reaction of each species to combinations of variables. The partial ecological profiles of the three Trichoptera species were constructed by taking the extremes of the values found when Trichoptera were present. The first comparison was with the water temperature(1), depth(2), velocity(3), discharge(4), transparency(5), altitude(6), distance from the water head(7), atmospheric pressure(8). The second took into consideration the parameters of the substance in solution: pH(9), conductivity(10), P(11) and M(12) alkalinity, total hardness(13), calcium-magnesium ratio(14), chloride(15), sulfate(16) and the third assayed the variables responsible for the quality of the water: phosphate(17), ammonia-nitrogen(18), nitrite-

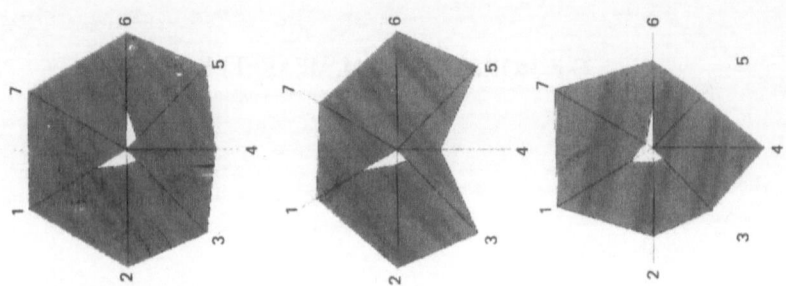

Fig. 1. Ecological profiles

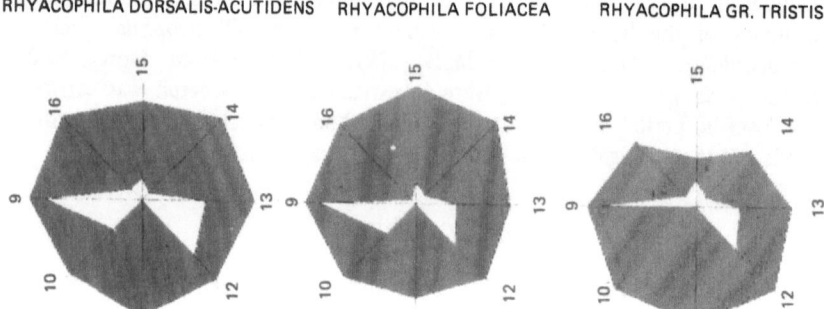

Fig. 2. Ecological profiles

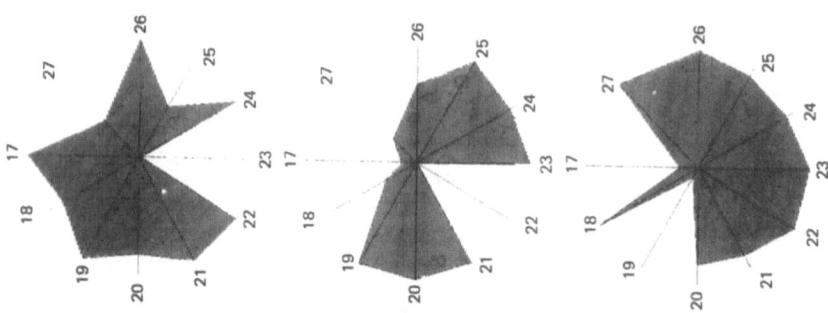

Fig. 3. Ecological profiles

228

nitrogen(19), nitrate-nitrogen(20), oxygen v.s.(21), oxygen deficiency(22), suspended solids(23), organic compounds(24), sulfide(25), B.O.D.5(26), surfactants(27).

A careful analysis of the overall range along the reference axis allows species variability to be determined for single parameters, as well as the stenoecious or the euryecious behaviour. Such a valuation can be calculated from the variations shown by the lined blocks of the ecological profiles.

The first result worthy of note is that, except in a few instances, there is high variability between the parameters. The second is the degree of homogeneity between the three ecological profiles. On the whole this is good for the first and the second profiles, whereas the parameters of the third profile, which characterize the water quality, reveal a high degree of variation. Consequently, there are no great differences in the ecological valence in the first group of parameters (Fig. 1). Due to its marked tolerance to all parameters, *Rhyacophila dorsalis-acutidens* is the species with the widest ecological profile. *Rhyacophila foliacea* is the stenothermic species which favours the coldest conditions and is found at higher altitude where the carrying capacity of the stream is lowest. *Rhyacophila* gr. *tristis* has the most quantitatively restricted ecological profile; it is seen only within a well-defined altitude band and only tolerates a narrow range of temperature, depth or velocity variations.

In the second comparison (Fig. 2), all three species manifest a tendency to stenoecious behaviour in relationship to the pH, the alkalinity and the total hardness, though it is more marked for the pH. In addition, chloride and calcium-magnesium ratio cause further reduction in the ecological valence of *Rhyacophila* gr. *tristis*, but are well tolerated by *Rhyacophila foliacea*. P alkalinity and sulphates result in euryecious behaviour in all three species.

The third group of parameters (Fig. 3) is responsible for the greatest difference of environmental behaviour in the three *Rhyacophila* species. *Rhyacophila foliacea* is the species with the most quantitatively restricted ecological profile. Qualitatively, no species is very susceptible to nitrite-nitrogen, oxygen % v.s., organic compounds and B.O.D.5. *Rhyacophila dorsalis acutidens* seems to be the most tolerant to the greatest number of the 11 parameters; its valence is only drastically reduced by suspended solids, surfactants and sulfide. *Rhyacophila foliacea*, on the other hand, is associated with low or absent values of phosphate, ammonia-nitrogen, oxygen deficiency and surfactants, while the *Rhyacophila* gr. *tristis* population is limited by the presence of phosphate, nitrite-nitrogen, surfactants and B.O.D.5.

These results should not be considered final, as they apply to only one river. The limits found for the parameters may be susceptible to compensation capacity variations within the profiles revealed. They are satisfactory in so far as they allow the ecological picture of the three species to be defined in a tract of a single river and, thereby, contribute to enlarging our knowledge of Trichoptera as biological indicators of the environment.

REFERENCES

Beck, W.M., Sew. Ind. Wastes 27: 227–458, 1955.

Cummins, K.W., Spec. publ. Pymatning lab. Fld. Biol. 4: 2–51, 1964.

Decamps, H., Annis Limnol. 3: 399–577, 1968.

Hynes, H.B.N., The ecology of running waters. Liverpool Univ. Press, 1960.

Verneaux, J., *In* La Pollution des Eaux Continentales, Eds Gauthier and Villars, pp. 229–285, Bordas, 1976.

Bournaud, M., Annis, Limnol. 16: 55–75, 1980.

MORPHOLOGICAL CHARACTERISTICS OF
LEPTODRUSUS BUDTZI ULM. IN THE
IMMATURE STAGES

G.P. MORETTI AND Q. PIRISINU

SUMMARY

The immature stages of *Leptodrusus budtzi* Ulm. endemic to Sardinia and Corsica are described for the first time. The taxonomic characteristic for certain and immediate identification of the larva is the clear field on the posterior two-thirds of the pronotum, which is delimited by a series of arches. The typical drusini hump is more pronounced than it is in other Italian drusini. These, other classification characteristic, the larval case and the pupa are shown in a series of designs.

INTRODUCTION

Ulmer classified a male specimen of drusinus, found by Peterson at Vizzavona in Corsica, as belonging to the *Potamorites* genus. Ulmer, however, doubted that the number of spurs was a sufficient taxonomic characteristic for certain identification of a genus. He named this specimen *Potamorites budtzi*. Mosely confirmed Ulmer's classification in 1930, but, in 1955, Schmid was of the opinion that *Potamorites budtzi* was, in fact, the genus type of *Leptodrusus* and he termed it *L. budtzi*.

Until 1964, neither the larva nor the pupa of *Leptodrusus budtzi*, which is endemic to Sardinia and Corsica, had ever been described. We decided to fill this gap and, after examining more than 100 specimens collected in Gallura, Sassari, between 1964 and 1966 produced a series of designs that clearly illustrate the main morphological features of the immature stages of the species.

DESCRIPTION

The *larval case* (Fig. 1b) is conical and only slightly tapered at the posterior apex. The overall length is 16–19 mm; the width at the anterior opening is 3–4 mm and at the posterior 2–2.1 mm. The case is compact, there is both dorsal and ventral flattening, it is fluted and more oblique at the anterior orifice than is the case of any other European drusini. Small, sometimes protruding, stones are used for constructing the case. Freed from its case the larva (Fig. 1a) measures 13–14 mm in length and 3–3.5 in width at the first abdominal segment. The head (Fig. 2c) is brownish, oval and extremely

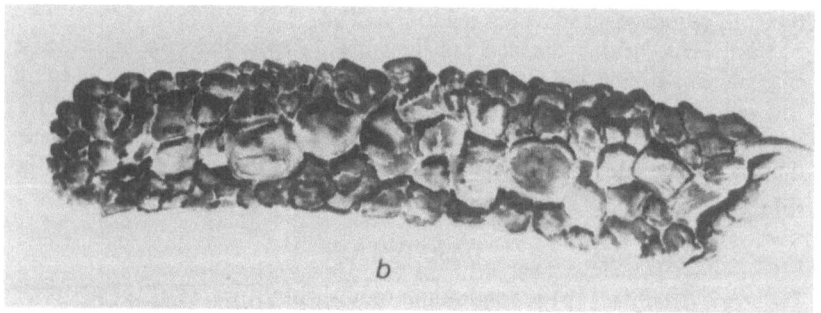

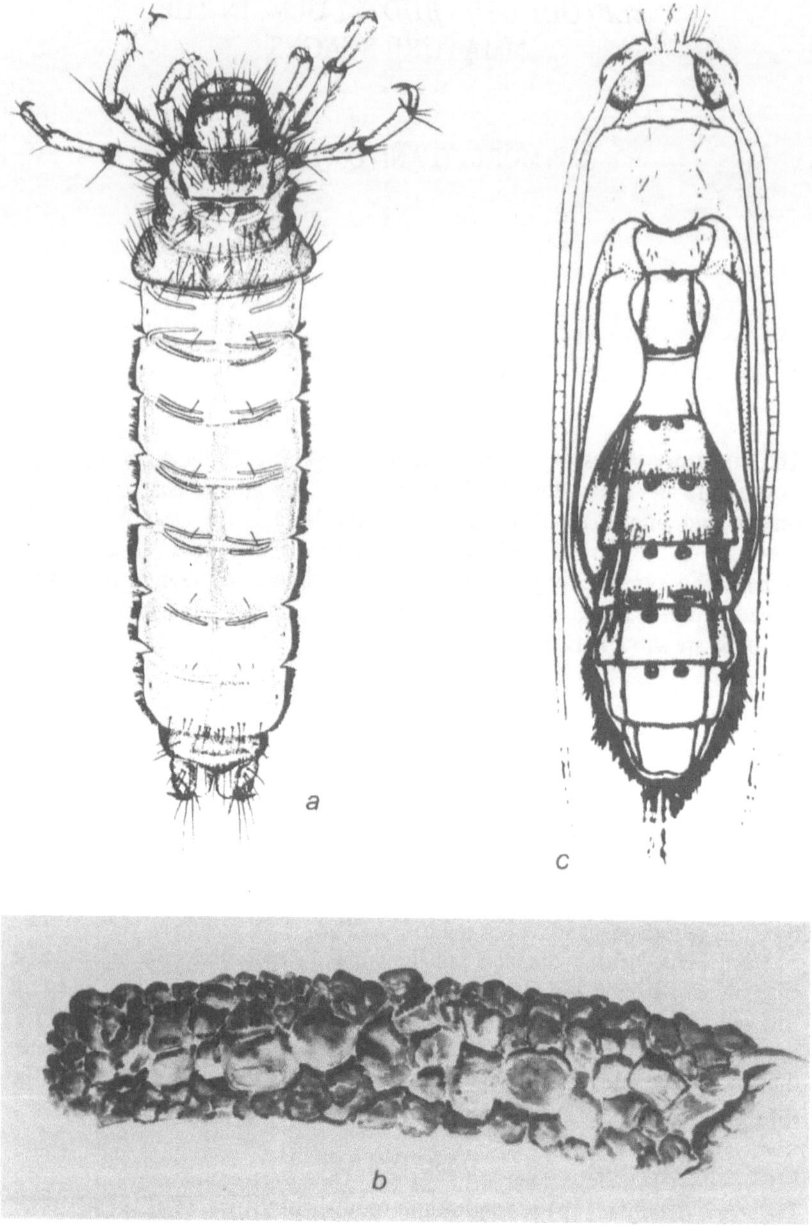

Fig. 1. Leptodrusus budtzi Ulm.
a – larva; b – larval case; c – pupa.

hypognate. The occipital region is lighter, but edged by a dark stripe. The typical drusini crest is absent. The eyes are located slightly behind the median periocular heart shaped marking, which suggests that the larva looks forward and upward. The labrum is chitinous, transversal and

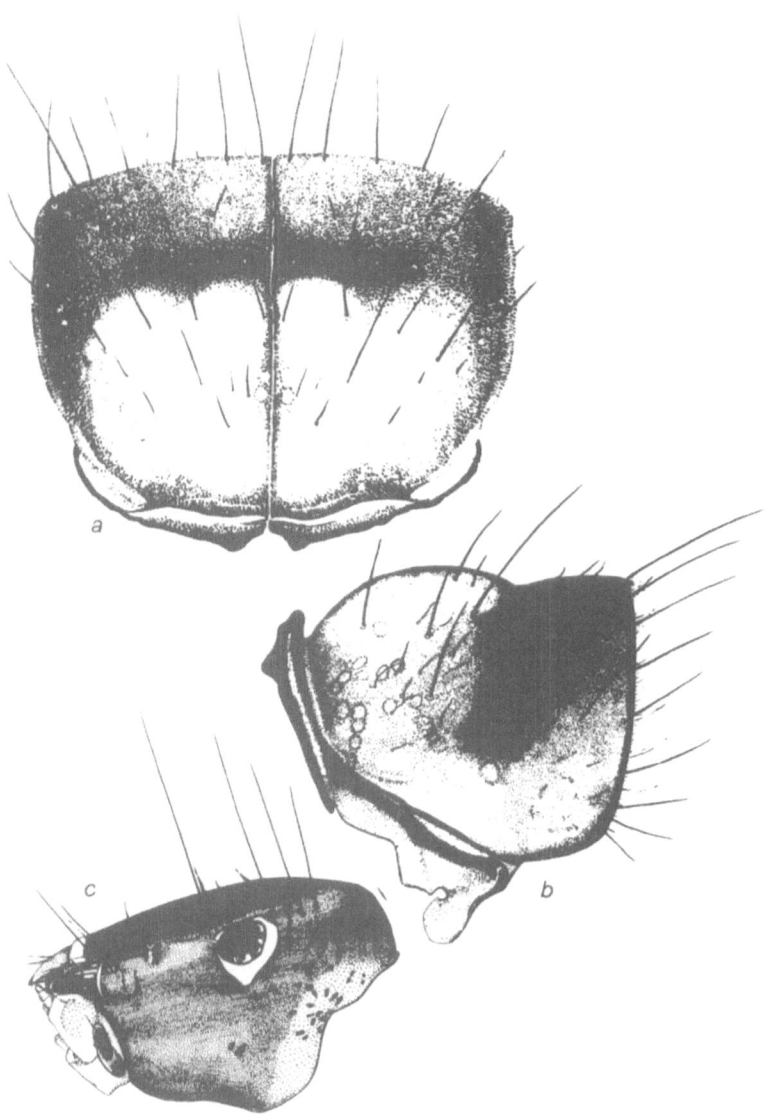

Fig. 2. Leptodrusus budtzi Ulm.
a – pronotum, dorsal view
b – pronotum, lateral view
c – head, lateral view

elliptical; the lateral convex lobes have tufts of yellow hairs which extend over the entire ventral surface. The dorsal surface of the mandibles are flat and the internal line is free of teeth, but there is a single undulation and an apical tooth curved like a spoon. The labial lobe terminates in a fleshy bulb and is carpeted with short hooked setae. The stipes are trapezoid and

233

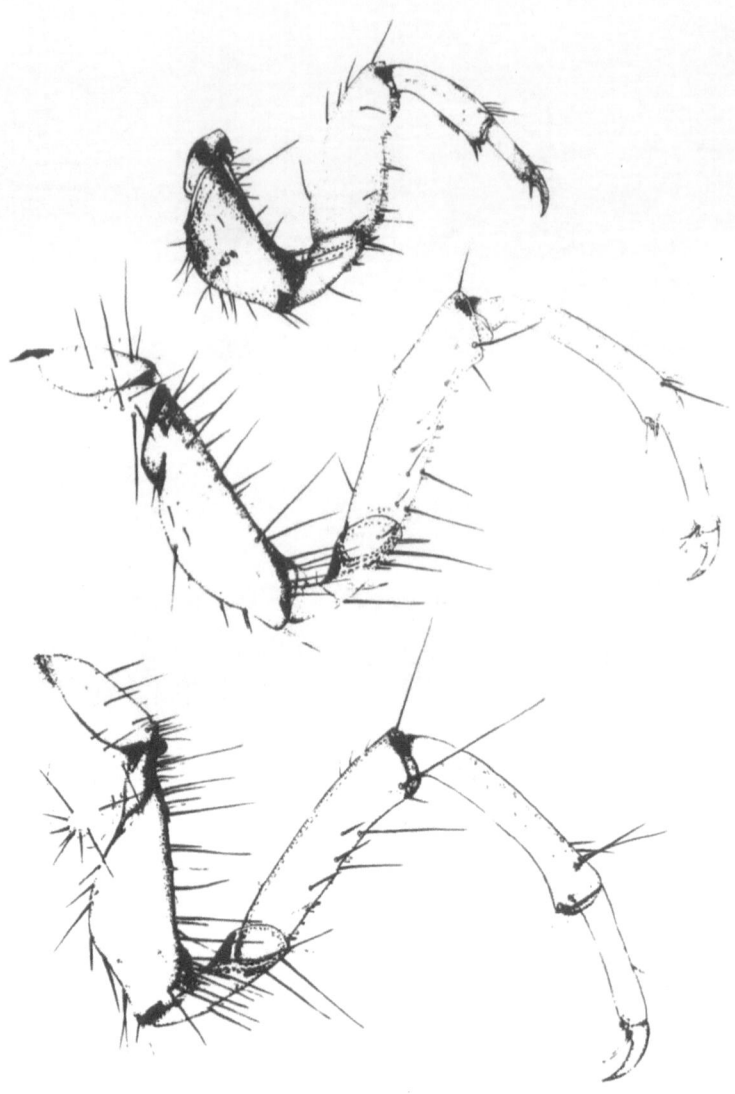

Fig. 3. Leptodrusus budtzi Ulm.
legs, right side.

less sclerified than the cardum. The whole of the maxillary labium cuticle forms a characteristic gently rippled sculptured border.

The *pronotum* (Fig. 2a, b) is the most important distinguishing taxonomic feature, it is quadrangular, deeply convex, well sclerified, the same colour as the head in the anterior third, but lighter in the posterior two-thirds, where there is a large brownish-yellow zone limited anteriorly by a series of arches. The anterior margin is fringed with pale flexible cilia between which

234

eight tiny spines stick out from each side. The transversal furrow which demarks the division between the anterior third and the posterior two-thirds is indented, particularly the darker medial tract. The posterior field forms the pronounced hump typical of the drusini.

The *mesonotum* is larger than the pronotum and the two adjacent plaques form a rectangular plate. There is trapezoid darkening at the anterior margin, which narrows towards the base, and two comma-like marks on a light lateral, posterior field. The lateral posterior margin is very dark and has a pronounced black border.

The fleshy *metanotum* is almost rectangular and has anterior median sclerites. The half-moon lateral plaques are well sclerified, while the triangular posterior ones are less so.

The legs are yellowish with the dorsal face of the trochanter tending to red. The borders are blackened at the joint (Fig. 3). The coxa of *the right anterior leg* is cuneiform and conical. The trochanter has two apical spurs, one larger than the other, in front of which there is a tuft of yellow setae and two wide-based median spurs among the tiny spines that fringe the internal edge of the femur, which is one of the main distinguishing characteristics. The tibia is narrower than the femur. It is equipped with a toothbrush of tiny spines, which are repeated on the tarsum, and two apical spurs.

The right median leg anterior profile of the coxa is straight and the posterior convex. There is a tuft of pale flexible setae below the apical spur on the trochanter. Small spines arranged like a comb run along the proximal two-thirds of the ventral femur and there are two apical spurs. A few tiny spines are seen on the tarsum and the pretarsal claw is equipped with a robust basal spine.

The right posterior leg coxa is similar to that of the median leg, but the pale yellow setae below the spur are absent. The femur is shorter and the tibia is folded back at the joint.

The abdomen is mainly cylindrical. The first segment is darker than the others and is the widest part. The wart-like protubrances are domed laterally and more pointed dorsally and have setae spaced around them. There are less setae on the other segments and, from the second to the sixth, there are two postsegmental lateral median setae. The lateral line, which is formed by a blackish-brown fringe, begins at the lateral posterior margin of the second segment and ends at the eighth. Presegmental and postsegmental single gills are observed on the second to seventh segments, but on the last there are only presegmental gills.

The large *anal claw* is formed by a roughly trapezoid fixed part, the dorsal proximal border of which is black. The small anal claw forms a right angle with the segment on which it is embedded and there is a spine on its edge.

The pupa (Fig. 1c) has an elliptical transversal head with large hemispherical protuding brownish-violet eyes. The filamentary antenna are usually made up of sixty-nine segments. They are longer than the body and curve back dorsally and recurve up at the apex. There is a group of thin black robust hooked setae on the pale squarish fleshy labrum. The mandible has a robust almost conical black base which is tapered in the lower half and toothed on the internal edge. The pterothecae have a slightly sclerified border, the posterior ones are considerably larger and wider than the anterior ones. About thirty small teeth are seen on the posterior lobes of the first abdominal tergum. The case hooking plates are found from the fourth to seventh segments and the hooks, themselves, are arranged in a fairly regular pattern. Two small presegmental and two large post segmental plaques are located on the fifth segment. The case is formed by cutting the larval house at the end. The overall length is 19.4 mm, the width 4.5 mm. The extremities are plugged with tiny stones and little pebbles bond together by a strong silky open weave web. The larvae are found in slow running streams with low carrying both near to and far from the source.

REFERENCES

Giudicelli, J., Thèses, Fac. Sci. Univ. - Marseille, 104–106, 1968.
Moretti, G.P., Boll. Soc. Entom. It., 68: 20–24, 1937.
——, Mem. Soc. Entom. It. 19, 1940.
Mosley, M.E., Eos Rev. Esp. Entom. 6: 147–184, 1930.
——, Eos 8: 165–184+IV–V, 1932.
——, and Berland, Ann. Soc. Entom. France. 105, Pt I, 111–144, 1936.
Petersen, P., Entom. Medd. 10: 25–26, 1913.
Schmid, F., Inst. Royal Science. Nat. Belgique. Mem. X Ser. 55, 1956.
Ulmer, G., Entom. Medd. 10: 17–19, 1913.

TAXONOMIC AND MORPHOLOGICAL CHARACTERISTICS OF THE *SERICOSTOMA ITALICUM* MORET. and *SERICOSTOMA PEDEMONTANUM* McL. LARVAE

G.P. MORETTI AND G. SPINELLI BATTA

SUMMARY

The morphological and taxonomic features which distinguish the fifth larval stage of *Sericostoma italicum* Moret. from that of *Sericostoma pedemontanum* McL. are shown in a table which facilitates distinction between the two species. *S. italicum* Moret. is endemic to Central Italy; *S. pedemontanum* McL., which is not accepted by all investigators, is found throughout Europe. Microscope studies on the sclerified parts of the head, thorax, legs and anal claws revealed a number of morphological characteristics which are more than sufficient for the differentiation of the larvae of these two species.

INTRODUCTION

Sericostoma italicum Moret. was found for the first time in a Central Italian stream (the Acilia, Lazio) by Castellani in 1939 and was, then, determined by one of us, Moretti, who presented the first taxonomic morphological description of the male during the 2nd International Symposium on Trichoptera at Reading in 1977 (Moretti and Cianficconi, 1977). This paper describes the morphological and taxonomic characteristics of the mature larva. In order to establish the distinguishing features of *Sericostoma italicum* Moret., an endemic Central Italian species, it was compared with *Sericostoma pedemontanum* McL. which is widely distributed in the running waters of Central Italy and is, therefore, easily collected. There is still disagreement on this species as some investigators consider it to be a synonym of *S. personatum* Spenc.. Both are rheophyls and prefer the fast running streams and brooks of the rhithron. Although omnivorous, they are mainly vegetarian and investigations on their gut contents have revealed remains of bryophyte leaves together with scrapes of insect larvae. The following description emphasizes the more important distinguishing characteristics.

Proc. of the 3rd Int. Symp. on Trichoptera, ed. by G.P. Moretti
Series Entomologica, Vol. 20. © 1981, Dr W. Junk Publishers, The Hague

DESCRIPTION

Table of distinguishing characteristics for the mature larvae of:
Sericostoma italicum Moret. and *Sericostoma pedemontanum* McL.

Sericostoma italicum Moret.

Larval case: 14 mm. in length, 2,9 mm wide at anterior orifice; curved cylinder, constructed in a mosaic of small stones that stick out and make the surface rough; external posterior diameter of the silky open-weave web 1,8 mm., interior orifice aperture 0,6 mm.

Head capsule (dorsal view) (Fig. 1–2): 1,8 mm. in length, 1,8 mm. in width, tawny on the upper face, clypeus wide and slightly paler in the centre; numerous indistinct markings, distinct only on the occipital border (Fig. 2b); pale ocular marking; small eyes, frontoclypeal setae long and strong.

Head capsule (ventral view) (Fig. 2c): slightly testacean in the perioccipital hypogenal region; two small barely visible lateral pregular marks.

Labrum: large, poorly sclerified not tapered at the border with respect to the posterior condyles; two large, pale sabre-like distinct setae on the anterior margin, no dark border on anterior margin; slightly convex.

Mandibles: hollowed like a spoon, large, short and squat, with five robust teeth and a tuft of blonde hairlike on the internal face.

Maxillary-labium: short, well sclerified only slightly protruding maxillary-palps.

Sericostoma pedemontanum McL.

Larval case: 15 mm. in length, 3 mm. wide at anterior orifice; less cylindrical, only slightly curved, constructed in a smooth and regular mosaic of perfectly juxtaposed tiny pebbles; external posterior diameter of the silky open-weave web 2 mm., interior orifice aperture 0,5 mm.

Head capsule (dorsal view) (Fig. 2d): 1,5 mm. in length, 1,4 mm. in width, more smokey on the upper face, clypeus has the same colour but narrower distinct, oblong, transversal markings on all the upper face and the lateral genae (Fig. 2e); pale ocular marking; large eyes, frontoclypeal setae shorter and less strong.

Head capsule (ventral view) (Fig. 2f): white in the lower perioccipital region; two small lateral pregular marks contrast with white background.

Labrum: smaller, pale, well sclerified, tapers back with respect to the condyles, two poorly developed setae on the anterior margin, dark border on anterior margin; very convex.

Mandibles: hollowed like a spoon, smaller, the five teeth are less robust.

Maxillary-labium: longer, less sclerified, slender maxillary-palps.

238

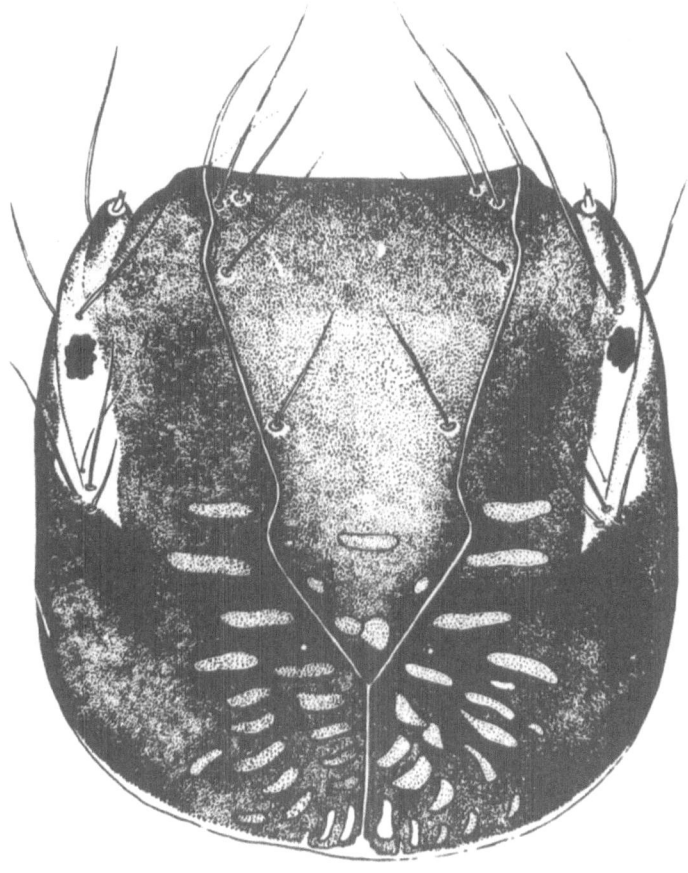

Fig. 1. Sericostoma italicum Moret. larva: head

Pronotum: rectangular with very elongated anterior lateral vertex; well sclerified with equidistant, sloped forward setae on the anterior border; an area of long setae sticking straight up covers the anterior half of the sclerite; two brown marks on the median suture, several posterior lateral brown markings on the sclerite.

Mesonotum: wide, with dark lateral borders covered with setae; a few distinct markings on the anterior dorsal surface.

Pronotum: rectangular, protruding anterior, lateral vertex; darker and less sclerified;

the same;

two dark brown marks on the median suture, posterior lateral brown markings on the sclerite are more distinct.

Mesonotum: narrower, no darkening at the lateral borders covered with setae; a few ill-defined brown markings on the anterior dorsal surface.

239

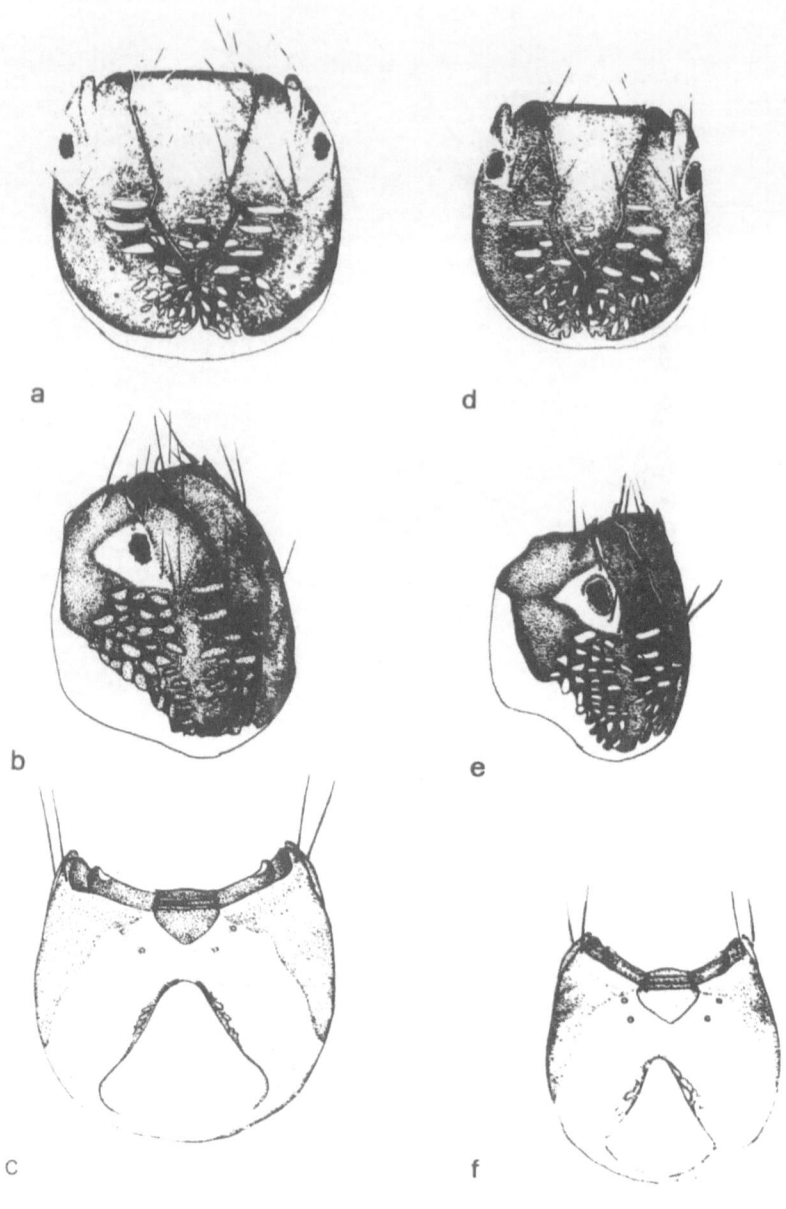

Fig. 2. Sericostoma italicum Moret. larva: head, a) dorsal view, b) lateral view, c) ventral view; *Sericostoma pedemontanum* McL. larva: head d) dorsal view, e) lateral view, f) ventral view.

Metanotum: pale and transparent with two tuft of lateral and two series of transversal setae.

Metanotum: the same

Anterior leg: squat, strong coxa, well dilated femur; tibia and tarsum with convex external border, abundance of long flexible setae on all parts; a series of pale spurs on the trochanter (8); two large .spurs on the femur; two small pointed spurs on the tarsum; a single spur on the tibia; short wide-based pretarsal claw with basal spine.

Anterior leg: short, strong coxa, well dilated femur; tibia and tarsum with external border convex;

the same
the same

the same

Median leg: brown, elongated and finer than the anterior, fringed with long flexible setae; a spur on the tibia; long scythe-like tarsal claw with spine at base.

Median leg: pale brown, long and fine, less fringed with setae; the same; shorter tarsal claw with spine at base.

Posterior leg: brown, longer and finer than the median; fringed with long flexible setae; no spurs; long curved tarsal claw with proximal, basal spine.

Posterior leg: lighter, longer and finer than the median; covering of setae scarce; no spurs; the tarsal claw is fine and shorter.

Anal claw: intermedial sclerite has a very blackened internal border; three hooks on external sclerite; fleshy anal lobe with few long thick black setae.

Anal claw: the intermedial sclerite border is less dark; the same the same

As will be seen, the comparative system set out above facilitates the identification of those taxonomic characteristics which distinguish *Sericostoma italicum* Moret. from *Sericostoma pedemontanum* McL. = *personatum* Spenc.?

NOTE: The denomination for the cuticolar formations are those of Lepneva (1966)

REFERENCES

Despax, R., *In* Traitè de Zoologie ed. P.P. Grassé, Masson, 10(I): 143–149, 1951.
Fischer, F.C.J., Trichopterorum Catalogus, Nieder. Entom. Vereen. Amsterdam, 2: pp. 214–254, 1970.
Giudicelli, J., Entom. Nice, 46: 201–212, 1978.
Hickin, N.E., *In* Caddis Larvae, Hutchinson, London, p. 476, 1967.

Hiley, P.D., *In* Proc. of the First Int. Symp. on Trichoptera, ed. Junk, The Hague, pp. 21–24, 1974.

Lepneva, S.G., Fauna of the U.S.S.R. Trichoptera. I. Isr. Pr. Sc. Transl. Jerusalem, pp. 672–679, 1971.

Lestage, J.A., Les larves et numphes aquatiques des insectes d'Europe, Bruxelles I: 918–931, 1921.

Moretti, G.P., Cianficconi F., *In* Proc. of the Second Symp. on Tichoptera, ed. M.I. Crichton, Junk, The Hague, pp. 7–30, 1977.

Ulmer, G., Die Susswasserfauna Deutschlands. Trichoptera. Verlag Gustav Fischer, 1909.

Wiggins, G.B., Larvae of the North American Caddisfly Genera (Trichoptera) Univ. Toronto Press, 1975.

242

TRICHOPTERA IN THE INTESTINAL CONTENT
OF CERTAIN FISH SPECIES

G.P. MORETTI, C. CORALLINI SORCETTI
AND P. GATTAPONI

SUMMARY

As Trichoptera are often observed in the intestinal contents of fish, a research was carried out in the three Central Italian lakes: Bolsena, vulcanic, Piediluco, an artificially regulated water system, and Trasimene, laminar. Various sizes and ages of 12 fish species were assayed in all seasons of the year. The Trichoptera larvae and pupae found in the food contents belonged to 24 taxa. Their incidence differed from lake to lake and the highest aliquot, with respect to the total content, was found in the fishes of Lake Piediluco.

INTRODUCTION

As Trichoptera are a natural food of many fish, we studied 12 fish species in the three Central Italian lakes: Bolsena (Viterbo), Piediluco (Terni) and Trasimene (Perugia).

Lake Bolsena, a typical "caldera" lake situated in the Volsini Mounts at 305 m above sea level, is roughly oval, has a surface area of 113 square kilometres, a maximum depth of 151 m and a periphery of 43 km.

Lake Piediluco is a small natural, but artificially controlled water for which reason its levels vary widely during the course of a day. It lies in a lacustral basin at 368 m above sea level, has a surface area of 1.52 km and measures 13 km around its rocky, indented shores.

Lake Trasimeme, a typical laminar lake standing at 258 m above sea level and with a maximum depth of 6.30 m, is the fourth largest lake in Italy. Its surface area of 124. 3 square kilometers and its perimeter of 53 km make it the most extensive lacustral water in the peninsular Italy. Its shores are deeply edged with reeds and hydrophytes that, usually, grow well into the waters, although here they become less dense. The marshiest zone is that situated on the South-East shores known as the valley.

MATERIALS AND METHODS

The more common fish species were chosen for this research. The seasons were taken into consideration when sampling, so that all ages and sizes of fish were included. The fishing techniques in the three lakes are similar and reasonably common. Net shape and gauge vary according to the type of catch.

Intestinal wet weight contents were examined for the foods of the different species. The digestive tract was removed whole, the contents weighed in mg and then stored in 5% acquous formaline in individual plastic containers.

RESULTS

The fish from Lake Piediluco were smaller than those of the other two lakes, no doubt the thermic chemical and turbulence factors resulting from the hydro-regulation were responsible for their slow growth.

The number and incidence of Trichoptera remains for each individual specimen from each of the lakes are listed in Table I. *Coregonus "forma hybrida"*, a perfect plankton phage, *Rutilus rubilio,* a cyprinide with pharyngeal teeth, and *Lepomis gibbosus* were the only species free of Trichoptera remains. The greatest number of fish examined came from Lake Piediluco. They represented 10 species and had the highest incidence of Trichoptera. *S. trutta fario, S. gairdneri, E. lucius* and *S. erythrophthalmus* proved to be the species which most frequently ingested these insects. Table II lists the Trichoptera found in the digestive tract of all fish specimens taken from the three lakes. Here, too, the results from Lake Piediluco are interesting as the remains of seven taxa were recovered from *S. erythrophthalmus* and five each from *S. trutta fario* and *L. cephalus.* The highest number from Lake Bolsena was three species in *T. tinca,* while the only prey from Lake Trasimene was *Leptocerus tineiformis.*

Table 1. Fish species and Trichoptera findings

FISH SPECIES	L. BOLSENA FISHES			L. PIEDILUCO FISHES			L. TRASIMENE FISHES		
	EXAMINED No.	WITH TRICHOPTERA No.	%	EXAMINED No.	WITH TRICHOPTERA No.	%	EXAMINED No.	WITH TRICHOPTERA No.	%
Salmo trutta fario L.				5	3	60.00			
Salmo gairdneri Rich.				2	1	50.00			
Coregonus "forma hybrida"	18	0	0.00						
Esox lucius L.	29	2	6.90	5	1	20.00			
Tinca tinca L.	2	1	50.00	13	3	23.00			
Leuciscus cephalus L.	13	0	0.00	36	4	11.11			
Scardinius erythrophthalmus L.	27	6	22.22	64	16	25.00	175	14	8.00
Rutilus rubilio Bp.				17	0	0.00			
Anguilla anguilla L.	28	4	14.28	22	1	4.54			
Atherina boyeri Risso	22	1	4.54						
Lepomis gibbosus L.				5	0	0.00			
Perca fluviatilis L.	33	3	9.09	67	8	11.94	40	0	0.00
TOTAL	172	17	9.88	236	37	15.68	215	14	6.51

244

Table II. Trichoptera found in the digestive tract of various fish

FISHES EXAMINED / LAKES STUDIED	L. BOLSENA	L. PIEDILUCO	L. TRASIMENE
Salmo trutta fario L.		Rhyacophila foliacea Moret. Hydropsyche pellucidula Curtis Plectrocnemia sp. Micrasema setiferum Pict. Chaetopteryx gessneri Mc L.	
Salmo gairdneri Rich.		Micrasema setiferum Pict.	
Esox lucius L.	Orthotrichia costalis Curt. Hydroptila aegyptia Ulm.	Hydropsyche instabilis Curt.	
Tinca tinca L.	Tinodes waeneri L. Oecetis furva Ramb. Leptocerus tineiformis Curt.	Rhyacophila sp. Orthotrichia costalis Curt. Agapetus sp.	
Leuciscus cephalus L.		Hydropsyche pellucidula Curtis Limnephilus sp. Lasiocephala basalis Kol. Mystacides azurea L. Mystacides longicornis L.	
Scardinius erythrophthalmus L.	Hydroptila aegyptia Ulm. Leptocerus tineiformis Curt.	Rhyacophila sp. Orthotrichia costalis Curt. Hydroptila sp. (non aegyptia) Agraylea sp. Tinodes waeneri L. Limnephilus flavospinosus Stein. Mystacides longicornis L.	Leptocerus tineiformis Curt.
Anguilla anguilla L.	Holocentropus sp. Leptocerus tineiformis Curt.	Orthotrichia costalis Curt. Hydroptila sp. (non aegyptia)	
Atherina boyeri Risso	Tinodes maclachlani Kimm.		
Perca fluviatilis L.	Orhotrichia costalis Curt.	Polycentropus flavomaculatus Pict. Limnephilus flavicornis Fbr. Ceraclea furva Ramb.	

The total number of Trichoptera found in the intestinal contents of the fish investigated was 7 for Lake Bolsena, 19 for Lake Piediluco and 1 for Lake Trasimene. The total taxa were 24 (Table III).

The fact that the *pabulum* of Lakes Bolsena and Trasimene favour pond-loving species, while the marshy, lotic and rheic environments of Lake Piediluco each invite its characteristic species (*Rhyacophila foliacea, Rhyacophila* sp. etc.), may explain both the greater number of taxa and the higher numbers of Trichoptera found in the intestinal contents of the fish taken from the last lake.

245

Table III. Trichoptera present in three lakes

SPECIES FOUND LAKES	L. BOLSENA	L. PIEDILUCO	L. TRASIMENE
1 - Rhyacophila foliacea Moret.		X	
2 - Rhyacophila sp.		X	
3 - Agapetus sp.		X	
4 - Orthotrichia costalis Curt.	X	X	
5 - Hydroptila aegyptia Ulm.	X		
6 - Hydroptila sp. (non aegyptia)		X	
7 - Agraylea sp.		X	
8 - Hydropsyche instabilis Curt.		X	
9 - " pellucidula Curtis		X	
10 - Plectrocnemia sp.		X	
11 - Polycentropus flavomaculatus Pict.		X	
12 - Holocentropus sp.	X		
13 - Tinodes maclachlani Kimm.	X		
14 - " waeneri L.	X	X	
15 - Micrasema setiferum Pict.		X	
16 - Limnephilus flavicornis Fbr.		X	
17 - " flavospinosus Stein.		X	
18 - Chaetopteryx gessneri·McL.		X	
19 - Lasiocephala basalis Kol.		X	
20 - Mystacides azurea L.		X	
21 - " longicornis L.		X	
22 - Oecetis furva Ramb.	X		
23 - Ceraclea fulva Ramb.		X	
24 - Leptocerus tineiformis Curt.	X		X
TOTAL	7	19	1

Larva and pupa are the preferred prey of fishes and the finding of intact *Chaetopterix gessneri* and *Ceraclea fulva* pupae lead to easy classification. The tiny stone and silk larval cases are not generally destroyed by the digestive processes; however, this is not so when they are constructed of leaf particles. The soft inner tissues are digested, while the cuticle and the sclerified parts are not. The larvae are often fragmented, as happens with *P. fluviatilis, S. trutta fario, S. gairdneri, L. cephalus* and *S. erythrophthalmus,* so that only chitine remains are found. The organic-matter-free cephalic capsule and the thoracic schlerites retain their distinctive pigmented designs intact, so, in this case, they can be identified systematically, sometimes even the species. The Trichoptera content varied from a few fragments to one or more Trichoptera and, at times, very large numbers were observed. Up to 30 *P. flavomaculatus* were counted in *P. fluviatilis* and a similar number of cases were seen in *S. trutta fario.*

The investigation on the presence of Trichoptera revealed a dietary pattern. Fish of more than six years showed a preference for large animal components, which younger fish are not yet able to ingest. *L. cephalus, S. erythrophthalmus* and *P. fluviatilis* seemed to prefer insect larvae, particularly those of the Trichoptera.

Only two species, *S. erythrophthalmus* and *P. fluviatilis,* where represented in the 215 fish taken for examination from Lake Trasimene. As previously

246

mentioned, the only finding was *Leptocerus* in *P. fluviatilis*, which, as far as the animal component of the diet is concerned, prefers *Palaemonetes antennarius* H.M. Edw..

COMMENTS

It is known from previous studies (Moretti et al., 1959; Cianficconi, 1959; Viganó, 1965) carried out on *Rutilus rubilio, Atherina boyeri* and *Lepomis gibbosus* from Lake Trasimene that only the last yields *Hydroptila aegyptia* and *Leptocerus tineiformis.*

Although the dietary habits of fish species are still under investigation, results already suggest that there is a close relationship between the composition of the lacustral basin and the composition of the available food resources.

REFERENCES

Cianficconi, F., Boll. Zool. 26: 607–613, 1959.
Losos, B., Zool. Listy. 25: 275–288, 1976.
———, Scripta Fac. Sci. Nat. Ujep. Brunensis, Biologia 1 (7): 31–46, 1977.
Moretti, G.P.; Gianotti, F.S. and Giganti, A., Riv. Biol. 51 (1): 3–38, 1959.
———; Gattaponi, P. and Corallini Sorcetti, C., Boll. Zool. 45: 230, 1978.
———; Cianficconi, F.; Corallini Sorcetti, C.; Gattaponi, P. and Tucciarelli, F. Boll. Soc. It. Biol. Sper. 55: 1288–1294, 1979.
———; Corallini Sorcetti, C. and Gattaponi, P., Boll. Zool. Suppl. 46: 154–155, 1979.
Viganó, A., Riv. Idrob. 5: 125–138, 1965.

TRICHOPTERA OF THE ISLE OF ELBA (TUSCANY ITALY)

G.P. MORETTI, F.S. GIANOTTI, M.I. TATICCHI
AND A. VIGANÓ

SUMMARY

During 1956–1958 a research on Trichoptera of the Isle of Elba was carried out. Fifty-three stations which included different biotopes were inspected and a total of 29 taxa were found. Certain environmental factors were assayed at the same time as the adult and immature stages were collected. The physical and chemical characteristics of the island waters reflect the geology of the Isle and have a very important influence on the Trichoptera distribution. Three new taxa (1 species and 2 subspecies) were found.

INTRODUCTION

The Isle of Elba is a meeting point of alpine and apennine systems, as some areas can be included in the tettonic alpine structure, while other parts are in the apenninic structure. The entire Monte Capanne (1019 m. above sea level) region and most of the West of the Isle of Elba belong to the first classification and are characterized by granodiorite and thermo-metamorphic rocks; whereas, East Elba, which is typified by regional sedimentary and metamorphic rocks, falls into the second category. Between them is an intensely cultivated arenary and marn plain. It is interesting to note that the famous iron (hematite) mines, once worked by the Etruscans, are situated in the East of the island, where extremely fractured and porous lime formations rise from the ground. These acted as collecting areas for molten mineralized liquids and they are similar to the Tuscan metalliferous chain. The Monte Calamita headland, where the outcrops are mainly due to quartz-biotic shists, is characterized by this lithological type of rock. The lithological variations are reflected in the hydrobiological features of the island and, therefore, in the distribution of Trichoptera biocenosis. Infact, in the East there are short and astatic water ways with a high degree of hardness that, at times, exceeds 50 French degrees; the pH is below 5 and there is also an abundance of iron salts on the Eastern slopes. However, on the North Western slopes the hardness ranges from 20 to 30 French degrees, the pH is between 7 and 8 and the iron salts are considerably less (0.1 or 1 mg/l). In West Elba there are more permanent springs, especially at high altitude, and the water courses are longer. The pH here is between 6 and 7, and the hardness never exceeds 10 French degrees. The 53 points shown in the map (Fig. 1) were inspected several times, between 1956 and 1958. The lack of stations in the centre is due to the fact that during the sampling period the water courses had dried out. The examined biotopes included hygropetrics, drips, springs, puddles, brooks and streams of varying discharge capacity.

RESULTS

Adult and immature Trichoptera were collected at all stations and, at the same time certain environmental factors (Table I), as well as the bottom type, current velocity (Km/h), relative humidity (%), air temperature (°C), water temperature (°C), dissolved oxygen (% s.v.), hardness (Fr. degree) and pH were assayed. All these are very important factors in selection of habitat by the immature Trichoptera stages. The physical-chemical and faunistic characteristics, also, varied greatly with the type of soil on which the water was found; bottoms were muddy, slimey, granitic or lime rocks, stoney, pebbly or rich in decaying vegetation. The environmental characteristics, as already mentioned, covered a wide range in the various sectors of the island. The stations with the lowest dissolved oxygen and pH, but with the highest hardness are those of the mine zones. The granitic regions have a slightly acid pH and a minimum total hardness of between 2 and 10 French degrees.

The zone least populated by Trichoptera, other than the water-poor central area, is the East of Elba, particularly from the 1st to 12th and 45th to 53th stations. Whereas the zone with the greatest number of species is the West, especially the Monte Capanne hygropetric biotopes which are generally located at more than 200 metres above sea level (No. 26, 33, 34 and 16, 17, 18, 20, 22, 24, 35, 36). A total of 29 taxa[1] (Tab. II) were recorded, some only in the immature stages. The reophil and spring species are considerably more in evidence; and this, no doubt, is due to the predominating biotope type of the island. Monospecific populations, sometimes of high density, are found in certain spring environments: *Helycopsyche sperata* at the 4th and 45th stations; *Tinodes maclachlani* at the 44th and 52nd stations. *Sericostoma pedemontanum* at the 12th and *Beraea ilvae* at the 23rd are monospecific puddle populations. The species which are well represented throughout the island are *Stactobia furcata, Hydropsyche instabilis, Tinodes maclachlani, Sericostoma pedemontanum, Helycopsyche sperata, Odontocerum albicorne.* Those found in all the island but with lower populations in the West, or which are limited to the Monte Capanne mass and to the Monte Calamita promontory are *Rhyacophila italica ilvana, Wormaldia occipitalis, Micropterna* sp., *Stenophilax mitis.* The dense population species that inhabit only the West of the island are: *Wormaldia variegata denisi, Plectrocnemia geniculata, Tinodes waeneri, Micrasema togatum, Potamophylax cingulatus, Lepidostoma hirtum, Crunoecia irrorata, Mystacides azurea, Bereae ilvae.* Finally, a few specimens of the *Catagapetus nigrans, Agapetus cyrnensis, Hydroptila* sp., *Hydroptila vectis, Agraylea* sp., *Diplectrona magna,*

---

[1] A specimen collected at Station No. 24 on 3/10/58, probably belong to the *Oxytrichia* Mor. genus, is not included owing to its very bad condition.

ISLE OF ELBA

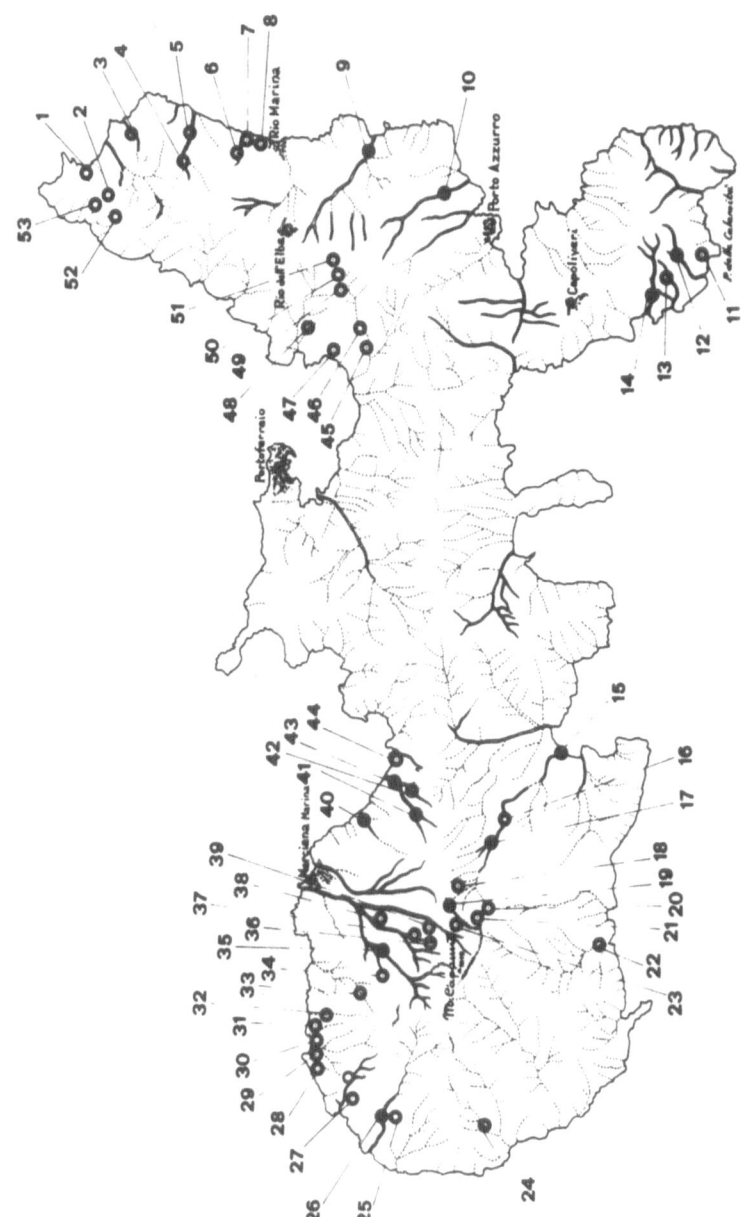

Fig. 1. Map of the Isle of Elba showing the 53 sampling stations.

251

Table I. Physico-chemical values registered at the 53 biotopes investigated, including certain environmental characteristics.

X = Stations with no Trichoptera.

Stat. without Trichoptera	Stat. No.	Altitude m sea level	ENVIRONMENT	Exposure	BED	DATE	R.U. %	AIR °C	WATER °C	O₂% S.V.	HARD-NESS	pH
X	1	30	mine dump (Fe mg/1 320)	E		6/8/56			29.0	51.0	310.0	2.6
	2	40	brook with pools	N	earth	28/8/57	65	24.5	18.5	93.4	28.0	7.0
						8/10/58		23.0	20.0	23.5	28.0	7.0
X	3	50	pool	NW	lime	28/8/57			20.0			7.5
	4	158	spring with pool	NE	rock musk	27/8/57	80	21.5	19.0	90.5	6.0	6.5
	5	50	ditch stones slow water (Fe 1 mg/1)	E	red lime	31/7/56			25.0	58.6	26.0	5.0
						6/8/56			23.5	67.8	18.0	
						27/8/57		24.5	22.0	81.4	32.0	4.5
X	6	60	ditch rich in water (Fe 10/mg/1)	SE	red lime stone	31/7/56			30.0	61.0	76.0	4.7
						27/8/57	70	26.0	20.0	91.5	22.5	6.5
X	7	86	pipe from mine	NE	sulphurous	27/8/57	75	26.0	20.5	75.1	40.0	7.0
X	8	3	drips (Fe 28 mg/1)	E	stones	31/7/56			19.0	30.0	108.0	3.1
	9	163	small falls, hygropetric pools			3/6/57			18.0	119.6	3.4	6.6
	10					13/7/56						
X	11	130	spring	SW	red earth	25.8.57	60	28.0	20.0	79.9	138.0	4.5
	12	170	residual pool (Fe 0.5 mg/1)	SW	lime vegetation	1/8/56			29.0	94.4	26.0	7.5
						4/6/57			18.2	126.4	14.1	6.5
						25/8/57		33.5	27.0	106.2	27.0	6.5
"	"	"	"	"	"	26/8/57			24.0	98.4	16.0	7.5
						26/8/57		25.0	23.5	0.0		6.5
						26/8/57	68	23.5	22.5	101.7	17.5	6.8
"	"		hygropetric	"	limestone	2/8/56						
						8/10/58			24.0	20.0	42.5	9.0	6.0
	13	200	hygropetric and pools (Fe 1 mg/1)	W	granite rock	1/8/56			22.5	70.0	13.0	7.0
						4/6/57			19.0	169.0	5.5	7.0
"	"		pool	"	lime and vegetation	25/8/57	37	33.5	22.5	78.3	5.0	6.5
						8/10/58		26.0	19.0	38.8	7.0	6.0
	14	49	hygropetric and elochrenic pools		rocks	1/6/57			21.3	101.7	4.5	6.0
						1/6/57			21.8	179.0	5.4	6.0
X	15	2	brook		stones	25/8/57	60	28.5	21.5	75.1	36.0	7.5
	16	77	small falls	SE	granite rocks	3/6/57			14.0	151.2	2.5	6.0
						24/8/57	95	23.0	19.0	101.0	3.0	6.0
	16	77	pool	"		24/8/57			19.5	83.6		6.0
						6/10/58		22.0	18.0	30.0	4.7	5.5
	17	250	pool under hygropetric (Fe 0.1 mg/1)	SE	musk	3/8/56			20.0	63.0	8.0	6.8
	18	640	hygropetric	SE	granite	22/8/57	68	23.5	15.0			5.9
	"	700	pool	S	granite	22/8/57	60	26.5	14.5	89.9	3.0	6.0
X	19	820	pool	N	granite and vegetation	22/8/57	52	22.0	14.0	92.0	2.5	6.0

Table I. (cont.)

X	No.		Habitat	Aspect	Substrate	Date						
	20	400	dripping spring	NE	granite	22/8/57	57	24.7	12.5	53.4	2.0	6.5
	21	630	canalized spring	NE	earth and vegetation	22/8/57	73	22.5	13.5	39.9	3.5	6.0
	"	"	"	"	"	3/10/58		18.0	14.0	14.5	3.2	6.0
	22	50	hygropetric	SW	granite	30/8/57			23.0	80.0	4.5	6.5
	"	"	ditch (c.v.2km/h)		granite rocks	30/8/57	60	25.0	22.0	87.2	4.0	6.0
						6/10/58		23.5	20.0	23.2	5.0	7.0
	23	750	pool (depth 3-4 cm)	N	granite and earth	22/8/57	60	19.5	15.5		2.0	6.0
	24	130	pools and hygropetric		granite	30/8/57						
						3/10/58		26.0	21.0	32.0	8.5	6.0
X	25	230	dripping spring (Fe 0 mg/l)	W	earth	28/7/56 20/8/57			20.0		7.5	7.0
	26	130	hygropetric (c.v.1.km/h)	SW	granite rocks	20/8/57		27.5	26.0	87.9	5.5	6.5
	"	"	pools (depth 20 cm)	"	leaves and detritus	20/8/57		27.5	28.0	103.9	6.5	6.5
	27	200	hygropetric (Fe 0.5-0.9 mg/l)	SW		28/7/56 20/8/57 30/8/57 3/10/58					10.0	7.0
X	28	100	small pool	W	lime	23/8/57						5.5
	29	40	pool (depth 30 cm)	N	granite and stones	23/8/57 3/10/58	55	30.0	21.0	65.0	7.5	7.0
	30	320	hygropetric and pools	N	limestone	23/8/57 3/10/58						6.3
X	31	6	pool	NW	black/lime	23/8/57						7.0
X	32	190	ditch		granite rocks	20/8/57 3/10/58			17.0	83.2	4.0	5.5
	33	340	hygropetric	NE		20/8/57 30/8/57 3/10/58		20.0	16.0	37.1	2.5	4.5
	34	375	ditch with pool (c.v.2.5km/h)	NE	granite rocks	28/7/56			21.5	65.0	4.0	6.0
						4/6/58			12.8	154.7	1.8	6.2
						19/8/57		20.0	18.0	57.9	5.0	6.0
						2/10/58		18.0	16.5	14.9	3.0	5.5
	"	"	pools (depth 20 cm)		stones and vegetation	19/8/57 31/8/57		20.0	18.0	53.9	5.0	6.0
	35	520	ditch	NW	limestone	21/8/57		20.5	10.1	96.7	2.5	6.5
	36	420	ditch	SW	granite	19/8/57	85	22.0	12.5	83.7	2.5	6.5
						4/10/58		18.0	15.0	29.4	4.0	6.0
	"	"	pools (depth 20 cm)	SW	stones and vegetation	19/8/57			12.5	75.4	2.5	6.5
	37	460	ditch			31/8/57						
						2/10/58		18.0	13.0	35.3	3.5	5.0
	38	250	Poggio	night collections								
	39	500	ditch		granite and musk	4/10/58		18.0	16.0	37.5	3.0	6.0
X	40	20	ditch			24/8/57						
	41	80	pools (depth 2 m) (Fe 0.1 mg/l)	NE	granite	27/7/56			24.0	65.0	8.0	7.1
						24/8/57	95	23.5	19.5	89.5	22.0	7.0

Table I. (cont.)

42	400	ditch with pools	N	limestone and vegetation	24/8/57	80	24.5	20.0	9.6	30.0	7.5
43	71	ditch	NE		3/6/57				11.4 105.1	12.2	7.0
44	100	ditch and pools	NE	granite	3/6/57						
45	40	spring	W	stones	29/8/57	42	32.0	20.5			7.0
					9/10/58		23.0	20.0	30.9	20.0	7.0
X 46		small pool		large-stones	29/8/57						7.0
X 47	30	pool	W	earth	29/8/57						7.5
X 48	86	spring with drips	W	rock	29/8/57	65	26.0	20.0	92.1	38.0	7.0
X 49	177	spring (Fe 0.1 mg/1)	SW	musk	31/7/56			23.0	66.0	34.0	7.7
					29/8/57			17.5	63.2	28.0	7.5
X 50	89	pool drips	SW	rock and earth	29/8/57			25.0	22.5 59.7	17.0	7.0
X 51	100	spring	E		29/8/57						6.5
52	155	canalized spring		little water	28/8/57						7.0
X 53	65	"	NE	" "	28/8/57	80	21.0	19.0			7.5

Table II. The 29 taxa found at the 53 biotopes arranged clockwise and numbered according to the direction of the slope.

0 = Immature stages.
+ = Adults.

Nₒ TAXA	SLOPE E										S													W			
STATIONS	1	2	3	4	5	6	7	8	9	10	11	12	13	14	15	16	17	18	19	20	21	22	23	24	25	26	27
1 Rhyacophila italica ssp.ilvana Moretti								o						+					o					o			
2 Catagapetus nigrans McL.											o																
3 Agapetus cyrnensis Mosely															o	o				o						o+	
4 Stactobia furcata Mosely											o+	+	+								o+			+		o+	
5 Stactobia moselyi Kimm																											
6 Hydroptila vectis Curtis																										o	
7 Agraylea sp.																											
8 Wormaldia occipitalis Pictet													+														
9 Wormaldia variegata ssp.denisi Moretti																				+				+			
10 Diplectrona magna Mosely																	o			o							
11 Hydropsyche instabilis Curtis							o	o							o+		o		o		o				o		
12 Hydropsyche pellucidula Curtis																					o				o		
13 Plectrocnemia geniculata McL.				o											o+	o			o	o					o		
14 Lype reducta Hagen																								+			
15 Tinodes maclachlani Kimmins		o					o				o+	o			o	o	+		o+		o				o	o	
16 Tinodes waeneri L.																											
17 Brachycentrus sp.																									o		
18 Micrasema togatum Hagen																									o		
19 Limnephilus gr.flavicornis Fbr.																	o										
20 Potamophylax cingulatus Steph.																									o		
21 Stenophylax mitis McL.									o						o		o										
22 Microptera sp.									o			o															
23 Lepidostoma hirtum Fbr.																								+			
24 Crunoecia irrorata Curt.																	o		o		o						
25 Mystacides azurea L.														+							o			+		o	
26 Sericostoma pedemontanum McL.	o+										o+				o	o			o+	o+	o+			+			+
27 Berara ilvae Moretti																+	o		o	o			+	+		o	
28 Helicopsyche sperata McL.	o+		+											+			o+		o	o+				+		o	
29 Odontocerum albicorne Scop.	+						o						+		o	o	o		o+		o			+		o	

Table II. (cont.)

N° TAXA \ STATIONS	SLOPE	N 28	29	30	31	32	33	34	35	36	37	38	39	40	41	42	43	44	45	46	47	48	49	50	51	52	53
1 *Rhyacophila italica ssp.ilvana* Moretti							o	+		o		+	+				o										
2 *Catagapetus nigrans* McL.								+																			
3 *Agapetus cyrnensis* Mosely																											
4 *Stactobia furcata* Mosely						o+																					
5 *Stactobia moselyi* Kimm						+																					
6 *Hydroptila vectis* Curtis						o+	o		o																		
7 *Agraylea* sp.						o																					
8 *Wormaldia occipitalis* Pictet															+											
9 *Wormaldia variegata ssp.denisi* Moretti							+		+	+																	
10 *Diplectrona magna* Mosely																											
11 *Hydropsyche instabilis* Curtis						o	o+	+	o+		+	+		o		o										
12 *Hydropsyche pellucidula* Curtis							o																				
13 *Plectrocnemia geniculata* McL.						o	o	o+		+		o															
14 *Lype reducta* Hagen						+	+																				
15 *Tinodes maclachlani* Kimmins		o				o	o+	o+	o		+	+		o		o			o				o			
16 *Tinodes waeneri* L.		o				+																					
17 *Brachycentrus* sp.																											
18 *Micrasema togatum* Hagen						o	o+		+	+																	
19 *Limnephilus gr.flavicornis* Fbr.		o																									
20 *Potamophylax cingulatus* Steph.		o					o	o		+																	
21 *Stenophylax mitis* McL.										+	+		o														
22 *Micropterna* sp.						o																					
23 *Lepidostoma hirtum* Fbr.																											
24 *Crunoecia irrorata* Curt.		o				+	+	o+			+																
25 *Mystacides azurea* L.		o				o+	+																				
26 *Sericostoma pedemontanum* McL.		o				+	o+	+	+																		
27 *Beraea ilvas* Moretti						o	+	+o		+																	
28 *Helicopsyche sperata* McL.						+								+													
29 *Odontocerum albicorne* Scop.		o				o	o+	o																			

Hydropsyche pellucidula, *Lype reducta*, *Brachycentrus* sp., *Limnephilus flavicornis* groups were noted at two or three western region stations.

The more interesting species from the taxonomic point of view are *Rhyacophila italica ilvana*, which has recently been described by Moretti; *Wormaldia*, which we classify as *variegata denisi* n.ssp., because of the

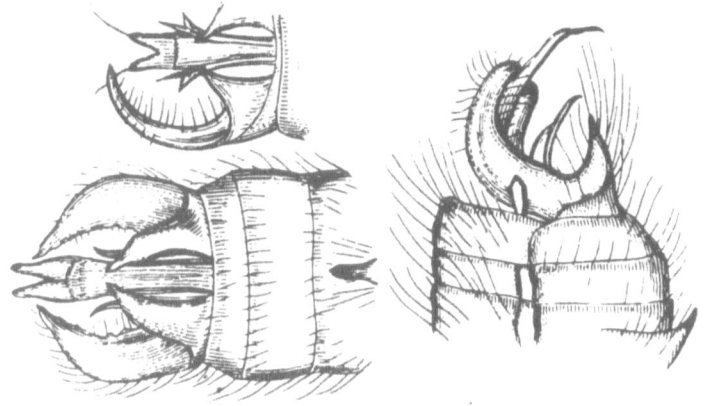

Fig. 2. H. sperata. Ventral process of the lower appendix of the genitalia. Note the two shapes of the appendix.

255

number of spines on the penis endotheca, not only varies from Kimmins' drawings in the shape of the 10th segment but, also, in the shape and position of the median spine of the penis endotheca. In *Hydropsyche instabilis* the extension of the 10th segment of the adult ♂ is longer than it is in *fulvipes* and the tracheogills are missing from the 7th larval segment. *Sericostoma pedemontanum* is typical, sensu Moretti; that is, there are two branches of different lengths, the upper always being longer than the lower, on the spina of the 10th segment. *Beraea ilvae* has 8 spines on the fleshy penis cover, as Moretti has recently shown. *Helycopsyche sperata* has two populations; one with a blunt triangular ventral process on the lower genital appendages (Fig. 2) and, the other, a low population, with the same process regularly pointed, as has already been reported for the Italian peninsula. Geonemically, *Agapetus cyrnensis* and *Woemaldia variegata* with a Sardinian-Corsican distribution; *Rhyacophyla italica*, with the exclusive presence of subspecies *ilvana; Catagapetus nigrans* and *Helycopsyche sperata*, which are strictly appenninic terms, are particularly worth of note. The presence of these species clearly demonstrates that, although the island has its own physionomy in the endemic terms (*Beraea ilvae, R. italica ilvana* and *Wormaldia variegata denisi* n. ssp.), it is a true meeting point for alpine and apennine systems. In conclusion, the wealth of exponents in an island of such modest dimensions (223.5 Kmq) is particularly worthy of note.

REFERENCES

Angelier, E., Vie et Milieu. Bull. Lab. Arago 5: 258–279, 1954.
Botosaneanu, L., Beitrage Zur Entomologie 6: 590–624, 1956.
Fischer, F.C.J., Nederl. Entom. Vereen, I-XV, 1960–73.
Giudicelli, J., Theses, Fac. Sc., Univ. Marseille, II–III: 1–437, 1968.
———, Ann. Fac. Sc., Univ. Marseille 43: 107–125, 1970.
Kumanski, K., Acta Zool. Bulg., Sophia 12: 58–66, 1979.
Lestage, J.A., *In* Les larves et nymphes aquatiques des insects d'Europe, pp 343–959, Lebegue, Bruxelles, 1921.
McLachlan, R., A Monographic Revision and synopsis of the Trichoptera of the European Fauna. Add. Suppl., Friedlander, Berlin, 1874–1884.
Moretti, G.P., Boll. Zool. Agr. Bachic. 10: 1–10, Univ. Milano, 1940.
———, Mem. Soc. Entom. Ital. 19: 259–291, 1941.
———, and Cianficconi, F., *In* Proc. of the 2nd Int. Symp. on Trichoptera, ed. M.I. Crichton, pp 7–30, Junk The Hague, 1977.
———, Cianficconi, F., Gianotti, F.S., Pirisinu, Q. and Viganó, A., Lav. Soc. Ital. Biogeor. I: 488–532, 1970.
———, Viganó, A. and Viganó Taticchi, M.I., *In* Proc. of the 1st Symp. on Trichoptera, ed. H. Malicky, pp. 87–92, Junk, The Hague, 1974.
Morton, K.J., Trans. Entom. Soc. 9: 75–82, 1893.
———, The Entom. Month. Mag. 70: 7, 1933.
Mosely, M.E., Eos, Rev. Esp. Entomol. 6: 147–184, 1930.
———, Eos, Rev. Esp. Entomol. 8: 165–184, 1932.
———, Proc. R. Ent. Soc., Lond. 6: 121–122, 1937.
Nielsen, A., Ann. Idrob., Suppl. 17: 255–631, 1942.
Perrin, M., Boll. Soc. Geolog. Ital. 94: 1929–1955, 1955.
Pirisinu, Q., Riv. Idrobiol. 9: 171–200, 1970.
———, Riv. Idrobiol. 13: 361–376, 1974.
Trevisan, L., Mem. Ist. Geolog., Univ. Padova 16, 1950.
Ulmer, G., Mittheil. Zool. Mus., Berlin 12: 266, 1926.
Viganó, A., Boll. mus. Zool. Univ. Torino 4: 25–32, 1974.

A PHYLOGENY AND CLASSIFICATION OF FAMILY-GROUP TAXA OF LEPTOCERIDAE (TRICHOPTERA)

J.C. MORSE

SUMMARY

Until now only four tribes have been defined in Leptoceridae, all in the sub-family Leptocerinae. Genera in Triplectidinae and most genera described since 1955 in Leptocerinae have never been assigned to tribes. Based on observations by Ulmer and on recent phylogenetic studies two of these four tribes (Mystacidini and Leptocerini) are paraphyletic groups. Consequently, as a conceptual framework for predictive biology and for future generic revisions, a new phylogeny and tribal classification are proposed for the family and its 45 currently recognized genera. The resulting eleven tribes in two subfamilies include six new family-group names.

INTRODUCTION

Over 900 species in 45 currently recognized genera have been described in the caddisfly family Leptoceridae from throughout the world. These vary considerably in size, color, structure, behavior, habitat, and distribution. Although much is known about many of these insects, few attempts have been made to synthesize this information into a broad conceptual framework. Such a phylogenetic classification is presented here in order to provide a theoretical foundation and a set of natural subunits for future revisions within the family and in order to present aquatic biologists with a hypothetical tool for projecting various ecological and physiological trends and relationships in this ubiquitous family.

Fig. 1 shows a hypothetical reconstruction of the evolution of the family, using simplified diagrams to depict some of the character changes through time relative to the ancestral character states shown at the base of the phylogeny. This figure illustrates the fundamental division of the Leptoceridae into two subfamilies, Leptocerinae Leach and Triplectidinae Ulmer. Monophyly for the Triplectidinae may be inferred by the loss of the primitive phallic parameres and marked reduction of the apical phalicata. Monophyly for the Leptocerinae may be inferred by the loss of one branch of the median vein (third cell) in the hindwing, by loss of the sectoral

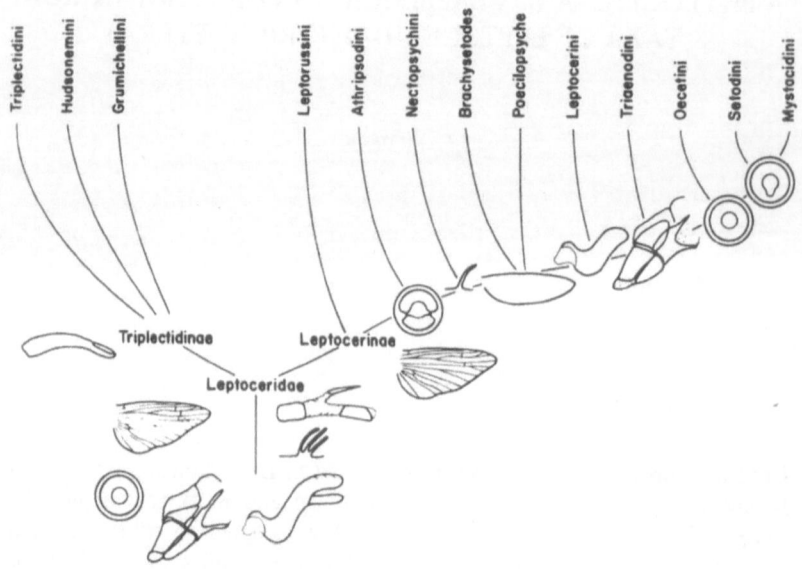

Fig. 1. Phylogeny of the tribes of Leptoceridae.

crossvein in the hindwing, and by reduction of the adult tibial spur formula from 2, 4, 4 to 2, 2, 4.

Mosely (1936) reviewed the subfamily Triplectidinae when only about half of our presently known 71 species had been described. Mosely and Kimmins (1953) reviewed the Australian-New Zealand portion of this fauna and Neboiss (1977) the Tasmanian species, adding most of the others to our present number. Neither they nor anyone else before now have grouped the genera into natural units.

Monophyly for the seven genera of Triplectidini can be hypothesized by their unique possession of an extra appendage which articulates at the base of each inferior appendage of the male genitalia (Mosely and Kimmins 1953, Fig. 134 ff.). Another four genera belonging to the new tribe Hudsonemini probably evolved from a single ancestor in which the male phallotremal sclerite developed into a pair of broad, vertical plates (Fig. 2). According to Ulmer (1955) and Flint (pers. comm.), *Leptocellodes* Ulmer is almost certainly a synonym of *Grumichella* Müller. Acting on those opinions, I recognize a new tribe Grumichellini in which the two included Neotropical genera share a synapomorphic 0, 2, 2 adult tibial spur count. The sister-group relationships of these three tribal lineages have not yet been resolved.

Fig. 2. Phallus of *Notalina fulva* Kimmins, left lateral view.

258

Based largely on his study of the immature stages, Silfvenius (1905) recognized three tribes in the subfamily Leptocerinae: Leptocerini, Mystacidini, and Oecetini. Ulmer (1955), also using larval and pupal characters, basically concurred with Silfvenius' classification and grouped the genera as follows:

Leptocerini Leach
    Leptecho
    Leptocerus (auctt. non Leach; = *Athripsodes + Ceraclea)*
    Leptocerodes
    Leptocella
    Nectopsyche
    Parasetodes
Mystacidini Burmeister
    Mystacides-Group
        Mystacides
        Tagalopsyche
        Setodes (including *Leptocerus* Leach)
        Trichosetodes
    Triaenodes-Group
        Adicella
        Erotesis
        Triaenodes
Oecetini Silfvenius
    Oecetis
    Oecetodella
    Paroecetis
    Potamoryza

In his discussion, Ulmer noted that the tribe Oecetini was a homogeneous group, but that the genera of Leptocerini were classified together on the basis of negative characters (*i.e.,* character states ancestral to those in the rest of the subfamily) and that Mystacidini was a highly heterogeneous assemblage of genera. Fig. 3 is a cladogram from Fig. 1 of the Silfvenius-Ulmer Leptocerinae

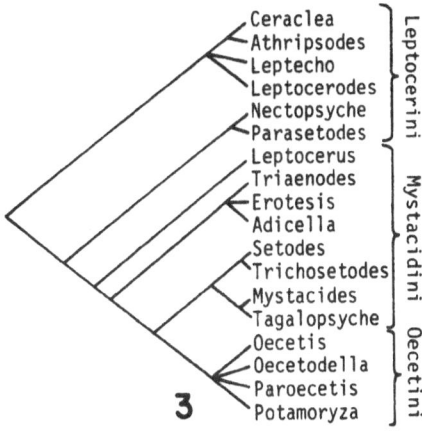

Fig. 3. Silfvenius-Ulmer classification of Leptocerinae genera represented in a cladogram from Fig. 1.

259

tribes and their genera. (Note that *Leptocella* now is considered a synonym of *Nectopsyche* and thus is not shown as a distinct lineage. Also observe that *Ceraclea* and *Athripsodes* [together = *Leptocerus auctt. non* Leach] and *Leptocerus* Leach are shown as distinct lineages, but in the tribes to which Silfvenius and Ulmer assigned them.) The Silfvenius-Ulmer tribe Oecetini is indeed monophyletic. However, their tribe Leptocerini (besides not including the type genus) is paraphyletic, as is Mystacidini. Except for the addition of the more recently defined tribe Athripsodini and the accomodation of modern generic synonyms, Wiggins' (1977, p. 46) classification followed the earlier ones, including the same two paraphyletic tribes.

You may well ask, "So what is the problem with paraphyletic taxa?" This philosophical issue is a matter of great controversy, often debated with vehemence in literature and in scientific meetings. Briefly, my position is as follows: (1) Paraphyletic taxa (groups which exclude selected descendant lineages) have very little utility for predictive biology. In other words, any prediction for ecological or physiological or other characters in the lineages of the paraphyletic group must also hold for each excluded lineage. (2) Conversely, if a derived state of some character is inferred for the ancestor of a particular excluded lineage, that inference is always relative to the more ancestral state of the character seen in the next antecedent sister lineage. Reference to a named group of arbitrarily selected lineages further antecedent is hardly necessary and potentially confusing. (3) By assigning to a descendant lineage a categorical rank equivalent to that of its paraphyletic ancestor group, the natural hierarchal classification is disrupted, thus unnecessarily complicating our essential biological information storage and retrieval system. A familiar example which illustrates these points is paraphyletic superfamily Rhyacophiloidea (Ross, 1967). Respectively, (1) no general predictions for as-yet-unknown biological characteristics of Rhyacophilidae and Glossosomatidae and Hydroptilidae can be made which must not also hold true for the excluded lineage Limnephiloidea, such that the term "Rhyacophiloidea" has very little predictive value. (2) On the other hand, the construction of tubular cases by larvae of the limnephiloid ancestor is seen as a derived behavior pattern relative to the construction of purse cases by larvae of Hydroptilidae. Reference to glossosomatid saddle cases or to rhyacophilid caselessness is interesting; but including one or both groups or the net-spinning lineages in a named taxon with hydroptilids is an unnecessary, arbitrary decision. (3) The fact that excluded, descendant superfamily Limnephiloidea has the same categorical rank as ancestral Rhyacophiloidea can be confusing for someone unfamiliar with trichopteran phylogeny. If such a person wants to assemble literature on feeding habits, for example, in Rhyacophiloidea, he may inadvertently omit consideration of the subject in limnephiloid species simply because they are not shown in classifications as belonging to a subunit of Rhyacophiloidea.

Adhering to these arguments, the following revised classification accomodates the new assessment of the evolution of the Leptocerinae shown in Fig. 1.

(A) In the ancestor at the node from which Athripsodini arose, (i) ninth sternal and phallic sclerotized strips developed in the genital chamber to brace

260

the male phallus basally (these structures are described in more detail by Morse, 1975), and (ii) adult tibial spurs were reduced to a 2, 2, 2 formula.

(B) In the Nectopsychini node ancestor, (i) larval gills lost the branched condition, (ii) one branch of the female median vein (3rd cell) was lost in the forewing, and (iii) adult tibial spurs were reduced to a 1, 2, 2 formula.

(C) The hindwings became narrower in the ancestor of the higher Leptocerinae. *Brachysetodes* and *Poecilopsyche* belong with these more apomorphic genera, but their sister-group relationships are not yet known.

(D) The ancestor at the Leptocerini node, (i) lost the apical segment (harpago) of the male inferior appendage and (ii) lost the adult midcranial sulcus.

(E) In the ancestor at the Triaenodini node, (i) the mesopleural katepisternum became constricted dorsally, obscuring the plesiomorphic suture which had closed it, and (ii) the male tenth tergites fused dorsally.

(F) In the ancestor at the node from which the Oecetini arose, (i) the male ninth sternal and phallic sclerotized strips were lost, superficially reclaiming the plesiomorphic, unbraced condition.

(G) In the ancestor at the Setodini-Mystacidini node, (i) a ventral phallic sclerotized strip developed to brace the male phallobase against the fused bases of the inferior appendages.

Returning now to the individual tribes of Leptocerinae, the very old, and now interestingly modified, monobasic genus *Leptorussa* constitutes the new tribe Leptorussini, whose ancestor apparently lost the apical segment (harpago) of the male inferior appendages (Mosely and Kimmins, 1953, Fig. 192).

Six genera assigned to Athripsodini arose from an ancestor in which the larva became much shorter and remarkably broad in the metathorax (Lepneva, 1966, Fig. 716A; Resh, 1976, Fig. 1).

The ancestor of the new tribe Nectopsychini lost the stems of the sector (radial sector *auctt.*) and median veins in the hindwing (Marlier, 1962, Fig. 89C).

The single genus *Leptocerus* constitutes the tribe Leptocerini. The tarsus of the larval middle leg became curved and its tarsal claw became markedly hooked in the ancestor of this genus (Wiggins, 1977, Fig. 9.2A).

Six genera assigned to the new tribe Triaenodini arose from an ancestor in which the cubital vein in the hindwing lost its apical fork 5 (Marlier, 1962, Figs. 89A and 89B).

The ancestor of the Oecetini underwent several larval and pupal modifications and the posterior branch of the median vein in the forewing came to be in unusually close association with the anterior branch of cubitus (Betten, 1934, Fig. 23).

The first branching of sector in the hindwing of the ancestor of the new tribe Setodini moved apically, beyond the fork of S_3 and S_4 (fork 2; or, the origin of the open discal cell moved apically beyond the radio-median crossvein *auctt.;* Marlier, 1962; Figs. 89H and 89K).

Two genera assigned to Mystacidini arose from an ancestor in which the ventral apex of the ninth segment became produced posteriorly (Kimmins, 1963, Figs. 56, 58, 80 and 82).

Thus, the 45 currently recognized genera of Leptoceridae may be classified in the following family-group taxa:

Triplectidinae Ulmer
    Grumichellini, new tribe
        Atanatolica, Grumichella
    Hudsonemini, new tribe
        Condocerus, Hudsonema, Notalina, Triplexa (= *Gracilipsodes* Sykora, new synonym)
    Triplectidini Ulmer
        Lectrides, Notoperata, Symphitoneuria, Symphitoneurina, Triplectides, Triplectidina, Westriplectes
Leptocerinae Leach
    Leptorussini, new tribe
        Leptorussa
    Athripsodini Morse and Wallace
        Athripsodes, Axiocerina, Ceraclea, Leptecho, Leptocerina, Leptocerodes
    Nectopsychini, new tribe
        Nectopsyche, Parasetodes
    Leptocerini Leach
        Leptocerus
    Triaenodini, new tribe
        Adicella, Allosetodes, Erotesis, Triaenodella, Triaenodes, Ylodes
    Oecetini Silfvenius
        Oecetinella, Oecetis, Oecetodella, Paroecetis, Pseudosetodes, Ptochoecetis, Setodellina, Setodina
    Setodini, new tribe
        Episetodes, Hemileptocerus, Setodes, Trichosetodes
    Mystacidini Burmeister
        Mystacides, Tagalopsyche
    Incertae sedis
        Brachysetodes, Poecilopsyche

Evidence has been presented elsewhere (Morse 1978) that the Leptoceridae are at least 65 million years old. Some further implications for the age of the family are evident from the phylogeny outlined above (Fig. 4). Although the sister-group relationships for the lineages of Triplectidinae have not been resolved, it is fairly obvious that this subfamily and the primitive leptocerine tribe Leptorussini represent an Australian-Neotropical biogeographic region distribution. No additional Australian species of Leptocerinae appear until the time of origin of Leptocerini. This implies that Australia was effectively isolated from other habitable land masses between the times of the Leptorussini node and the Leptocerini node and that most of the Australian Leptocerinae arose from Asian immigrants. If this were so, and if the recent estimates for the separation of Australia from Antarctica-South America are correct (Smith and Briden, 1977), then the Leptorussini node ancestor apparently lived at least 40 million years ago. Since Antarctica has been situated over the South Pole for the past 80 million years, the Australian fauna may have been effectively isolated by ice since the late Cretaceous Period. It is much more problematic to project when the Australian plate became close enough to Southeast Asia to begin receiving Leptocerinae immigrants.

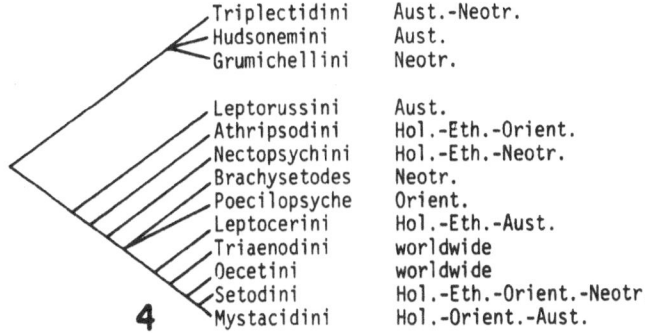

Triplectidini	Aust.-Neotr.
Hudsonemini	Aust.
Grumichellini	Neotr.
Leptorussini	Aust.
Athripsodini	Hol.-Eth.-Orient.
Nectopsychini	Hol.-Eth.-Neotr.
Brachysetodes	Neotr.
Poecilopsyche	Orient.
Leptocerini	Hol.-Eth.-Aust.
Triaenodini	worldwide
Oecetini	worldwide
Setodini	Hol.-Eth.-Orient.-Neotr.
Mystacidini	Hol.-Orient.-Aust.

4

Fig. 4. Biogeographic region distribution of tribes of Leptoceridae represented in a cladogram from Fig. 1. Aust. = Australian, Eth. = Ethiopian, Hol. = Holarctic, Neotr. = Neotropical, Orient. = Oriental.

ACKNOWLEDGEMENTS

Many useful discussions with the late H.H. Ross assisted the author in gathering evidence for the hypotheses expressed herein.

This study was supported in part by the National Science Foundation and in part by the U.S. Department of Energy's Savannah River National Environmental Research Park, Aiken, South Carolina.

This Technical Contribution No. 1839 is published by permission of the Director, South Carolina Agricultural Experiment Station.

REFERENCES

Betten, C., Bull. New York State Mus. 292: 1–576, pls. 1–67, 1934.
Kimmins, D.E., Bull. Brit. Mus. (Nat. Hist.), Entomol. 14: 263–316, 1963.
Lepneva, S.G., Fauna of the U.S.S.R., Trichoptera, Vol. II, 2, 1966.
Marlier, C., Ann. Mus. R. Afr. Centr., Sci. Zool. 8: 1–261, 1962.
Morse, J.C., Contr. Amer. Entomol. Inst. 11: 1–97, 1975.
———, *In* Proc. 2nd Int. Symp. Trichoptera, ed. M.I. Crichton, pp 199–206, Junk, The Hague, 1978.
Mosely, M.E., Trans. R. Entomol. Soc. London 85: 91–129, 1936.
———, and Kimmins, D.E., The Trichoptera (caddis-flies) of Australia and New Zealand. British Museum, London, 1953.
Neboiss, A., Mem. Nat. Mus. Victoria 38: 1–208, pls. 1–3, 1977.
Resh, V.H., Ann. Entomol. Soc. Amer. 69: 1039–1061, 1976.
Ross, H.H., Ann. Rev. Entomol. 12: 169–206, 1967.
Silfvenius, A.J., Acta Soc. Fauna Flora Fenn. 27: 1–168, 1905.
Smith, A.G. and Briden, J.C., Mesozoic and Cenozoic paleocontinental maps. Cambridge Univ. Press. Cambridge, 1977.
Ulmer, G., Arch. Hydrobiol., Suppl.-Bd. 21: 408–608, pls. 10–17, 1955.
Wiggins, G.B., Larvae of the North American caddisfly genera (Trichoptera). Univ. Toronto Press, 1977.

DISCUSSION

Moretti: Your classification is very interesting, but don't you think the larval forms would aid in the validation of Leptoceridae tribes and subfamilies? Morse: Yes, I certainly do. As I explained, several larval characteristics are utilized in this phylogeny; for example, the reduced gills of the Nectopsychini node ancestor, the broadened metathorax of the Athripsodini ancestor larva, and the highly modified foretarsi and claws of the Leptocerini ancestor larva. Characteristics from as many life history stages as possible should always be considered.

DISTRIBUTION OF TRICHOPTERA FAMILIES IN AUSTRALIA WITH COMMENTS ON THE COMPOSITION OF FAUNA IN THE SOUTH-WEST

A. NEBOISS

SUMMARY

This review gives a brief description of the three faunal provinces of Australia and characterises their caddis-fly faunas by describing family representation in each. A more detailed account of the caddis-fauna of south-west Western Australia is given; the concept of refuge area is invoked to account for its composition.

INTRODUCTION

The only previous attempt to analyse the distribution and relationships of Australian caddis-flies is that of Mosely and Kimmins (1953); the lack of subsequent studies is probably a direct consequence of the paucity of information available until now. The data used for this review are based on extensive collections accumulated during the last 10 to 15 years from localities throughout the entire continent at various seasons; they can therefore be regarded as reasonably representative.

The knowledge of Australian species and their distribution has increased dramatically with every survey conducted. This was illustrated by the results obtained in a recent study of Tasmanian caddis-flies where the recorded number of 18 species in 7 families was reached in 1936 (Mosely, 1936), 58 species in 13 families in 1953 (Mosely and Kimmins, 1953) but 157 species in 21 families in 1977 (Neboiss, 1977). Similarly, the present study of south-west Australian Trichoptera shows an increase from 18 species in 6 families to 43 species in 9 families (Neboiss, in preparation).

This increased knowledge of species has led to the elucidation of family relationships and amendments to certain groups have been necessary. Mosely and Kimmins (1953) recorded 17 families with 69 genera from Australia; since then some families have been subdivided, new families described and others replaced so that, at present, 24 families with 91 genera are recognized; their distribution is summarized in Table I. The total numbers given include several unpublished genera; the classification is based on the proposal by Ross (1967), with amendments by Schmid (1970), Malicky (1973) and Neboiss (1977).

Table I. Number and distribution of genera

Provinces / Families	Australian total	Eyrean province	Torresian province North-west	Torresian province North-central	Torresian province North-east	Bassian province South-east	Bassian province Tasmania	Bassian province South-west
Superfamily Rhyacophiloidea								
Glossosomatidae	2	–	–	–	1	2	1	–
Hydroptilidae	10	2	3	2	7	6	6	4
Hydrobiosidae	14	–	–	–	6	12	10	2
Superfamily Hydropsychoidea								
Philopotamidae	2	–	1	1	2	2	1	1
Stenopsychidae	1	–	–	–	1	1	1	–
Hydropsychidae	8	1	2	2	6	6	4	2
Ecnomidae	2	1	2	2	2	2	2	2
Polycentropodidae	7	–	1	1	4	4	3	2
Superfamily Limnephiloidea								
Chatamiidae	1	–	–	–	–	1	–	–
Tasimiidae	2	–	–	–	–	2	2	–
Limnephilidae	1	–	–	–	–	1	1	–
Oeconesidae	1	–	–	–	–	–	1	–
Kokiriidae	3	–	–	–	–	1	2	–
Plectrotarsidae	3	–	–	–	–	1	3	1
Conoesucidae	6	–	–	–	3	6	6	–
Antipodoecidae	1	–	–	–	–	1	–	–
Calocidae	3	–	–	–	3	3	3	–
Helicophidae	2	–	–	–	–	2	2	–
Odontoceridae	2	–	–	–	1	2	–	–
Atriplectididae	1	–	–	–	1	1	1	–
Philorheithridae	5	–	–	–	2	4	5	1
Helicopsychidae	1	–	1	1	1	1	1	–
Calamoceratidae	1	–	1	1	1	1	1	–
Leptoceridae	12	2	4	6	9	10	11	8
Total:	91	6	15	16	50	72	67	23
Number of families:	24	4	8	8	16	23	21	9

FAUNAL PROVINCES

To appreciate the relationships and geographic distribution of south-west Western Australian caddis-flies, it is necessary to look at the faunal provinces of Australia. Spencer (1896) proposed and characterised three such provinces (sub-regions) based on the presence of a limited group of animal species. Subsequently amendments and modifications have been proposed, but the basic concept of distinguishing the peripheral provinces from interior provinces has remained unchanged (Sloane, 1915; Iredale, 1929; Keast, 1959, 1961; Paramonov, 1959; Mackerras, 1970; Horton, 1973; Serventy and Whittell, 1976).

The development and distribution of vegetation, on which animal communities depend, is influenced by the interaction of temperature, rainfall and runoff conditions, and together with surface physiography they create certain climatic zones with distributional barriers and refuge areas (Keast, 1961; Koch, 1977). These conditions also govern the flow of water in streams and levels in lentic water bodies. Distribution of animal species is restricted by their needs and tolerances to numerous variables (Keast, 1959). Rainfall is one of the key factors which limits distribution by affecting the availability of suitable food.

In Australia the average rainfall gradually increases towards the margins of the continent and the isohyets show a somewhat concentric zone pattern. However, the low rainfall arid areas extend to the coast in the west and south central regions, creating an effective climatic barrier.

Eyrean province. The boundary between the large arid Eyrean province in the centre of the continent and the other two provinces, which occupy the higher rainfall, marginal zone correspond approximately to the 500 mm/annual isohyet (Fig. 1). The rainfall in the Eyrean province is low and irregular, both annually and seasonally. As a result, most of the province is desert, saltbush semi-desert or dry grassland, and also includes marginal areas of mallee and mulga scrub and some woodland (Gentilli, 1968).

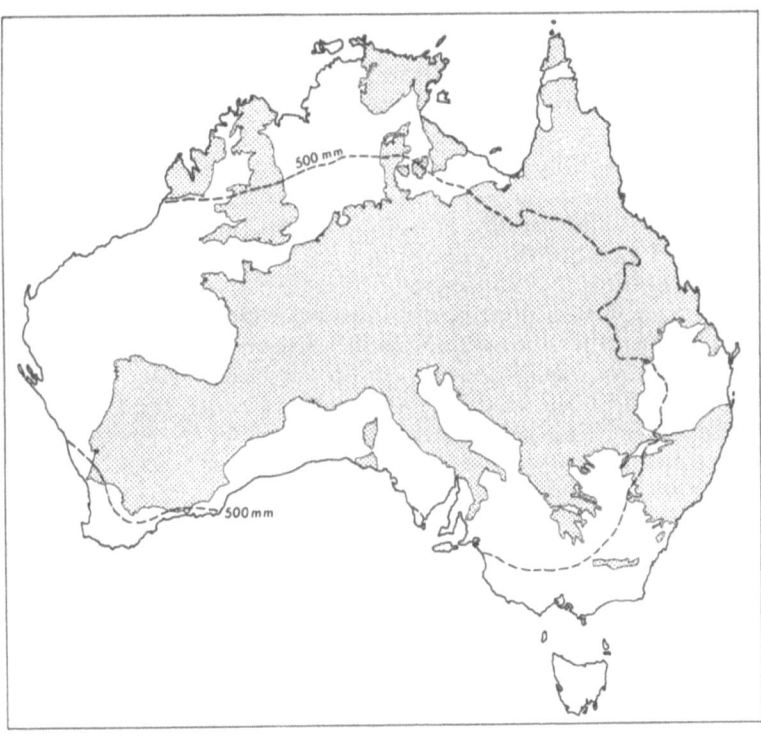

Fig. 1. Average annual rainfall 500 mm isohyet in Australia. Contour of Europe of the same scale included for comparison.

267

The insect fauna is rather restricted and not dominated by any particular element. It is composed of specialized and adapted offshoots derived from various groups (Mackerras, 1970). The majority of water bodies are either lentic or ephemeral in character, and consequently the aquatic insect fauna is adapted to these conditions.

Although it is the largest faunal province, its caddis-fly fauna is the most deficient in family and species diversity. Only 4 families, containing some 10 to 15 species, are known from this enormous area, larger than Europe. Among these the Leptocerid genera *Oecetis* and *Triplectides* dominate the entire fauna. Some species in the genus *Oecetis* are known to survive arid desert conditions in Central Australia. Two other families are also widely distributed, the Economidae (*Ecnomus*) and Hydropsychidae (*Cheumatopsyche*), although with only a few species each. The fourth family, Hydroptilidae is somewhat marginal in distribution and recorded from disjunct localities.

Torresian province. This region includes the north-western (Kimberleys), north-central and north-eastern coastal areas, which correspond to the distribution of northern woodlands and savannah with some tropical rainforests in the east (Gentilli, 1968). The moist, tropical and subtropical climate is dominated by a monsoonal rain pattern, with maximum precipitation in the summer months between November and April. The southern boundary, between the Torresian and Bassian province, is usually placed just south of the Clarence River in northern New South Wales. However, the change from northern to southern species in caddis-flies has been noticed in Townsville – Rockhampton area rather than further south. This has been illustrated in a recent study of the genus *Anisocentropus* (Neboiss, 1980a). The narrow stretch of dry savannah near Townsville was considered by Keast (1961) as a faunal barrier, and, no doubt, has a considerable influence on caddis-fly distribution. Kikkawa and Pearse (1969) and Horton (1973) demonstrated that this area more likely is the boundary between the Torresian and Bassian provinces.

The fauna, including the insects, is dominated by northern elements; the Papuan influence is more noticeable in the north-east.

The Torresian caddis-fly fauna is also dominated by the Leptoceridae, however it shows higher generic and species diversity and close affinities to the northern faunal elements. The genus *Oecetis* retains its rather dominant position with an increased number of species; the genus *Triplectides* constitutes only a small proportion of the fauna, but a more prominent place is shared by a number of *Triaenodes* and *Leptocerus* species.

Of the other families present, it appears that the majority of species are in Economidae (*Ecnomus* and *Ecnomina*), followed by Hydroptilidae (several genera), with a smaller number in Philopotamidae (*Chimarra*) and Calamoceratidae (*Anisocentropus*). There appear to be even fewer species in Hydropsychidae (*Cheumatopsyche*), Polycentropodidae (*Nyctiophylax*) and Helicopsychidae (*Helicopsyche*).

The eastern section of the province, with its rainforests, adds another dimension to the fauna. Families otherwise well represented in cooler, more southern localities, include Hydrobiosidae, Glossosomatidae, Stenopsychidae, Calocidae, Atriplectididae and Philorheithridae. To the warm adapted northern elements the recently discovered *Aethaloptera* (Barnard, in preparation) and *Hyalopsyche* (Neboiss, 1980b, in press) are now added. The family representation increases from 8 in the north-western and north-central section to 16 in the north-eastern section of the province.

Bassian province. This area with its cool, moist temperate climate includes the south-eastern coastal areas, Tasmania and a small isolated area in the south-western corner of the continent which has been regarded by some authors as part of the Eyrean province (Spencer, 1896), or as a separate province by others (Iredale, 1929). The vegetation types are dominated by wet and dry sclerophyll forests, some woodland and mountain moors, except for the south-western section, where the vegetation is almost entirely wet and dry sclerophyll forests, with some heathlands. Rainfall, although distributed throughout the year, has maximum falls in winter months (May to October). Permanent streams, as well as permanent and ephemeral lentic water bodies, are found throughout the province. The insect fauna is dominated by southern elements with strong trans-antarctic relationships; although cold adapted, many forms can tolerate relatively high temperatures during their immature stages (Mackerras, 1970).

The caddis-fly fauna of the Bassian province includes all 24 families known from Australia. However, the distribution within the province is not uniform and three separate regions can be recognized. The Bassian province proper, located at the south-east of the continent, supports 23 families, estimated to contain about 200 species. Only the family Oeconesidae is absent. To the south, the second region, Tasmania, has representatives of 21 families, about 160 species, of which 74% are endemic (Neboiss, 1977). The third region is the south-west Western Australia, from which only 9 families with 43 species are known (Neboiss, in preparation). Here also about 70% of species are endemic. Common to all three regions are the trans-antarctic faunal elements the Hydrobiosidae and Philorheithridae. Most of the Bassian province is regarded by Keast (1961) as a refuge area. The south-east, with its more favourable climate supports the richest caddis-fly fauna in Australia, whereas the south-west shows considerably lower family and species diversity.

Refuge areas and distributional barriers. The refuge area concept and faunal barriers have been used widely to explain the presence and diversity of various animal groups — Australian birds by Keast (1961), Tabanid flies by Mackerras (1962) and scorpions by Koch (1977). Refuge areas have more hospitable conditions and therefore attract and support certain animal populations. These conditions are usually found in mountainous areas, which have relatively high rates of precipitation, richer, more diverse vegetation and generally are more suited for habitation than the surrounding arid land. Almost all major refuge areas, as described by Keast (1961) in his study of Australian birds, are found within, and occupy extensive areas of Torresian and Bassian faunal provinces (Fig. 2).

Mackerras (1962) discussed the effect of climatic conditions during Pleistocene glacial phases, which resulted in shifting and compression of climatic zones in equatorial direction. This period provided a broad, fertile

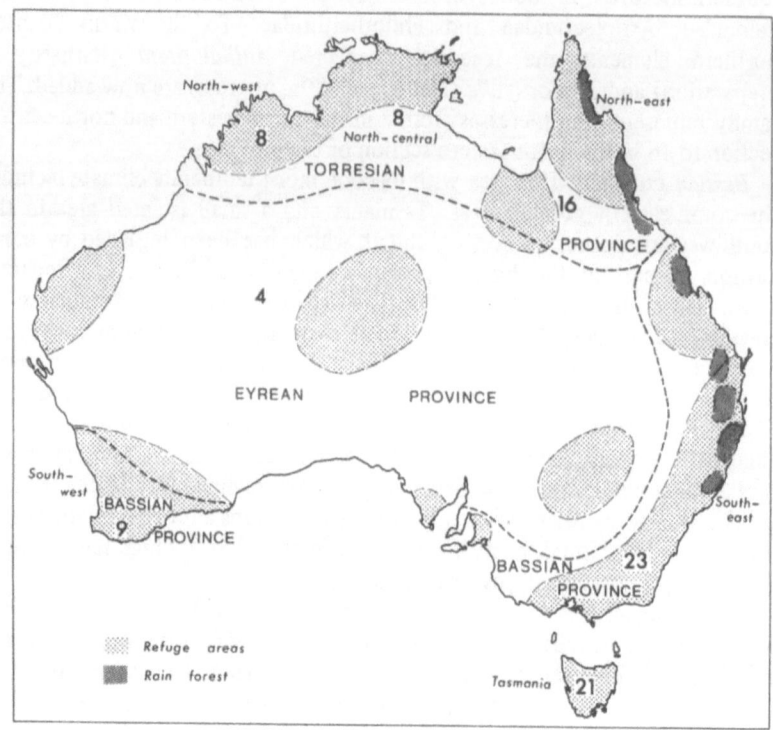

Fig. 2. Faunal provinces of the Australian region. Trichoptera families indicated by large numerals; shaded areas show refuge areas and rainforests as described by Keast (1961).

corridor between the south-eastern and south-western parts of Australia which may have allowed free movement of species in both directions; a narrow, effective desert barrier persisted in the north. During the post-glacial period the aridity extended southward and finally, by reaching the sea in the south created the present effective barrier between east and west.

The composition of south-western fauna. The area here referred to as the south-western section of the Bassian province, is part of the south-western refuge area. Undoubtedly, the surrounding tracts of arid land have restricted movement of certain animal groups and are a strong factor determining the present composition of the south-western caddis-fly fauna.

Representation of the superfamily Rhyacophiloidea is limited to two families – Hydroptilidae and Hydrobiosidae; the family Glossosomatidae is absent. The Hydroptilidae are known by four genera, one of which, based on a single species and found at the extreme south-west, is endemic. Three species in other genera are also endemic and only two are widespread, with distributions ranging across the southern part of the continent. The Hydrobiosidae is represented by two species, one in each subfamily; both are endemic and confined to cool, fast running streams.

Of the five families in the superfamily Hydropsychoidea, four are represented in the south-western fauna by 15 species; the family Stenopsychidae

270

is absent. The families Philopotamidae and Polycentropodidae with two species each, are confined to the cool, fast running coastal streams as are the Hydrobiosidae. The family Ecnomidae is well represented with 9 species, of which 2 in the genus *Ecnomus* are known from many localities in other parts of the continent; the 7 species in the genus *Ecnomina* are endemic and rather restricted in distribution. The family Hydropsychidae with two species in two genera: *Cheumatopsyche* and *Smicrophylax*, the former is widespread, the latter endemic.

The superfamily Limnephiloidea has a very limited family representation where out of 16 only 3 families are present. In addition, two families, Plectrotarsidae and Philorheithridae are known only by a single, rare and local species. The third family Leptoceridae, on the other hand, has a far higher diversity with 18 species in 9 genera, some of which have distributions extending across the entire continent.

ACKNOWLEDGEMENTS

The author is grateful for the extensive material made available for study by many individuals and institutions, particularly to the Australian National Insect Collection, Canberra and the Australian Museum, Sydney. Valuable information was obtained from the expedition material collected at many North-western, Northern and Central Australian localities and donated to the National Museum of Victoria by Mr and Mrs Max Moulds of Sydney and Mr J. Blyth of Melbourne.

REFERENCES

Gentilli, J., Sun, Climate and Life, Jackaranda Press, Brisbane and Melbourne, 1968.
Horton, D., Syst. Zool. 22: 191–195, 1973.
Iredale, T., Rep. Australas. Ass. Advmt. Sci. 1928: 244–245, 1929.
Keast, A., *In* Biogeography and Ecology in Australia, eds A. Keast,; R.L. Crocker, and C.S. Christian, pp 115–135, Junk, The Hague, 1959.
———, Bull. Mus. comp. Zool. Harvard. 123: 305–495, 1961.
Kikkawa, J. and Pearse, K., Aust. J. Zool. 17: 821–840, 1969.
Koch, L.E., Rec. West. Aust. Mus. 5: 83–367, 1977.
Mackerras, I.M., *In* The Evolution of Living organisms, Roy. Soc. Victoria, Melbourne, University Press, 1962.
———, Insects of Australia, CSIRO. Melbourne, University Press, 1970.
Malicky, H., Handbuch der Zoologie. 4(2)2/29. Walter de Gruyter, Berlin, 1973.
Mosley, M.E., Proc. Zool. Soc. London. 1936: 395–424, 1936.
——— and Kimmins, D.E., The Trichoptera (caddis-flies) of Australia and New Zealand, British Museum of Natural History, London, 1953.
Neboiss, A., Mem. Natl. Mus. 38: 1–208, 1977.
———, Aust. J; Mar. Freshwater Res. 31: 193–213, 1980a.
———, Archiv. für Hydrobiol. 1980b, (in press).
Paramonov, S.J., *In* Biogeography and Ecology in Australia, eds A., Keast, R.L. Crocker and C.S. Christian, pp 164–191, Junk, The Hague, 1959.
Ross, H.H., Ann. Rev. Ent. 12: 169–203, 1967.
Schmid, F., Mem. Ent. Soc. Canada 66: 1–230, 1970.

Serventy, D.L. and Whittell, H.M. Birds of Western Australia 5th edit., University of Western Australia Press. Perth, 1976.

Sloane, T.G., Proc. Roy. Soc. Vic. 28: 139–148, 1915.

Spencer, W.B., Report on the Work of the Horn Scientific Expendition to Central Australia, Vol. 1, Dulau and Co, London, 1896.

ON THE EVOLUTION OF THE PHALLUS AND OTHER MALE TERMINALIA IN THE HYDROPSYCHIDAE WITH A PROPOSAL FOR A NEW GENERIC NAME

A. NIELSEN

SUMMARY

The plesiomorphic structure of the phallus is discussed. In *Cheumatopsyche* and in *Hydropsyche silfvenii* Ulmer some plesiomorphic characteristics are probably found, whereas the apparently more simple conditions in other species of *Hydropsyche* are considered as the end result of an evolution.

Structures, interpreted as epiproct and paraprocts, are described in the two forms mentioned, and in *Cheumatopsyche* also the cerci. The latter are of very rare occurrence among male holometabolous insects.

The plesiomorphic structure of the trichopteran aedeagus is discussed.

The subgenus *Ceratopsyche* (Ross and Unzicker, 1977) is raised to generic status. It includes the European species *Hydropsyche silfvenii* Ulmer and *H. nevae* Kolenati.

McFarlane (1976) and Ross and Unzicker (1977) have proposed to use phallic structures as a means of separating genera within the subfamily Hydropsychinae. I approve of this procedure, with the addition that other terminalia, as well as female structures, should also be included.

The probably plesiomorphic structure of the phallus in Trichoptera (Nielsen, 1957a) as well as in most other holometabolous insects (Nielsen, 1957b) is that it consists of a phallobase, an aedeagus (phallicata Morse) and one pair, exceptionally two pairs (1957a, pp. 104–05), of parameres (endothecal processes Morse). The distal end of the phallobase is invaginated as an eversible endotheca, from the bottom of which the aedeagus arises, flanked by the parameres. In the majority of Trichoptera, however, the phallus is undivided, and this is the case in *Hydropsyche*.

In this genus (Nielsen, 1957a, pp. 66–67) the gonopore is in the bottom of a cavity, flanked by two pairs of immovable lips, smaller internal ones and larger external ones (in 1957a, p. 66, Fig. 36 labelled 4 and 5). McFarlane considers this a primitive condition. Like Ross and Unzicker I think that it is the end result of an evolution, "a simplicity arisen by specialization" (1957a, p. 63).

Probably plesiomorphic conditions in the Hydropsychidae are found in *Cheumatopsyche* and in *Hydropsyche silfvenii* Ulmer, though in the former the whole aedeagus, in the latter its dorsal branch (see below) is entirely reduced. In *Cheumatopsyche* (Fig. 1) the gonopore is flanked by a pair of meso-laterally movable, slightly asymmetrical valves. Each valve carries basally on its mesal side a solid, sclerotized knob (phallotremal sclerite R. & U.). The right knob is smaller than the left one; its height especially decreases rapidly in a dorsal direction. Like Ross and Unzicker I consider the valves as parameres. No doubt

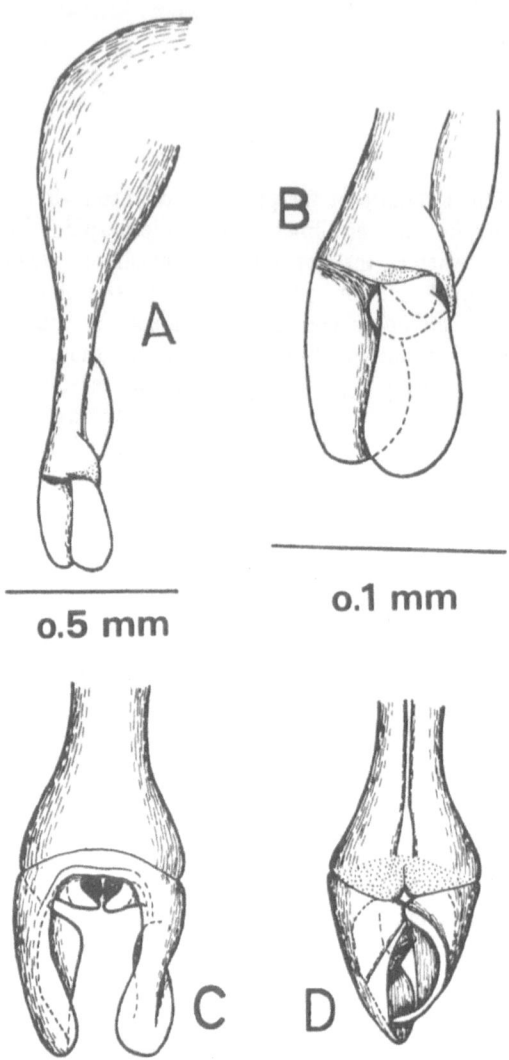

o.5 mm

o.1 mm

Fig. 1. Cheumatopsyche lepida. Phallus as seen from the right side (A). Distal end of phallus as seen from the right side (B), in a dorsal (C) and in a ventral view (D). In D the parameres are "adducted".

the valves themselves are homologous with the external lips in *Hydropsyche*, the knobs with the internal lips. Distally, on the ventral side of the phallobase, there is a thin, vertical lamella.

I have seen preparations in which the valves were abducted at 90°, but the condition shown in Fig. 1C probably is the normal resting position. In this position the meso-basal knobs touch each other. Due to the fact that the knobs touch each other in repose and to their asymmetry, an attempt at adduction of the valves results in a rotation of the latter (Fig. 1D).

In *Hydropsyche silfvenii* Ulmer the structure of the phallus (Fig. 2) is rather complicated. Distally on the dorsal side there is a pair of roughly globular processes, to some degree movable in a vertical plane by means of simple ball-and-socket articulations. This movement is effected by the retractor aedeagi through a pair of cuticular reinforcements in the ventral walls of the ejaculatory duct, which is seen also in other species of *Hydropsyche*. The same no doubt is the case in *Cheumatopsyche*, though at least in *Ch. lepida* Pictet the reinforcements are indistinct. I consider these processes as parameres. Proximally to them there is a large, very thin-walled and probably extensible bladder. Its proximal end is drawn out into a pair of horns, each of which carries a short, conical, heavily sclerotized process distally. This may be a second pair of parameres or it may be a structure *sui generis*.

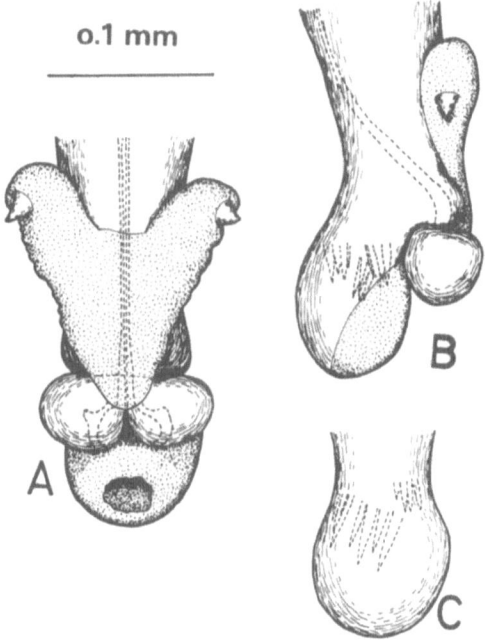

0.1 mm

Fig. 2. Ceratopsyche silfvenii. Distal end of phallus in a dorsal view (A) and as seen from the left side (B). Extreme distal end of phallus in a ventral view (D).

275

The ventral side of the phallobase projects beyond the parameres and the gonopore. On the distal end of the membranous dorsal side of this projection a somewhat irregular opening is seen. It leads into a sac, in the bottom of which three groups of very strong spines (Fig. 2B and C shown by broken lines) are seen. The sac no doubt is eversible, never to be retracted again.

A comparison with *Agapetus* (1957a, pp. 23–24) and *Rhyacophila* (pp. 18–20) suggests that the ventro-distal projection represents a ventral branch of the aedeagus, the dorsal branch of which has been reduced. Previously (1957a, pp. 151 and 156) I have put forward the theory that the aedeagus of the hypothetical primeval Trichopteran was divided into a dorsal and a ventral branch, the former carrying the gonopore. The peculiar, so-called titillator in the Hydroptilidae (1957a, pp. 74–75 and 80–81) may very likely, be interpreted as a ventral aedeagal branch.

On the posterior corners of the genital segments in *Cheumatopsyche* (Fig. 3) there is a pair of slender processes, which I consider as paraprocts. Each carries a small number of short, but very strong, slightly curved setae dorso-distally. Subdistally and meso-ventrally, there is a small number of a little longer and straight, but also strong, setae. Moreover, a sensillum

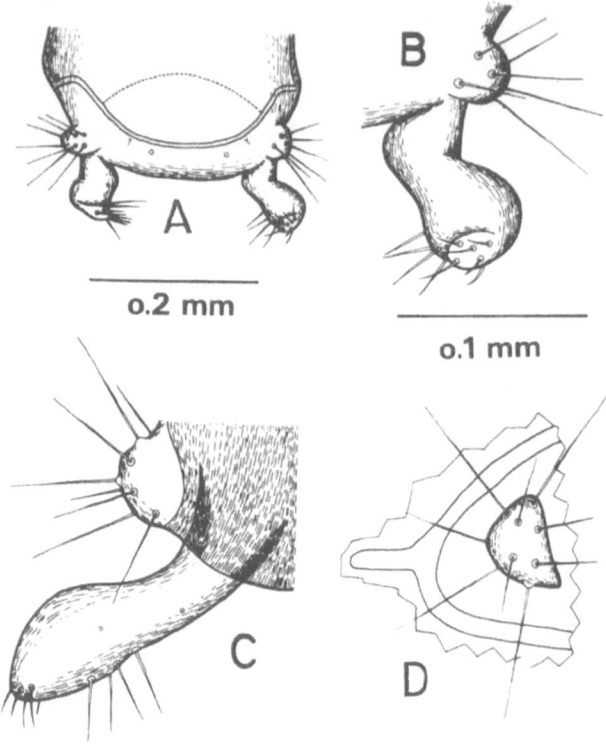

o.2 mm

o.1 mm

Fig. 3. Cheumatopsyche lepida. Posterior end of genital segments in a dorsal view (A). Right paraproct and cercus in a dorsal view (B). Left paraproct and cercus in a mesal view (C). Left cercus in a latero-dorsal view (D; anterior direction to the right).

campaniformium is present on both the dorsal and the ventral side. Between the two paraprocts the posterior margin bulges a little backward as a very low tongue, here considered as the epiproct. Except for a pair of sensilla campaniformia and a pair of tiny setae it is devoid of sensilla.

Again, between the epiproct and the paraprocts there is a pair of low, rounded processes, situated on areas framed by cuticular reinforcements and carrying setae similar to but a little longer than those subdistally on the paraprocts. I think that these processes represent cerci, a very rare feature in male holometabolous insects. Tjeder (1954) has interpreted similar processes in some Megaloptera and Neuroptera as cerci.

In *Ch. lepida*, at least, each harpago carries on its distal end two setae like, but a little longer than, those distally on the paraproct.

In *H. silfvenii* (Fig. 4) the posterior end of the genital segments is divided into three short, blunt processes, an unpaired dorsal one and a pair of laterals, which I interprete as epiproct and paraprocts, resp. The dorsal keel (in 1957a labelled 2) is very broad, triangular, with a slight longitudinal reinforcement. It is densely clothed with small, backward directed microtrichia. It carries on its anterior end a number of small sensilla campaniformia and on its lateral sides three pairs of very long setae. Otherwise the dorsal side s.str. of the genital segments is devoid of sensilla. On the distal end of the harpago there are rather a lot of very close-set and small setae.

In the structure of the phallus *Hydropsyche silfvenii* Ulmer resembles *H. nevae* Kolenati and *H. etnieri* (Schuster and Talak, 1977) as well as a

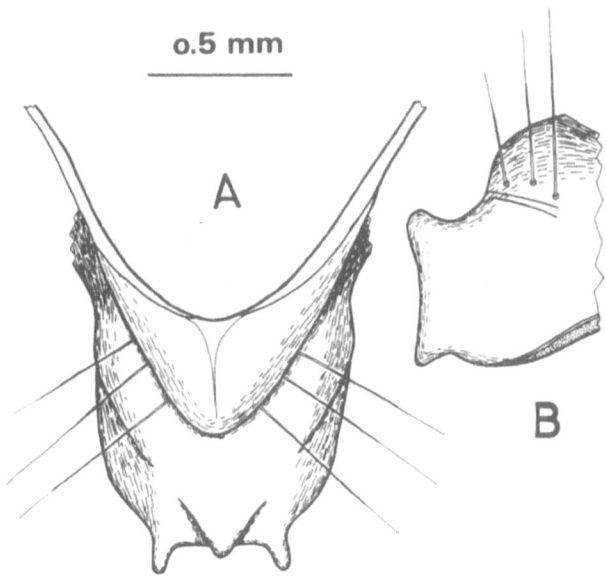

Fig. 4. Ceratopsyche silfvenii. Upper half of genital segments in a dorsal view (A). Posterior end of genital segments as seen from the right side (B; anterior direction to the right). Setae omitted except for the three pairs of long ones.

277

number of North American species, listed by Ross (1944, pp. 96–99) as belonging to the *H. bifida* group. Ross and Unzicker included this group in the genus *Symphitopsyche* (Ulmer, 1907), though as a separate subgenus, *Ceratopsyche.* I do not think it is justified to include the *bifida* group in the genus *Symphitopsyche,* and hence the name *Ceratopsyche* has to be given generic status. The generic name of *Hydropsyche silfvenii* Ulmer (and *H. nevae* Kolenati) will thus be *Ceratopsyche* (Ross and Unzicker, 1977).

Judged by a figure (Ross and Unzicker, Fig. 10) the dorsal branch of the aedeagus may not be reduced in the type species of *Symphitopsyche* (*Hydropsyche mauritiana* Maclachlan, 1873). If true, *Symphitopsyche* is more plesiomorphic in this respect than other hydropsychids.

REFERENCES

McFarlane, A.G., J. R. Soc. New Zealand 6: 23–35, 1976.
Nielsen, A., Biol. Skr. Dan. Vid. Selsk. 8: 5, 1957a.
———, Ent. Medd. Kbh. 28: 27–57, 1957b.
Ross, H., Nat. Hist. Surv. Div. Illinois 23: 1, 1944.
——— and Unzicker, J.D., J. Georgia Ent. Soc. 12: 292–312, 1977.
Tjeder, B., Ent. Medd. Kbh. 27: 23–40, 1954.

DISCUSSION

Botosaneanu: You have found a very interesting morphological characteristic in the male of *Hydropsyche silfvenii*. But I do not believe that this is a reason to establish a new genus, and I believe that Ross and Unzicker were wrong when they made a new genus for the *Antillean Hydropsyche*. I ask myself: is there a genus containing more than one species, which could not be further divided in 2, or several, genera, using some morphological characteristic? Certainly not! *Hydropsyche* could also be divided in many genera: think, for instance, of the Asiatic species belonging to the "orientalis-group"! But I see no reason for doing it, *Hydropsyche* is a "good" genus, characterized by the habitus, the wing venation, the antennae, the shape of male and female genitalia, the appearance of the larva. It would be so easy to divide *Rhyacophila* in, say, 100 different genera or in 50, or in 20, but it would be a pity to do so. We should be as happy to work with genera as unartificial as possible, such as *Hydropsyche,* or *Rhyacophila,* or *Drosophila.*

Nielsen: Opinions differ. Like McFarlane I consider genital structures as more reliable than minor differences in wing venation. I find it justifiable to establish genera and subgenera in an attempt to elucidate phylogeny I especially wanted to point out some plesiomorphies. I think that the genera *Rhyacophila* and *Limnephilus* ought to be subdivided.

In the latter there is a very great ecological diversity.

BIBLIOGRAPHY OF TRICHOPTERA

A.P. NIMMO

SUMMARY

History and progress of work on the bibliography are outlined.

In 1972, I corresponded with F.C.J. Fischer of Holland, concerning the desirability of working on a bibliography of Trichoptera from 1961 onwards; to continue on from where his 'Trichopterorum Catalogus' left off.

Fischer welcomed the idea, remarking that he was himself then working on such a bibliography, to complement the 'Catalogus' (which is essentially taxonomic in nature). The bibliography was to deal with Trichoptera up to 1960 inclusive. Unhappily this project appears to have died with Fischer; apparently no trace of it has yet been found.

Fischer also mentioned that Higler (elsewhere in this volume) was to continue the 'Catalogus' beyond 1960.

I started accumulating references in 1972, and now estimate, roughly, that I may have as many as 3,000 entries in my card file, for the period 1961—80. Any mention of Trichoptera, at whatever taxonomic level, in any field of inquiry, is currently accepted, including popular accounts.

Biological Abstracts, Entomology Abstracts, Aquatic Sciences and Fisheries Abstracts, Bioresearch Index, and Zoological Record are regularly searched for relevant entries. Entries are also obtained from wherever else I may find them.

Acquisition of reprints or photocopies of all relevant publications is continuous; many earlier ones have still to be obtained. Beside other reasons, it is necessary to have these for preparation of the index. Cooperation of authors has been excellent.

Hoping that it will, eventually, be published, the bibliography will be laid out as follows. An introductory section will be followed by a numbered listing of all entries arranged alphabetically by author, and chronologically within author. Also, following each entry will be abbreviated references to known abstracting publication entries (e.g. BA69-55555 = Biological Abstracts, Vol. 69, entry 55555), which will allow speedy consultation of an abstract if the paper itself is not immediately available.

Proc. of the 3rd Int. Symp. on Trichoptera, ed. by G.P. Moretti   279
Series Entomologica, Vol. 20. © 1981, Dr W. Junk Publishers, The Hague

An index will complete the bibliography. This index will be thoroughly cross-indexed, to allow access to any entry from any relevant approach. Index reference to entries under author and date will be through the numbering of entries mentioned above.

The first volume will embrace literature from the period 1961 to the year of the then current volume of Zoological record (probably about 15 years). Following volumes would probably embrace 5-year periods.

It is anticipated that this work will be of interest to workers in disciplines other than simply Trichopterology; primarily aquatic studies generally.

The continued cooperation of authors, in obtaining reprints or copies of publications is earnestly solicited.

TRICHOPTERA DISTRIBUTION PATTERN DIFFERENCES FOUND BY SWEEPING, BEATING AND LIGHT TRAPS AT THREE SOUTHERN BOHEMIAN SITES

K. NOVÁK

SUMMARY

An investigation on the incidence of caddis flies was carried out between May and October in three different geographical locations in Southern Bohemia by sweeping, beating and light traps. The results were: Site 1 — total species from sweeping and beating 34, light trap 38, species found only by sweeping and beating 11, light trap only 15. The figures for the other sites were: Site 2 — 20, 36, 3, 19; Site 3 — 36, 50, 3, 17.

INTRODUCTION

Light traps are very effective for investigating Trichoptera incidence and the samples collected in these traps have greatly contributed to our knowledge of caddis fly incidence at many localities. Although most Trichoptera are attracted to light (Malicky, 1973), the degree of attraction varies with individual species. Crichton (1960, 1971, 1974) made a valuable contribution to this investigation with his detailed analysis of samples from light traps in Great Britain, and he also examined the differences between light and suction traps (Crichton, 1965); while Svensson (1972, 1974) compared light trapping with other methods of collecting adults. As we had the opportunity of investigating caddis fly incidence at three sites in Bohemia, not only by the classic methods of sweeping, beating and sighting, but also with light traps, we were able to make comparisons between the various methods.

METHODS AND COLLECTING SITES

Horská Kvilda in the Bohemian Forest (Šumava Mts.) at the altitude of 1050 m, which is an area of peat bogs and small brooks where the head-waters of the River Vltava are found, was chosen as our first site. The second site was the River Malše near the village of Roudná in the lowlands of southern Bohemia (altitude 535 m) and the third a carp pond area in the Třeboň Basin (altitude 423 m).

Proc. of the 3rd Int. Symp. on Trichoptera, ed. by G.P. Moretti
Series Entomologica, Vol. 20. © 1981, Dr W. Junk Publishers, The Hague

A Minnesota-type light trap with a white mercury lamp was used for taking samples every three to five days from May to October. At the first site, it was placed in a peat-bog meadow about 500 m from water; at the second, also in a meadow, about 100 m from the water's edge and, at the third, in a wet, periodically flooded meadow about 300 m from a carp pond. Sweeping and beating were carried out at monthly intervals at all three sites from May to October and supplemented by the collection of individual adult specimens.

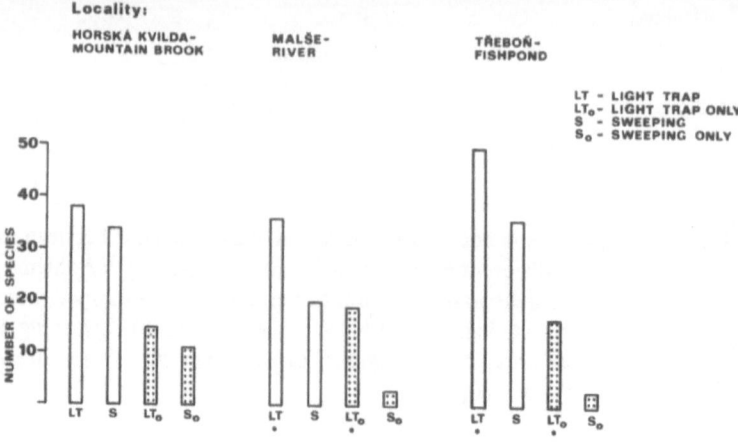

Fig. 1. Numbers of species collected at individual localities by sweeping, beating and light trapping.

RESULTS

Horská Kvilda: Altogether 34 species of caddis flies were found by sweeping and beating at the first locality. The most abundant were *Apatania fimbriana* Pictet (592 specimens), *Drusus annulatus* Stephens (221 specimens), *Agapetus fuscipes* Curtis (184 specimens) and *Allogamus auricollis* Pictet (102 specimens). Thirty eight species were caught in the light trap in 1974 and 35 in 1975. The most abundant species in the light trap in 1974 were *Drusus annulatus* Stephens (394 specimens), *Potamophylax latipennis* Curtis (389 specimens), *Ecclisopteryx madida* McLachlan (204 specimens) and *Allogamus uncatus* Brauer (143 specimens). In 1975, they were *Allogamus auricollis* Pictet (277 specimens), *Potamophylax latipennis* Curtis (176 specimens), *Drusus annulatus* Stephens (167 specimens), *Limnephilus coenosus* Curtis (145 specimens) and *Allogamus uncatus* Brauer (139 specimens). In all, 1618 specimens were collected by beating and sweeping 1541 were caught in the light traps in 1974 and 1188 in 1975. Eleven species – *Rhyacophila praemorsa* McLachlan (6 specimens), *Agapetus fuscipes* Curtis (184 specimens), *Philopotamus ludificatus* McLachlan (2 specimens), *Micrasema longulum* McLachlan (1 specimen), *Silo pallipes* Fabr. (6 specimens), *Drusus trifidus* McLachlan (30 specimens), *Acrophylax zerberus* Brauer (1 specimen), *Pseudopsilopteryx zimmeri* McLachlan

282

(1 specimen), *Chaetopteryx villosa* Fabr. (80 specimens), *Chaetopterygopsis maclachlani* Stein (44 specimens) and *Parachiona picicornis* Pictet (54 specimens) — were collected by sweeping and beating only.

According to Crichton (1974), although *Agapetus fuscipes* has never been found by the Rothamsted Insect Survey in light traps in England, it is an abundant species. However, at Horská Kvilda, *Agapetus fuscipes* was one of the most abundant species; the others were *Chaetopteryx villosa* Fabr. *Parachiona picicornis* Pictet and *Drusus trifidus* McLachlan. These species seem to fly only short distances and this may explain why they were not caught in the trap situated 500 m from water. *Apatania fimbriata* Pictet, the most abundant species collected by sweeping, was also rarely found in the light traps (5 specimens). On the other hand, 15 species never collected by sweeping and beating were caught in the light trap. They were mainly species of the Limnephilidae family and had, probably, flown there from distant places. *Halesus rubricollis* Pictet, *Halesus digitatus* Schrank and *Stenophylax permistus* McLachlan were reasonably abundant; *Hydropsyche contubernalis* McLachlan, *Hydropsyche pellucidula* Curtis and *Drusus discolor* Ramb. were also found.

Malše: At the second locality, the river Malše, 20 species were collected by sweeping and beating and 36 species were trapped. The most abundant among those collected by sweeping and beating were *Polycentropus flavomaculatus* Pictet (98 specimens), *Psychomyia pusilla* Fabr. (94 specimens) and *Hydropsyche pellucidula* Curtis (74 specimens). *Hydropsyche pellucidula* Curtis (6015 specimens), *Psychomyia pusilla* Fabr. (2285 specimens) and *Ceraclea dissimilis* Steph. (936 specimens) were the most frequent in the light trap. Three species *Athripsodes bilineatus* L., *Athripsodes commutatus* Rostock and *Beraeodes minutus* L. — were found by sweeping only. Nineteen species were caught only in the light trap, again mostly representatives of the Limnephilidae that fly long distances. Most of the *Athripsodes* and *Mystacides* species occurred more often in samples obtained by sweeping, whereas *Ceraclea dissimilis* Steph. was far more frequent in samples from the light trap. The total number of specimens caught in the light trap at this locality exceeded 10,000, while only 345 specimens were collected by sweeping and beating.

Meadow near a carp pond: Thirty six species were collected by sweeping and beating at this site. *Mystacides longicornis* L. (156 specimens), *Oecetis lacustris* Pictet (115 specimens) and *Athripsodes aterrimus* Steph. (108 specimens) were the most abundant. Fifty species were caught in the light trap; the most abundant of them were *Oecetis ochracea* Curtis (1991 specimens), *Oecetis lacustris* Pictet (164 specimens) and *Hydropsyche angustipennis* Curtis (156 specimens). Three species — *Anabolia furcata* Brauer, *Notidobia ciliaris* L. and *Oligotricha striata* L. — were found by sweeping only; however, all were rare at this site. On the other hand, some species of Leptoceridae frequently collected by sweeping, in particular *Athripsodes aterrimus* Steph., *Mystacides longicornis* L. and *Trienodes bicolor* Curtis, were seldom found in the light trap. Seventeen species were caught only in the light trap, most of them belonging to the Limnephilidae; they occurred sporadically and had apparently flown in from distant places.

CONCLUSIONS

The results show that, in all cases, more species and more individual specimens are collected with light traps than by sweeping and beating, and confirm that no all Trichoptera species are equally drawn to light; some only slightly, others not at all. This is true of certain representatives of the genera *Athripsodes* and *Mystacides*, and, as has already been pointed out by Crichton (1974), also to *Agapetus fuscipes* and some other species living in springs (*Parachiona picicornis* Pictet, *Drusus trifidus* McLachlan). Naturally, the ability to migrate, especially to fly long distances to light traps situated far from water, also plays a certain role. Some species which fly reluctantly or only short distances (*Chaetopteryx villosa* Fabr., *Chaetopterygopsis maclachlani* Stein, *Anitella obscurata* McLachlan) are rarely found in samples from light traps, although they are otherwise abundant. Therefore the siting of a light trap is of prime importance, especially its distance from water, as Svensson (1972, 1974) has pointed out.

It can, on the whole, be said that light traps are most suitable for studying Trichoptera in major rivers, ponds and lakes, where they yield much more extensive samples than sweeping and beating. The variations in the different sampling methods are not as pronounced in mountain streams and springs. Finally, methods may be usefully combined to obtain as complete a picture as possible on caddis fly distribution patterns.

REFERENCES

Crichton, M.I., Trans. R. ent. Soc. Lond. 112: 319–344, 1960.
———, Proc. R. ent. Soc. Lond. (A) 40: 101–108, 1965.
———, J. Zool. Lond. 163: 533–563, 1971.
———, Proc. of the First Int. Symp. on Trichoptera, pp 147–158, Junk, The Hague, 1974.
Malicky, H. Handbuch der Zoologie IV. Bd, Arthropoda, 2. Hälfte Insecta. Nr. 29. W. de Gruyter, Berlin, pp 1–114, 1973.
Svensson, B.V., Oikos 23: 370–383, 1972.
———, Oikos 25: 157–175, 1974.

WHY DO *POTAMOPHYLAX CINGULATUS* (STEPH.) (TRICHOPTERA) LARVAE AGGREGATE AT PUPATION?

C. OTTO AND B.S. SVENSSON

SUMMARY

In stream-living *Potamophylax cingulatus*, current velocity was found to be an ultimate factor in guiding settlement of the larvae prior to pupation, areas of high current velocity being avoided. In areas suitable for pupation, i.e. in a low current regime, the larvae formed aggregations only under certain stones. Most pupae were found in large ($>$ 40 inds.) aggregations, where also hatching success was highest. Infestation by a chironomid larva, *Polypedilum fallax* Joh., peaked at intermediate aggregation size. We suggest pupal aggregations to be formed in order to reduce the risk of becoming a victim of the chironomid.

INTRODUCTION

During spring and early summer the locomotory activity of the lotic *Potamophylax cingulatus* (Steph.) larvae is markedly increased (Otto, 1971; 1975). The elevated activity, mainly in the form of drift but also upstream movement, leads to increased densities in areas of hard bottom where pupation takes place. Within these areas the larvae aggregate under certain objects, at the same time as many similar objects are rejected (Otto, 1976). Similar pupal aggregations have been described for *Potamophylax luctuosus* (Pill.) (Tobias, 1967) and *Potamophylax latipennis* (Curt.) (Campbell and Meadows, 1972). The reasons for aggregative behaviour prior to pupation can be divided into abiotic and biotic ones. However, if abiotic factors, viz. microhabitat differences with respect to e.g. illumination and current velocity, alone are responsible for the aggregative behaviour, the aggregations should not be as large as are often observed. Stones on and partially inserted into the bottom are often bordered all round by pupae fastened on the stone close to the bottom. Since current velocity differs between the upstream and downstream side of a stone and since the velocity tolerance therefore must be wide, current seems hard to accept as a proximate factor guiding settlement. Besides, in lake dwelling *P. cingulatus*, aggregation formation has been observed (J. Brittain, pers. comm.). Also in the laboratory, *P. latipennis* formed aggregations in a lentic environment, and consequently current could be excluded as a factor determining the formation of aggregations (Campbell and Meadows op. cit.). The fact that some pupae, often as a consequence of settling on each other, are situated outside the vertical projection of a stone i.e. in an area of high illumination indicates that the light regime is of subordinate importance. In *Potamophylax* (*Stenophylax*) sp. Scott (1958) found it unlikely that a physical factor governed settlement, and he postulated that last instar larvae attached their cases in the immediate vicinity of already fastened cases leading to the formation of 'clusters'.

Since the pupae are immobile for several weeks, they are vulnerable to predators and parasites. Gallepp (1974) put forward the hypothesis, that caddisfly larvae pupate in aggregations in order to reduce the efficiency of predators or parasites. The fastening of cases on the undersides of stones and roots can be an adaptation which make the pupae less accessible to predatory birds and fish. In a previous study (Otto, 1975) these predators did not use pupae as food. Besides, if the settlement on the undersides of stones is of prime importance in reducing predation, it does not explain the formation of aggregations. The small space available under stones is obviously not an impediment for invertebrate predators and parasites to reach their victims. Formation of aggregations might lead to a decreased efficiency of the searching parasite (cf. Wilson, 1975), and, if the aggregation is found, a simultaneous occurrence of a large number of pupae eventually result in a 'selfish herd' or 'swamping' effect (Hamilton, 1971; Wilson, 1975). Alternatively, if the size of an aggregation is not large enough to result in a parasite being swamped, the functional response by a parasite should result in an increased infestation rate with aggregation size (cf. Hassell et al., 1977).

In the present field study we have analysed the distribution of pupae of *P. cingulatus* with respect to 'recruitment area', current velocity and incidence of stones utilized for pupation. The results are discussed in the light of knowledge of the infestation by a 'parasitic' chironomid.

MATERIAL AND METHODS

The material of this study was collected in the Stampen stream, South Sweden, described by Malmqvist et al. (1978). Two adjoining sections of the stream were investigated. In the downstream section the distribution of pupae was analysed. The section was divided into small riffles and pools, the bottoms of the pools being soft and consequently unsuitable for pupation. Since by far most of the locomotory activity during spring – early summer is directed downstream (Otto, 1971), the length of a pool situated upstream of a stony riffle was used as a measurement of 'recruitment area'. When possible each riffle (1–6 m long) was divided into three parts with respect to current velocity. Thus, the uppermost part was in the low current velocity regime, all stones of which were covered with fine particulate organic matter. The stones of the intermediate and downstream parts of the riffles were not covered with organic matter, and the current velocity peaked in the downstream part. The intermediate and downstream parts were of similar lengths. In each part of the riffles all stones, which on account of previous experience were judged to be suitable for pupation, were turned and the absence of pupae or the number of pupae present was noted.

In the upstream section of the stream a total number of 2282 pupae, representing different aggregation sizes (1–10, 11–20, 21–30, 31–40 and >40 inds.), were collected. Hatching success as well as the presence of parasites with respect to aggregation size was controlled in the laboratory. The pupae were collected in the end of August, when all larvae had pupated, and the pupae already hatched were added to those subsequently hatching in the laboratory.

RESULTS

The distribution of pupae on 268 stones taken in the upper parts of the riffles were analyzed. It differed significantly ($P < 0.001$) from a Poisson distribution. Since the variance was large in relation to mean, we concluded that the distribution of pupae under the stones was highly clumped.

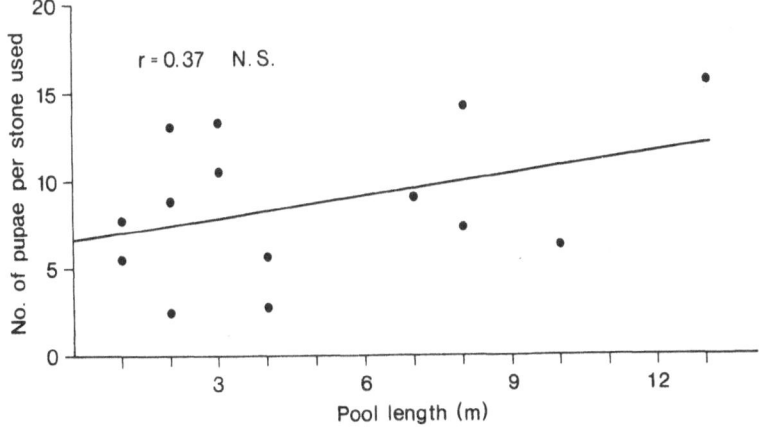

Fig. 1. Number of pupae per stone used for pupation in each riffle with respect to pool length (m) upstream the riffle.

The relative abundance of pupae in the riffles did not show any significant correlation to pool length (recruitment area) upstream of the riffle (Fig. 1). Thus, if the number of larvae is positively related to pool length, many larvae apparently do not stop at their first contact with a pupation site (riffle). Within each riffle the most pupae by far were found in the slow current i.e. the upstream part (Table 1) and the number of pupae steadily declined parallel to current velocity increments in the downstream direction.

Table I. Distribution of *P. cingulatus* pupae within a riffle with respect to current velocity and number of stones available. Through the upstream, intermediate and downstream part of a riffle current velocity steadily increased. n refers to the number of riffles.

| Part of riffle | No. of pupae | | | Total no of stones available | Total no of stones used | Per cent of stones used |
	X̄	S.D.	n			
Upstream	125	118	14	299	176	59
Intermediate	22	32	10	182	38	21
Downstream	3	4	8	140	8	6

Thus, current velocity acts as an ultimate factor in guiding the settlement. In the uppermost part of a riffle some 60 per cent of the stones were used for pupation, while only 6 per cent were used in the downstream part (Table 1).

With respect to aggregation size most pupae were found in aggregations consisting of more than 40 individuals. More individuals formed small aggregations (1–10 pupae) than aggregations of intermediate size (11–40 pupae) (Fig. 2a). Hatching success within aggregation sizes ranged between 93.6 and 97.8 per cent. Pupae from intermediate aggregation size (21–30 inds.) suffered the highest mortality, while small as well as large aggregations had higher hatching success (Fig. 2b). By dissecting those pupae not hatched, depending on aggregation size, 1.4–3.2 per cent of the total number of pupae were found infested by a chironomid larva, *Polypedilum fallax* Joh. High infestation rates were found in aggregations of intermediate size (11–30 inds.) (Fig. 2c). The chironomid killed and consumed its host, and, if not entering the pupal case only as a predatory larva but obligately associated to its host for development, it should be assigned parasitoid. Since the chironomid has not been recorded in Europe previously, it' is probably uncommon, indicating that parasitoid is an adequate designation. Only one chironomid was found in each infested pupa. By adding mortality due to the parasitoid to hatching success of the pupae, it was found, that these factors could explain only 96.8 and 97.8 per cent of the pupae in aggregations of 21–30 and 31–40 individuals, respectively. In small and large aggregations the figures reached by this addition was 98.5–98.9 and 99.4 per cent, respectively. Thus, some other mortality factor(s) also peaked at intermediate aggregation size, but in these remaining dead pupae no indications of parasitoids were found.

DISCUSSION

On a coarse scale, current velocity apparently restricted the area suitable for pupation. On the other hand, within a pupation area, i.e. in a low current regime, current velocity seemed insignificant with respect to pupation site. On the average, only about 60 per cent of the stones in a suitable area was used for pupation, but nothing indicated that those stones not used were in a bad position with respect to current. The fact that not all stones were

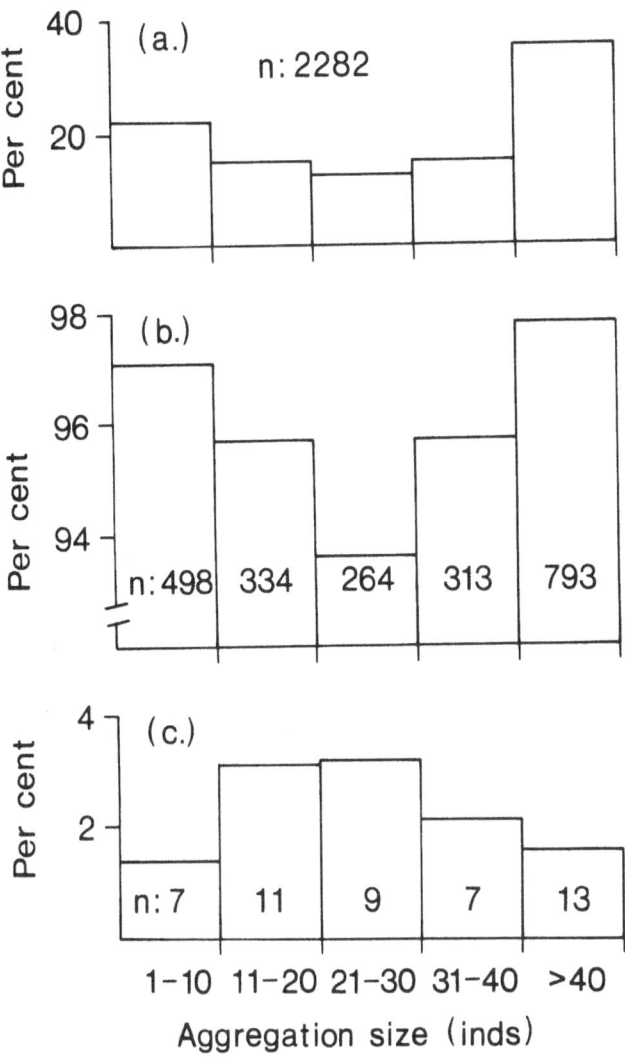

Fig. 2(a). Distribution of pupae with respect to aggregation size. (b). Hatching success with respect to aggregation size. (c). Infestation by chironomids with respect to aggregation size.

used should rather be viewed as an expression of the aggregate behaviour of the larvae prior to pupation. Concluding that abiotic factors within wide limits are unimportant with respect to settling, we have restricted our field of interest to biotic ones. As pointed out earlier, predation by birds and fish on the pupae was not observed. It is probable that the pupae, confined to the undersides of stones, escape predation by vertebrates. No invertebrate except the chironomid was found associated to the pupal case. In *Brachycentrus occidentalis* Banks, Gallepp (1974) found pupae being cannibalized by larvae, but cannibalism was not observed here. Gallepp (op. cit.) also found a chironomid larva, *Eukiefferiella* sp., in pupae of *B. occidentalis*. However, this chironomid seemed to damage by crowding in the pupal case, not by eating the pupae. Recently Parker and Voshell (1979) found pupae of hydropsychids to be infested by parasitic larvae of *Cardiocladius* sp., another chironomid. The infestation rate by this chironomid was as high as 61 per cent. The only biotic agency so far known to affect pupal mortality in *P. cingulatus* is the larval chironomid *P. fallax*. A commensalistic chironomid has been found in *P. latipennis* (Dratnal, 1979).

Do *P. cingulatus* larvae form their aggregations prior to pupation as a consequence of the attacks by *P. fallax?* In birds incidence of predation and brood parasitism has been shown to be much lower in large colonies or at high prey abundances (e.g. Patterson, 1965, Robertson 1973; Clark and Robertson, 1979). However, much of the reduced predation/parasitism in large bird colonies is caused by behavioural group responses (e.g. early warning, bunching, mobbing) which do not apply to immobile pupae. On the other hand a 'selfish herd' or 'swamping' effect (Hamilton, 1971; Wilson, 1975) should also work in pupae. Thus, also in *P. cingulatus* high density could be regarded as a way to reduce the effects by a parasitoid. In reducing the effect of the parasitoid, from the above reasoning, large aggregations should be most favourable, and in fact most pupae were found in these aggregations where hatching success also peaked. Intermediate aggregations suffered from the highest infestation rates suggesting that swamping effects did not occur here. Intermediate aggregations, because of their size, were probably more easily found than single pupae and small aggregations. Since intermediate aggregations were comparatively highly infested and had a low hatching success, why are they formed? A reasonable explanation might be the limited space available under most stones. Thus, we propose that, given space enough, intermediate aggregations would have been large ones, but addition of pupae is stopped when the aggregation reaches the borders of the stone. Single pupae-small aggregations were not heavily infested, and their hatching success was next to that of large aggregations. However, it should be remembered, that if aggregations were not formed, pupal distribution would be much more even. Thus, if alone, the pupa would run a many times higher risk of being encountered by the random search of a parasitoid (cf. Cullen, 1960; Wilson, 1975). In closure, we propose that a pressure exerted by a parasitoid might be the selective force behind the formation of aggregations in *P. cingulatus* pupae, the limits of which are set by the physical environment.

ACKNOWLEDGEMENTS

We thank Prof. O. Saether who kindly identified the chironomid. Dr J. Brittain commented on the manuscript.

REFERENCES

Campbell, J.I. and Meadows, P.S., J. Zool. Lond. 167:133–141, 1972.

Clark, K.L. and Robertson, R.J., Behav. Ecol. Sociobiol. 5:359–371, 1979.
Cullen, J.M., *In* Proc. 7th Int. Orn. Congr. Helsinki, pp. 153–157, 1969.
Dratnal, E., Bull. Acad. Pol. Sci. Biol. 27:183–193, 1979.
Gallepp, G.W., Ecology 55:1283–1294, 1974.
Hamilton, W.D., J. Theor. Biol. 31:295–311, 1971.
Malmqvist, B.; Nilsson, L.M. and Svensson, B.S., Oikos 31:3–16, 1978.
Otto, C., Oikos 22:292–301, 1971.
——, Oikos 26:159–169, 1975.
——, Oikos 27:93–100, 1976.
Parker, C.R. and Voshell, J.R., Environ. Entomol. 8:808–809, 1979.
Patterson, I.J., Ibis 107:433–459, 1965.
Robertson, R.J., Ecology 54:1085–1093, 1973.
Scott, D., Arch. Hydrobiol. 54:340–392, 1958.
Tobias, W., Oikos 18: 55–75, 1967.
Wilson, E.O. Sociobiology – the New Synthesis. Harvard University Press, Cambridge, Massachusetts, 1975.

DISCUSSION

Moretti: In Italy *P. cingulatus* pupae form large aggregations as a protection against predation by trout, as isolated pupae are an easy prey. *P. cingulatus* pupae also protect themselves by hiding under stones in running waters?

Otto: *P. cingulatus* pupate under stones etc. where it is hard for trout to reach them. We do not think aggregation is selected for by trout predation. Besides, in previous studies, we have not found trout to feed on *P. cingulatus* pupae.

Jalon: Have you determined the minimal stone size for larval aggregates at pupation?

Otto: About 5 cm, but it very much depends on the position of the stone in relation to the soft (e.g. sand) substrate.

THE INFLUENCE OF RESERVOIR DISCHARGE ON BENTHIC FAUNA IN THE RIVER TER, N.E. SPAIN

N. PRAT

SUMMARY

Water discharge and temperature seem to have been responsible for variations in the stone-dwelling benthic fauna of the River Ter, N.E. Spain during 1974–1975. The river is fed by the discharge from the Susqueda Reservoir 10 kilometres upriver from the sampling site. Floods in September, 1975 affected the numbers of stone-dwelling invertebrates. Because of high discharge between March and July, 1974, there was greater development of benthic fauna than in the same period of 1975 when discharge was minimal. Water fed from the reservoir hypolimnion in 1975 caused a drop in water temperature in the river and the extended growth period, as well as the emergence of a single generation, of *Psychomyia pusilla* was probably a response to the low temperatures. There is evidence to suggest that there was a partial emergence of a second generation in 1974.

INTRODUCTION

Studies on aquatic insects and their role in river ecology have been badly neglected in Spain. However, there is a growing interest in the systematic study of aquatic insects, particularly Trichoptera (Jalon, 1977; Viedma and Jalon, 1980) and their importance in river ecology (Prat, 1980; Prat et al., 1979). In this paper we report data on the composition and population dynamics of an invertebrate community, with special reference to insects, during the years 1974 and 1975. A detailed study was carried out on the life cycle of *Psychomyia pusilla*.

Study area

The River Ter in North East Spain flows along a 208 km course from the Pyrenees to the Mediterranean. The sampling station is in the middle reach of the river, 225 m above sea level and 10 km downstream from the Susqueda Reservoir, which has a total capacity of 233 Hm^3 and is contained by a 125 m high dam wall. For most of the year the river volume is determined by the discharge from this reservoir.

Proc. of the 3rd Int. Symp. on Trichoptera, ed. by G.P. Moretti
Series Entomologica, Vol. 20. © 1981, Dr W. Junk Publishers, The Hague

MATERIAL AND METHODS

Separate samples were taken from the upper and lower surfaces of stones with quite different composition in the middle of the river (Prat, 1979, 1980). Only those samples which covered 100 cm^2 of the upper surface of the stones are reported in this paper. All material was fixed in 10% formol, then later, in the laboratory, it was stirred with distilled water, filtered through a 250 μ net and separated under a stereoscopic microscope at × 10. Samples were taken monthly or every two months during 1974 and the end of 1975, from March to August, 1975 they were taken every two weeks. Water and air temperatures were always recorded and, on occasions, certain other physico-chemical parameters.

RESULTS

River discharge records are available at the E1 Pasteral reservoir 3 km above the sampling site (Fig. 1). As will be seen from Fig. 2, the daily mean discharge in cubic metres shows great seasonal variation between the two years. 1974 was a dry year with little snow or rain fall in autumn and winter and, in consequence, the water level in the Susqueda Reservoir was low in the spring of 1975 and so, therefore, was the discharge. However, snow runoff from the Pyrenees caused high discharge between March and July, 1974. In contrast, 1975 was a wet year with heavy rains in June that were, probably, responsible for the high discharge in July. The autumn was, also, very rainy and there was extensive flooding in September (Fig. 2). The human regulation of the river flow through the discharge from the Susqueda Reservoir explains the fluctuations.

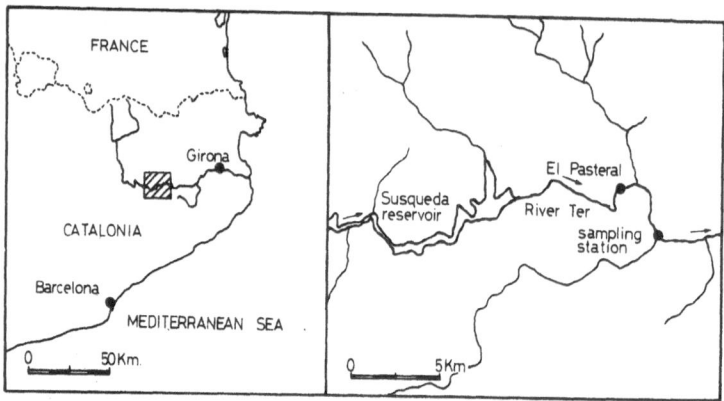

Fig. 1. Site of sampling station in the River Ter downstream from the Susqueda Reservoir.

294

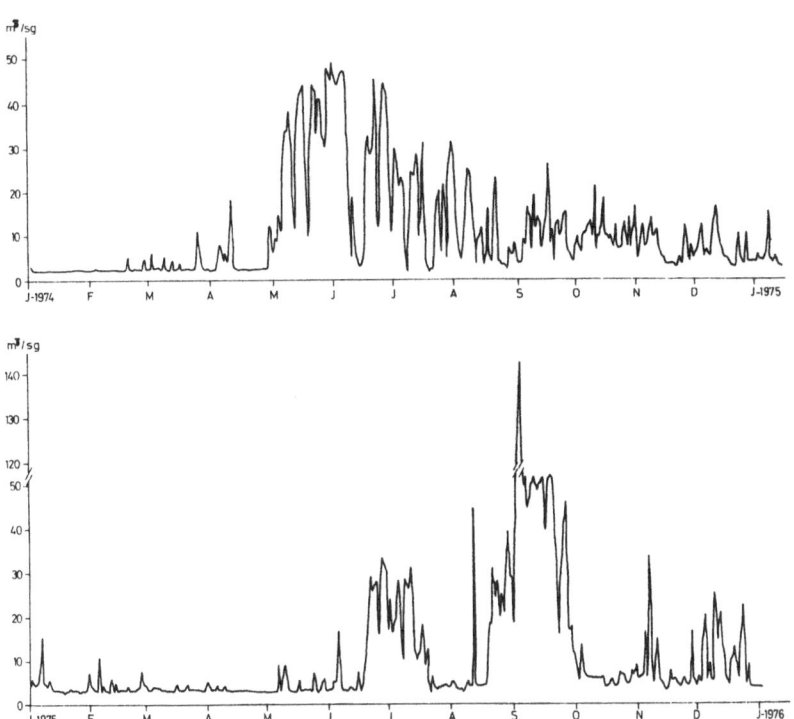

Fig. 2. Daily mean discharge for 1974 and 1975 at El Pasteral Dam.

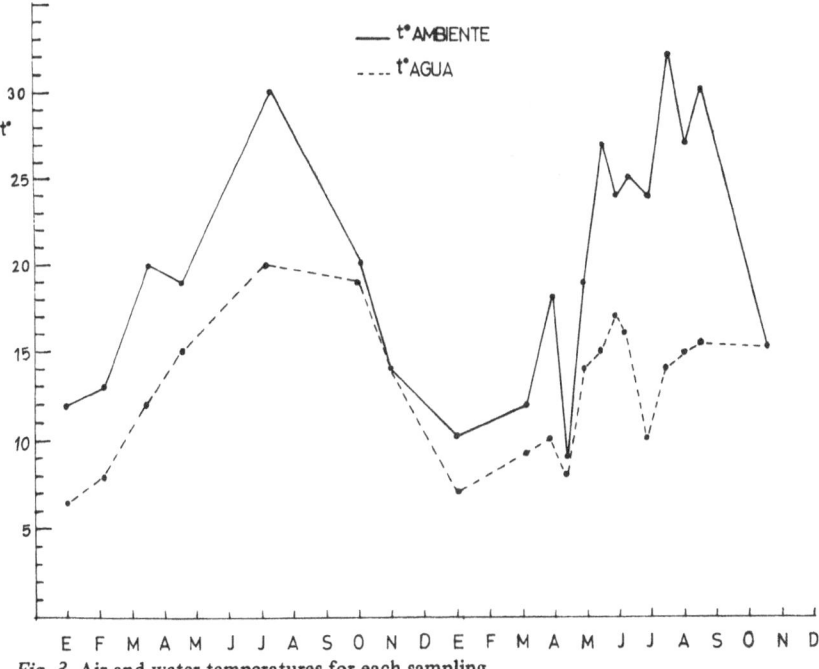

Fig. 3. Air and water temperatures for each sampling.

Temperature

The air and water temperatures for each sampling day are presented in Fig. 3; daily temperatures are not available. In the summer of 1974, the water temperature reached 20°C and seems to have been high throughout the summer. In the summer of 1975 water temperatures were lower and at the beginning of July a low 10°C was registered.

As the water-feed from the Susqueda Reservoir is at the level of the hypolimnion, temperatures can be below average. Both water discharge and turnover rate were high between March and August, 1974 and this could have resulted in warmer waters descending to the hypolimnion. Until the end of June, 1975 discharge was minimal, therefore, there was less exchange between hypolimnetic and surface waters, so that the temperature of the discharge water would have been very low in July. The level of the water outlet to the turbines of the hydro-electric power station was raised nearer the surface in 1974 and, naturally, this, also, influenced water temperatures throughout the year.

Nutrients

Nitrite, nitrate and phosphate concentrations can be very high (maximums 65, 525 and 142 µgr/1) in sampling areas owing to the hypolimnetic origin of the water and the influx of untreated waste water from villages and farms. There is an abundance of dissolved organic matter (DOM) and particulate organic matter (POM). Primary production is high and as there are no severe pollution problems, the oxygen supply is plentiful. There is ample food for the development of an abundant benthic fauna.

Macroinvertebrates on stones

Larvae of *Psychomyia pusilla* (Trichoptera) and certain Chironomidae (Diptera) genera, especially *Orthocladius, Eukiefferiella* and *Cricotopus*, dominated the benthic fauna on the upper surface of the stones at the sampling station. The two groups formed between 60% and 95% of the total fauna and, at times, *Psychomyia* represented as much as 47 and 78% of the total individuals (Prat, 1980). Another Trichoptera, *Hydroptila*, was abundant in certain periods, especially in March, 1974, when 19% of the total fauna were larvae of a *Hydroptila* species. The high numbers seemed to be associated with the degree of *Cladophora* development on the stones; for in 1975 when *Cladophora* was scarce, the percentage of *Hydroptila* was low. Other invertebrates present on the stones were Ephemeroptera (*Baetis, Ephemerella, Caenis*), Trichoptera (*Hydropsyche, Rhyacophila* and *Tinodes*), Diptera (Simuliidae, Tipulidae, Empididae) and Mollusca (*Ancylus*), but they were never plentiful.

296

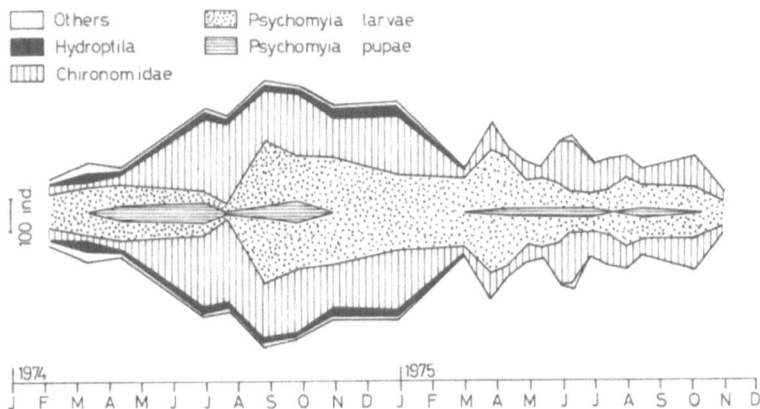

Fig. 4. Changes in total number of specimens on the upper 100 cm² stone surface in the River Ter. The relative importance of each group is shown.

Changes in composition and numbers of individuals

Hydroptila was relatively important only in 1974, when *Cladophora* developed profusely and water temperatures were high. In 1975, when water temperatures were lower and *Cladophora* development poor, *Hydroptila* did not form an important part of the benthic fauna. Parallel changes were noted in certain *Chironomidae* (Fig. 4).

In September, 1974, 429 *Psychomyia* were counted in the total 846 specimens that made up the mixed community on the upper 100 cm² surface. Before the flood in the same month of 1975, 321 specimens were recorded, but after, in November, only 138 could be found; while in the November of 1975, 620 were collected.

Life cycle of Psychomyia pusilla

The five *Psychomyia pusilla* larval stages are separated by measuring the head width. Because of the mesh gauge employed, stages 1 and 2 are difficult to separate and stage 1 may be under represented in the sample. Pupae first appeared in March and were present until October in 1974 and until September in 1975. A comparison of the distribution of the larval stages in the two years is shown in Fig. 5. The pupae incidence was high from March to June, 1974, when they made up 50% of the total *Psychomyia*. By July, practically all populations were pupating or in the last larval stages, but few larvae remained on the stones in August and the first generation seems to have petered out at this time. One month later, in September, numerous first and second stage larvae were collected together with others in the last larval or early pupal stages. During October and November, larvae were in one of the first three stages, but there was also a good stock of pupae in October. The September and October temperatures were as high

297

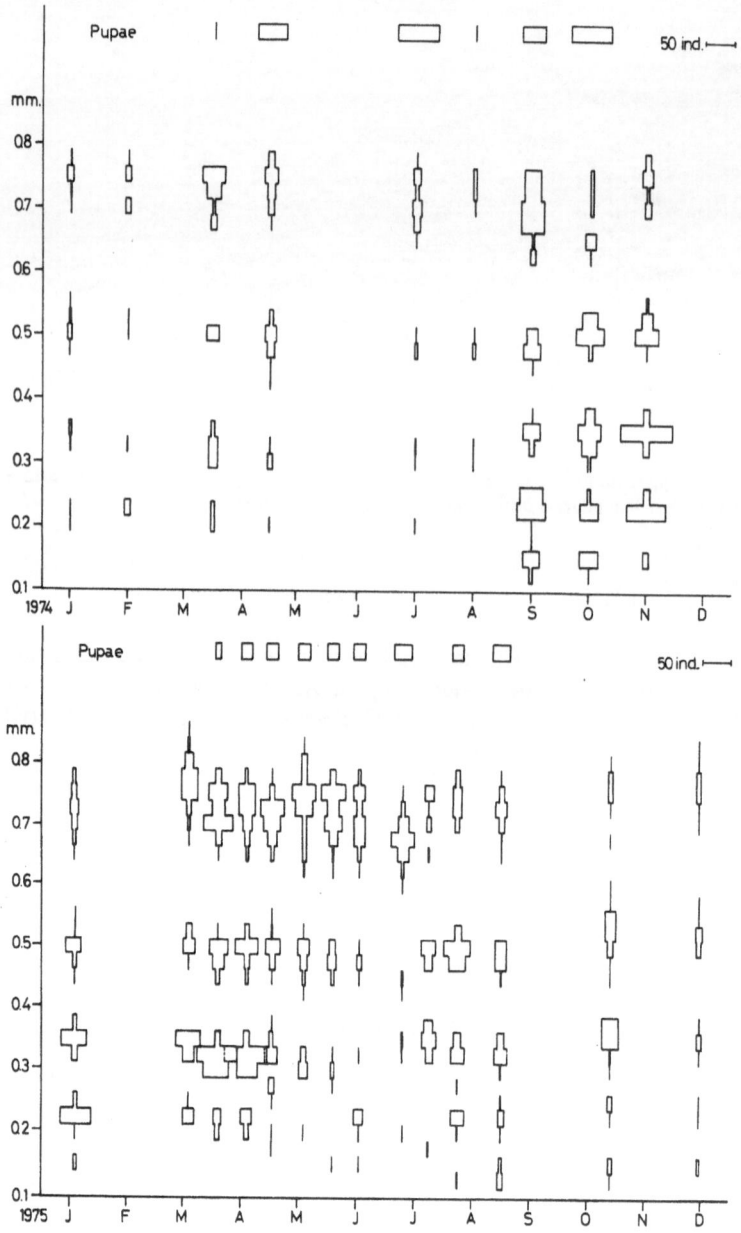

Fig. 5. Psychomyia pusilla larval stages for two successive years. Ordinates equal width of head capsule in mm.

as those of summer and in these conditions, even when larvae are absent, a second generation can develop very quickly, although only part of this generation emerges before November.

The bi-monthly samples taken in 1975 give a better picture of the

298

evolution throughout the year. There was a first generation between March and July (Fig. 5). As in the previous year, most specimens were at the final larval stage or pupating by July, although the pupae did not form as high a percentage of the population. No pupae were found when water temperatures dropped to around 10°C in July and all larval stages were scarce. The number of larvae and pupae was already reduced before the September flood, after it subsided no pupae could be found on the stones. The low growth rate of *Psychomyia* in 1975, as well as the absence of pupae in July of the same year, can be attributed to the lower water temperatures when the life cycle of the first generation had not been completed and when there was little chance of a second generation developing. Pupae found at the end of July may be the metamorphosized larvae of the first generation, although the development of some larvae from egg masses hatched in summer cannot be excluded.

DISCUSSION

Many factors, such as discharge, current velocity, temperature, food supply, heterogeneity of the substrate etc., influence the distribution and composition of river benthic fauna (Cummins, 1975). If certain of these factors remain constant, it would seem resonable to attribute changes in distribution and composition to those factors which undergo variations. As specimens were always collected from the same stone substrates at the same sites in the centre of the stream, where the current was fastest, during sampling in the River Ter and, as food was plentiful, it could be hypothesized that the modifications observed in incidence and composition of the benthic populations during the two-year study period were the result of variations in discharge and temperature.

Discharge plays a major role in river ecology. The result of extensive flooding can be disasterous for benthic fauna (Kohmann, 1980) while regulated water discharge has been shown to enhance the biomass of benthic populations (Ward and Short, 1978). The greater discharge from the Susqueda Reservoir in 1974 would have been expected to favour high numbers of stone-dwelling macroinvertebrates, and the September flood to reduce them in October and November.

It has been suggested that temperature is one of the factors responsible for the distribution and numbers of river larvae (Ide, 1935; Carlsson, 1977) and this has been demonstrated in controlled temperature investigations on Trichoptera (Anderson, 1978; Roux, 1979). Life cycles are temperature dependent, while the emergence of imagos is related to both absolute temperature and to temperature change patterns (Flannagan, 1978). A given species can complete one or more life cycles within a given period of time according to temperature (Hynes, 1970) and this accounts for the shorter life cycles of the aquatic insect populations in Southern Europe (Thibault, 1971).

The temperature in the River Ter would seem to affect both the composition and the life cycles of aquatic insects. It would, also, seem to have

299

been responsible for the scarce summer growth of *Cladophora* in 1975 and, in consequence, for the limited development of *Hydroptila* and certain *Chironomidae*. In that year there was no second generation of *Psychomyia*, whereas, in 1974, part of the larvae born in the summer seem to have emerged in autumn.

The effects of the regulation of discharge into the River Ter by the Susqueda Reservoir are similar to those described for other impoundments (Ridley and Steel, 1975). The river temperature in thrown out of balance when the river is fed from the hypolimnion of a reservoir (Hannan and Young, 1974), because the temperature range is narrowed; the winter and autumn temperatures being higher than normal (Short and Ward, 1980); the summer ones lower.

We propose that discharge and temperature are the principal factors responsible for composition and growth differences in the stone-dwelling benthic organisms in this reach of the River Ter. Furthermore, we suggest that the controlled discharge from the Susqueda Reservoir influences both factors for many kilometres downstream.

ACKNOWLEDGEMENTS

I should like to thank Dr R. Margalef for his helpful criticism and for reviewing the text; Anna M. Domingo for the drawings and the Hidroeléctrica de Catalunya for providing data on reservoir discharge. Sampling was done when the author was a Ministerio de Educación y Ciencia postgraduate fellow.

REFERENCES

Anderson, N.H. *In* Proc. 2nd Int. Symp. on Trichoptera, ed. M.I. Crichton, pp. 317–329, Junk, The Hague, 1978.
Carlsson, M.; Nilsson, L.N.; Svensson, B.; Ulfstrand, S. and Wotton, R.S., Oikos 29: 229–238, 1977.
Cummins, K.W., *In* River Ecology, ed. B.A. Whitton, pp. 170–191, Blackwell, London, 1975.
Flannagan, J.F., *In* Proc. 2nd Int. Symp. on Trichoptera, ed. M.I. Crichton, pp. 183–197, Junk, The Hague, 1978.
Hannan, H.H. and Young, W.J., Hydrobiologia 44:177–207, 1974.
Hynes, H.B.N., The Ecology of Running Waters. Liverpool, United States, 1970.
Ide, F.P., Publ. Ont. Fish. Res. Lab. 50:1–76, 1935.
Jalon, D.G.de, Annl. Limnol. 13:221–226, 1977.
Kohmann, F., Spixiana 3:91–97, 1980.
Prat, N., Quad. Ecol. Apl. 4:87–107, 1979.
——, Col. loquis Soc. Cat. Biologia 12:27–34, 1980.
——, Bautista, I.; Gonzalez, G. and Puig, M.A., *In* El Patrimoni Natural Andorra, ed. R. Falch, pp. 261–309, Ketres, Barcelona, 1979.
Ridley, J.E. and Steel, J.A., *In* River Ecology, ed. B.A. Whitton, pp. 565–587, Blackwell, London, 1975.
Roux, C., Fresh. Biol. 9:111–117, 1979.
Short, R.A. and Ward, J.V., Can J. Fish. Aquat. Sci. 37:123–127, 1980.
Thibault, M., Ann. Hydrobiol. 2:241–274, 1971.

Viedma, M.G.de and Jalon, D.G.de, Aquatic Insects. 2:1–12, 1980.
Ward, J.V. and Short, R.A., Verh. Int. Verein. Limnol. 20:1382–1387, 1978.

DISCUSSION

Terra: Was *pusilla* the only species of *Psychomyia* you found in the River Ter? Have you found *P. ctenophora* in any other type of habitat, particularly in larger rivers with slower current?

Prat: *P. pusilla* was the only species identified in this study.

Garcia de Jalon: I find, as Dr Terra does, that *P. pusilla* inhabits the upstream reach of a river and *P. ctenophora* the downstream.

LES LARVES DE TRICHOPTÈRES DE LA RIVIÈRE LLOBREGAT (CATALOGNE, ESPAGNE). DISTRIBUTION LONGITUDINALE ET RELATION AVEC LA QUALITÉ DE L'EAU.

M.A. PUIG, I. BAUTISTA, M.J. TORT ET N. PRAT

SUMMARY

The longitudinal distribution of Trichoptera along the River Llobregat (Catalonia, NE Spain) seems to be explained by the different physico-chemical conditions of the water. The higher zone (only one sampling station) is characterised by more varied and very different Trichopteran fauna than the lower ones. The lower reaches are very polluted and no Trichoptera can be found there. In the intermediate zone, the Trichopteran fauna changes from higher to lower sampling stations. In the higher intermediate zone, cleaner and colder water favours the dominance of *Rhyacophila evoluta*, the polluted lower intermediate zone is dominated by *Hydropsyche exocellata*. Four species of *Hydropsyche* are present in the river, *H. pellucidula* is found only at station 56 (the higher sampling station), and *H. angustipennis*, *H. exocellata* and *H. contubernalis* appears succesively from high to lower river bed. *H. exocellata* is the most resistent species and the only Trichoptera inhabiting the highly polluted lower intermediate zone.

INTRODUCTION

Dans notre pays il n'y a guère de travaux sur les insectes aquatiques et sur leur importance sur l'écologie des fleuves (Prat et al., 1979).

Ce mémoire constitue une partie de l'étude écologique des communautés bentoniques des réseaux hydrographiques de deux fleuves catalans: Llobregat et Besós. Les données appartiennent à trois prélèvements effectués pendant l'été et l'hiver 1979 et le printemps 1980. Dans l'ensemble de ce travail nous présentons une description de la distribution de Trichoptères tout au long de l'axe fluvial du Llobregat, ainsi que les mesures et références qui semblent liées à la vie des espèces.

MATERIEL ET METHODES

De la totalité des 109 stations de prospection, 18 ont été choisies pour cette étude, étant situées tout au long de l'axe fluvial du Llobregat, depuis la source jusqu'à l'embouchure du fleuve (Fig. 1).

Pour chaque station nous avons tenu compte de la conductivité, la température, le pH, et au laboratoire nous avons déterminé les mesures des phosphates, nitrites, nitrates, d'oxygène dissous, alcalinité, chlorures, sulfates, silicates et le poids sec, ainsi que la mesure de la quantité de quelques cations.

La récolte des larves a été réalisée qualitativement en choisissant les prélèvements destinés à obtenir le plus d'espèces possible, plutôt qu'un grant nombre d'individus pour quelques unes. Les larves ont été récoltées à la main sur des surfaces de pierres, macrophites, mousses, etc. Les échantillonnages étaient ensuite filtrés à travers un filet de 250 microns de diamètre et fixés. C'est au laboratoire que s'est effectué le tri et la classification de toutes les larves de Trichoptéres.

Situation du cours d'eau étudié

Le bassin du Llobregat, qui a une surface de 4.948 Km2, est situé au Nord-Est de la Péninsule Ibérique. Le fleuve Llobregat naît à la 'Sierra del Cadi' (Pré-Pyrénées), environ à 1500 m d'altitude et atteint son embouchure dans la Méditerranée à Barcelone; il a une longueur de 156 Km.

Par son profil (Fig. 1C) on constate que sa pente est faible, à l'exclusion des 20 premiers kilomètres de son parcours.

RESULTATS

Caractéristiques physico-chimiques

Les terrains traversés par le Llobregat sont calcaires, et dans quelques zones (Balsareny, station 68) il traverse des gisements de chlorures. Pour cette raison, dès sa source, la minéralisation de ses eaux est déjà très importante; mais c'est à partir de la station 68 que se produit une augmentation remarquable de la concentration des chlorures qui deviennent chaque fois plus élevés jusqu'à son embouchure (Fig. 2).

L'augmentation de la contamination des eaux du fleuve est aussi graduelle à cause de la très grande utilisation de ses eaux par les villages et les usines des alentours.

À partir des stations 68 et 103 il se produit une grande augmentation de la quantité de chlorures, phosphates et nitrogènes, et après la station 91 on constate une augmentation considérable de différents contaminants (organiques et inorganiques) qu'on peut cbserver dans l'accroissement des phosphates et ammonium (Fig. 3) accompagnés d'une diminution de la concentration d'oxygène dissous (Fig. 4).

Les apports du fleuve Anoia, effluent qui reçoit sur sa droite le Llobregat, les déversements des villages et l'accumulation industrielle sur les rives, sont les principaux agents de cette grande dégradation du fleuve.

304

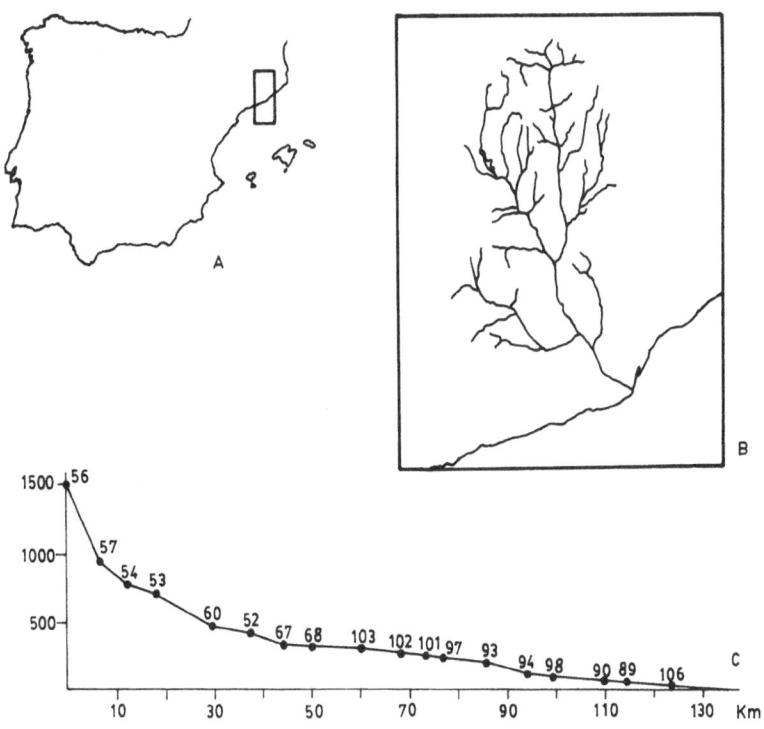

Fig. 1. A, Situation du bassin du Llobregat dans la Péninsule Ibérique; B, Carte du Llobregat et de ses affluents; C, Profil du Llobregat.

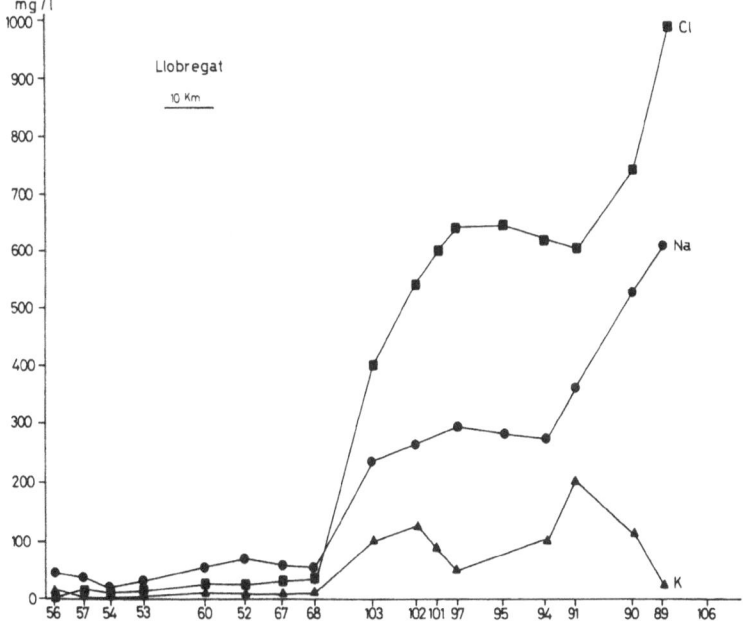

Fig. 2. Profil en ion chlorure, sodium et potassium du Llobregat en aval. (Décembre 1979)

305

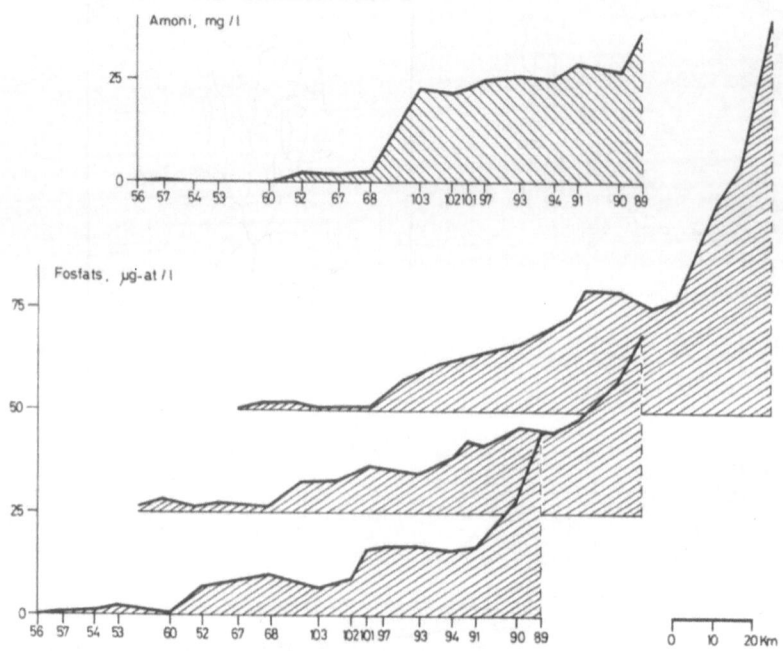

Fig. 3. Evolution de la teneur en ammonium (en haut) et en phosphates (en bas) en aval du Llobregat. (Août et Décembre, 1979 et Avril, 1980)

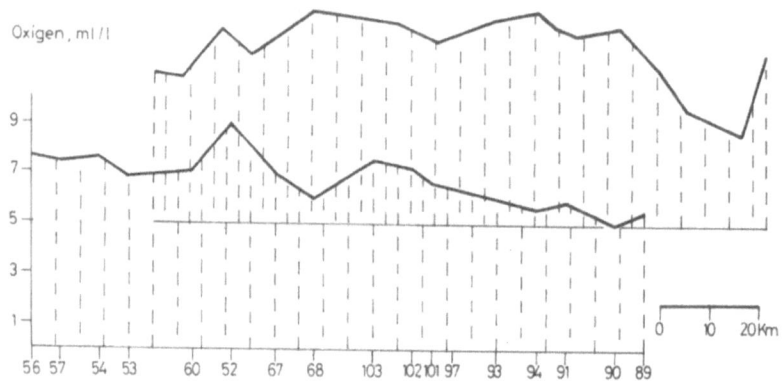

Fig. 4. Evolution de la teneur en oxygène du Llobregat en aval. (Août et Décembre, 1979)

Distribution des Trichoptères tout au long du Llobregat

Nous avons utilisé les données de l'hiver pour la représentation des espèces de Trichoptères tout au long du fleuve, étant donné que, pendant cette époque, la plus grande partie des espèces se trouve en phase larvaire dans l'eau.

De cette distribution spécifique on peut s'apercevoir de l'identité particulière de la station 56 par rapport aux autres, la forte pente, les eaux rapides, propres et froides sont des habitats excellents pour le développement de différentes espèces de Trichoptères comme: *Limnephilus sp.*, *Tinodes dives*, *Micrasema morosum* et *Hydropsyche pellucidula* qu'on trouve seulement dans cette station, et avec une répartition plus grande dans le Llobregat où l'on trouve aussi *Rhyacophila evoluta*.

Rhyacophila evoluta reste importante jusqu'à la station 68, c'est-à-dire, avant l'entrée du déversement du sel des gisements de Balsareny et au début de la zone avec forte contamination organique (Fig. 3). Dans cette zone de distribution de *Rhyacophila evoluta*, d'autres espèces de Trichoptères commencent à apparaître ainsi que différents Hydropsychidae ou *Hydroptila*. La présence de *Hydroptila* peut être dévalorisée à cause de la relation directe de cette espèce avec *Cladophora* (Prat, 1980), étant donné que ces organismes se développent sur les algues, presque absentes en hiver. Au mois de Mai 1980, on trouve la présence de ce Trichoptère dès la station 53 jusqu'à la station 101, en même temps qu'une apparition de l'algue.

On trouve une succession des différentes espèces de *Hydropsyche* tout au long du fleuve, avec l'apparition d'abord de *H. angustipennis*, ensuite de *H. exocellata* et finalement *H. contubernalis*, séquence semblable à celle qui a été trouvée par d'autres auteurs en France (Verneaux et Faessel, 1976). On constate la présence de *H. exocellata* jusqu'à la station 94, à partir de laquelle on vérifie une véritable élimination de ses larves, par la grande augmentation de la contamination des eaux (Fig. 3).

La station 68 doit être soulignée, car elle marque les limites pour la distribution de Trichoptères, associés aux eaux froides et propres, et qui sont en concordance avec la limite de distribution des Plécoptères et de quelques Ephéméroptères, liés à un type d'habitat d'eaux plus propres et froides comme *Ecdyonurus*.

Entre les stations 68 et 91 les communautés de *Hydropsyche* se développent très abondamment, par la coexistence de plusieurs espèces, cas particulièrement connu dans ce genre (Hildrew, 1978).

On doit souligner aussi la présence remarquable de *Psychomia pusilla* à la station 103; on pourrait penser qu'il s'agit d'un reste d'une plus grande distribution de cette espèce, maintenant limitée à cet espace écologique à cause du changement des conditions du fleuve dans son dernier parcours, parce que c'est une espèce très fréquente dans les régions plus proches à l'embouchure des fleuves (Decamps, 1967).

La dominance relative des différentes espèces peut s'observer sur la Fig. 5, où l'on constate la grande diversité de la faune de Trichoptères à la source du fleuve (station 56). La dominance de *Rhyacophila evoluta* jusqu'à la station 68 sur les autres Trichoptères (spécialement *H. angustipennis*), l'apparition de *H. exocellata* à la station 60 et la dominance de cette dernière espèce tout le long du fleuve sont évidentes. On s'en rend également compte, à partir de la station 102 et surtout de la station 97. Cette augmentation est favorisée par l'eau en provenance du fleuve Cardoner qui débouche dans le Llobregat sur la rive droite et porte une charge de sel et de contamination urbaine et industrielle très importante, mais qui

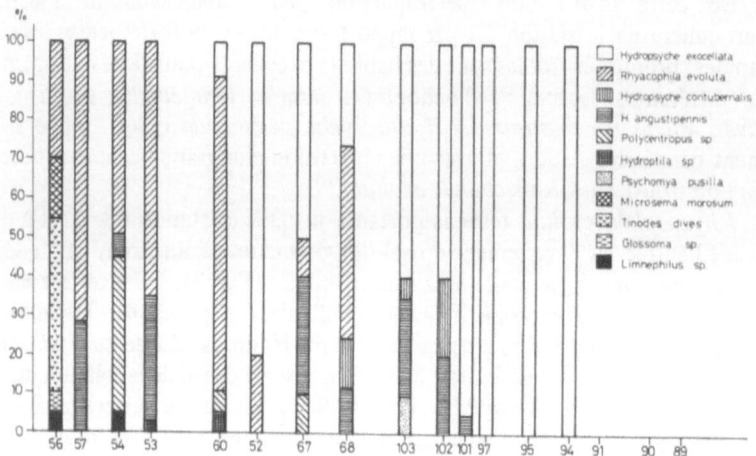

Fig. 5. Importance des différentes espèces dans le benthos, en pourcentage du nombre total d'individus récoltés, dans les stations d'échantillonnage.

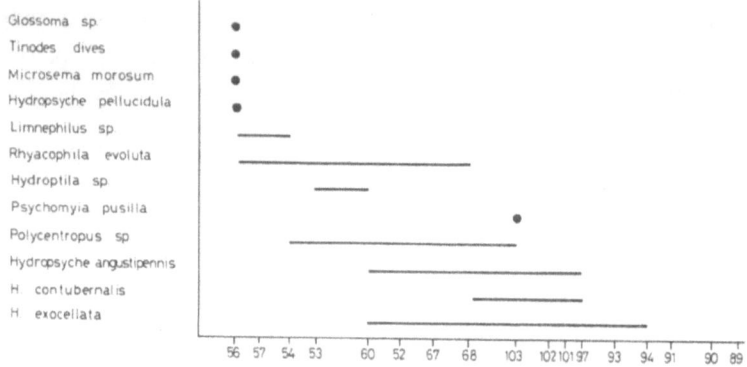

Fig. 6. Répartition altitudinale des Trichoptères du Llobregat.

n'a pas de conséquence significative sur les taux d'oxygène dissous (Fig. 4). C'est dans ces conditions que *H. exocellata* devient l'espèce dominante.

DISCUSSION

Avec les données disponibles actuellement, on peut apprécier l'identité très différente de la station 56 (la plus haute) et les autres. La plus grande partie des espèces qui se trouvent à cette station, ne se retrouvent pas aux autres (Fig. 6). Aussi la zone comprise au-dessous de la station 91 se détache-t-elle parmi les autres par l'absence de Trichoptères en conséquence de la forte pollution.

Dans le cours intermédiaire du fleuve, on peut relever une différence

308

entre la section supérieure du fleuve, jusqu'à la station 68 et la section inférieure, surtout après la confluence du Cardoner, c'est-à-dire, au-dessous de la station 102. La zone supérieure se caractérise par la présence d'eaux ayant un degré de contamination modérée, diluée à cause du débit du Llobregat et de la faible concentration de population et industrielle. Dans ces conditions on peut évaluer le développement d'une communauté constituée par *Rhyacophila evoluta* et *Hydropsyche angustipennis*, avec *Polycentropus* abondant dans les zones à eaux calmes (Edington, 1963), et l'apparition de *Hydropsyche exocellata*. Cette espèce devient la plus importante à cause de sa résistance aux changements des conditions de l'environnement, fait qui est plus évident à partir de la station 101. On pourrait toujours penser que le plus grand pourcentage de *H. exocellata*, dans les stations intermédiaires (52, ou 67), peut être en rapport avec un degré plus élevé de la pollution dans ces stations, déterminée par des activités humaines.

On est obligé de signaler la présence dans tout le cours d'eau de petits barrages qui favorisent la présence des espèces filtrantes sur d'autres espèces avec alimentation différente (Short & Ward, 1980).

REMERCIEMENTS

Nous remercions sincèrement la "Diputacion de Barcelona" pour l'aide économique qui a permis la réalisation de cette étude.

REFERENCES

Decamps M.H., Ann. Limnol. 3:101–176, 1967.
Edington, J.M., Proc. zool. Soc. Cand. 143: 281–300, 1963.
Hildrew A.G., *In* Proc. 2nd Int. Symp. on Trichoptera, ed. M.I. Crichton, 269–281, 1978.
Prat N., Col. loquis Soc. Cat. Biologia 12:27–34, 1980.
Prat N.; Bautista I.; Gonzalez G. et Puig M.A., *In* El Patrimoni natural Andorrà, ed. Folch, pp. 261–309. Ketres éditora Barcelona, 1979.
Short, R.A. et Ward J.V., Can. J. Fish. Aquat. Sci. 37: 123–127, 1980.
Verneaux J. et Faessel B., Ann. Limnol. 12:7–16, 1976.

RESPONSES OF THE SERICOSTOMATID CADDISFLY *GUMAGA NIGRICULA* (McL.) TO ENVIRONMENTAL DISRUPTION

V.H. RESH, T.S. FLYNN, G.A. LAMBERTI, E.P. McELRAVY, K.L. SORG, AND J.R. WOOD

SUMMARY

Life history features of the sericostomatid caddisfly *Gumaga nigricula* (McL.) are described and include: a firm gelatinous egg mass that has a 'figure-8' outline; laboratory-reared larvae that molt as many as 14 times; field-collected larvae whose instars cannot be distinguished by head capsule measurements; pupae that are aggregated along stream margins; and adults, larvae, and pupae that serve as paratenic hosts of a merminthid nematode. Density of larvae and pupae in a northern California spring seepage (Hopland, CA, USA) was 157.6 ± 78.9 individuals $/158$ cm^2, or $\approx 10,000/$m^2. Following severe drought and complete loss of habitat, *G. nigricula* age structure in the Hopland spring changed from that of a multiple cohort population to that of a single cohort population. Population densities and faunal dominance of *G. nigricula* in a stream habitat (Big Sulphur Creek, The Geysers, CA) were higher following drought conditions (which resulted in reduced flow but not complete habitat loss) than following years of normal and above-normal precipitation. A recolonization study in Big Sulphur Creek indicated that hyporheic and upstream movements are more important in habitat recolonization than drift or aerial oviposition. In a long-term geothermal energy development area (The Geysers, CA, USA) numbers of macroinvertebrate species decrease with the addition of thermal, silt, and chemical contaminants, but both density and faunal dominance of *G. nigricula* increase along this gradient. Although *G. nigricula* exhibits considerable potential for use as an indicator of water quality, taxonomic and biometric considerations require further clarification.

INTRODUCTION

Over the past several years, we have examined the effects of environmental disturbance on aquatic macroinvertebrate communities in several northern California streams. An interconnecting link among these studies has been the presence of populations of the sericostomatid caddisfly *Gumaga nigricula* (McL.). This paper presents an overview of life history features of this species and examples of its response to specific environmental disruptions.

SITE DESCRIPTIONS

Our studies on *G. nigricula* have been concentrated in two types of habitats, springs and streams, that occur in the northern California Coastal Range, an area characterized by cool, wet winters and hot, dry summers. The particular spring discussed in this paper is located at the University of California Hopland Field Station (Mendocino Co., CA, USA). The spring seepage is 30 m in length from its near-constant temperature source ($16° \pm 1°C$; $\bar{x}$ discharge = 0.2 1/s) to its downstream reaches where water temperature fluctuations approach ambient air temperatures (-2 to $+40°C$). The stream habitat discussed is Big Sulphur Creek, a second-order stream located at The Geysers (Sonoma Co., CA, USA).

LIFE HISTORY

G. nigricula eggs are contained within a firm gelatinous covering which, when unfolded, has a 'figure-8' outline. Each mass contains from 78–358 eggs.

The number of larval instars in either spring or stream habitat cannot be distinguished by head width measurements (Fig. 1A, B) unlike many other caddisfly larvae (e.g., Mackay, 1978). Winterbourn (1978) has reported a similar inability to distinguish larval instars by head capsule measurements in a New Zealand sericostomatid caddisfly population of *Olinga feredayi* (McL.). These findings may be related to our observation that individual larvae of laboratory-reared *G. nigricula* have molted as many as 14 times. Similarly, C. Denis (personal communication) has observed that laboratory-reared larvae of *Sericostoma galeatum* Rambur have molted 10 times.

Prior to pupation, full grown larvae of *G. nigricula* form extensive aggregations, usually along shoreline margins (e.g., Waters and Resh, 1979, Fig. 2). In 69 individual clumps examined along a 100-m stretch of Big Sulphur Creek shoreline, clump size varied from 3 to 1,410 individuals but the majority of clumps contained less than 200 organisms.

Although adult flight in Big Sulphur Creek occurs from May to September, most adults emerge during the spring. Adult flight information is not available for the Hopland spring, but pupae have been found year-round.

Poinar et al. (1976) have reported that larvae, pupae, and adults of *G. nigricula* from the Hopland spring contain infective stage juveniles of the merminthid nematode *Pheromermis pachysoma* (von Linstow). *G. nigricula* serves as a paratenic host for this parasite of the yellowjacket, *Vespula pensylvanica* (Saussure).

Populations can reach very high densities. In March 1976, extensive sampling of the Hopland spring (64 158-cm² quadrats with a modified Surber sampler) indicated that although an average of 9 macroinvertebrate species occurred in each sample, *G. nigricula* comprised > 80% of all individuals collected. Densities ranged from 19–404 larvae and pupae ($\bar{x}$ = 157.6 ± 78.9)/158 cm² or $\approx 10,000/m^2$. This is among the highest densities yet recorded for any species in the caddisfly suborder Integripalpia.

312

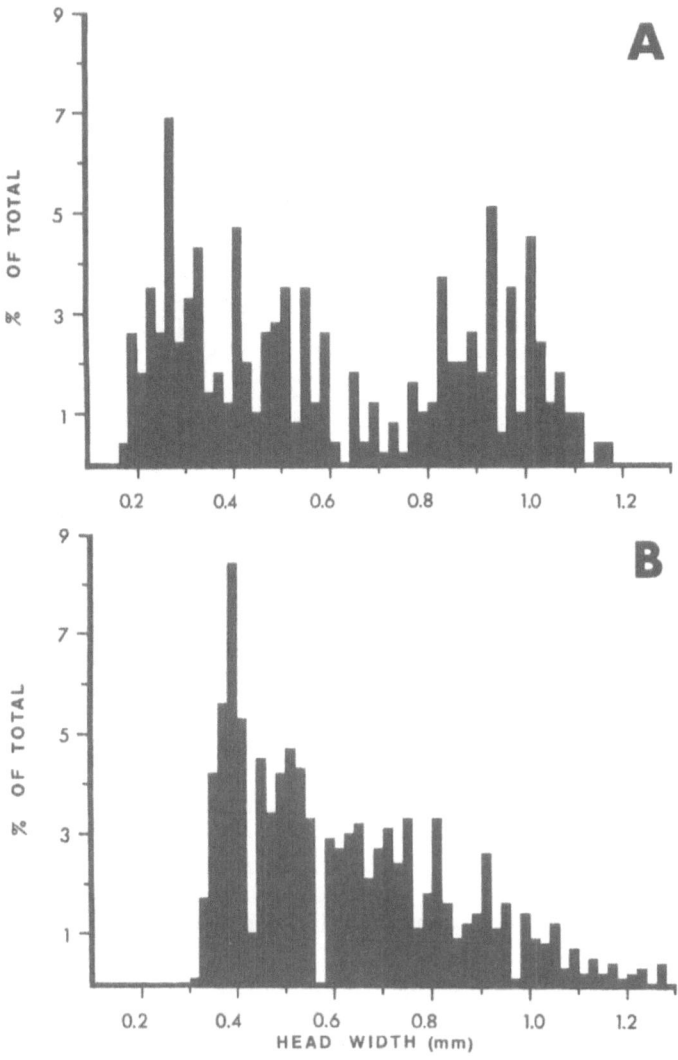

Fig. 1. Frequency distribution of *G. nigricula* headwidth measurements from (A) the Hopland spring and (B) Big Sulphur Creek.

RESPONSE TO ENVIRONMENTAL DISRUPTION

Habitat loss and recovery. Based on a year-long study (November 1975–October 1976) of *G. nigricula* at the Hopland spring, we determined that during each month of the year early instar larvae (i.e., head width < 0.50 mm, Fig. 1A) comprised > 50% of the population present, whereas pupae comprised 1–6% of the population. This age structure is typical of a multiple cohort population. However, during both the summer of the year (1976)

313

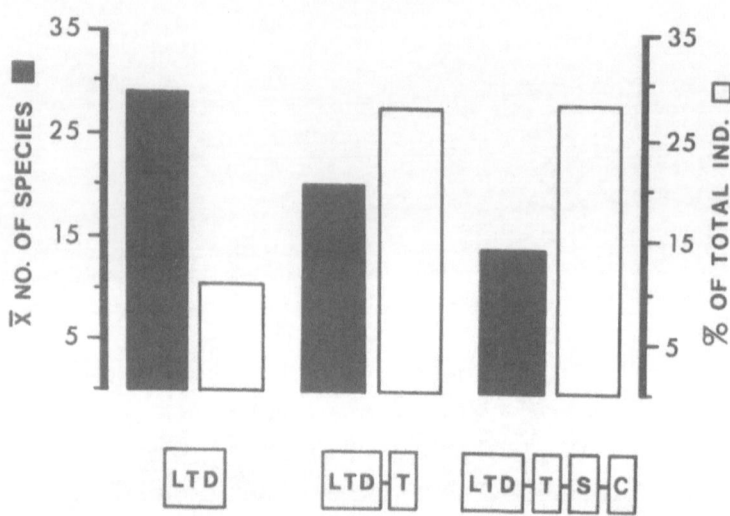

Fig. 2. Mean number of macroinvertebrate species per Surber sample (shaded bars) compared with relative abundance of *G. nigricula* [as % of total macroinvertebrates collected per site (open bars)] at three sites along Big Sulphur Creek (see text for further explanation).

when these estimates were made and the following summer (1977), precipitation (45.6 and 40.7 cm, respectively) was lower than the minimum annual rainfall recorded during any of the previous 25 years ($\bar{x} = 88.9$ cm, 55.6–153.4 range). Consequently, during early summer of 1977, the Hopland spring went completely dry and flow did not resume until autumn of 1977. No live individuals were found when sub-surface areas of the dry spring were examined nor have either field or laboratory tests indicated that eggs or larvae can survive dessication. The adults are short-lived (5–8 days). Thus, all evidence indicates that no *G. nigricula* individuals survived this prolonged loss of habitat (but see comment by L. Botosaneanu following literature citations).

By spring of 1978, following a winter of heavy rains (133.8 cm), the permanent habitat had returned. Although no specimens of *G. nigricula* were found in April 1978 collections, *G. nigricula* was present in April 1979 collections but the population was almost entirely composed of full grown larvae and pupae, with early instar larvae comprising < 2% of the population. By July 1979, early instar larvae comprised > 99% of individuals present with densities approximating pre-drought levels. A comparison of pre- and post-drought population age structure indicates that a shift occurred from a multiple cohort to a single cohort pattern following the loss and subsequent recovery of the spring habitat. Age structure determined in 1980 followed that of 1979, indicating a continuation of the single cohort population pattern.

Severe Drought and Washout Conditions. The Big Sulphur Creek *G. nigricula* population was also markedly affected by the 1975–77 drought but total

314

habitat loss did not occur. Precipitation during 1975–77 was lower (1975–76: 54.7 cm, 1976–77: 48.9 cm) than normal (25 yr $\bar{x}$: 146.2 cm) but heavy rains occurred during the following year (1977–78: 216.4 cm). In 1978–79, rainfall was closer to normal levels (97.9 cm). Thus, over the period 1975–79, Big Sulphur Creek experienced drought (1975–77), washout (1977–78), and near normal rainfall conditions (1978–79). Corresponding *G. nigricula* densities in May (when the seasonally heavy rainfall ends) were: 1977 (following drought): $\bar{x} = 4698/m^2$; 1978 (following washout): $\bar{x} = 65/m^2$; 1979 (following normal precipitation): $\bar{x} = 398/m^2$. Reduced spate activity during the 1975–77 drought years may have resulted in lower larval mortality in the overwintering population. In contrast, more severe spate activity following the 1977–78 rainy season may account for the lower densities in May of 1978 than 1979. When numbers of *G. nigricula* are expressed as a percentage of total macroinvertebrates collected, the relative abundance of *G. nigricula* is greater during drought conditions (34%) than compared to normal precipitation (3%) or washout (1%) conditions.

Recolonization. Caddisflies and other aquatic insects may colonize habitats by: 1) downstream drift, 2) upstream movements, 3) aerial oviposition, and 4) hyporheic movements. Since there was apparently no upstream, downstream, or hyporheic source for recolonization of the Hopland population, we presume that adults migrated from other *G. nigricula* populations that had survived the drought (cf. L. Botosaneanu comment). Recolonization mechanisms were also examined in a section of Big Sulphur Creek during the spring of 1978. Traps designed after those of Williams and Hynes (1976) were used to distinguish the influence of each of the above recolonization sources. The colonization pattern for *G. nigricula* indicated that hyporheic movements accounted for 50% of total recolonization, upstream movements 37%, aerial oviposition 8%, and drift 5%.

Geothermal Energy Development and Operations. Big Sulphur Creek is the main drainage of The Geysers Geothermal Resource Area, the largest development of this type currently operational worldwide. Surber sampler (0.093-m² quadrats) collections were made at sites along a section of this stream where long-term geothermal energy development has occurred. Comparison of a site without additional inputs (LTD), with a site having a thermal (T) input ($\bar{x}$ annual $\Delta T = +4°C$), and a site having thermal ($\bar{x}$ annual $\Delta T = +5°C$ from LTD site), silt (S), and chemical (C, e.g., sulfates, ammonium, heavy metals) inputs indicates that numbers of macroinvertebrate species decrease (Fig. 2). Conversely, both the density of *G. nigricula* and its dominance of the macroinvertebrate fauna (Fig. 2) increase along this gradient.

Laboratory Bioassays. Russell et al. (in press) examined the influence of oil shale retort water on *G. nigricula* in laboratory bioassay experiments and determined that even at the highest concentrations tested, no demonstrable effects on survival or activity patterns of this species were evident. In contrast, ammonium carbonate concentrations of 4.52 mM to which *G.*

315

nigricula showed no response, resulted in significant larval mortality for the limnephilid caddisfly *Dicosmoecus gilvipes* (Hagen).

DISCUSSION

Rosenberg and Wiens (1976) and Resh (1979) discussed three criteria for determining potential use of aquatic insects as indicators of environmental change due to human interference. These criteria will be examined in terms of known information about the biology of *G. nigricula:*

1) *Taxonomic soundness* – i.e. there should be no disagreements on the validity of the species and it should have reasonably distinct characteristics for ease of identification. There are two described species of *Gumaga* from North America, *G. nigricula* and *G. griseola* (McL.). A third species (*G. okinawaensis* Tsuda) is known from Japan. However, at least two morphological types of larvae are known to occur for *G. nigricula.* Larvae of this species from Big Sulphur Creek have head and pronotum characteristics that are very distinct from those of the Hopland spring population (Fig. 3). However, both G.W. Wiggins and D.G. Denning (personal communication) indicate that adults from both populations are *G. nigricula.* A series of specimens from each population has been placed in the Royal Ontario Museum and the California Insect Survey collections.

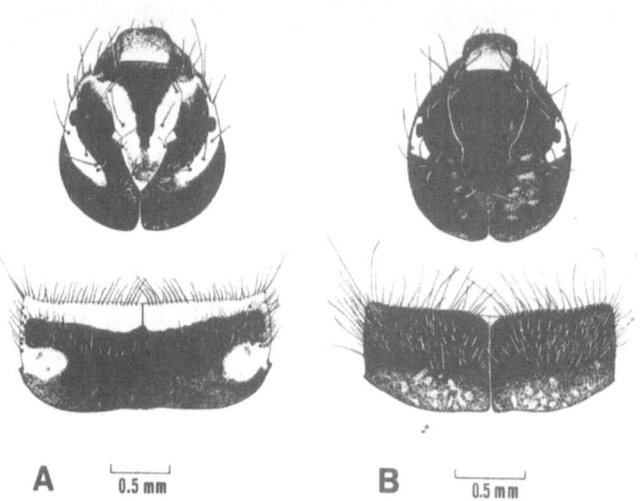

A 0.5 mm   **B** 0.5 mm

Fig. 3. Larval head capsule and pronotum of *G. nigricula* from (A) Big Sulphur Creek and (B) the Hopland spring.

2) *Widespread distribution* – i.e. it or an ecologically similar 'sister' species should have a broad zoogeographic distribution in order that it may be used predictively over wide geographic areas. *G. nigricula* is known from Arizona, California, and Oregon, although it seems to occur most commonly in California coastal streams and small tributaries in the Sierra Nevada mountain range.

3) *Abundance* – i.e. it must be numerous in the community before impact occurs and the population response to the impact can be either positive or negative. As found in the Hopland spring population ($\bar{x} = 10,000$ individuals/m^2), *G. nigricula* often reaches very high densities. Increased population densities accompanying chemical, thermal, and silt inputs to Big Sulphur Creek would be an example of a positive population response to environmental impact.

To the above three criteria, we feel that an additional category should be added: 4) *Biometric considerations* – i.e. the population can be quantitatively sampled. *G. nigricula* has a particularly useful life history feature in this regard. The presence of its distinctive egg mass enables the potential size of the total population to be determined. Almost all previous population studies of aquatic insects have relied on early instar densities to determine maximum population size, which invariably results in an underestimate. Since newly hatched larvae may be particularly sensitive to environmental perturbations, an ability to quantitatively sample the egg stage is a major advantage in assessing such potential effects. Pre-pupal movements and subsequent pupal aggregations of *G. nigricula* could bias population estimates since a sampling design for estimating larval abundances would probably fail to accurately estimate pupal densities. However, this can be solved by developing a two stage sampling design in which larval and pupal densities are determined separately.

It is apparent that the effective use of *G. nigricula* as an indicator organism will require additional clarification of two criteria: taxonomy (criterion 1) and biometrics (criterion 4). In practice, these two criteria are inevitably the ones that must be given detailed consideration before the indicator organism potential of many species of caddisflies and other aquatic insects can be fully realized.

ACKNOWLEDGEMENTS

We thank R. Nowierski, S. Balling, and W. Tozer for field and laboratory assistance, and Drs G.B. Wiggins and D.G. Denning for taxonomic determinations. The research leading to this report was funded by the Office of Water Research and Technology, USDI, under the Allotment Program of Public Law 88–379, as amended and by the University of California Water Resources Center, as part of Office of Water Research and Technology Project No. A-063-CAL and Water Resources Center Project UCAL-WRC-W-519.

REFERENCES

Mackay, R.J., Ann. Ent. Soc. Am. 71:499–509, 1978.
Poinar, G.O.Jr., Lane, R.S. and Thomas, G.M., Nematologica 22:360–370, 1976.
Resh, V.H., *In* Ecological Diversity in Theory and Practice, eds J.F. Grassle; G.P. Patil; W.K. Smith and C. Taillie, pp. 241–253, Int. Coop. Publish. House, Fairland, Maryland, USA, 1979.

Rosenberg, D.M. and Wiens, A.P., J. Fish. Res. Bd. Can. 33:1955–1963, 1976.
Russell, P.P., Resh, V.H. and Flynn, T.S., *In* Proc. Ist EPA Oil Shale Symp.: Sampling, Analysis, and Quality Assurance (In press).
Waters, W.E. and Resh, V.H., *In* Contemporary Quantitative Ecology and Related Ecometrics, eds G.P. Patil and M. Rosenzweig, pp. 569–617, Int. Coop. Publish. House, Fairland, Maryland, USA, 1979.
Williams, D.D. and Hynes, H.B.N., Oikos 27:265–272, 1976.
Winterbourn, M.J., N.Z.J. Zool. 5:157–169, 1978.

DISCUSSION

Botosaneanu: I am very glad to see that you are interested in the problems of the hyporheal, so often neglected by the North American workers. One personal observation was made in Rumania, in a stream with a rather large and extremely isolated population of *Notidobia ciliaris*, a genus somewhat related to *Gumaga*. During a very dry and hot summer, the stream completely dried up and the *N. ciliaris* population disappeared. Using the Karaman-Chappuis method I succeeded in finding a few stages I–III larvae in the hyporheal at a depth of about 0.8 m in the gravel. The following year there were some young larvae in the stream. The year after the adults could be caught again. This shows the extremely important role of the hyporheal in stream biology.

Resh: This is a very useful comment. If in fact early instar larvae did survive the dry period deep within the hyporheic, a single cohort population would have resulted following habitat recovery. Our hyporheic examination of the Hopland spring was not as detailed as the one you described and it is possible that even though we did not find any early instar larvae, they could have been present.

318

TRICHOPTERA OF WESTERN SWITZERLAND

C. SIEGENTHALER

SUMMARY

During the past four years, two new distributions and several first recordings have been found in Western Switzerland, which can be divided into several zoogeographical zones with the Jorat appearing as a distinct entity. It is suggested that the Plateau and the Alps be considered as separate zones.

INTRODUCTION

Most studies on Trichoptera in Western Switzerland were carried out at the turn of the century, consequently most of the insects in Swiss museums date from that period to about 1945. I therefore thought it worthwhile to restudy this area and complete the data. This was done by
(a) establishing a faunistic list of Trichoptera, something that had not previously been done;
(b) analysing the distribution of the species with respect to the surroundings: geology, climate, vegetation, type of river, quality of water, etc. . .

TWO NEW DISTRIBUTIONS

It is risky to state the extinction or the rarefaction of a species. I have, however, been successful in ascertaining that some species have a wider distribution than was previously thought.

As a point of reference for the distribution of Trichoptera, I consulted Limnofauna Europaea (Illies, 1978) and although its zoning is not entirely satisfactory for local study (see below), I shall quote it as it is currently used in Europe.

Limnofauna Europaea gives two zoogeographical zones for Western Switzerland, Nos 4 and 8 (Map 1). Zone 4, running from east to west, includes the Alps, Prealps, Jorat and the Plateau, while Zone 8 covers the Jura west of its 1000 m limit. Based on this zoning, I found two new distributions of Trichoptera in Western Switzerland:

(1) *Rhyacophila aurata* Brau., which had not been recorded in the Swiss Jura (Zone 8), was caught on numerous occasions in two distinctly separated locations (Map 2);

(2) *Tinodes pallidulus* McL., which would be expected to inhabit the Plateau (Zone 4) but had never been collected there, was found in two locations in this area (Map 3).

ZOOGEOGRAPHICAL ZONING BASED ON TRICHOPTERA

Other Swiss entomologists before me attempted to lay down an insect based zoogeographical zoning for Western Switzerland. In 1968, both J. de Beaumont and W. Sauter put forward a proposal for such a zoning; the first was based on the general insect population, the second on Lepidoptera. They had in common the fact that they considered the Alps and the Plateau separately, and this in my view is a valid distinction.

The zoning proposed by Illies is remarkably useful for Europe in general; however, when working on a local basis, I found it insufficiently detailed. I do not believe that the inclusion of the Alps and the Plateau in a single zone is valid. The geology, climate and vegetation of these two regions are distinctly different and, therefore, so are the insect populations. *Drusus discolor* Ramb. (Map 4), for example, only inhabits the Alps, whereas *Oecetis ochracea* Curt. lives only on the Plateau (Map 5).

Having established a distribution map for each species I collected, I would suggest that entomologists who wish to study Swiss fauna complete Illies' zoning. Therefore, I would recommend an alternative zoning map for Western Switzerland which partly retains the proposals of J. de Beaumont and W. Sauter (Map 6).

It is not my purpose to insist on details of local interest in this paper, nevertheless, I consider the Jorat worthy of special mention as the Trichoptera in this region are specific. For instance, *Rhyacophila fasciata* Hag. can only be found in this part of Western Switzerland (Map 7) and not in the surrounding Plateau, in spite of its very wide European distribution. This and other data lead me to suggest that the Jorat be considered as a separate entity, although it was not identified as such by J. de Beaumont and W. Sauter who were mainly working on other insect orders.

These results, when added to future investigations, should lead to a fairly comprehensive picture of distribution of Trichoptera in relation to their environment in Western Switzerland.

REFERENCES

De Beaumont, J., Mitt. schweiz. ent. Ges. 41: 323–329, 1968.
Illies, J. et al., Limnofauna Europaea. Fischer Verlag, 1968.
Sauter, W., Mitt. schweiz. ent. Ges. 41: 330–336, 1968.

OVERWINTERING STRATEGIES IN SOME NORWEGIAN CADDISFLIES

J.O. SOLEM

SUMMARY

Aquatic stages of many invertebrates living in shallow areas of lakes and pools may at wintertime be found completely surrounded by ice. Papers remarking on this phenomenon were published as early as in the last decade of the 18th century. Field investigations and temperature measurements in lakes and pools in Central Norway have revealed that e.g. the larvae of the phryganeid *Agrypnia obsoleta* may be completely embedded by ice for a period of 6 months and survive temperatures down to -10 to $-11°C$ in the bottom substrate layers where the specimens stayed.

In laboratory freezing experiments, *A. obsoleta*, as expected, survived being kept for several weeks in iceblocks at temperatures at about $-2°C$, while e.g. *Phryganea bipunctata* was dead after a similar treatment. For two *P. bipunctata* larvae no supercooling point was found, while two *A. obsoleta* larvae had supercooling points of $-3°C$. Other species showed similar patterns. The temperature point where freezing occurred for *A. obsoleta* is not the most important thing about this, as *A. obsoleta*, in shallow pools in nature, has survived temperatures down to -10 to $-11°C$. The importance is the difference in the ability to survive sub-zero temperatures between *P. bipunctata* and *A. obsoleta*. Two different mechanisms seem to operate, and the one operating in *A. obsoleta* is undoubtedly a key factor mechanism for the larvae to survive in many habitats. These few in number experiments show that some caddis larvae may withstand sub-zero temperatures and do not freeze when the water freezes, while other freeze when the surrounding water freezes. Caddis larvae collected in spring in pools that have been frozen solid during wintertime or in frozen areas of the littoral zone of lakes are *A. obsoleta*, *Nemotaulius punctatolineatus*, *Oecetis ochracea*, and *Molanna albicans*. Other species like *Asynarchus contumax*, *A. lapponicus*, *Limnephilus stigma*, and *Grammotaulius signatipennis* inhabit temporary vernal pools and spend the winter as eggs or young larvae inside the gelatinous matrix.

DISCUSSION

Nielsen: I have, on occasions, found adult *Chaetopteryx* quite accidentally in the middle of winter and very near streams. They were active at a few degrees centigrade below zero. I had, probably, disturbed them in their winter quarters. I have found abundant numbers of first instar *Halesus* larvae in early spring. Perhaps, one cannot exclude hibernation in the egg stage.

Solem: I, also, have found *Chaetopteryx villosa* adults in the middle of winter. Considering that *C. villosa* is normally a late autumn species, I think, that because of slow development close to $0°C$, it may spend much of the winter in the egg stage. There is no similar data for *Halesus*, as it flies earlier in the summer and egg development is likely to be faster.

PRELIMINARY INTERPRETATIONS OF THE DISTRIBUTION OF HYDROPSYCHIDAE IN A REGULATED RIVER

J.A. STANDFORD AND J.V. WARD

SUMMARY

In pristine, headwater segments of the Gunnison River (Colorado, U.S.A.), *Arctopsyche grandis* dominated Trichoptera biomass, but was replaced by *Hydropsyche* and *Cheumatopsyche* species in downstream reaches. Hydropsychids did not inhabit tailwaters below dams, but were present in very high densities 30–130 km downstream from the last dam. The distributional pattern was attributed to temperature and availability of particulate organic carbon.

INTRODUCTION

Tailwater segments (i.e. those areas immediately downstream from on-channel dams) of rivers regulated by hypolimnial releases from deep reservoirs are often characterized by a constant, cold thermal regime, armored substrate, and a reduced diversity of benthic fauna in comparison to pre-impoundment conditions. Benthic production may vary but generally is also depressed (Ward and Stanford, 1979). In natural streams, Vannote et al. (1980) have suggested that organisms tend to segregate functionally along the longitudinal continuum of the river and that the pattern of segregation may be altered or 'reset' in a downstream or upstream direction by side flow (i.e. inflow of tributary streams). We suggest that a similar reset mechanism should, then, be evident in a regulated stream if one considers the regulated stream on the basis of an 'interrupted' continuum. That is, at some point downstream from the severe stress imposed in a tail-water segment, the river might be expected to regain some portion of its biotic integrity.

We have begun a long-term study of the entire reach of a 6th order, regulated river. We report here early findings on the distributional responses of indigenous Hydropsychidae to stream regulation; these findings indicate the reset mechanism proposed above may actually exist.

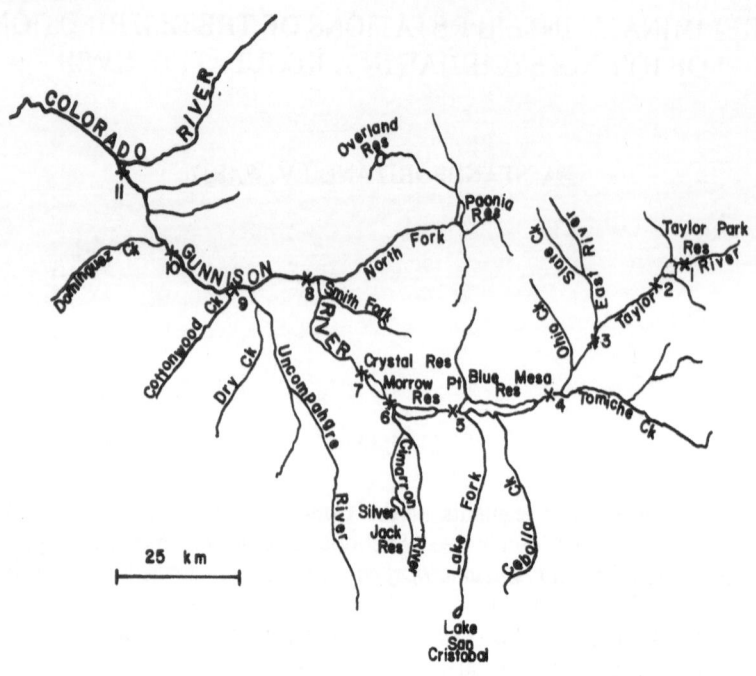

Fig. 1. The Gunnison River, Colorado, U.S.A. and sampling sites (x).

STUDY AREA

The Gunnison River drains 20,533 km² in west-central Colorado. The river flows westerly from the Continental Divide for nearly 300 km to join the Colorado River. Prior to impoundment, it was a rhythron stream from headwaters downstream ca. 170 km into a deep (830 m), high-gradient canyon segment. Below the confluence with the North Fork and Uncompahgre Rivers (Fig. 1), the Gunnison was a turbid, increasingly salty, and warm-water river (Table I). Thus, upstream from the canyon segment, productive populations of benthic insects, notably the plecopteran, *Pteronarcys californica,* and several hydropsychid trichopterans, supported a substantial trout fishery; downstream from the gorge, unique, indigenous cyprinids (e.g. *Gila robusta* and *Ptychocheleilus lucius*) and catostomids (e.g. *Catostomas latipinnis* and *Xyrauchen texanus*) characterized the riverine fauna (Wiltzius, 1978). No historical data on benthos in this lower river segment exists, other than qualitative observations since the 1950's by Dr Jack A. Stanford.

Four on-channel reservoirs have been built on the Gunnison River and its main headwater tributary, the Taylor River (Fig. 1). The first, Taylor Park Reservoir, was completed in 1937. The others, located in the upstream portion of the canyon segment, were completed between 1965 and 1977. All are deep (i.e. 50 m or more) reservoirs with hypolimnial drains.

324

Table I. Summary of pre- and post-impoundment thermal regima measured at selected sites along the Gunnison River continuum.

Station No.	Km from Headwaters	Post-Impoundment Mean Annual Degree Days (Annual Thermal Range, °C)	Pre-Impoundment Mean Annual Degree Days (Annual Thermal Range, °C)
1	18	1950[a] (0–15.0)	1950[a] (0–15.0)
2*	24	1000[a] (2.5–7.2)	2000[a] (0–15.0)
3	54	–	–
4	81	2550[a] (0–18.8)	2547[b] (0–18.8)
5*	115	2323[c] (3.3–11.1)	2650[a] (0–19.0)
6*	130	–	–
7*	144 '	1429[d] (0–8.8)	2895[a] (0–20.0)
8	195	–	–
9	228	4059[d] (3.3–21.6)	3606[a] (0–24.0)
10	271	–	–
11	290	3432[d] (0–23.3)	4132[e] (0–26.6)

*Tailwater area.
[a]Estimated from fragmentary field data and interpolation of all literature data.
[b]Calculated from 1964–65 data (Wiltzius, 1966).
[c]Calculated from 1973 data (Wiltzius, 1978).
[d]Calculated from 1978 data (Water and Power Resources Services, unpubl.).
[e]Calculated from 1952–1965 data (Wiltzius, 1978).

METHODS

Serial sets of replicated (3) benthic samples were obtained monthly at eleven sites (Fig. 1) along the river continuum. Trichoptera larvae were quantitatively collected by hand-cleaning substrata in a 0.5 m^{-2} area and subsequently retaining dislodged specimens in a large net (150 μm mesh size) held immediately downstream (Hauer, 1980). Adults were collected at each site using sweep nets and a night light. Samples were preserved in 10 percent formalin and shipped to the University of Montana Biological Station (UMBS) for analysis. Taxonomic designations, presently only preliminary, were based on consideration of preimpoundment records, Wiggins (1977) and other pertinent literature, and comparisons to a reference collection of Montana trichopterans verified by G.B. Wiggins. Biomass estimates of trichoptera larvae in samples were obtained volumetrically using a 10 ml graduated cylinder.

Physicochemical measurements (i.e. temperature, conductivity, pH, and alkalinity) were obtained with field meters and on-site titrations. Grab samples of water were obtained at each site for laboratory (UMBS) analysis of organic carbon constituents (Oceanography International Inc., TOC system) and major ions (Dionex Ion Chromatograph). Continuously recorded temperature data were kindly supplied by Water and Power Resources Service.

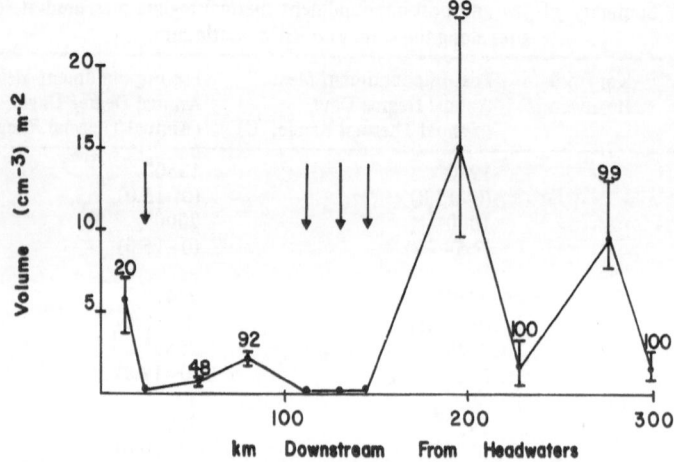

Fig. 2. Standing crops of Trichoptera at eleven sites on the Gunnison River, Colorado, in September, 1979. Means are connected, bars represent ranges in three samples, numbers at top of bars indicate percent of total biomass made up of Hydropsychidae, and arrows indicate location of impoundments.

RESULTS AND DISCUSSION

Hydropsychids were present in all samples, and were the most abundant functional group (Cummins, 1973) at the majority of the collecting sites. Although absolute values varied, the pattern illustrated in Fig. 2 emerged in sample sets collected in all seasons. At Station 1 *Brachycentrus similis* and *Glossosoma* sp. outnumbered *Arctopsyche grandis;* but, the latter species dominated in biomass, owing to its large size and two year life history. *A. grandis* was also the dominant species at Station 3 in terms of biomass, but *Lepidostoma* sp. and smaller net-spinners (i.e. *Cheumatopsyche* sp., *Hydropsyche* spp.) were numerically abundant. *Hydropsyche* and *Cheumatopsyche* were dominant in biomass and numbers at Station 4. Caddis larvae (all species) were virtually absent from samples taken in tailwater areas. Beginning ca. 30 km downstream from the last dam, the hydropsychids (excluding *A. grandis*) were again present and reached high densities at Stations 8–10. These species were the same ones that occurred at Station 4. In addition, *Psychomyia flavida* was occasionally observed in benthic samples at the downstream sites. In samples taken on subsequent dates biomass at Station 9 was consistently higher than shown in Fig. 2, indicating a general downward trend from a maximum biomass at Station 8 to a minimum (for the lower river segment) at Station 11.

Hydropsychid species at these downstream sites may produce more than one generation per year, as several instars were present in each sample. We collected adults of each species in September–October 1979, and again in April–August, 1980. At Stations 8–10 large swarms of adults were observed in September, 1979. Only a few adults were collected at the upstream sites (i.e. 1–4) and all species, except *A. grandis*, appeared to be univoltine.

The distribution pattern in the upstream areas probably remains similar to historical conditions. The negative effects of Taylor Park Reservoir are likely diluted (or reset) by side flow from the East River and Tomichi Creek. Conditions at Station 3 and, especially, 4 are likely close to pristine. High standing crops of hydropsychids in the lower segment of the river have only eventuated since construction of the canyon dams, however.

Table II. Mean concentrations of particulate (POC) and dissolved (DOC) organic carbon measured of 11 sites along the Gunnison River Continuum during 1979–80.

Station No.	Km from Headwaters	POC (Mg 1^{-1})	DOC (Mg 1^{-1})	N
1	18	0.43	0.73	4
2*	24	0.24	1.97	5
3	54	0.23	1.92	6
4	81	0.45	1.79	5
5*	115	0.16	2.15	5
6*	130	0.18	2.12	5
7*	144	0.21	2.19	5
8	195	0.40	2.36	5
9	228	0.57	2.44	4
10	271	0.74	3.31	4
11	295	0.74	4.88	5

*Tailwater Area

We believe that the explanation for the success of the hydropsychids in the lower river lies in an understanding of temperature, and the spiraling (Webster and Patten, 1979) or organic carbon as a function of distance from the downstream dam. In tailwater areas significant depression of the pre-impoundment thermal regime has occurred. Also, the hypolimnial discharges carry comparatively less particulate organic carbon (Table II) presumably due to mineralization of allochthonous and autochthonous particulates as they sediment through the water column of the deep reservoirs. Moderation of these apparently severe 'constancy' effects (Ward and Stanford, 1979) occurs progressively downstream from Crystal Dam. A favorable thermal regime is reconstructed by solar insolation and agglutination of particulates increases spatially from the dam. These effects, coupled with less turbidity and naturally high total organic carbon pool (and its associated nutrient cycles) would tend to favor collector organisms like the hydropsychids at some point along the regulated portion of the stream continuum. In the Gunnison River that point appears at Station 8, some 51 km from Crystal Dam. This natural reset mechanism implies an ecosystem subsidy response to a stress (Odum et al., 1979) that is reflected in the life history strategies of an indigenous group of aquatic insects. Our investigations on the Gunnison River and in regulated streams elsewhere are designed to test the reality of this implication.

ACKNOWLEDGEMENTS

We thank J.D. Coulter, B. Martinson and L.S. Stanford for assistance in

327

field and R. Hauer, K. Thomas and J. Stuart for help with the manuscript. This work was supported in part by the U.S. Environmental Protection Agency.

REFERENCES

Cummins, K.W., Ann. Rev. Entomol. 18:183–206, 1973.
Hauer, F.R., Ph.D. Dissertation, North Texas State University, Denton, Texas, 1980.
Odum, E.P., Finn, J.T. and Franz, E.H., Bioscience 29: 349–352, 1979.
Vannote, R.L., Minshall, G.W.; Cummins, K.W.; Sedell, J.R. and Cushing, C.E., Can. J. Fish. Aquat. Sci. 37:130–137, 1980.
Ward, J.V. and Stanford J.A., The Ecology of Regulated Streams, p. 398, Plenum Press, New York, 1979.
Webster, J.R. and Patten, B.C., Ecological Monographs 49, 1:51–72, 1979.
Wiggins, G.B., Larvae of the North American caddisfly Genera (Trichoptera), pp. 401, University of Toronto Press., 1977.
Wiltzius, W.J., Preimpoundment investigations of the Curecanti Unit. Upper Colorado River Storage Project, Colorado Dept. Game, Fish and Parks, p. 70, 1966.
Wiltzius, W.J., Some factors historically affecting the distribution and abundance of fishes in the Gunnison River, pp. 215, Colorado Division Wildlife. 7f. Collins Co., 1978.

DISCUSSION

Smith: I have found the *Arctopsyche grandis* larva without a dorsal head stripe in Idaho and Washington; it is, evidently, widespread.

A PROGRESS REPORT ON HYDROPSYCHIDAE FROM THE IVORY COAST: CHARACTERS FOR THE SPECIFIC IDENTIFICATION OF LARVAE AND POPULATION DYNAMICS OF FOUR ABUNDANT SPECIES.[1]

B. STATZNER

SUMMARY

Characters for the specific identification of hydropsychid larvae from the Ivory Coast are given for the following genera: *Cheumatopsyche, Polymorphanisus, Aethaloptera, Amphipsyche, Protomacronema, Macronema, Leptonema*. Especially the genus *Cheumatopsyche* is represented by different morphological groups and a systematic revision of this genus seems to be necessary in the future.

In the N'Zi River, a temporary running water habitat during the study period, the population dynamics of three *Cheumatopsyche spp.* and *Aethaloptera dispar* were followed over 15 months. Densities and life cycles of the species were influenced by the discharge pattern. Undoubtedly their larval lives lasted longer than is generally assumed for insects of tropical rivers, and probably only 2 or 3 generations per year occurred.

INTRODUCTION

The ecology of tropical running water invertebrates is not well known and this especially holds true for taxonomic groups which are difficult to identify. Such a difficult group are the tropical representatives of the Hydropsychidae, particularly when larvae are considered. While new characters for the larval identification of this family from temperate zones were recently published (e.g. Statzner, 1976a; Boon, 1978; Mackay, 1978; Schuster and Etnier 1978), our knowledge on tropical forms remains much more incomplete (Scott, 1975). Thus the characters determined after the morphological examination of numerous specimens and then used for identifying hydropsychid larvae in zoogeographical and ecological studies from the Ivory Coast (West Africa) are briefly demonstrated in this paper.

The lack of basic taxonomic data naturally causes a lack of ecological data, and our knowledge, particularly on the population dynamics of running water invertebrates near the equator, is nearly zero. From Africa we have best

---

[1] The field studies were carried out during a period of WHO consultantship in the Onchocerciasis Control Programme.

Proc. of the 3rd Int. Symp. on Trichoptera, ed. by G.P. Moretti
Series Entomologica, Vol. 20. ©1981, Dr W. Junk Publishers, The Hague

information on the ecology of the *Simulium damnosum* complex, due to its role as vector of human onchocerciasis (e.g. Le Berre 1966). Other than this well studied group, less comprehensive data on the phenology of other invertebrates from permanent running water habitats have been published (e.g. Cridland, 1958; Ramanankasina, 1973; Statzner, 1976b; Zwick, 1976; Lehmann, 1979). The development rate of the total fauna is considered in permanent streams after treatment with DDT (Hynes and Williams, 1962) or in temporary streams after the start of flow (Harrison, 1966; Hynes 1975). From most of these studies it can be concluded that life cycles of benthic invertebrates are rather short. The size/instar analysis of several macroinvertebrate species from a temporary running water habitat in the Ivory Coast indicated that life cycles lasted much longer, examples of this will be demonstrated in this paper.

METHODS AND THE STUDY AREA

In 1977/78 the hydropsychid fauna of about 100 running water localities scattered over the Ivory Coast was sampled. In addition, 5 rapids of the N'Zi River were visited monthly to obtain data on the population dynamics of abundant species. At one rapid, Surber (area: 16 x 17 cm; mesh: 320–200 μm) and core samples were taken from a gravel/pebble substratum.

Qualitative samples (mesh: 200 μm) were taken from all substrata. In the laboratory the material was handled as mentioned in Statzner (1975a, b). Larvae were assigned to imagines via pupae (larval exuvia from a pupal chamber with a pupa possessing a well developed genital). Larval instars were separated according to head capsule width.

The N'Zi, flowing from north to south through savannah, is a tributary of the Bandama River. During the study period discharge of the N'Zi decreased to zero from January 77 until April 77. In May 77 flow started again, a second period of low discharge was followed by the flood in the main rainy season, after that discharge decreased again from October 77 until March 78, when it reached zero in the northern part of the N'Zi. Thus the N'Zi was a temporary running water habitat during the study period with one (southern part) or two (northern part) periods without flow, during which surface water remained solely in the pools. Water temperature, which reached maximum values around the periods without flow, and transparency were closely related to the discharge pattern.

While the middle section remained as a control, the northern and southern part of the N'Zi was treated with chlorphoxim, a product of Bayer, from June until December 77. This was done to obtain data on the effect of this material on non-target fauna if this insecticide is applied against *Simulium damnosum*.

CHARACTERS FOR THE SPECIFIC IDENTIFICATION OF LARVAE

The following characters are the most useful ones in the separation of hydropsychid larval instar V. In the same locality most of the characters can be used to differ between species also in larval instars IV and III. The intensive study of three *Cheumatopsyche spp.* of the N'Zi demonstrates, that the characters stridulatory ridges on the pleurae and secondary setae on the head can also be used for instar II. First instar larvae were separated according to the head capsule width (Mackay, 1978). In *Aethaloptera* the character of the setae on the foretrochantin is even useful for the separation of larval instar I. Below only characters for larval instar V are considered.

Diplectroninae, one subfamily of the Hydropsychidae, were not found in this study. From the Hydropsychinae only one genus is represented by more than one species in the material, which is *Cheumatopsyche* (sensu Kimmins 1963). Ten different larval types were found, the most common ones are assigned to different imagines; thus these ten types are assumed to be distinct species. With the help of the following characters: posterior prosternites, anterior margin of frontoclypeus, and secondary setae on the head, these species can be grouped as in Fig. 1. Additional characters for the specific identification are the stridulatory ridges, the submentum, the foretrochantin, and the form of the prosternal sclerite(s).

The well developed posterior prosternites in *C. digitata*, which were not mentioned by Gibbs (1973), and *Cheumatopsyche sp. I* and *II*, makes that up-to-now-important character useless for purposes of separating *Hydropsyche* from *Cheumatopsyche*. This and particularly the isolation of *C. albomaculata* and *Cheumatopsyche sp. VIII* from the other species by several characters (Fig. 1; further characters: secondary setae on head dark and arranged in a semicircle, head as wide as long, width of the frontoclypeus at the tentorial pits equals that of the anterior margin) suggests great non-homogenity in the recent genus *Cheumatopsyche* and requires a systematic revision of that group.

The Macronematinae, the third subfamily of the Hydropsychidae, are represented in the material by more than one species in all genera known from Africa. Larvae of 2 *Polymorphanisus* spp. differ mainly in the coloration of the head (dark design + or −) and the kind of setae on the foretrochantin. *Aethaloptera* (2 spp.) larvae can be identified by the use of the anterior margin of the frontoclypeus, the submentum, setae on the foretrochantin and on the ventral sclerites of abdominal segment IX. Larvae of *Amphipsyche* (2 spp.) can be distinguished by the stridulatory ridges, submentum, form of the ventral and dorsal sclerites on abdominal segment IX and setae of the ventral ones. The 2 larval types of *Protomacronema* are separated by characters of the submentum and the dorsal sclerites on abdominal segment IX. From the genus *Macronema*, 4 spp. occurred in the larval material. They differ mainly in the form of the head and the submentum, the stridulatory ridges and the setae on the ventral sclerites of abdominal segment IX. *Leptonema* larvae (3 spp.) are easy to identify using the characters stridulatory ridges, form of forecoxae, and setae on abdomen.

This brief presentation of characters most useful for differential analysis is hoped to enable the separation of african larvae of this genera as to facilitate the identification in this family in temperate regions, where characters used for this purpose are known to be often variable.

POPULATION DYNAMICS OF FOUR ABUNDANT SPECIES OF THE N'ZI RIVER

This section deals with the species *Cheumatopsyche sp. III* and *VII, C.*

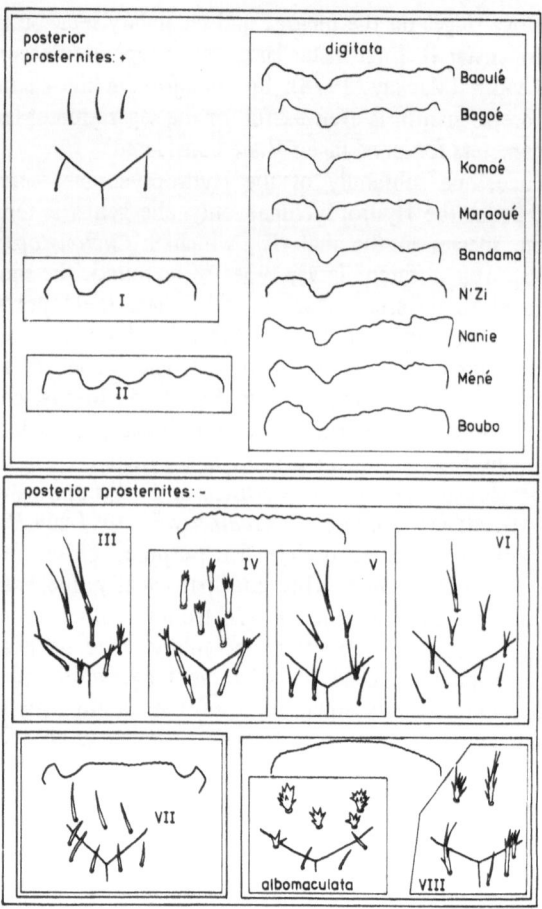

Fig. 1. Larval characters of *Cheumatopsyche* (sensu Kimmins, 1963). The ten species are grouped according to the morphology of the ventral sclerites on the prosternum, the form of the anterior margin of the frontoclypeus, and the secondary setae on the head (region at the back of the frontoclypeal sclerite). For *C. digitata* the variability of the form of the anterior margin of the frontoclypeus is shown from different streams from the Ivory Coast.

digitata, and *Aethaloptera dispar.* Their abundance was influenced by the discharge pattern, as is shown examplary for two species on a rapid from the control section of the N'Zi (Fig. 2). The density of *Cheumatopsyche sp. III* and *C. digitata* increased each time discharge decreased. An exception was found in December when the condition of the fauna in the rapid indicated that poisoning had occurred, probably due to activities of local fishermen. A pattern as in Fig. 2 was found in *Cheumatopsyche sp. VII* and the sum of the benthic fauna. *A. dispar* behaved different from this pattern since it was not able to survive the period without flow and was not captured there before November. Because *A. dispar* was eradicated during the insecticide application in the southern part of the N'Zi and it

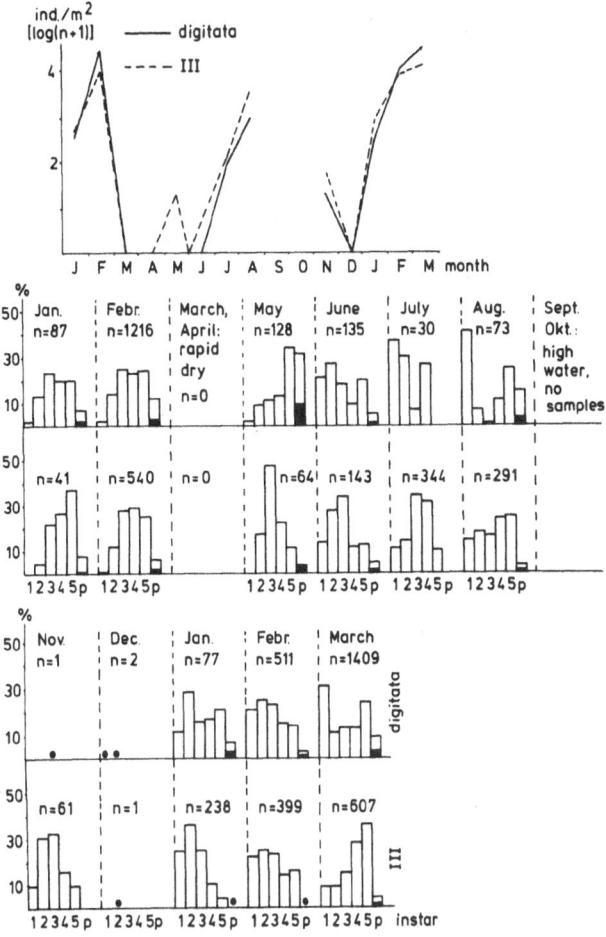

Fig. 2. Cheumatopsyche sp. III and *C. digitata* at Ouokoukro, a rapid from the control section of the N'Zi River (January 1977–March 1978). Above: Density on a gravel/pebble substratum. Below: Percentage instar-distribution in qualitative samples (1–5: larval instars; p: pupae, black part of p: prepupae). If samples contained only few specimens, they are presented directly by points.

survived the dry season in streams with current, the distance that must be overcome by imagines to recolonize the control section of the N'Zi by eggs was at least 80 km.

A. dispar larvae build tubes which protrude into the current and changed the character of the substratum to some extent. Thus the density of this species is expected to influence that of other benthic macroinvertebrates by producing places with reduced current.

Core samples taken near the places of surber samples demonstrate that the abundances of hydropsychids in the hyporheic habitat changed as they did at the surface of the substratum. In core samples from the dry rapid as in the filtrate of interstitial water obtained from a deep hole, no hydropsychids were found.

As with density, the life cycles of the four species under study were influenced by the discharge pattern. This was valid for all five rapids studied when the period of insecticide application was excluded. An example of the instar distribution in the same two species from the same rapid mentioned above is given in Fig. 2. Both species occurred mainly as later instars before the rapid dried up. Less than a fortnight after flow resumed, later instars predominated again; at other rapids this was true even a week after flow had started. In the period from June until August younger instars occurred more or less pronounced in higher percentages. The spate then made sampling impossible. Afterwards only *Cheumatopsyche sp. III* was found in larger numbers and had later instars in lower percentages. In December only few specimens were captured, which is the expected result of poisoning (cf. above). At the second rapid from the control area of the N'Zi, late instars were found in higher percentages in December and populations became older there until March 78 in both species. Thus the picture in Fig. 2, where young instars were more frequent from January until March 78 than in the same period of the previous year, seems to be a result of the assumed poisoning of the rapid and probably reflects some density related effects on development.

Based on the results from all five rapids studied the following conclusions can be drawn for *Cheumatopsyche sp. III* and *C. digitata:* A period of low reproduction activity was found before flow stopped and larvae occurred as late instars. Immediately after flow started the instar percentage distribution indicated no reproduction, but a survival of the period without flow by the larvae. This is in contrast to results of Harrison (1966) and Hynes (1975). Both authors assumed that the recolonization of dry rapids took place by eggs in *Cheumatopsyche*, without studying the instar distributions in the populations.

It is not yet clear where the larvae survived in the N'Zi. Larvae were missed in the interstices of dry rapids but one larva was found in a core sample from sand in a pool. Once, a few larvae were found alive in a groove with stagnant water, probably few days after flow had stopped there. This indicated a survival of larvae in the remaining pools, although no further larvae were found there. However, in the N'Zi the area of the rapids was tiny in relation to that of the pools, i.e. if larvae from the rapids dispersed over the pools after flow had stopped the larval density in the pools was very low and therefore larvae were missed there.

The instar distribution in the period before flow stopped in connection with the survival of larvae in the period without flow indicated that the latest instars from June were quite old, probably of an age of 6 months. From June until August probably a second generation occurred. What happened during the spate is completely unknown, but probably a second period of increased reproduction occurred at the end of the high water period. If rapids remained undisturbed the populations became older from November 77 until March 78.

On the rapids where the chlorphoxim application was stopped in December 77, the first pupae occurred within four weeks whereas both species were nearly eradicated before. This suggested that single specimens developed within 4 weeks. However, during that period the areas with current decreased distinctly on the rapids. This led to the above mentioned concentration of larvae on the rapids, i.e. single later instars could be more easily overlooked in December 77 than in January 78.

Summing up those results the average development of both *Cheumatopsyche sp. III* and *C. digitata*, undoubtedly lasted longer than can be assumed after the study of most of the literature mentioned in the introduction, and probably only 2 or 3 generations per year occurred. This is valid too for *Cheumatopsyche sp. VII*, which developed in a way similar to that mentioned above.

Before flow stopped in 1977 *A. dispar* showed an instar distribution similar to that of *Cheumatopsyche*. In November 77 it occurred again at one control rapid, where few specimens of the first three instars were found. In December 77 pupae were recorded from that rapid and up to March 78 all larvae reached the last larval or the pupal instar. This suggested a long larval life also for this species, although the material was more fragmentary due to the inability of the larvae to survive without flow.

Thus the dynamics of the discharge influenced that of the population in a way that highest larval densities were reached by older populations, whereas low larval densities occurred when smaller instars predominated, an interesting result particularly in connection with problems related to the determination of the productivity of such populations.

ACKNOWLEDGEMENTS

Thanks are due to the staff of the hydrobiological laboratory of the O.R.S.T.O.M. in Bouaké, to members of the onchocerciasis control programme, and to my laboratory assistents in Bouaké for much help. Dr Vincento H. and Kay Resh corrected the english manuscript which is gratefully acknowledged.

REFERENCES

Boon, P.J., *In* Proc. 2nd Int. Symp. Trichoptera, ed. M.I. Crichton, pp. 165–173, Junk, The Hague, 1978.
Cridland, C.C., J. Trop. Med. Hyg. 61:3–7, 1958.
Gibbs, D.G., Dtsch. Ent. Z.N.F. 20:363–424, 1973.
Harrison, A.D., Arch. Hydrobiol. 62:405–421, 1966.
Hynes, H.B.N. and Williams, T.R., Annls Trop. Med. & Parasit. 56:78–91, 1962.
Hynes, J.D., Freshwat. Biol. 5:71–83, 1975.
Kimmins, D.E., Bull. Br. Mus. Nat. Hist. Ent. 13:117–170, 1963.
Le Berre, R., Mémoires O.R.S.T.O.M., Paris, 17:204 pp., 1966.
Lehmann, J. Spixiana, Suppl., 3:1–144, 1979.
Mackay, R.J., Annls Entomol. Soc. Am 71:499–509, 1978.
Ramanankasina, R.E., C.R. Acad. Sc. Paris, Sér. D 277:1513–1515, 1973.
Schuster, G.A. and Etnier, D.A., Res. Rep. Off. Res. Develop. U.S. Environ. Protect. Agen., ser. Environ. Monitor.-EPA-600/4–78–060: 128 pp., 1978.
Scott, K.M.F., Proc. I. Congr. Ent. Soc. Sth. Afr.:41–52, 1975.
Statzner, B., Zool. Anz. 193 (1974):382–398, 1975a.
———, Arch. Hydrobiol. 76:153–180, 1975b.
———, Ent. Germ. 3:265–268, 1976a.
———, Arch. Hydrobiol. 78:102–137, 1976b.
Zwick, P., Int. Revue ges. Hydrobiol. 61:683–697, 1976.

OCCURRENCE OF THE GENUS *HYDROPSYCHE* IN THE NORTH AMERICAN GREAT LAKES

J.L. SYKORA, B.G. SWEGMAN AND J.S. WEAVER III

SUMMARY

Two North American species, *Hydropsyche recurvata* Banks and *H. separata* Banks, are known to inhabit the Great Lakes. While *H. recurvata* is widely distributed in all of the lakes, *H. separata* seems to be absent in Lakes Superior and Michigan, occurs rarely in Lakes Huron and Ontario but is dominant in Lake Erie. It is suggested that this distribution of the genus *Hydropsyche* in the Great Lakes is the result of glacial and related geological events. Similarly, the distribution of several *H. separata* populations occuring in North America may have also been caused by the glaciation. The taxonomy and ecology of *Hydropsyche* adults and larvae inhabiting lakes is also briefly discussed in this paper.

INTRODUCTION

Members of the family Hydropsychidae normally inhabit fast flowing portions of streams and rivers, and their occurrence in lakes is less common. According to Ulmer (1909) *Hydropsyche pellucidula* (Curtis) inhabits both lowland streams and lakes of continental Europe. Hickin (1967) confirms that unlike most members of the family Hydropsychidae *H. pellucidula* is found in lakes and ponds. Crichton et al. (1978) support this statement and acknowledge that this species is a common Hydropsychid found throughout Britain not only in ponds and lakes but also in rivers. There is also some evidence that *H. angustipennis* (Curtis) inhabits the wave-zones of large lakes. Wichard and Unkelbach (1974) collected this species in two crater lakes in Rhineland.

Two North American species – *H. recurvata* Banks and *H. separata* Banks are known to inhabit lakes. In this paper the distribution and zoogeography of these two species are discussed.

MATERIALS AND METHODS

Unpublished records on the distribution of *H. recurvata* and *H. separata* were obtained from Dr Richard W. Bauman, Bringham Young University, Provo, Utah; Dr Oliver S. Flint, Jr, U.S. National Museum of Natural History, Washington, D.C.; Dr David R. Barton and Dr H.B.N. Hynes, University of Waterloo, Ontario, Canada, and Dr Glenn B. Wiggins, Royal Ontario Museum, Toronto, Ontario, Canada. Additional information were obtained from several published sources listed in the reference section.

Our own material was collected in Lake Erie and Lake Ontario. Larvae were picked by hand from stones. In addition, 'black light' trap was used for collection of adults. The trap was operated from shortly before dusk through the night. All the collected specimens were preserved in 70% alcohol.

The following material was obtained:

Hydropsyche recurvata

Lake Ontario, Somerset, N.Y., June–September 1972–73, 38 ♂, 56 ♀

Hydropsyche separata

Lake Erie, Cleveland, Ohio, August 17, 1974, 26 larvae
Lake Erie, Cleveland, Ohio, July 12, 1972, 115 ♂, 216 ♀
Lake Erie, Cleveland, Ohio, June 14, 1973, 3 ♂, 3 ♀
Lake Erie, Presque Isle, Pa., July 10, 1971, 4 ♂, 5 ♀

RESULTS AND DISCUSSION

Taxonomical notes: The structure of male and female genitalia of *H. recurvata* was sufficiently described by Ross (1944). On the other hand, the taxonomical identity of *H. separata* was not always clearly defined. Ross and Spencer (1952) suggested that this species was a synonym of the European species *H. guttata* Pictet. In a more recent paper Ross (1965) considers *H. guttata* and *H. separata* to be distinct species indicating evolutionary change after intercontinental isolation. This author finds the two species to be remarkably similar except one slight but constant difference in male genitalia. Unfortunately Ross (1965) failed to provide any details concerning this characteristic which may be used to separate these two closely related species. Smith (1979) after consultation with Dr Malicky concluded that *H. guttata* and *H. separata* are two distinct species. They also noted a close morphological resemblance of *H. separata* male genitalia to those of *Hydropsyche bulgomanorum* Malicky. In addition, Smith (1979) has observed different number of spurs in males of *H. guttata* and *H. separata* without making any specific comments. The males of the latter species in

338

H. separata, front tibia, male

Fig. 1. Hydropsyche separata Banks, front tibia, male.

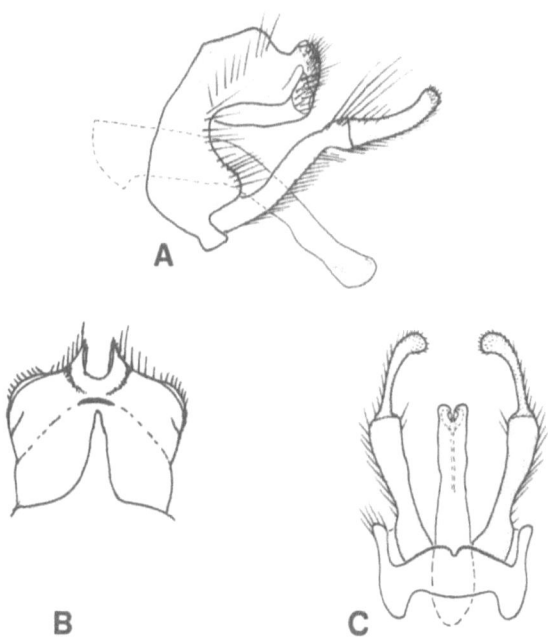

A

B   C

Fig. 2. Hydropsyche separata Banks, male genitalia. A, lateral view; B, dorsal view; C, ventral view.

our collection have only one spur on the front tibia whereas the second spur is reduced to a rudimentary knoblike structure (Fig. 1). The male genitalia of *H. separata* are very similar to those of *H. bulgomanorum* but differ in the shape of the dorsal section of the ninth segment and slightly in the morphology of aedeagus and claspers (Fig. 2). The larva of *H. separata* was sufficiently described by Smith (1979).

339

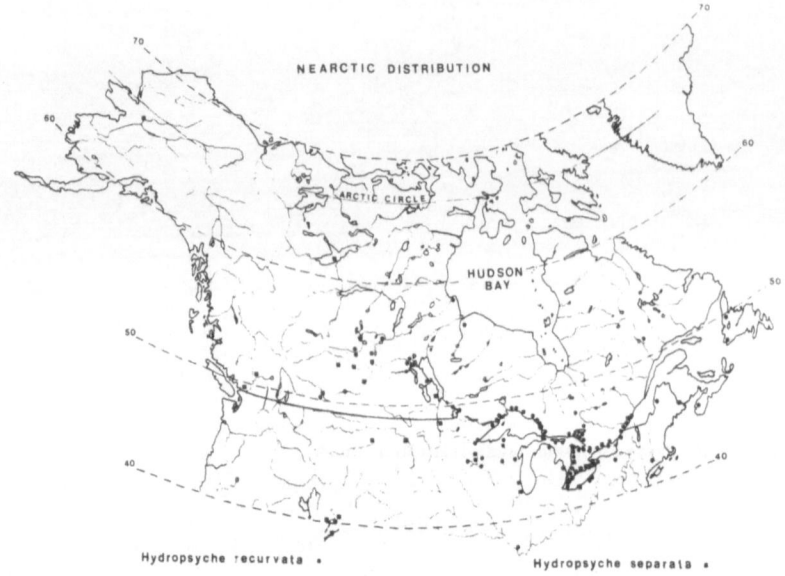

Hydropsyche recurvata •   Hydropsyche separata •

Fig. 3. Distribution of *H. recurvata* and *H. separata* in North America.

Distribution of H. recurvata *and* H. separata: *H. recurvata* seems to be widely distributed in one continuous band across boreal North America (Fig. 3), and occurring from Newfoundland west to Alberta. This distribution pattern is not interrupted by any distinct gaps. This species is very common in the Great Lakes receding through Wisconsin and Minnesota to Manitoba, Saskatchewan and Western Canada. It inhabits fast flowing, cold streams in the North with infrequent occurrence in Lake Erie and very common appearance in Lakes Michigan, Ontario, Huron and Superior. According to Barton and Hynes (1978a) the larvae of *H. recurvata* were collected in Lake Erie in a narrow strip of gravel and cobbles which extends about 5–10 cm offshore along most of the central basin. The larvae of this species were found in the wave zone of the Great Lakes by Barton and Hynes (1978a), most frequently in Lake Ontario and Lake Huron. We have recorded numerous adults in light traps set up at Somerset, N.Y. near Lake Ontario shore during summer 1972–73. Marshall (1939) recorded *H. recurvata* in light trap material from Gibraltar Island (Western basin of Lake Erie). Based on our adult and immature stages collection from Lake Erie this species is not common in the central basin of Lake Erie. Masteller and Flint (1979) collected 2 ♂ and 1 ♀ of *H. recurvata* in West Springfield, Pa. located on Lake Erie shore whereas adults of *H. separata* collected at the identical site were more numerous (19 ♂, 1 ♀).

It has been observed that *H. recurvata* has a long flight period. Nimmo (1966) recorded a flight period for this species to last from June through August. Marshall (1939), and Leonard and Leonard (1949) also indicated a long flight period for *H. recurvata*.

The distribution of *H. separata* is also transcontinental. However, based

340

on published and unpublished records this species appears in four rather distinct populations. A Great Lakes population is centered in the central basin of Lake Erie with some scattered records from Lake Ontario and Lake Huron. This species is rather rare in the latter two lakes. Another population is located in northern Great Plains extending from northeast Utah, possibly as far north as Great Slave Lake. The Yukon Valley is inhabited by the third group. The fourth population is located in the Columbia River basin and may be isolated from the Plains population by the Rocky Mountains. Whether or not the western population is indeed disjunct is not yet clear and requires more material from the northwest Canadian provinces, i.e. Alberta, British Columbia, and Yukon. More records are also needed from Alaska. An interesting feature of the present distribution is that the Great Lakes population and Plains population, which seem to be disjunct, may presently be isolated by a biotic barrier, i.e. a dominant population of *H. recurvata* in the western Great Lakes with which *H. separata* cannot compete effectively. It is possible that these disjunct populations of *H. separata* may represent different taxonomical entities. However, to support this statement would require additional morphological studies and possibly electrophoretic analysis and/or some other techniques used in modern taxonomy.

The present distribution of *H. separata* is probably the result of events surrounding the last glacial period in North America and the pattern of retreat of the ice. First, it can be assumed that this species was widely distributed across boreal N. America prior to Wisconsin glaciation. With the advance of ice, the species was forced southward. In the Great Plains, retreat probably occurred along water courses which flowed north to Hudson Bay. Lemke et al. (1965) reported that 'In pre-glacial time the Cheyenne River in South Dakota and all streams north of it drained into Hudson Bay. Possibly even the ancestral Bad and ancestral White Rivers also drained northward into the Cheyenne instead of south' (Fig. 4). Presently these rivers flow south to the Mississippi.

With the retreat of ice, this species had again recolonized lakes and drainages south of the glacier. Thus, the streams draining now to the Mississippi River still maintain a sizeable population of *H. separata*.

The Alaskan population might have existed during all this period (glaciation) as the Yukon Valley in Alaska has never been glaciated (Pewe et al. 1965). It would be interesting to find out if this Alaskan population has any affinities to some of the Siberian *Hydropsyche* species or if *H. separata* also occurs across from the Behring strait. The Columbia River population may be a remnant from any one of several 'nunataks' which existed in the far west. ('Nunatak' is a small area within the glaciated land which was not covered with ice).

The most interesting distribution pattern is demonstrated by the Great Lakes population of *H. separata*. The type and paratypes of this species were collected in 1905 near Westfield (N.Y.) which is located on Lake Erie. Therefore, it can be assumed that *H. separata* was living in Lake Erie before the major influx of pollution and its occurrence would be attributed to the natural causes such as geomorphological changes

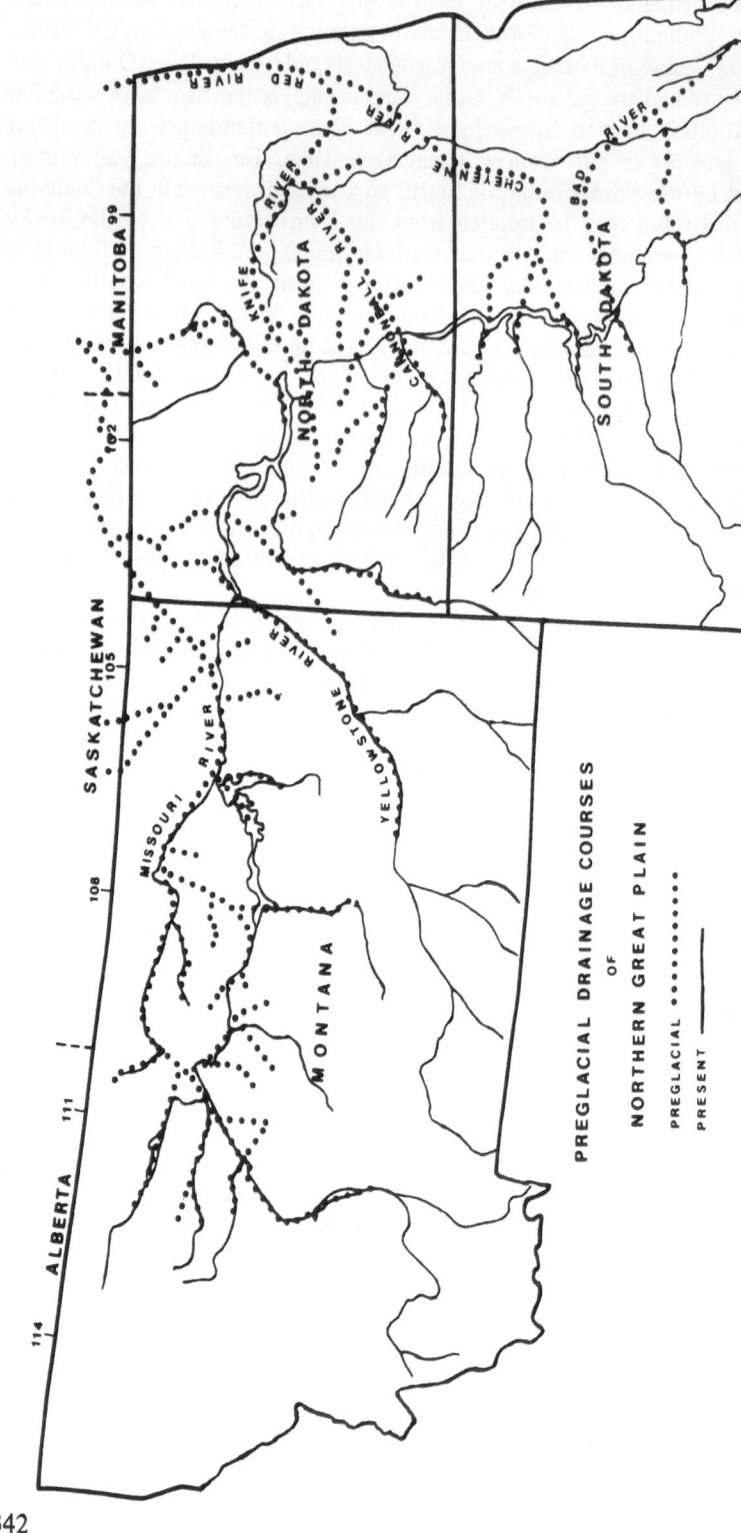

Fig. 4. Major preglacial and present drainage courses in Montana, North Dakota and South Dakota (Redrawn from Lemke et al., 1965).

during and after the glaciation or to the natural biological and ecological conditions.

It is possible that during the glaciation the Great Lakes population of *Hydropsyche* was isolated south of the ice. Upon retreat of the ice, *H. recurvata* may have recolonized the western Great Lakes quickly (from south of the glacial) through the old southwest drainages of early Lake Michigan and early Lake Superior. Possibly, early Lake Ontario was colonized by *H. recurvata* through old Trent Valley drainage. At certain stages in the later part of the Wisconsin glaciation, Lake Erie was isolated from Lake Huron, and water from this lake flowed directly into Lake Ontario (Hough, 1965 – Fig. 5). Therefore, *H. separata* may have been isolated in Lake Erie whereas *H. recurvata* bypassed this lake when colonizing Lake Ontario.

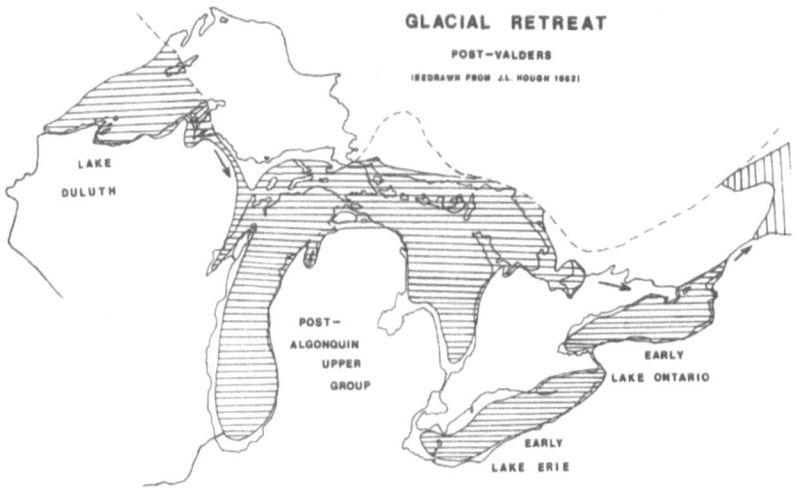

Fig. 5. Glacial retreat in the Great Lakes area (redrawn from J.L. Hough, 1962).

Another alternative is that *H. separata* was one time more widely distributed in the eastern Great Lakes, but has been more recently out-competed by *H. recurvata*. On the other hand, it could be that the final explanation is a combination of both of these ideas.

Thus, the events leading to the present occurrence of *Hydropsyche* species in Great Lakes cannot be defined with certainty. However, the final distribution is certainly remarkable (Fig. 6). Especially the isolation of *H. separata* in Lake Erie and its dominance in this water body is unusual and should be a subject of further investigations. Of interest is also the presence of limited population of *H. separata* in Lakes Ontario and Huron which could be possibly attributed to a relatively recent colonization effort of this species. It is also remarkable that the *Hydropsyche* fauna of Great Lakes is limited to two species which are able to thrive and prosper in this environment otherwise hostile to many rheophilic organisms. *Hydropsyche* as a net-spinning caddisfly needs fairly high current velocities for survival. Under winter ice cover this action is limited and/or non-existent. According to

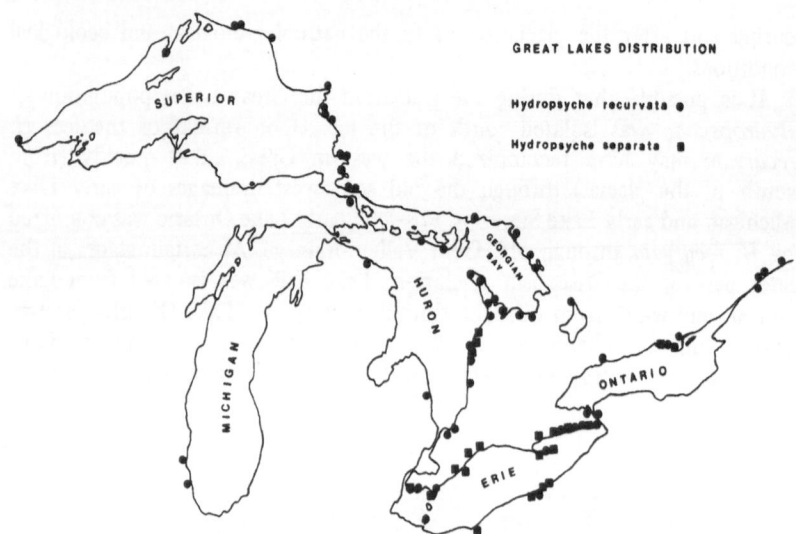

Fig. 6. Distribution of *H. recurvata* and *H. separata* in the Great Lakes area.

Barton and Hynes (1978b) most of the wave-zone invertebrates of Lake Erie overwinter offshore and re-enter the wave-zone by drifting via lake currents. However, overwintering refuge was not found for both *Hydropsyche* species living in Lake Erie and the origin of the larvae in the wave zone remains unknown.

It is, therefore, quite possible that the ability to survive fall and spring storms, ice scouring conditions and inactive winter periods in a refuge is the principle factor responsible for distribution of *Hydropsyche* species in the Great Lakes. This adaptation may be again related to the survival during glaciation periods.

ACKNOWLEDGEMENTS

We are extremely grateful to Drs Richard W. Baumann of the Bringham Young University; Glenn B. Wiggins of the Royal Ontario Museum; and Oliver S. Flint, Jr, of the U.S. National Museum of Natural History for supplying distribution records from these institutions. Dr Flint also kindly permitted examination of specimens in his curatorial care. We would also like to thank Dr Andrew P. Nimmo, for supplying copies of needed literature. Finally, very special thanks are extended to Drs David R. Barton and H.B.N. Hynes of the University of Waterloo, Ontario for supplying detailed un-published distributional data for the Great Lakes region.

REFERENCES

Barton, D.R. and Hynes, H.B.N., J. Great Lakes Res. 4:27–45, 1978a.
——, J. Great Lakes Res. 4:50–56, 1978b.
Crichton, M.I.; Fischer, D. and Woiwood, I.P., Holoarctic Ecology. 1:31–45, 1978.

Hickin, N.E., Caddis Larvae. London pp. 476, 1967.

Hough, J.L., Geology of the Great Lakes. University of Illinois Press, Urbana pp. 313, 1958.

Lemke, R.W.; Laird, W.M.; Tipton, M.J. and Lindvall, R.M., *In* The Quaternary of the United States, eds H.E. Wright and D.G. Frey, pp. 922, Princeton Univ. Press, 1965.

Marshall, A.C., Ann. Ent. Soc. Am. 32:665–688, 1939.

Masteller, E.C. and Flint, O.S. Jr., Great Lakes Ent. 12:165–177, 1979.

Nimmo, A.P., Quaestiones Entomologicae 2:217–242, 1966.

Pewe, Troy, L.; Hopkins, D.M. and Giddins, J.L., *In* The Quaternary of United States, eds H.E. Wright and D.G. Frey, pp. 922, Princeton Univ. Press, 1965.

Ross, H.H., Ill. Nat. Hist. Soc. Bull. 23:236 pp, 1944.

——, *In* The Quaternary of the United States. eds. H.E. Wright and D.G. Frey, pp. 922, Princeton Univ. Press, 1965.

—— and Spencer, G.J., Ent. Soc. Brit. Col., Proc. 48:43–51, 1952.

Smith, D., Pan-Pacific Entomol. 55:10–20, 1979.

Ulmer, G., *In* Die Süsswasserfauna Deutschlands, Vol. 5–6, ed. Brauer, pp. 326, 1909.

Wichard, W. and Unkelbach, G., Decheniana. 126:407–413, 1974.

DISCUSSION

Badcock: Do the disjunct populations of *Hydropysche separata* show any signs of subspeciation? Have you studied this morphologically and physiologically at all? You say that hydropsychids occur in lakes and ponds in Britain. I have not found this to be so. What is your evidence for it?

Sykora: We have not yet studied the disjunct populations of *H. separata* Hickin (1967) reported that *H. pellucidula* inhabits lakes and ponds in Britain.

Badcock: I have been unable to substantiate this. I wrote to Dr Hickin asking about it, but he did not reply. When I met him, he could not tell me, so it may be a doubtful record.

Williams: I can assure you that *Hydropsyche* was present in Lake Ontario at least 100,000 years ago. I have larval sclerites of this species from sediments of the last interglacial at Toronto, Ontario, Canada.

Sykora: This is very useful information which should be published as soon as possible.

Ward: Is there a difference in the structure of the larval net and retreat in lakes compared to streams?

Sykora: According to my limited observation, there is no noticeable difference in the larval net construction. For more information on this subject you may refer to Barton and Hynes (1978a, b).

CYCLES BIOLOGIQUES DES HYDROPSYCHIDAE ET D'UN POLYCENTROPODIDAE (TRICHOPTERA) DANS LE RHÔNE EN AMONT DE LYON[1]

H. TACHET ET M. BOURNAUD[2]

SUMMARY

Seven species of larvae of Hydropsychidae (*Cheumatopsyche lepida*, *Hydropsyche siltalai, H. ornatula, H. pellucidula, H. exocellata, H. dissimulata* and *H. contubernalis*) and 1 species of Polycentropodidae (*Neureclipsis bimaculata*) were studied in the Rhône river for 4 years by mean of artificial substrates. Their life cycles are very stable during the whole period in spite of very high variations of the discharge in the river. *C. lepida, H. siltalai, H. ornatula* and *N. bimaculata* are univoltine, *H. pellucidula, H. exocellata, H. dissimulata* and *H. contubernalis* seem bivoltine. In the smallest species (*C. lepida, H. contubernalis*) the overwintering instar is the 5 th. The relations with the water temperature, the discharges and dam clearing are discussed. *N. bimaculata* is characteristic of low discharges.

INTRODUCTION

Dans le cadre de recherches sur le marcobenthos du Rhône, nous avons plus particulièrement étudié les larves de Trichoptères filtreurs et particulièrement les Hydropsychidae qui constituent un des groupes dominants (Bournaud et al, 1978). Au niveau de la station étudiée, 7 espèces d'Hydropsychidae peuvent coexister: *Cheumatopsyche lepida* (Pict.) 1834, *Hydropsyche siltalai* Döhler 1963, *H. pellucidula* (Curt.) 1834, *H. ornatula* Mc L. 1878, *H. exocellata* Dufour 1841, *H. dissimulata* Kum. et Bots 1974 et *H. contubernalis* Mc L. 1865. Le Polycentropodidae: *Neureclipsis bimaculata* (L.) 1758 est parfois également abondant.

Nous avons abordé l'étude de la coexistence de ces différentes espèces en examinant la structure de leur cycle biologique et ses variations en fonction des principales composantes du milieu.

---

[1] Structure et fonctionnement des écosystèmes du Haut-Rhône français – no. 26.
[2] E.R.A. – C.N.R.S. – no. 849 – Ecologie des eaux douces, Université LYON-I, F. 69622 VILLEURBANNE CEDEX.

Les prélèvements ont été effectués d'avril 1975 à mars 1979 sur la rive droite du canal de Jonage (portion rectifiée du Rhône) à Jons, 20 km en amont de Lyon et 10 km après l'arrivée de la rivière Ain. Bien que partiellement aménagé, le Rhône présente à cet endroit les mêmes caractéristiques que dans sa partie encore sauvage (Perrin et Roux, 1978; Bournaud *et al*, 1978).

Les prélèvements on été réalisés à l'aide de substrats artificiels (Roux *et al*, 1976) déposés sur le fond, près de la rive, pendant 3 semaines. En général 2 substrats ont été immergés simultanément chaque mois. Cependant de mai à novembre 1977 un seul substrat a été immergé à la fois à raison de un par semaine. Nous disposons ainsi de 24 prélèvements en 1975, 24 en 1976, 33 en 1977, 17 en 1978, 6 début 1979 soit 104 prélèvements totalisant 14204 individus appartenant aux 8 espèces citées (Fig. 1). Le tri a été effectué sur tamis de 0,5 mm jusqu'en mai 1977, puis sur tamis de 0,3 mm. De ce fait les effectifs des larves de stade 1 sont sousestimés. Les larves d'*Hydropsyche* n'ont pu être identifiées spécifiquement qu'à partir

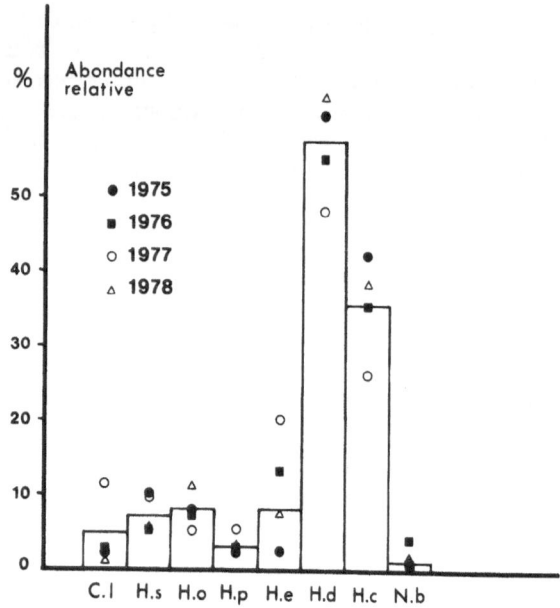

Fig. 1. Abondance relative des 8 espèces rencontrées
C.1. = *Cheumatopsyche lepida*
H.s. = *Hydropsyche siltalaï*
H.o. = *H. ornatula*
H.p. = *H. pellucidula*
H.e. = *H. exocellata*
H.d. = *H. dissimulata*
H.c. = *H. contubernalis*
N.b. = *Neureclipsis bimaculata*

du stade 3, en utilisant les travaux de Sedlak (1971), Szczesny (1974), Hildrew et Morgan (1974), Verneaux et Faessel (1976), Statzner (1976) et Boon (1977). La distinction entre *H. dissimulata*, *H. contubernalis* et *H. exocellata* n'a pu être faite qu'à partir de la mise au point de Malicky (1977) sur les imagos.

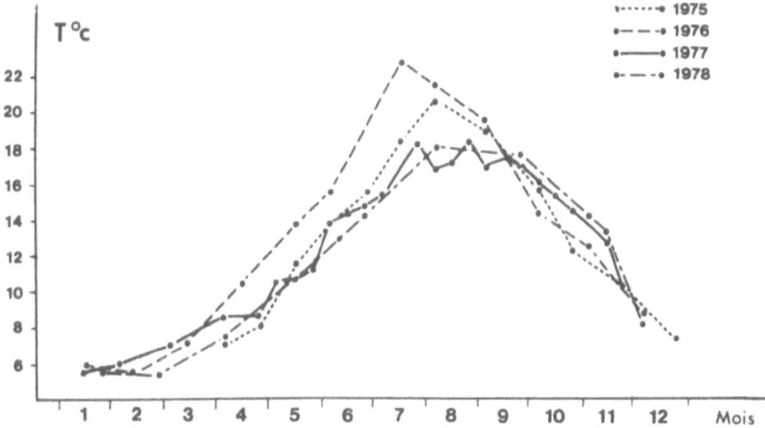

Fig. 2. Température moyenne de l'eau pendant les 21 jours de colonisation de substrats artificiels.

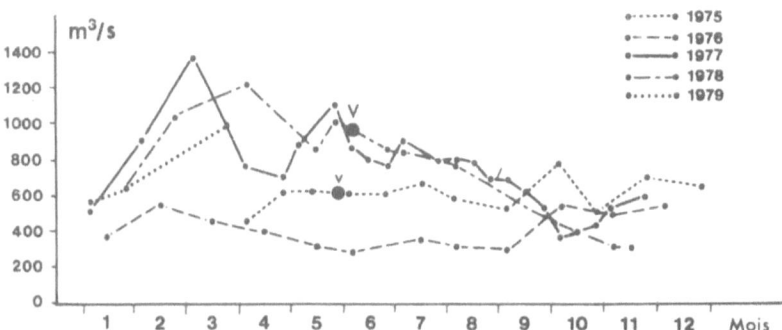

Fig. 3. Débit moyen pendant les 21 jours de colonisation des substrats artificiels. V, v = vidanges de barrages.

Les données sur la température de l'eau (Fig. 2) et le débit (Fig. 3) proviennent des relevés effectués en continu par la Compagnie Nationale du Rhône et l'Electricité de France.

Entre avril 1975 et mars 1979 il y a eu deux vidanges de barrage: la première correspond à la vidange complète du barrage de Verbois et partielle de Génissiat (du 24 au 30 mai 1975); la deuxième plus importante correspond à la vidange complète de Verbois et Génissiat (du 6 au 15 juin 1978).

Afin de hiérarchiser l'importance des ressemblances entre stades et entre espèces, nous avons utilisé l'*analyse factorielle des correspondances* (Benzecri 1973) sur l'ensemble des données. Les regroupements ci-après ont été opérés d'après les résultats de cette analyse. Les données concernant le débit et la température ont été introduites en *variable supplémentaire* (inactives). Le rapprochement de ces variables supplémentaires avec certains éléments des populations étudiées nous a guidés pour rechercher les effets de ces variables de milieu. Ces effets ont été testés au moyen de la méthode de l'information réciproque (Legendre et Legendre, 1979), au niveau d'un tableau de contingence. Un indice de similarité S entre la variable biologique et la variable de milieu (toutes deux divisées en modalité avec présence − absence) mesure l'intensité de la relation entre les deux. Cette relation est testée statistiquement au moyen d'un test χ^2.

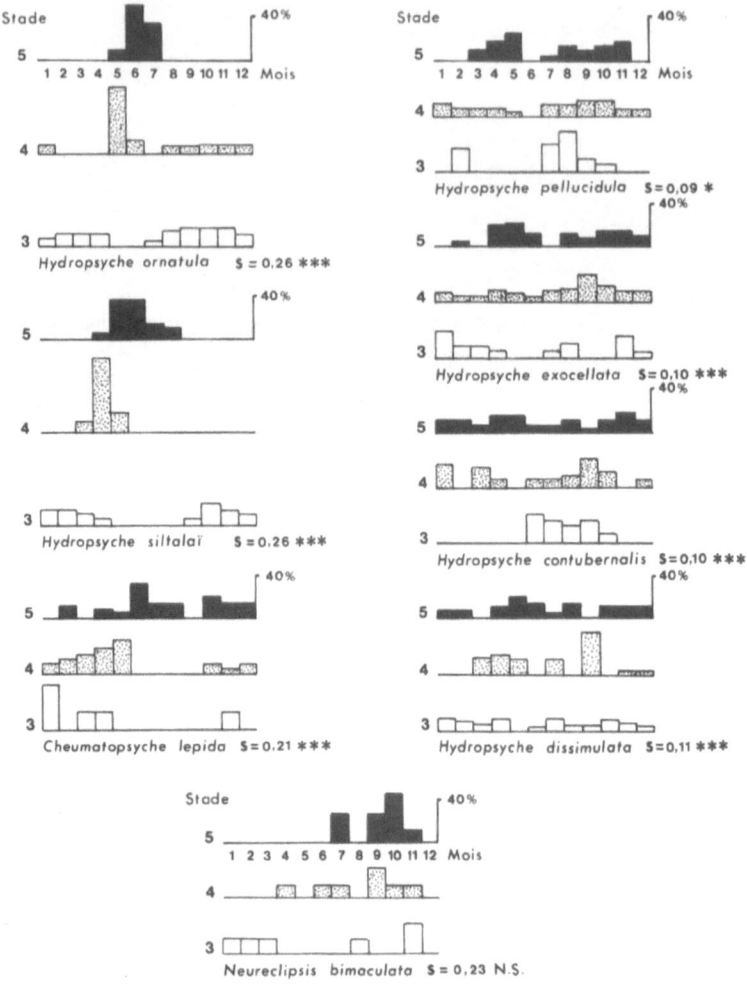

Fig. 4. Répartition mensuelle des dominances des stades larvaires, en pourcentage par rapport au nombre total de relevés où le stade larvaire considéré est dominant. S = indice de similarité: la relation mois − dominance des stades larvaires est significative à 5% (*), 1‰ (***) ou non significative (NS).

LES CYCLES BIOLOGIQUES

Pour comparer les cycles des différentes espèces nous avons utilisé la dominance de chaque stade larvaire par rapport aux autres. Les prélèvements où un stade donné est dominant sont répartis dans les 12 mois de l'année suivant une distribution en pourcentage mensuel (Fig. 4). Par exemple, les 11 prélèvements où le stade 4 de *H. siltalaï* est dominant se répartissent ainsi: 1 en mars (9%), 8 en avril (73%) et 2 en mai (18%). Cette présentation permet de suivre l'évolution des cohortes sur la moyenne des 4 années considérées. L'indice de similitude S entre le mois de prélèvement et le stade dominant permet de juger de la régularité du cycle et de ses altérations possibles: décalages dans le temps, étalement plus ou moins grand de l' apparition et de la présence des stades.

On peut ainsi distinguer 2 types d'espèces (Fig. 4). Celles dont chacun des stades ne domine qu'à une époque précise de l'année (*N. bimaculata, C. lepida, H. ornatula* et *H. siltalaï*) ont un indice de similitude S entre mois et stades relativement élevé (de 0,20 à 0,26). Celles dont au moins les stades 4 et 5 ont une dominance échelonnée sur toute l'année, avec 2 ou 3 maximums plus ou moins prononcés (mai, août et octobre—novembre pour le stade 5) ont un indice de similitude plus bas (entre 0,09 et 0,11). Il s'agit de *H. pellucidula, H. exocellata, H. contubernalis* et *H. dissimulata*.

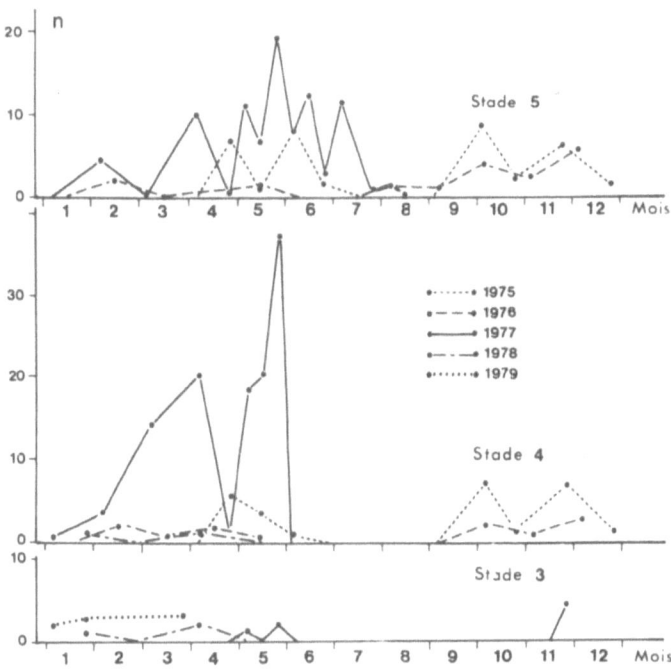

Fig. 5. Cheumatopsyche lepida: Nombre n d'individus par prélèvement pour les stades larvaires 3, 4 et 5.

Groupe des espèces à développement saisonnier limité dans le temps:
C. lepida, H. siltalaï, H. ornatula (Figs 5 à 8).

Ces espèces développent leur stade 5 au printemps (Fig. 4) et les imagos volent à partir de juillet (captures au piège lumineux). *C. lepida* est la plus précoce, puisqu'elle hiverne même sous la forme de stade 5 (Fig. 5). Puis viennent *H. siltalaï*, dont le pic de stade 5 est en général en mai (Fig. 6) et *H. ornatula* dont le pic de stade 5 est début juin (Fig. 7). Ces 3 espèces s'ordonnent dans le même sens vis-à-vis de leur préférendum de température, *C. lepida* ayant le préférendum le plus bas (Fig. 8). Après une disparition

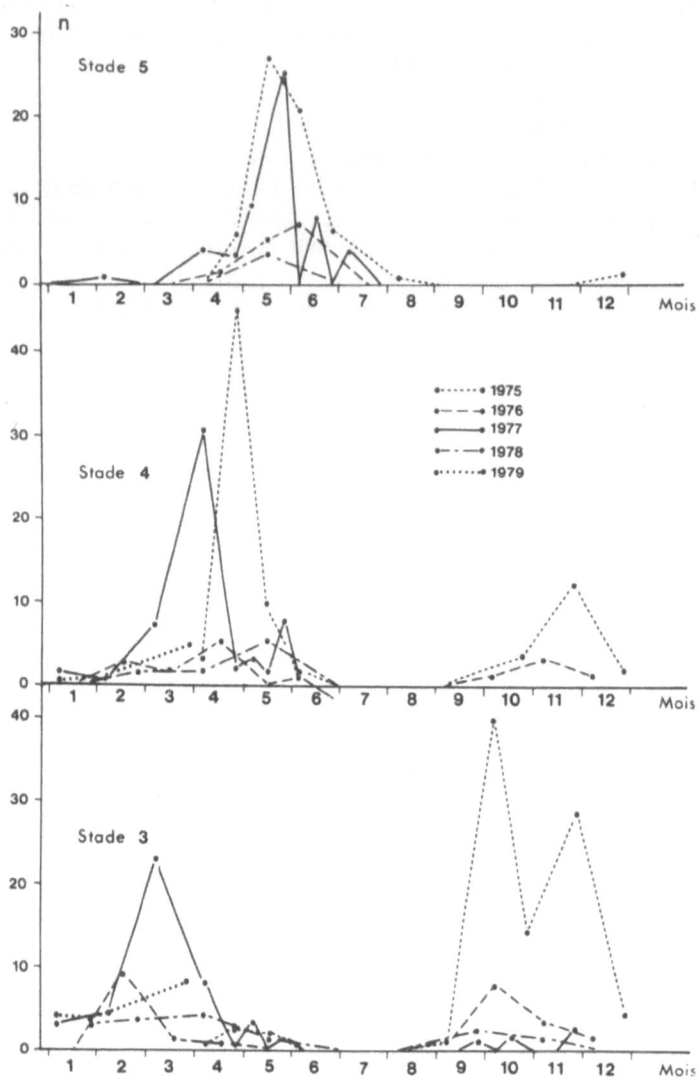

Fig. 6. Hydropsyche siltalaï, idem fig. 5 —

352

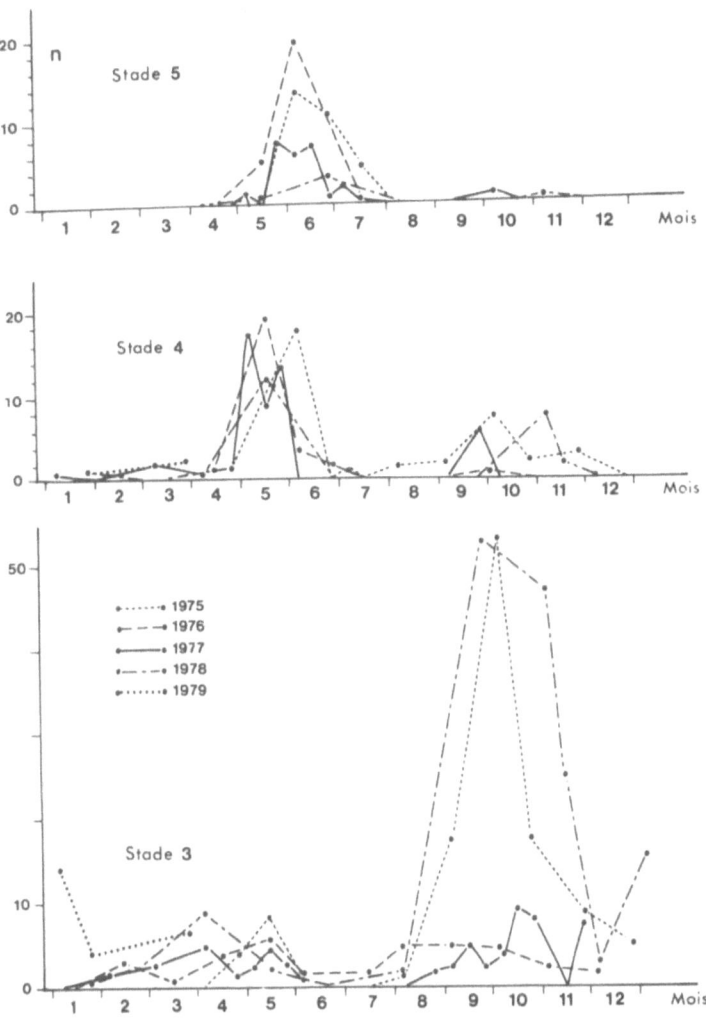

Fig. 7. *Hydropsyche ornatula*, idem fig. 5 –

quasi totale des stades 3, 4 et 5 en été, des effectifs parfois très importants (stade 3 et 4) réapparaissent en automne (Figs 5, 6 et 7), d'autant plus précocement que l'espèce a été tardive au printemps et a un préférendum termique plus élevé: en août pour *H. ornatula*, en septembre pour *H. siltalaï*. Le temps de nymphose et de vol des imagos est donc plus long pour les espèces à préférendum froid, ce qui pourrait correspondre à une diapause pendant la période chaude estivale, rappelant celle observée chez les Limnephilidae Stenophylacini à imagos cavernicoles. En effet, *H. siltalaï* continue son développement plutôt plus rapidement dans la Tyne, rivière britannique pourtant un peu plus fraîche que le Rhône (Boon et Shires 1976, Boon 1979). L'écart entre le maximum

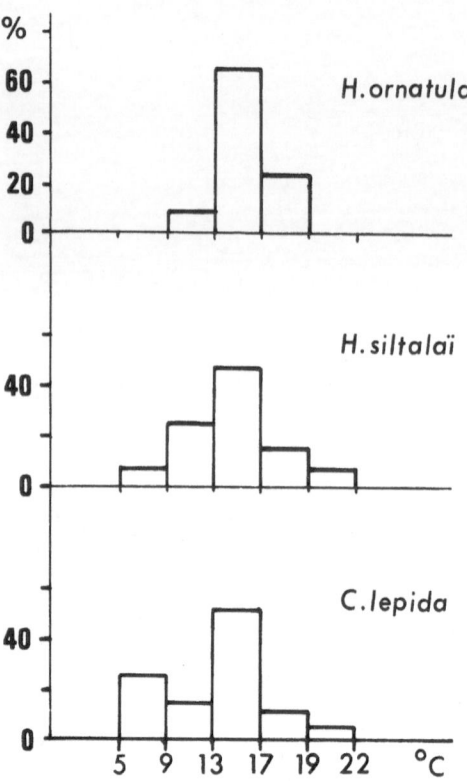

Fig. 8. Dominance du 5ème stade larvaire en fonction de la température de l'eau, en pourcentage par rapport au nombre total de relevés où le stade 5 est dominant.

du stade 5 au printemps et le maximum du stade 3 d'automne y est en effet de 3, 5 mois alors qu'il est d'au moins 4 mois dans le Rhône.

Pour *C. lepida* les stades 3, 4 et 5 ont tendance à persister en hiver, mais très irrégulièrement suivant les années (Fig. 5). Cette espèce apparaît comme liée au débit: les effectifs du stade 5 sont en effet d'autant plus élevés que le débit du jour du prélèvement est élevé (Fig. 9). S'agit-il d'un lessivage dû aux forts débits amenant sur le substrat artificiel des individus provenant de l'affluent amont (l'Ain)? Le problème est le même pour les stades 3 d'*H. siltalaï* qui sont d'autant plus abondants que le débit maximum pendant les 3 semaines de colonisation est élevé (Fig. 9).

Les stades 3 et 4 de *H. siltalaï* et *H. ornatula*, très nombreux en 1975 (et 1978 pour *H. ornatula*), ont tendance à diminuer en decémbre et janvier, sans qu'apparaissent pour autant des stades 5. La croissance hivernale semble donc inexistante et les stades 4 et 5 apparus en automne voués à la disparition en hiver. Les stades 3 apparaissant au printemps ne peuvent alors provenir que de la croissance des jeunes stades (1 et 2), non dénombrés, qui doivent persister tout l'hiver.

Ces 3 espèces sont donc univoltines, ce qui est conforme aux travaux antérieurs pour *H. siltalaï* (Hynes 1961, Elliott 1968, Crichton et Fisher

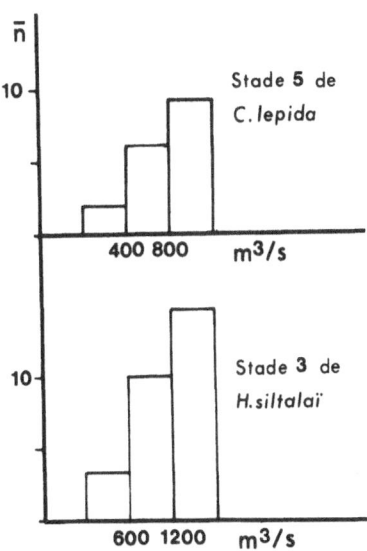

Fig. 9. Effectif moyen, $\bar{n}$, par prélèvement en fonction du débit du jour du relevé (*C. lepida*, stade 5) et du débit maximal pendant les 21 jours de colonisation (*H. siltalaï*, stade 3).

1978, Boon 1978 et 1979, Hildrew et Morgan 1979) et pour *C. lepida* (Brindle 1965, Lapchin et Neveu 1979). Mais les variations de leurs effectifs sont considérables suivant les années (Fig. 5 à 7), surtout pour les stades 3 en automne, sans que l'on puisse attribuer ces fluctuations à l'hydraulicité ou à la température (Fig. 2 et 3). Tout au plus peut-on remarquer que les stades 3 de *H. ornatula* semblent avoir été favorisés par les vidanges de barrages en juin 1975 et juin 1978 (Fig. 7), ceux de *H. siltalaï* favorisés seulement par celle de juin 1975 et défavorisés par celle beaucoup plus importante de juin 1978 (Fig. 6).

Il ressort de ces observations que *C. lepida*, *H. siltalaï* et *H. ornatula*, espèces assez différentes du point de vue taxonomique et écologique, sont très homogènes dans leurs cycles biologiques.

Groupe des espèces à développement saisonnier étalé: H. pellucidula, H. exocellata, H. dissimulata, H. contubernalis.

Ces 4 espèces ont une dynamique assez semblable, surtout par la présence de leurs stades 4 et 5 durant toute l'année (Fig. 4), mais avec des effectifs beaucoup plus importants en automne (Fig. 10). Leurs dynamiques sont suffisamment proches pour que l'on puisse la résumer globalement pour les 4 espèces par les moyennes géométriques des effectifs de stade 5 par panier (Fig. 11).

3 maximums apparaissent dans le développement des stades 5 (Fig. 4 et 11): un en mai, un en août et un en octobre-novembre. Ces maximums peuvent correspondre à des périodes de prénymphose, puisque les imagos sont capturés aux pièges lumineux sans interruption de juin à octobre. La croissance du stade 3 au stade 5 est visible sur la figure 4. Cette évolution

355

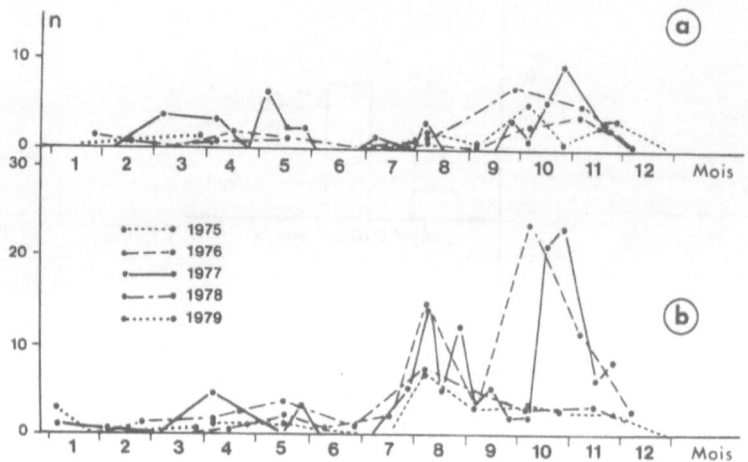

Fig. 10. Effectif n par prélèvement (5ème stade larvaire)
a = *Hydropsyche pellucidula*
b = *H. exocellata*

peut être interprétée comme suit. La croissance hivernale aboutit à une génération d'imagos émergeant en mai-juin. Les pontes de ceux-ci fournissent une génération à développement rapide en juillet-août, responsable du pic de stades 5 en août. Les imagos qui en sont issus sont responsables de la génération et du pic plus étalé d'automne. Ce cycle à allure de bivoltinisme se retrouve effectivement pour chacune des 4 espèces (Fig. 4). La figure 10 montre la grande différence d'effectifs (stade 5) entre la génération de printemps et celles d'été et d'automne, surtout pour *H. contubernalis* et *H. dissimulata*. La croissance est rapide et très bien synchronisée en été: le pic de state 5 est toujours dans la *1ère décade d'août* pour les 4 espèces et les 4 années. Par contre la génération d'automne est plus étalée: le pic des stades 5 se situant en octobre ou novembre. Nous ne savons pas dans quelle mesure ces stades 5 donnent lieu à des métamorphoses tardives qui expliqueraient la diminution des effectifs de stade 5 en hiver, mais cette diminution peut s'expliquer surtout par une mortalité hivernale importante. La proportion de stades 5 en période hivernale est de plus en plus grande dans la succession *H. pellucidula, H. exocellata, H. dissimulata, H. contubernalis* (Fig. 12). L'hivernage sous forme de jeunes stades est donc important surtout chez *H. pellucidula* et *H. exocellata*. Mais le synchronisme des pics de stades 5 chez les 4 espèces est encore souligné par les courbes de la figure 12.

Mises à part les modalités d'hivernage que l'on vient de souligner, ainsi que des effectifs plus faibles chez *H. pellucidula* et *H. exocellata* (Fig. 10), les 4 espèces de ce groupe présentent peu de différence. *H. pellucidula* est la seule non strictement potamique. Le rapport des effectifs *H. exocellata/ H. pellucidula* tend à augmenter avec la température de l'eau (Fig. 13), ce qui est à mettre en relation avec la plus grande importance du pic d'été chez *H. exocellata*.

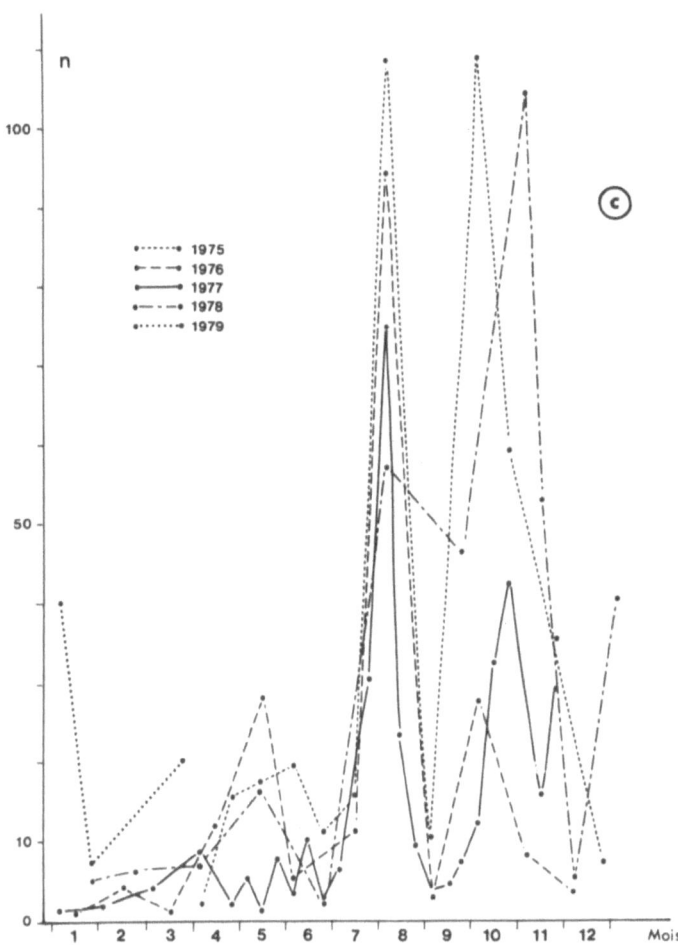

Fig. 10. Effectif n par prélèvement (5ème stade larvaire)
c = *H. dissimulata*

L'effet des crues ou du débit est peu marqué (Fig. 10). On peut enfin remarquer (Fig. 10) que les générations d'automne (stade 5) sont beaucoup plus abondantes en 1975 et 1978, (années des vidanges de barrages), chez *H. contubernalis, H. dissimulata* et à un moindre degré chez *H. pellucidula.* C'est l'inverse chez *H. exocellata.*

Ces 4 espèces apparaissent donc comme bivoltines. Seule *H. pellucidula* est bien connue de ce point de vue: bivoltine seulement en France à une température estivale de 16–18°C (Lapchin et Neveu 1979), univoltine en Pologne et en Grande-Bretagne avec une température estivale de 10–17°C (Szczesny 1975, Crichton et Fisher 1978, Boon 1978 et 1979, Hildrew et Edington 1979). *H. contubernalis* est bivoltine en Grande-Bretagne,

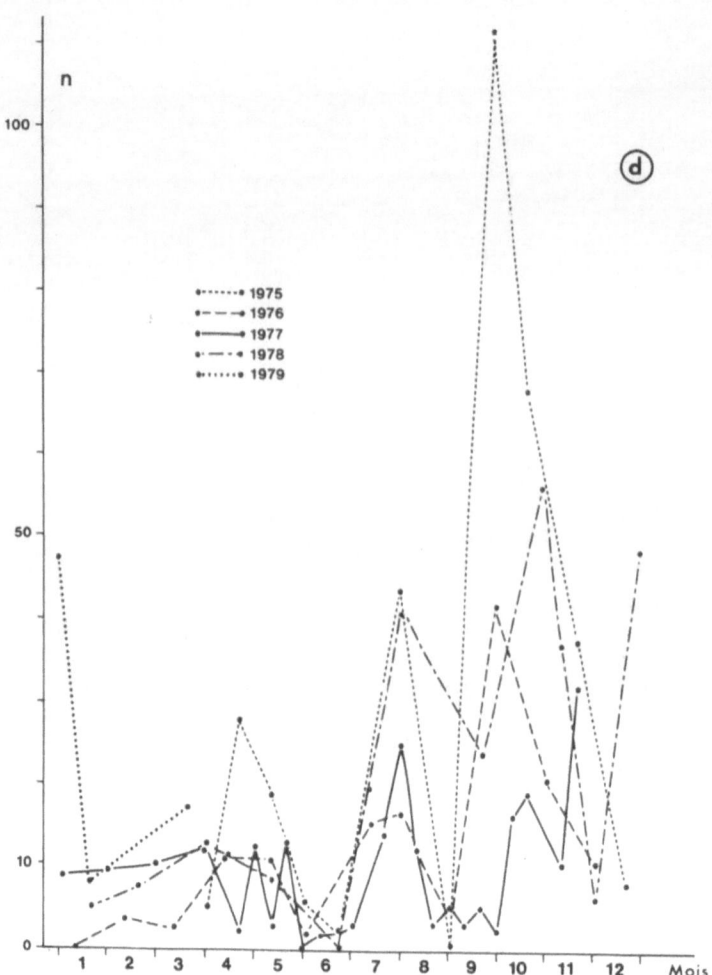

Fig. 10. Effectif n par prélèvement (5ème stade larvaire)
d = *H. contubernalis*

d'après les récoltes d'imagos (Crichton et Fisher 1978). *H. dissimulata* et *H. exocellata* n'ont pas été étudiées.

Neureclipsis bimaculata (Fig. 4 et 14)

Ce Polycentropodidae à larve carnivore n'apparaît presque uniquement qu'en été 1976 et en automne 1978 (Fig. 14), soit les 2 périodes de débits les plus faibles, inférieurs à 400 m3/s (Figs 3 et 15). Il s'agit bien d'une espèce du potamon lent. Ceci est confirmé par sa présence beaucoup plus continue dans le Rhône à Lyon, où le fleuve est ralenti par le barrage de Pierre-Bénite. A la différence des espèces précédentes, l'époque de

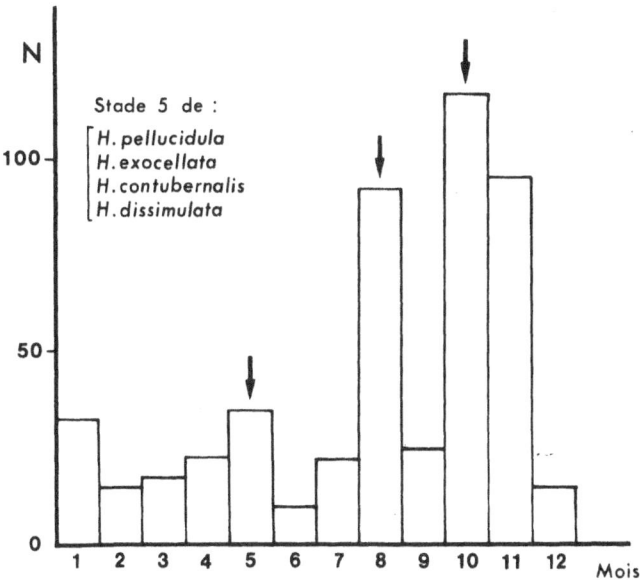

Fig. 11. Effectif N par prélèvement (moyenne géométrique) des 5èmes stades larvaires des 4 espèces ensemble: (*H. pellucidula, H. exocellata, H. contubernalis,* et *H. dissimulata*) en fonction du mois de l'année. Les flèches indiquent les 3 maximums communs aux 4 espèces.

son développement maximal varie beaucoup suivant les années: début septembre en 1976 à Lyon, juillet août en 1976 à Lyon et Jons (Fig. 15), octobre en 1978 à Jons (Fig. 14): ces 3 périodes correspondent aux plus faibles débits de l'année considérée (Fig. 3). Cette espèce potamique est donc un indicateur particulièrement net des faibles débits.

DISCUSSION

Pendant ces 4 années d'études qui intègrent des variations de milieu souvent considérables (débit passant de 190 à 2500 m3/s, température comprise entre 5 et 25°C, matières en suspension variant de quelques milligrammes à plusieurs grammes) nous avons pu constater une grande stabilité dans les cycles biologiques des espèces étudiées. En effet:

– les abondances relatives des différentes espèces ont trés peu varié (Fig. 1): *H. dissimulata* est resté l'espèce la plus abondante toujours suivie de *H. contubernalis.*

– à l'exception de *N. bimaculata,* dont l'apparition sur les substrats est étroitement liée aux variations de débit, toutes les autres espèces présentent un cycle de développement souvent très régulier, sans rapport net avec le régime des crues ou la température. Nous avons ainsi pu distinguer un groupe d'espèces monovoltines (*H. siltalaï, H. ornatula* et *C. lepida*) et un groupe d'espèces bivoltines (*H. pellucidula, H. exocellata, H. dissimulata* et *H. contubernalis*).

359

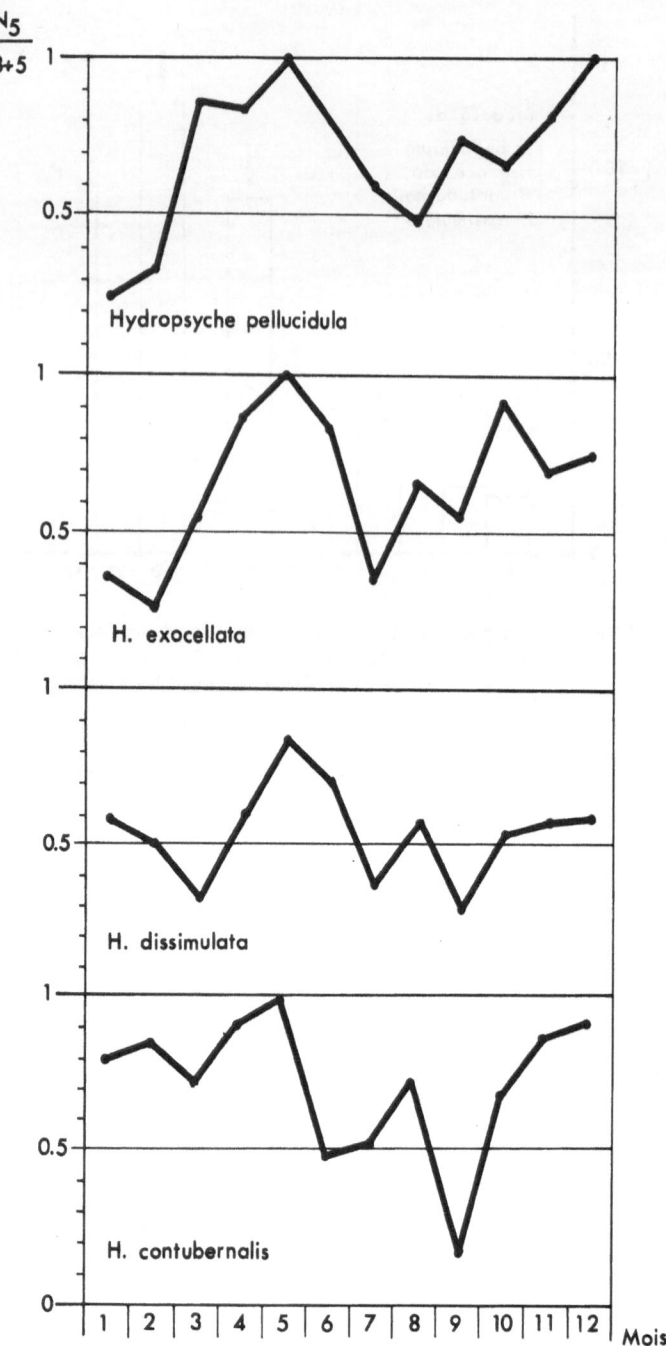

Fig. 12. Proportion des effectifs des stades 5 (N_5) par rapport aux effectifs des stades 3 et 5 réunis (N_{3+5}). La montée de la courbe au-dessus de 0,5 montre une prédominance du stade 5, sa descente au-dessous de 0,5 une prédominance des stades 3.

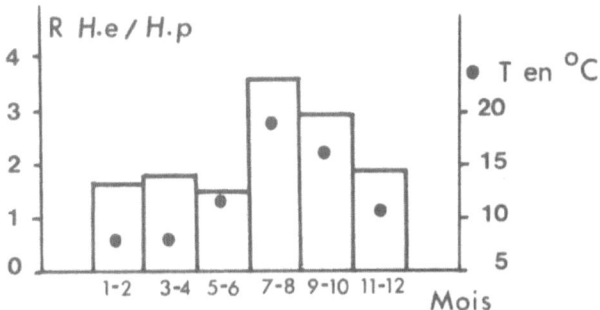

Fig. 13. Rapport des effectifs de *H. exocellata* sur ceux de *H. pellucidula* en fonction du mois de l'année et de la température de l'eau.

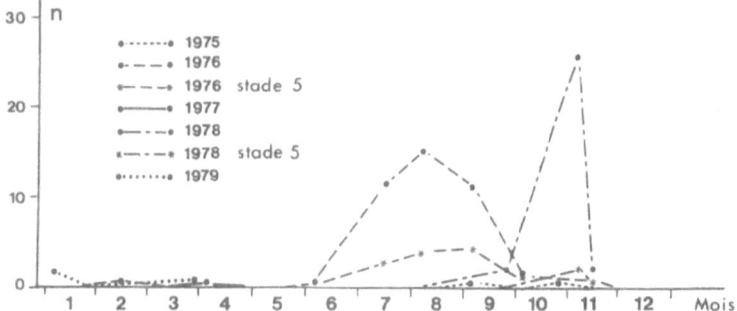

Fig. 14. *Neureclipsis bimaculata*: effectifs n par relevé, stades 3, 4 et 5 confondus.

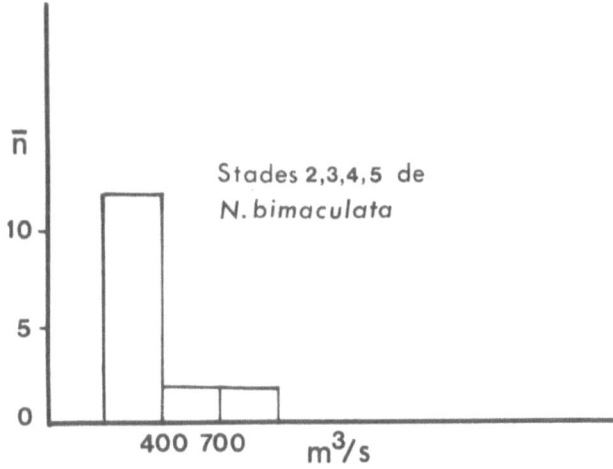

Fig. 15. Effectif moyen n̄ par relevé de *N. bimaculata* (stades 3, 4 et 5 confondus) en fonction du débit moyen pendant les 21 jours de colonisation des substrats artificiels.

Parmi les causes de ce mono ou bivoltinisme, l'action de la température peut être invoquée pour certaines espèces. C'est ainsi que *H. siltalaï* et *H. pellucidula* sont monovoltines en Grande Bretagne et en Pologne (cf. § 3) où les températures estivales sont basses. Dans le Rhône, ces 2 espèces bien que soumises, au moins théoriquement, aux mêmes conditions thermiques, réagissent différemment: *H. siltalaï* reste monovoltine tandis qu' *H. pellucidula* paraît s'adapter aux températures plus élevées et devient bivoltine, comme c'est le cas également pour la basse Nivelle (Lapchin et Neveu 1979).

Il y aurait donc aussi des espèces à monovoltinisme strict, telles que *H. siltalaï* et *C. lepida* et des espèces à bivoltinisme facultatif telle *H. pellucidula*. Le passage du monovoltinisme au bivoltinisme dépend éventuellement de la température.

H. contubernalis qui, avec *H. dissimulata*, est la plus petite espèce d' *Hydropsyche* rencontrée dans le Rhône est bivoltine comme en Grande-Bretagne (Crichton et Fisher, 1978).

Mc Kay (1979) a observé également que les petites espèces sont toujours bivoltines et hivernent plutôt sous forme de stade 5.

L'existence d'espèces apparemment strictement monovoltines, d'espèces mono ou bivoltines selon les circonstances et d'espèces plus strictement bivoltines pose le problème de leur déterminisme. Le facteur température n'est pas le seul à intervenir, le facteur nourriture est sans doute important à condition de considérer tous les stades larvaires et examiner si les éléments rencontrés dans le tube digestif ont réellement une valeur nutritive. Il est enfin nécessaire de ne pas oublier que la vie larvaire qui représente effectivement en durée les 9/10 du cycle n'est pas nécessairement le moment le plus crucial de ce cycle. Le problème de différences dans les niches écologiques de ces espèces sympatriques est posé. Il est encore difficile à résoudre du fait que l'écologie de *H. dissimulata* et *H. exocellata* était encore inconnue, celle de *H. contubernalis* et *H. ornatula* très mal connue. Il faudra notamment examiner les abondances et cycles de ces espèces dans d'autres stations du bassin du Rhône, afin de contrôler les tendances thermophiles de *H. exocellata* par rapport à *H. pellucidula* (Fig. 13), de *H. ornatula* par rapport à *H. siltalaï* et *C. lepida* (Fig. 8), la régularité d'apparition de *C. lepida* et *H. siltalaï* au moment des forts débits, et enfin l'effet dynamisant, voire eutrophisant, des vidanges de barrage, accompagnées de taux élevés de matières en suspension principalement sur *H. ornatula* et *H. dissimulata*, à un degré moindre sur *H. contubernalis* et *H. siltalaï*.

REMERCIEMENTS

Nous remercions tout particulièrement Madame Jazdzewska et Mademoiselle Dessaix, qui effectuèrent les relevés de terrain, respectivement en 1977 et 1978.

REFERENCES

Benzecri, J.P., L'Analyse des Donnés 2. L'analyse des Correspondances. Dunod, Paris, 1973.

Bon, P.J., *In* Proc. 2nd Intern. Symp. Trichoptera, ed. M.I. Crichton, pp. 165–173, Junk, The Hague, 1977.

———, Hydrobiologia 57:167–174, 1978.

———, Arch. Hydrobiol. 85:336–359, 1979.

——— and Shires, S.W., Freshw. Biol. 6:23–32, 1976.

Bournaud, M., Bull. Mens. Soc. Linn. Lyon 40:196–211, 1971.

———; Chavanon, G. et Tachet, H., Verh. Int. Ver. Limnol. 20:1495–1493, 1978.

Brindle, A., Ent. Rec. 77:148–159, 1965.

Caspers, N., Verh. Int. Ver. Limnol. 20:2617–2621, 1977.

Crichton, M.I.; Fisher, D. et Woiwod, I.P., Holarctic Ecology 1:31–45, 1978.

Elliott, J.M., J. Anim. Ecol. 37:615–625, 1968.

Fey, J.M. et Schumacher, H., Ent. Germ. 4:1–11, 1978.

Floessner, D., Limnologica 10:123–153, 1976.

Fremling, C.R., Res. Bull. Iowa State Univ. Sci. Techn. 483:855–879, 1960.

Hildrew, A.G. et Edington, J.M., J. Anim. Ecol. 48:557–576, 1979.

——— et Morgan, J.C., J. Ent. (B) 43:217–229, 1974.

Hopkins, C.L., N.Z.J. Mar. Freshwat. Res. 10:629–640, 1976.

Hynes, H.B.N., Arch. Hydrobiol. 57:344–388, 1961.

Kaiser, P., Int. Rev. Ges. Hydrobiol. 50:169–224, 1965.

Lapchin, L. et Neveu, A., Annls Limnol. 15:139–153, 1979.

Legendre, L. et Legendre P., Ecologie numérique, Masson, Paris, pp. 197 et 247, 1979.

Mackay, R.J., Can. J. Zool. 57:963–975, 1979.

Mc Cullough, D.A. et Minshall, G.W., Ecology 60:585–596, 1979.

Malicky, H., Z. Arbeitsgem. Ost. Ent. 29:1–28, 1977.

Matzdorf, F., Ent. Ber. 2:73–79, 1964.

Perrin, J.F. et Roux, A.L., Verh. Int. Ver. Limnol. 20:1494–1502, 1978.

Rhame, R.E. et Stewart, K.W., Trans. Am. Ent. Soc. 102:65–99, 1976.

Roux, A.L.; Tachet, H. et Neyron, M., Bull. Ecol. 7:493–496, 1976.

Schröder, P., Beitr. Naturk. Forsch. SüdwDtl. 35:137–148, 1976.

Sedlak, E., Acta Ent. Bohmoslov. 68:185–187, 1971.

Sprules, W.M., Univ. Toronto Stud. Biol. 56:1–88, 1947.

Statzner, B., Ent. Germ. 3:265–268, 1976.

Szczesny, B., Polsk. Archiv. Hydrobiol. 21:387–390, 1974.

———, Acta Hydrobiol. 17:35–51, 1975.

Verneaux, J. et Faessel, B., Annls Limnol. 12:7–16, 1976.

Wallace, Y.B., Ann. Ent. Soc. Amer. 68:463–472, 1974.

Williams, N.E. et Hynes, H.B.N., Oikos 24:73–84, 1973.

DISCUSSION

Botosaneanu: Comme profane dans ce genre de recherches, je n'ai pas bien compris la manière dont vous avez utilisé les pariers remplis de pierres plates. Si je vois correctement, le rôle majeur dans le peuplement en faune de ces milieux artificiels est joué par le drift; dans ces conditions pensez-vous que la méthode utilisée puisse vraiment donner une idée correcte de ce qui se passe dans la rivière?

Tachet: Les substrats artificiels utilisés, pendant 21 jours sont en effet alimentés en partie probablement importante par le drift. Mais nous avons pu constaté que le peuplement de ces substrats n'est différent ni qualitativement ni quantitativement, sur le plan des espèces et des stades larvaires, de prélèvements effectués directement sur le fond à l'aide d'une drague (résultats récents non encore publiés).

Higler: Est ce que vous avec pensé à la possibilité 'de cohortes' dans le cas des espèces potamiques au lieu des deux générations? nous ne croyons pas que ces espèces sont bivoltines et pensons que les prélèvements avec des substrats artificiels offrent une sélection qui rend l'interprétation dangereuse.

Tachet: Nous n'avons pas tenu compte des stades larvaires 1 et 2 non suffisamment échantillonnés. Il n'y a pas d'effets particuliers du débit sur les espèces du 2e groupe (avec 3 maximums de stade 5). Cette absence de relation, ainsi que la répétition des maximums du stade 5 aux mêmes dates les 4 années, nous a invité à considérer ces espèces comme bivoltines. Mais il existe certainement aussi une continuité d'apparition de cohortes au long de l'année.

Moretti: *Hydropsyche dissimulata* a aussi chez nous 2 générations en Italie Centrale. *Neureclipsis bimaculata* présente aux émissaires des lacs de Lombardie une émergence très forte pendant la décharge des lacs. Vous avez raison d'étudier le trophisme des lacs pendant la vidange.

Tachet: Cette espèce n'apparaît dans le Rhône qu'au moment de débits exceptionnellement bas. On ne peut donc pas la mettre en relation avec les vidanges de barrage.

THE PRINCIPLES OF CASE BUILDING BEHAVIOUR IN TRICHOPTERA

C. TOMASZEWSKI

SUMMARY

It is hypothesized that the caddis larva's portable case is the result of tactile demands which originated from its soil-dwelling ancestors. The external stimulus that triggers the building activity is thought to be lack of touch.

When the wart-like protuberances on the 1st segment were compressed in an experimental research, building activity ceased. It is, therefore, suggested that these are the superior mechanoreceptors responsible for triggering building activity.

INTRODUCTION

The portable cases of the caddis larvae, although constructed almost entirely of foreign material, are an integral part of their bodies. Together they form a functional unity, for without the case the larva could neither live nor develop. The problems of case building are, therefore, both phylogenic and ontogenic.

Ross (1964, 1967), Malicky (1973) and Tomaszewski (1973) investigated the evolutionary aspect of this problem and Tomaszewski, also, drew attention to its original causes. There is an ever increasing number of ontogenic papers which mainly deal with the methods of building, the plasticity and the repairing of the cases by the different species. However, they offer no answer to the question 'Why does the larva build a case and what triggers the building mechanism?'. Some investigators have attempted an answer and, in 1965, Merrill demonstrated that removal of the anal hook and hair sensills of the 9th and 10th segments prolongs the case building activity, but that resection of the nerve cord between the 6th and 7th ganglia causes the activity to cease. From these experiments, Merrill concluded that the impulses emitted by the posterior ganglia pass to the supposed anterior control mechanism and stimulate building activity, whereas sensory contact of the anal hooks and hair sensills with the case inhibits the activity. These findings have been confirmed by Marstaller (1969), Hansell (1973, 1974a, b) and Smart (1974). However, Marstaller doubts

Merrill's interpretation, because when an additional cut is made at the 4th and 5th ganglia, the building activity is not fully inhibited. Furthermore, Branch (1922) reported that the anal hooks of Limnephilidae are innervated by the 7th and 8th ganglia. Marstaller, therefore, considers that building activity is a function of the whole nervous system and his hypothesis is supported by the fact that when its posterior portion is destroyed by irradiation, the larva continues building. Smart's (1974) experiments suggest that the case length is more dependent on the presence, or absence, of anal claws than of hair sensills. The only function of the anal hook, in my opinion, is that of attaching the larva to the case at the posterior inside and if the larva cannot do this, it cannot measure the length of the case and so continues building. The question on the causes of case building, therefore, remains open.

The investigations mentioned deal with internal stimulation which must be triggered by an external stimulus; so one may well ask: 'What is the external stimulus, how is it perceived and where are the superior receptors located?'.

The case structure is multifunctional; it provides streamlining, ballast rigidity, buoyancy, camouflage, internal water circulation and protection from predators and other dangers (Wiggins, 1977). So, maybe, the next question should be: 'What is the original and most basic function of the case that may have determined its building?'. Hologenic data are used in an attempt to answer this question.

Tomaszewski (1973) hypothesizes that the process of adaptive evolution in the caddis larva can be visualized as a cycle of changes from terrestrial ancestors, through life in water to the larva of certain contemporary species that have been driven to live on the land. The cycle turns in time to generate a wave on the co-ordinate system. Fig. 1 depicts a single hologenic turn; on the first peak, the hypothetical soil-living ancestors, which, probably stuck silk linings to their tunnels in the substrate, are seen. Research carried out by Marstaller (1969) and Tachet (1978) suggest that spinning and building behaviour are independent and, in this case, building is secondary to

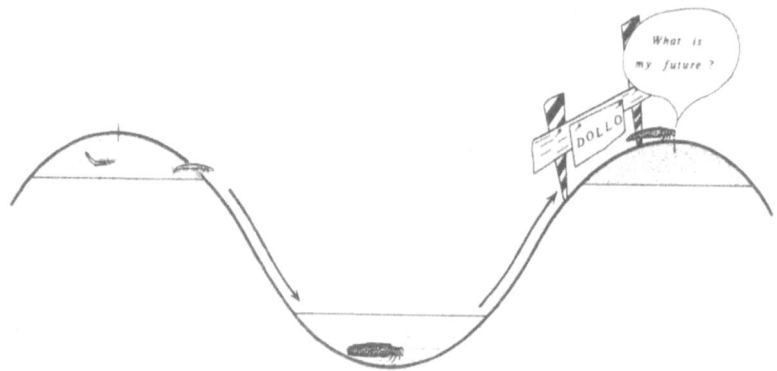

Fig. 1. The hologenic theory of the adaptive evolution of the caddis larva. One turn of the circle generates a sinus wave.

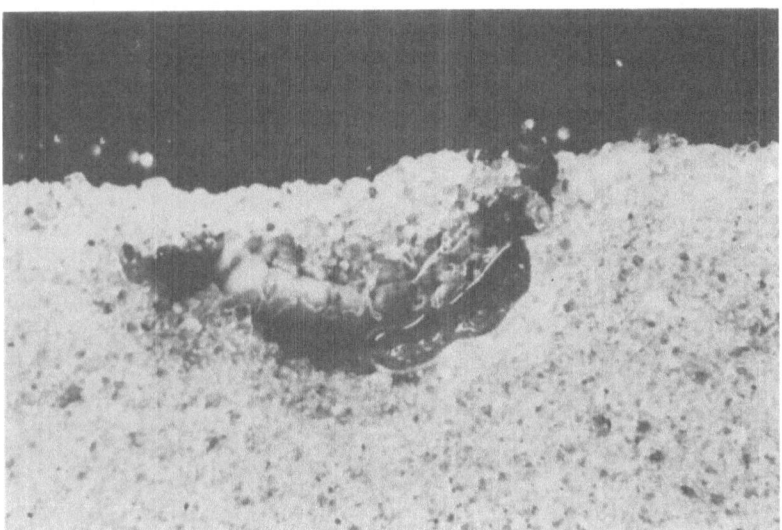

Figs. 2 and 3. A *Limnephilus flavicornis* larva burrows into the substrate. (Phot. M. Jankowska).

spinning behaviour. The passage from soil to lotic waters saw a new group of insects – the Trichoptera – arise. On the descending line two family branches – Hydropsychoidea and Rhyacophiloidea – evolved in cool waters (adapted from Ross, 1956) and building behaviour originated, while the evolution of the Limnephiloidea branch was associated with a passage into warm lenitic waters (adapted from Ross, 1956) and thence, in the larva

367

of some species – *Enoicyla pusilla* (Burm.), *Ironoquia parvula* (Banks) (Flint, 1958) and *Philocasca demita* (Ross) (Anderson, 1967) – secondarily, through lotic waters to the land. A new post-Trichoptera group of insects may evolve from these species in the future.

It would, therefore, seem safe to assume that the ability to build a portable case had its origins in lenitic waters, as when Limnephiloidea larva are denuded, their characteristic reaction is to bury themselves in the substrate (Figs 2, 3) and this, I think, shows regression to the way of life of their hypothetical ancestors. In its first phase the larva surrounds its body and satisfies tactile demands. The lack of touch is most probably the external stimulus which triggers building activity and, thus, the original case building behaviour.

EXPERIMENTAL DATA

A large number of hair sensills are distributed in various patterns over the body of the larva. If lack of touch stimulates building activity, then there should be a superior touch receptor. I decided to investigate the abdominal wart-like protuberances on the first segment which, it is generally thought, have the function of attaching the anterior of the larva to the inside of the case. The function attributed to them is questionable because they are too soft for this purpose. When the larva is out of the case they can be retracted, but not when it is inside. This suggests that they have another role, that of tactile receptors.

MATERIALS AND METHODS

Compression girdles made from a small plastic tube were used to exert pressure on the 1st segment abdominal protuberances (Fig. 4). Thirty *Limnephilus flavicornis* (Brau.) and 20 *Limnephilus borealis* (Zett.) at the last instar, caught in flood waters of the Brzostowka river in Glowno, Central Poland, were used for the experiments.

RESULTS

The results are reported in Tables I and II. The numbers in columns 2 and 3 show burrowing and case building times on the first day after the case was removed; they are given for comparison only. Column 5 lists non-building time in larva with compression girdles in place, which is, also, the time girdles were left on; the differences between the specimens are, therefore, due to chance. The behaviour of the larva after the girdles were slipped off is recorded in column 6. It will be seen from columns 5 and 6 that while the girdle exerted pressure on the warts, there was no case building activity. The longest non-case-building times observed were 14 days in *L. flavicornis* and 15 days in *L. borealis*.

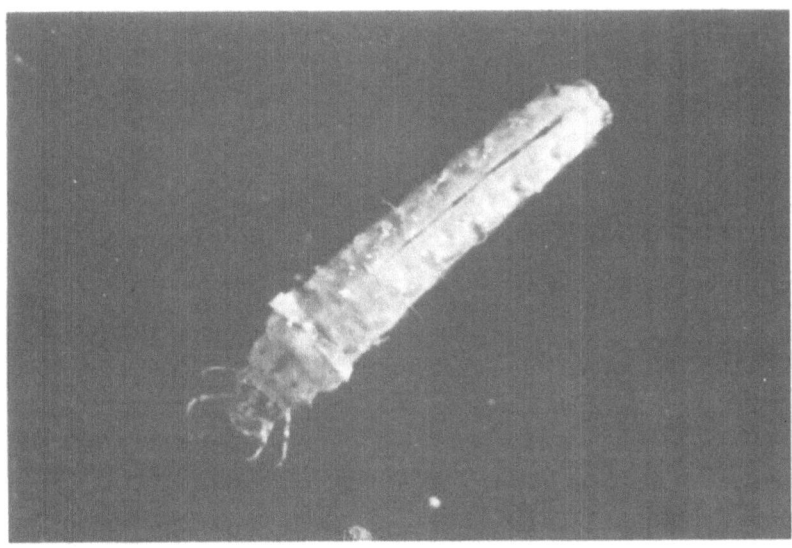

Fig. 4. Limnephilus flavicornis larva with compression girdle on 1st abdominal segment — it has slipped slightly. (Phot. M. Jankowska).

DISCUSSION

The case is a multipurpose structure, the basic function which determines its building is the satisfaction of the tactile demands and the external trigger mechanism is the lack of touch. The superior mechanoreceptors which receive the tactile information are the wart-like protuberances on the 1st segment, which, also, probably play a role in controlling case weight and length as the larva grows. Presumably, the shape of the case depends on the shape of the protuberances and their innervation system, but this is a subject for future research. By stretching out from the case with the help of its legs, while remaining hooked to the inside by the anal claws, the larva brings the abdominal protuberances into contact with the anterior rim and this is, probably, the way the larva measures its case. Hansell (1974a) reported on stretchreceptors, but did not examine them in the Trichoptera.

It can, therefore, be concluded that the abdominal protuberances are responsible for transmitting the input of touch information on the stretch maximum to the central nervous system and that this dictates when building activity ceases, and that contact of the protuberances with the case wall satisfies tactile demands.

ACKNOWLEDGEMENTS

I wish to thank Mgr. M. Jankowska for her help in carrying out the experiments.

Table I. Case making and 'non-case building activity' in *Limnephilus flavicornis* (Brau.) larvae

No.	Time of case building activity on the 1st day after denudation		Larvae with compression girdles on the 1st abdominal segment		
	Burrowing into the substratum (min.)	Case building (hours)	No.	Time of 'non case building' activity (24 hours)	Activity after slipping off the girdles (after time mentioned in the column 5)
1	2	3	4	5	6
1	10	8	1	4	The larva buried into the substratum and died after 2 days
2	15	4	2	7	
3	20	5	3	11	
4	20	4	4	12	Burrowing and case building
5	5	6	5	5	
6	20	7	6	10	
7	15	7	7	12	
8	8	5	8	–	The larvae died after 3 hours (probably injured)
9	15	6	9	–	
10	20	5	10	9	
11	20	5	11	3	
12	5;8;15	8;7;9	12	1	Burrowing and case building
13	20	1,5	12	0	
14	20	3	14	0	
15	5	6	15	5	
16	20	5	16	14	After 2 weeks the larva emerged from the water onto the surface. The larva re-entered the water and after a few days died
17	5;5	8;5;10	17	1	Burrowing and case building
18	5	4	18	–	Death after a few days (probably injured)
19	10	5	19	–	

Table I. continued

	10:5:15	8:5:9			
20	15	6	20	4	Burrowing and case building
21			21	8	
22	5	7	22	14	After 2 weeks the larvae started to build their pupal cases immediately, but did not burrow into the substrate and pupated. The case material was stuck to the girdles and the cases were fastened to the substrate with silk
23	35	5	23	14	
24	5	3	24	5	Burrowing and case building
25	35	6	25	10	
26	10	5	26	14	As larvae Nos 22 and 23
27	45	4	27	7	Burrowing and case building
28	25	3	28	5	
29	5	4	29	–	Death after few days (probably injured)
30	25	5	30	4	Burrowing and case building

Table II. Case making and 'non-case building activity' in *Limnephilus borealis* (Zett.) larvae

No.	Time of case building activity on the 1st day after denudation		Larvae with compression girdles on the 1st abdominal segment		
	Burrowing into the substratum (min.)	Case building (hours)	No.	Time of 'non case building' activity (24 hours)	Activity after slipping off the girdles (after time mentioned in column 5)
1	2	3	4	5	6
1	30	3,5	1	0	
2	35	2,5	2	5	
3	20	1,5	3	0	Burrowing and case building
4	25	3	4	5	
5	15	5,5	5	11	
6	10	5,5	6	0	The larva started to build its case immediately, but did not burrow into the substratum
7	10	5	7	5	Burrowing and case building
8	15	4	8	1	
9	5	5	9	–	Death probably due to injury when pulling off the girdle
10	15	4	10	9	
11	40	5	11	2	
12	15	4,5	12	0	
13	15	7	13	4	Burrowing and case building
14	20	5,5	14	10	
15	15	7	15	6	
16	15	3	16	15	
17	45	5,5	17	–	
18	20	9	18	–	Death probably due to injury when pulling off the girdle
19	20	4	19	–	
20	35	9	20	12	Burrowing and case building

REFERENCES

Anderson, N.H., Ann. ent. Soc. Amer. 60:320–323, 1967.

Branch, H.E., Ann. ent. Soc. Amer. 15:256–275, 1922.

Flint, O.S., Jr, J.N. York ent. Soc. 66:59–62, 1958.

Hansell, M.H., Behaviour 46:141–153, 1973.

——— Anim. Behav. 22:133–144, 1974a.

——— *In* Proc. 1st int. Symp. on Trichoptera, ed. H. Malicky, pp. 181–184, Junk, The Hague, 1974b.

Malicky, H., Handb. Zool. 4, 29:1–114, 1973.

Marstaller, R., Zool. Jb. Physiol. 75, 1:76–98, 1969.

Merrill, D., J. Exp. Zool. 158:123–130, 1965.

Ross, H.H.; Evolution and Classification of the Mountain Caddisflies, Univ. Illinois Press, Urbana, 1956.

——— Am. Zoologist 4:209–220, 1964.

——— Ann. Rev. Ent. 12:169–206, 1967.

Smart, K., *In* Proc. 1st int. Symp. on Trichoptera, ed. H. Malicky, pp. 185–186, Junk, The Hague, 1974b.

Tachet, H., Biol. Behaviour 3:97–112, 1978.

Tomaszewski, C., Acta zool. cracov. 18:311–398, 1973.

Wiggins, G.B., Larvae of the North American Caddisfly Genera (Trichoptera), Univ. Toronto Press, 1977.

ALTITUDINAL DISTRIBUTION AND ABUNDANCE OF TRICHOPTERA IN A ROCKY MOUNTAIN STREAM

J.V. WARD

SUMMARY

The larvae of 30 + species of Trichoptera in 10 families were collected from rubble riffles in a Colorado flowage, from alpine tundra to the plains. Mean larval density exhibited a general increase downstream, from 338 larvae per m² in the alpine tundra stream to 3967 larvae per m² at the plains location. Limnephilidae occurred over the entire altitudinal gradient and was the only family at the upper two sites. Only Limnephilidae and Rhyacophilidae were collected above 3000 meters. Some species were restricted to the headwaters, others to middle or lower reaches, whereas a few species occurred over a wide range of altitude. A sharp faunal break was apparent between the lower foothills and the plains. Seven species of *Rhyacophila*, which exhibited overlapping distribution patterns in the middle reaches, all occurred sympatrically on riffles at 3109 and 2816 meters elevation. Diversity and abundance patterns of Trichoptera are associated with gradients in physico-chemical parameters and biotic factors.

INTRODUCTION

The East Slope of the Cordilleran in Colorado exhibits an especially steep altitudinal gradient, thus providing excellent opportunities to investigate factors which control the diversity, composition, and abundance of stream organisms. Although longitudinal faunal patterns in stream systems have been thoroughly examined in certain regions (Illies and Botosaneanu, 1963; Hynes, 1970; Hawkes, 1975), remarkably little previous work has been conducted in Colorado. Dodds and Hisaw's (1925) pioneering study of South Boulder Creek employed only qualitative sampling techniques, was limited to the late June to early September period, and did not include physico-chemical data. Other studies in Colorado have also been limited in sampling period, altitudinal gradient, or taxonomic scope (Knight and Gaufin, 1966; Elgmork and Saether, 1970; Mecom, 1972; Allan, 1975).

Wiggins and Mackay (1978) presented a generalized view of the distribution of Nearctic Trichoptera along the longitudinal stream profile. Studies outside

North America which specifically deal with longitudinal zonation of Trichoptera include Décamps (1967), Statzner (1975), and Moretti and Mearelli (1978).

St. Vrain Creek, Colorado, traverses an elevation gradient of nearly 1900 m from its glacier-fed source to the plains. Data on the abundance and distribution patterns of lotic Trichoptera as a function of altitude were collected during a comprehensive study of stream zonation which will be reported elsewhere.

STUDY AREAS AND METHODS

Middle St. Vrain Creek originates in glacial cirques on the east side of the Continental Divide in northern Colorado. The stream originates in alpine tundra (3414 m a.s.l.) and drops 1870 m to the plains over a distance of only 54 km. The lakes of the region have been described in detail (Pennak, 1958). Mecom (1972) made qualitative collections of Trichoptera in the lower portion of the study stream, and Ward (1975) examined bottom fauna-substrate relationships at a single location on a major tributary.

Benthic samples were collected from rubble riffles at eleven locations (Fig. 1). Six Surber samples (each enclosing 929 cm^2) were taken from rubble riffles at each site, three with a regular mesh net (720 μm) and three with fine mesh (240 μm). Data from the regular mesh samples are reported herein.

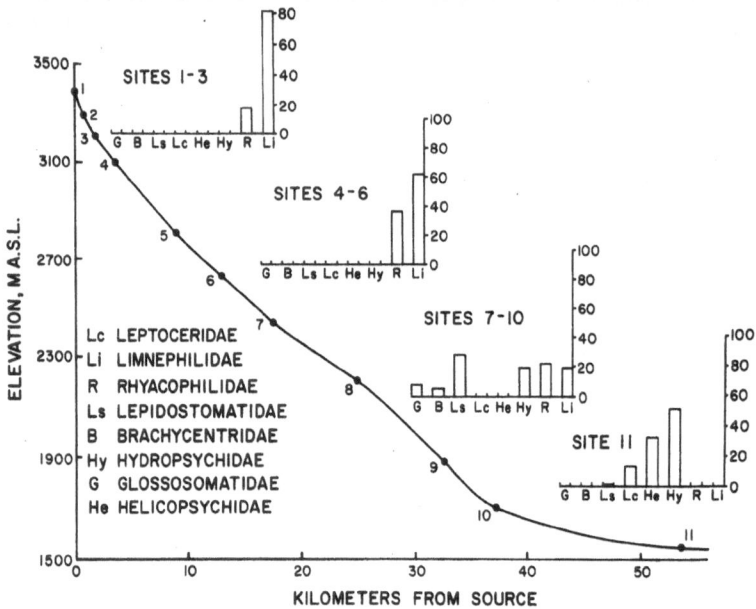

Fig. 1. The relative contribution (mean % composition by density) of common families to total Trichoptera as a function of altitude in St. Vrain Creek, Colorado. Site locations are indicated by Arabic numerals on the longitudinal stream profile. Values < 1% are not shown.

376

From the Montane Limnological Zone (2633 m) to the plains, sampling was conducted monthly from June 1975–May 1976. Upper sites, which were reached by hiking trail, were only accessible during summer because of the deep snowpack. Sites 4 and 5 were accessible by June, Sites 2 and 3 by July, but sampling at Site 1 was not possible until August due to an especially heavy snowpack. For this reason, a second summer of sampling was conducted during which Site 1 was accessible by July. Because the open season is so short at high elevations, it is unlikely that any common species were missed despite the restricted sampling period at upper sites.

Because immatures of some groups of trichopterans are poorly known, especially in Western North America, adults were collected with sweep nets to provide additional taxonomic data. Some species designations are therefore presumptive, based upon the distribution of larvae and adults, and the great likelihood of St. Vrain Creek having a similar trichopteran fauna as South Boulder Creek (Dodds and Hisaw, 1925) immediately to the south. A variety of factors, including the remoteness of the area and the comprehensive nature of the study, precluded definitive associations of immatures and adults by rearing.

Detailed data on physico-chemical parameters, periphyton, and macrophytes were collected but will be referred to only briefly.

RESULTS AND DISCUSSION

Trichoptera from 10 families, represented by 22 genera and at least 33 species, were collected (Fig. 2). The larvae of Sites 1 and 2 were limited to 5 species, one of which (*Imania tripunctata ?*) comprised over 98% of the caddisfly fauna at those locations. The number of species increased to 11 at Site 4 (3109 m) and remained relatively constant over the remaining altitudinal gradient. Mean larval density exhibited a general increase downstream with 338 organisms per m^2 in the tundra stream (Site 1) and 3967 organisms per m^2 at Site 11 on the plains.

The degree to which the common families contributed to the total caddisfly fauna varied as a function of altitude (Fig. 1). Sampling sites are grouped according to Pennak's (1958) limnological zones (alpine, Sites 1–3; montane, Sites 4–6; foothills, Sites 7–10; plains, Site 11). Only Limnephilidae and Rhyacophilidae occurred above 3000 m. Limnephilidae comprised the entire trichopteran fauna at Sites 1 and 2 (a single adult male *Rhyacophila alberta* was found at Site 2). Limnephilids, the only family to traverse the entire altitudinal gradient, were abundant in the Alpine and Montane Limnological Zones, but were rare or absent below Site 7 in the lower foothills and plains.

Rhyacophila larvae first appeared at Site 3 and reached maximum relative abundance (87% of total caddisflies) and diversity (7 spp.) at Site 4 (3109 m). No Rhyacophilidae were collected from the plains stream. Lepidostomatidae, Brachycentridae, Hydropsychidae, and Glossosomatidae were restricted to the lower half (Sites 5–11 or 6–11) of the longitudinal stream profile. Nectopsychidae and Hydroptilidae were collected only at Sites 10 and 11.

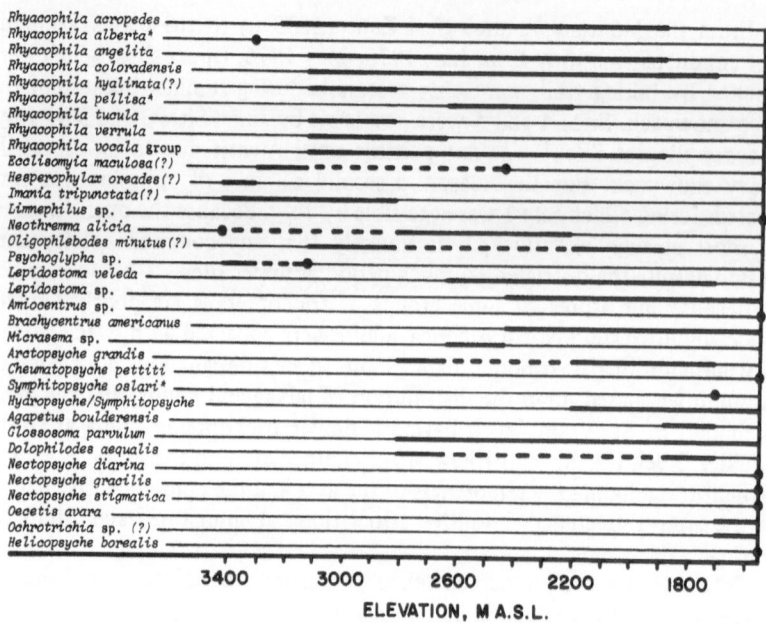

Fig. 2. Altitudinal distribution (dark horizontal lines and solid circles) of Trichoptera in St. Vrain Creek, Colorado. Asterisks indicate adults without associated larvae.

Helicopsychidae were restricted to the plains location where *Helicopsyche borealis* comprised 33% of caddisflies numbers, exceeded only by Hydropsychidae (52%).

The species distributions observed in this study are presented in Fig. 2. *Rhyacophila acropedes*, *R. coloradensis*, and *Glossosoma parvulum* exhibited the widest distributions, each occurring at seven contiguous sites. The immatures of seven species in five families were collected from only one location; all were restricted to Site 11 on the plains (Fig. 2). Two additional species were collected only from the plains and lowermost foothills site. Three limnephilids were restricted to headwater locations. Several species were restricted to the middle reaches of St. Vrain Creek. This is somewhat different than the altitudinal distribution of Ephemeroptera, the general pattern of which is the addition of species in the downstream direction without the loss of those occurring at higher elevations (Ward and Berner, 1980).

Dodds and Hisaw (1925), who collected over the same altitudinal range in a drainage system immediately to the south of St. Vrain Creek, reported the distribution of 48 species of caddisflies, including lake and stream forms. Elimination of lentic species and synonymies reduces that number to 29, the larval distributions of which are indicated for only 11 species (compared to 30 + larvae from St. Vrain Creek). Their restricted sampling period (late June to early September) probably accounts for the smaller number of species, especially at lower elevations. In addition, keys are now available which enable better discrimination of larval congeners (e.g., Smith, 1968; Haddock, 1977). Nonetheless, distribution patterns and faunal composition were

378

generally similar in upper and middle reaches of St. Vrain and South Boulder Creeks.

Elgmork and Saether (1970) conducted an intensive four-day study in mid-July of the portion of North Boulder Creek above the tree line during which four lotic trichopteran species were collected, *Rhyacophila* sp., *Brachycentrus* sp., *Pseudostenophylax edwardsi*, and *Hesperophylax oreades*. Of these species only *H. oreades* was collected from the alpine zone of St. Vrain Creek. The presence of numerous small lakes in the North Boulder Creek drainage undoubtedly modified thermal and trophic conditions of the stream, which may account for the range extensions observed.

Mecom (1972) reported only 11 species of caddisflies in qualitative collections from the St. Vrain Creek basin, the lower sampling stations of which corresponded to the middle and lower reaches examined in the present study. Since Mecom's collections included sampling stations over a similar elevation gradient, the greater number of species collected in the present study supports the use of quantitative sampling techniques even for determining species present. All species reported by Mecom were also found in this study except *Hydropsyche occidentalis*.

The genus *Rhyacophila* was represented by nine species, two of which (*R. alberta* and *R. pellisa*) were identified only from adult material (Fig. 2). All seven larval species occurred sympatrically on riffles at 3109 and 2816 meters elevation. From 4–5 congeners occupied four other sampling sites. Other investigators (e.g. Thut, 1969; Short and Ward, 1980) have also reported the sympatric occurrence of species of *Rhyacophila*, but the mechanisms of niche segregation remain to be elucidated.

Three species of the leptocerid genus *Nectopsyche* occurred sympatrically at the plains location. *N. gracilis* and *N. stigmatica* were abundant and exhibited similar seasonal patterns, whereas *N. diarina* was rare.

The abundance and distribution of lotic invertebrates has been attributed to a variety of factors (Hynes, 1970), many of which vary as a function of altitude and thus may be responsible directly or indirectly, for zonation patterns. Mackay and Wiggins (1979) delineated the following as major determinants of caddisfly distribution in lotic habitats: stream size, temperature, current, substrate (organic and inorganic), chemical factors, and food.

Some factors did not greatly vary between sites on St. Vrain Creek. Dissolved oxygen values were near saturation at all locations. Variations in current and substrate were reduced by restricting sampling to rubble riffles.

Several physico-chemical parameters varied significantly with altitude. Stream size and associated changes in habitat diversity may be important since Site 1 is a first-order stream, whereas the plains stream is fifth-order. The pH increased from near neutrality in the headwaters to 8.1 on the plains. Total dissolved salts which averaged less than 10 mg/l at Site 1 increased to nearly 200 mg/l at Site 11. Suspended matter and hardness also increased downstream. However, many physico-chemical parameters varied only slightly from Sites 1–10, but increased greatly from the lower foothills to the plains location (i.e., from Sites 10–11).

Mean summer temperatures increased from 2.3°C to 20.8°C over the altitudinal gradient. The thermal regime undoubtedly is a major factor determining the distribution and abundance of stream fauna (Dodds and Hisaw, 1925; Hynes, 1970; Ward and Stanford, 1979).

Submerged aquatic angiosperms were restricted to the plains location. The dense beds at Site 11 provide microhabitats, current refugia, and case building materials not present at other sites. Various mosses and liverworts comprised the macrophytes at mountain stream sites.

Allochthonous organic inputs and *in situ* autochthonous production vary in quality and quantity over the gradient of altitude. Combined with changes in amounts and size composition of suspended organic particles, a wide range of trophic conditions is provided (Wiggins and Mackay, 1978).

Predation pressure and competitive interactions are also expected to vary with altitude, although few studies have dealt with the influence of these phenomena in determining the distribution of Trichoptera or other stream invertebrates in North America (McIntire and Colby, 1978). The presence of congeners may, for example, influence the niche breadth and therefore the altitudinal range of a species.

CONCLUSIONS

Based upon relative abundance of larvae, the following faunal assemblages are apparent: (1) an exclusively limnephilid community comprised primarily of *Imania* at Sites 1 and 2; (2) a limnephilid-rhyacophilid community at Site 3; (3) a diverse rhyacophilid community containing a few limnephilids at Site 4; (4) a return to a limnephilid community now dominated by *Neothremma alicia*, with some rhyacophilids at Sites 5–7; (5) a rhyacophilid-lepidostomatid community at Site 8; (6) a lepidostomatid-hydropsychid community at Site 9; (7) a lepidostomatid-hydropsychid-glossosomatid-brachycentrid community at Site 10; and (8) a hydropsychid-helicopsychid-nectopsychid community at Site 11 on the plains.

Some species were restricted to the headwaters, others to middle or lower reaches, whereas a few species occurred over a wide range of altitude. Representatives of the family Limnephilidae occurred over the entire altitudinal gradient, but no species traversed more than 7 of the 11 sampling stations.

A sharp fauna break was apparent between Sites 10 and 11, coincident with the abrupt change in physico-chemical parameters and the addition of aquatic angiosperms. Seven species in five families were collected from the plains stream, but did not occur at mountain stream sites. In addition, six species present at Site 10 were not found at Site 11 during the year's study.

The gradient of environmental conditions which occurs over relatively short distances in the Front Range of the Colorado Rocky Mountains provides exceptional opportunities, not only for investigating factors controlling patterns of diversity and abundance of lotic organisms, but also for developing and testing models of predation, competition, and niche segregation.

ACKNOWLEDGEMENTS

Appreciation is extended to Dr R.W. Pennak, University of Colorado, for suggestions regarding the manuscript. Drs D.G. Denning, J.D. Haddock, S.D. Smith, and G.B. Wiggins provided valuable taxonomic expertise. S.P. Canton greatly assisted with laboratory analyses.

REFERENCES

Allan, J.D., Ecology 56:1040–1053, 1975.

Décamps, H., Annls. Limnol. 3:399–577, 1967.

Dodds, G.S. and Hisaw, F.L., Ecology 6:380–390, 1925.

Elgmork, K. and Saether, O.R., Univ. Col. Stud. Ser. Biol., 31: 1–55, 1970.

Haddock, J.D., Amer. Midl. Nat. 98:382–421, 1977.

Hawkes, H.A., *In* River Ecology, ed. B.A. Whitton, pp. 312–374, Blackwell Sci. Publ. Oxford, 1975.

Hynes, H.B.N., The Ecology of Running Waters, pp. 555, Univ. Toronto Press, 1970.

Illies, J. and Botosaneanu L., Mitt. Int. Verein. Theor. Angew. Limnol. 12:57 p, 1963.

Knight, A.W. and Gaufin, A.R., J. Kansas Ent. Soc. 39:668–675, 1966.

Mackay, R.J. and Wiggins, G.B., Ann. Rev. Entomol. 24:185–208, 1979.

McIntire, C.D. and Colby, J.A., Ecol. Monogr. 48:167–190, 1978.

Mecom, J.O., Hydrobiologia. 40:151–176, 1972.

Moretti, G.P. and Mearelli, M., Riv. Idrobiol. 17:137–186, 1978.

Pennak, R.W., Verh. Internat. Verein. Limnol. 13:264–283, 1958.

Short, R.A. and Ward, J.V., Southwest. Nat. 25:23–32, 1980.

Smith, S.D., Ann. Entomol. Soc. Amer. 61:655–674, 1968.

Statzner, B., Arch. Hydrobiol. 76:153–180, 1975.

Thut, R.N., Ann. Amer. Entomol. Soc. 62:894–898, 1969.

Ward, J.V., Ecology 56:1429–1434, 1975.

—— and Berner, L., *In* Advances in Ephemeroptera Biology, eds J.F. Flannagan and K.E. Marshall, pp. 169–177, Plenum Publ. Corp. New York and London, 1980.

—— and Stanford, J.A., *In* The Ecology of Regulated Streams, eds J.V. Ward and J.A. Stanford, pp. 35–55, Plenum Publ. Corp., New York and London, 1979.

Wiggins, G.B. and Mackay, R.J., Ecology 59:1211–1220, 1978.

LARVAL BEHAVIOR AND DISPERSION OF *PYCNOPSYCHE LUCULENTA* (BETTEN) DEMONSTRATED BY A UNIQUE TAGGING METHOD (LIMNEPHILIDAE: TRICHOPTERA)[1]

J.S. WEAVER III AND J.L. SYKORA

SUMMARY

In order to observe the larval behavior and dispersion of *Pycnopsyche luculenta* (Betten)(Limnephilidae: Trichoptera), 367 final instar specimens were tagged with numbered pins. The specific location of each recovered larva was observed on a monthly basis from February through August, inclusively. The distribution and habitat preferences in a small springbrook were noted during periods of active mobility and pupation.

INTRODUCTION

Many aquatic entomologists are concerned with the recognition of the specific habitat preferences of benthic macroinvertebrates. This interest lies in the fact that if aquatic insects are to be considered useful indicators of pollution or other environmental alterations, then scientists must be familiar with the aquatic conditions which species prefer. The process of obtaining this information is complicated by the vast diversity of aquatic insects and their relation with various habitats which are difficult to define. Furthermore, different instars of a species may exhibit various behavior or habitat preferences, which may be affected by seasonal effects.

The purpose of this investigation was to observe the behavior and movement of tagged *Pycnopsyche luculenta* (Betten) larvae. This species is endemic to small woodland streams and springs of eastern North America. Larvae are common in the winter, spring and early summer, pupae and adults towards the end of the summer.

STUDY LOCALITY

The study area was Maul Spring (Fig. 1) which is located in the headwaters of Powdermill Run, Westmoreland County, in the Allegheny Mountains of

---

[1] This Technical Contribution No. 1846 is published by permission of the Director, South Carolina Agricultural Experiment station, Clemson University.

southwestern Pennsylvania. The spring maintains a relatively constant temperature of $9° \pm 1°C$ throughout the year; average discharge is 0.05 m³/sec. The source of the springs is limnocrene, forming a pool area 6 m in diameter with a maximum depth of 50 cm. The resulting springbrook is 6 m wide and relatively shallow with a maximum depth of 6 cm. The spring issues from strata consisting of sandstone. The water is well-oxygenated at the source with the dissolved oxygen saturation ranging from 85 to 95%. The pH was slightly acidic, averaging 6.8. The spring was surrounded by trees, predominantly beech (*Fagus grandifolia*), sugar maple (*Acer saccharum*), black oak

Fig. 1. Maul Spring.

(*Quercus velutina*), tulip poplar (*Liriodendron tulipifera*) and hemlock (*Tsuga canadensis*). The springbrook was unique in having an abundance of several aquatic bryophytes. The water moss (*Oxyrhynchium riparioides* and the leafy liverwort *Chiloscyphus pallescens* were the most common aquatic plants, occurring throughout the year.

Fig. 2 shows some of the major habitats of Maul Spring. The pool area at the source was deepest in the middle where water issued from the strata. Detritus accumulated in this area and also on the right bank near the base of the large hemlock tree at the edge of the pool. This tree was a useful point of reference as it marked the end of the pool area above and the beginning of the riffle area below it. Other areas in the pool were relatively free of loose sticks due to the pattern of surface run-off above the spring which resulted from spring thaws and rain storms, and which carried detritus from these areas. Even during high flow conditions the water level of the spring rose only about 5 cm. Thus, the effects of drift due to high water velocity were

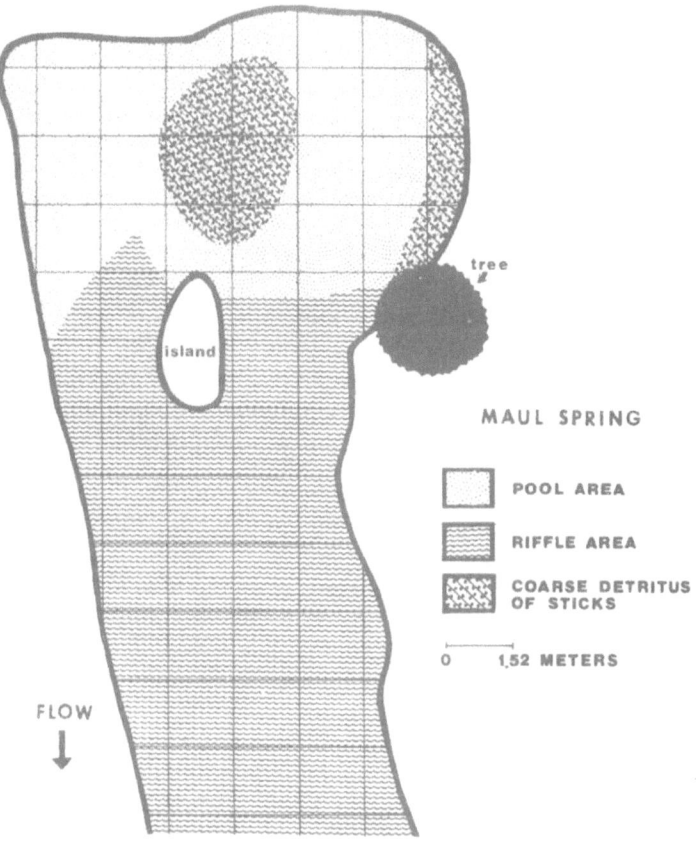

Fig. 2. Diagram of habitats in Maul Spring.

minimal. This was verified by observing that a few loose sticks tagged in the springbrook did not move appreciably during the study. The substrate in the pool area was primarily loose sand and gravel, while the riffle was primarily rocks and smaller stones.

METHODS AND MATERIALS

The larval case of this species is constructed with two lateral sticks which are joined with smaller pieces of bark or sticks. Tagging pins, (numbered map pins) were placed posteriorly in one of the major lateral sticks of the case (Fig. 3). Each larva was tagged individually and returned to the location from which it had been taken in the springbrook. A total of 367 last instar larvae of *Pycnopsyche luculenta* were tagged: 45 in February, 30 in March, and 292 in April (Fig. 4). In this figure each dot represents a tagged individual in the square of the grid where it was recovered, because clumping would

385

Fig. 3. A tagged *Pycnopsyche luculenta* larva.

FEBRUARY - APRIL
Tagged Larvae

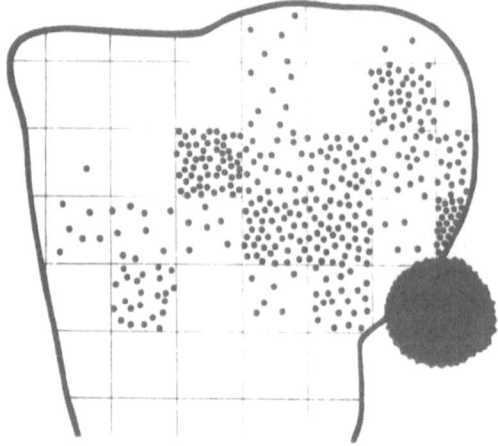

Fig. 4. February – April, positions of tagged *Pycnopsyche luculenta* larvae.

have caused too much overlapping of dots if the exact positions had been depicted.

The springbrook was searched for tagged larvae on a monthly basis. Individuals were tagged and/or recovered on the following dates: 19 February, 19 March, 14 April, 21–22 May, 22 June, 19 July, and 27 August, 1975. The

386

number and location of each recovered specimen were recorded, then it was returned immediately to the position from which it was taken. By using this method, it was not necessary to preserve the larvae, and the population was not threatened or altered by oversampling.

A grid was superimposed over the study area. In the field, each point of intersection of the grid was defined by a stick vertically inserted into the substrate. In order to record the location of each tagged larva, the nearest stake was used as the point of origin, the exact position defined by horizontal (x) and vertical (y) coordinates. Each square in the grid was 1.52 m long and represented an area of 2.31 m^2.

RESULTS

The majority of individuals tagged were found in the pool area in places where loose sticks were prevalent. Larvae were uncommon in parts of the pool which did not accumulate loose sticks, and almost none were found in the riffle area.

During a 24 hour period in May, the distances traveled by individuals were observed. The locations of individuals in the most heavily populated area in the middle of the spring were recorded. The following day, the entire spring was searched for tagged larvae. The change over the 24 hour period for 65 larvae is shown in Fig. 5. The bar graph in this figure shows the frequencies of distances traveled by individuals. There was a substantial distance covered during this period. The average distance traveled per larva was 1.77 m/24 hours. There seemed to be a tendency for individuals either to disperse from the middle of the pool and aggregate near the right bank, or to remain in the middle of the pool. In either case, they were found in areas where sticks were common. Many of the larvae found during May were chewing on the bark of sticks. It was a month of active feeding and mobility.

In May 213 tagged larvae were found, 58% of the original number tagged (Fig. 6). Also shown is the number of individuals in each square of the grid. Most of the individuals were aggregated in the pool. Only a few larvae traveled into the riffle area in May.

The 129 tagged individuals that were recovered in July and August (35% recovery) are shown in Fig. 7. This figure also shows the number of individuals in each square of the grid. Most of the individuals recovered at this time were pupating. Of those found pupating, 83 were found attached to stones, 34 were attached to sticks, and 4 were found burrowed in the substrate. However, 8 individuals were freely crawling larvae. The population at this time was more evenly dispersed throughout the pool and riffle areas.

In Fig. 8 the exact positions and routes of six tagged individuals were noted throughout this study. The upper right diagram of Fig. 8 shows larva #273 (white circle) which was tagged in April near the right bank and found in May, at which time it was eating an acorn. Acorns which fell into the spring would swell, splitting the outer pericarp of the seed. These split acorns appeared to be considered a delicacy by larvae, and as many as three individuals were found together eating the same acorn. Larva #32 (Fig. 8, black

MAY

Distances Traveled by Tagged *Pycnopsyche luculenta*

Larvae Within 24 Hours

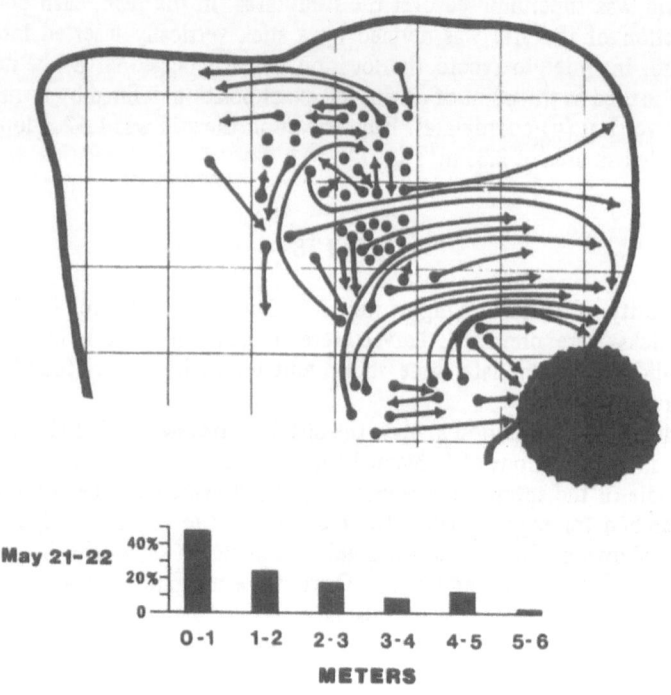

Fig. 5. Change during 24 hour period in May of tagged *Pycnopsyche lucuenta* larvae.

circle) was tagged in February, and was recovered in March and again in May. By June, it had burrowed in the substrate, and by July it had pupated, attached under a stone.

Larva #113 (Fig. 8, black circle, lower right diagram) was tagged in April, recovered May 21 and 22, and in July was found attached under a stone, pupating. Larva #55 (Fig. 8, black circle, lower right) was tagged in February, recovered in March and April, and in June was attached to a stick. In August, it was found pupating, burrowed into the substrate.

Larva #249 (Fig. 8, white circle, left diagram) was tagged in April, recovered in May, and in June was found burrowed shallowly in the substrate. However, in July it was found crawling freely. In August it was found attached to a rock, pupating. Larva #95 (Fig. 8, black circle, left diagram) was one of the best long distance movers, covering a distance of 13.7 m between May and July. It traveled from the middle of the pool to the riffle area, where it was attached to a stone, pupating.

388

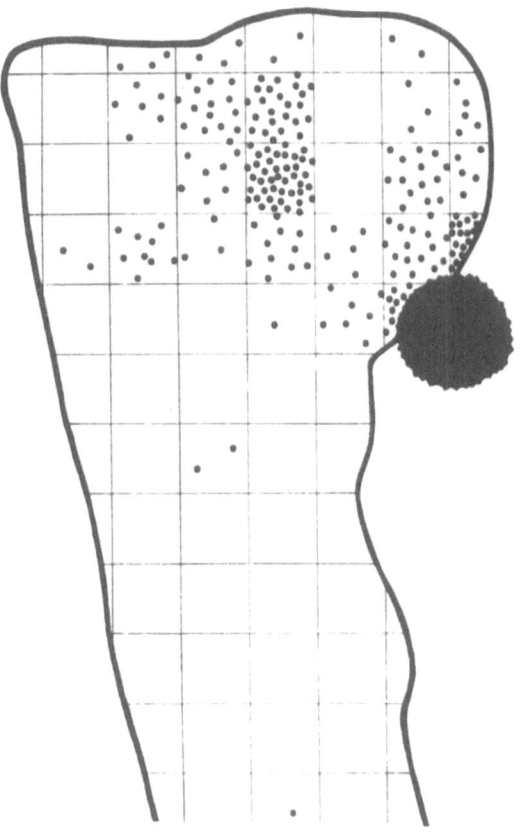

Fig. 6. May, recovery of tagged *Pycnopsyche luculenta* larvae.

CONCLUSIONS

The results show that *Pycnopsyche luculenta* larvae during the last instar are quite ambulatory, increasing in activity from February to June. During this time they are also feeding actively, and aggregate in quiescent pool areas where sticks are common. It was noted that the larvae may temporarily attach to stable objects or burrow in the substrate, and a month later be found attached to a different stone or stick. This occurred frequently enough to suggest that larvae may, after consuming food, rest for a short while, at which time they may secure themselves temporarily to hold their position. Local anglers sometimes call these larvae 'two sticks'. The two longitudinal sticks of the case are reminiscent of the travois used by the American Plains Indians to carry heavy items. Similarly, the two sticks may serve as runners of a sled,

389

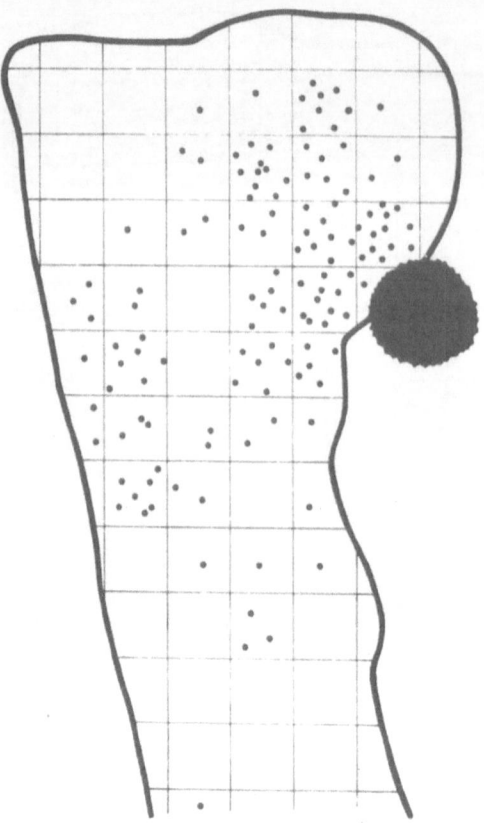

Fig. 7. July – August, routes of six tagged *Pycnopsyche luculenta* individuals.

and suggest that the design of the case was selected due to the active, mobile nature of this species.

In July and August the population became more dispersed, as some burrowed in the substrate and attached to small stones, while others attached to rocks or large sticks. The main objective of a larva at this time may be to locate a stable pupation site. To accomplish this, some larvae choose stable rocks in the riffle area; others, in the pool, attach to sticks or burrow in the substrate.

To show the general trend of movements for the population, individuals were grouped according to distances traveled. These monthly trends are compared in Fig. 9, which shows the frequencies of distances traveled per month for individuals collected on consecutive months throughout the study. In March and April frequencies are more evenly distributed than in later months. Perhaps this indicates that larval movement is more sporadic at this time. In May and June many individuals were collected on consecutive

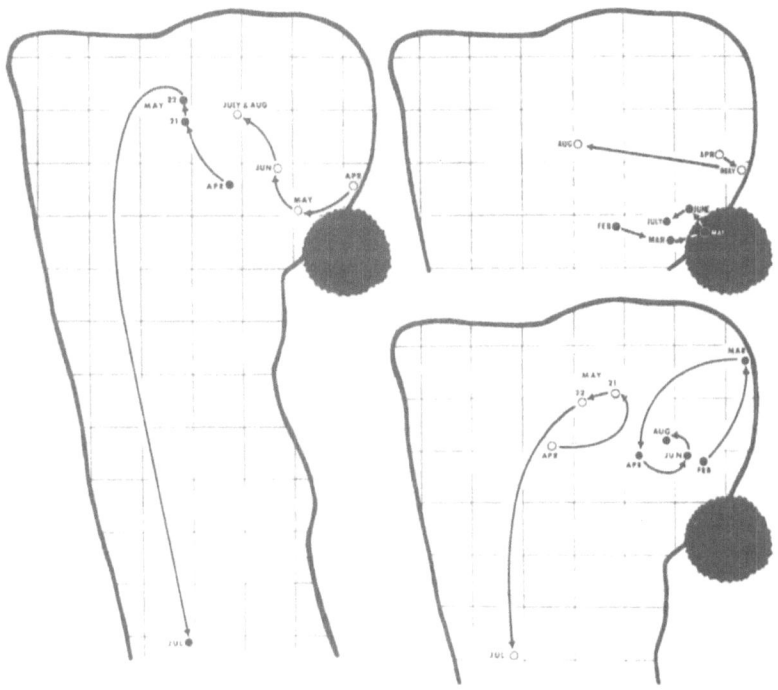

Fig. 8. February – August, routes of six tagged *Pycnopsyche luculenta* individuals.

months and the frequencies for these months approaches a normal distribution about the means. In July and August the average distances traveled by individuals is much less than in previous months. This is due to the increasing number of stationary pupae as the summer progresses.

Table I summarizes the average distances traveled by individuals per month. There is an increase from 2.41 m in March to a peak of 3.06 in June. Then there is a rapid decline to 1.84 m in July, followed by the lowest value of 1.17 m in August. This seems to indicate that the population steadily increases its mobile activity until the time of pupation.

ACKNOWLEDGEMENTS

The authors wish to thank Larry Garver and Diane E. Ashton, who helped record data in the field. We also extend our gratitude to Dr John C. Morse, Clemson University, for reviewing the manuscript.

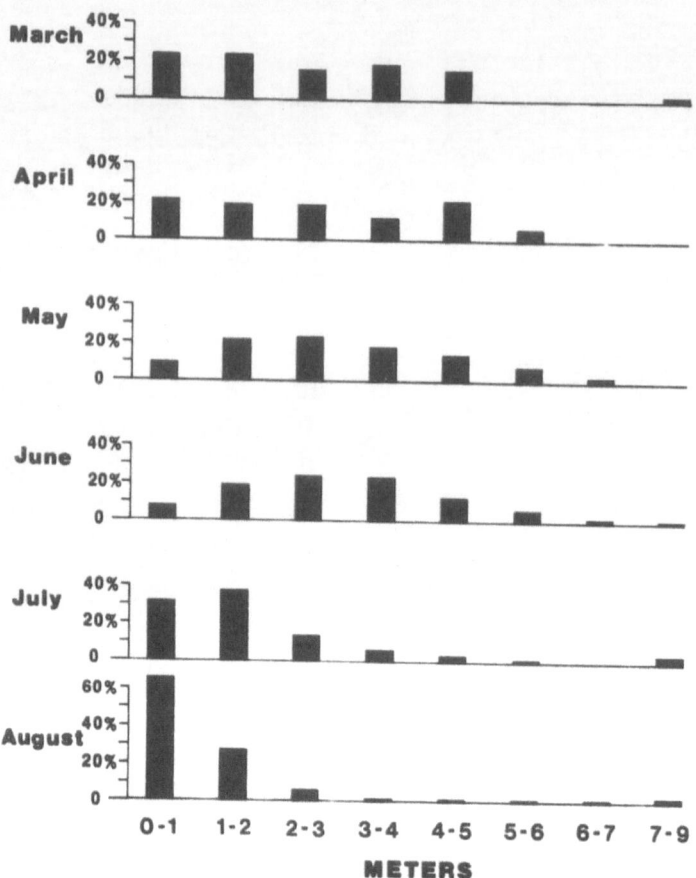

Fig. 9. Frequencies of distances traveled by tagged *Pycnopsyche luculenta* individuals.

392

Table I. Mean distances traveled by *Pycnopsyche luculenta* larvae.

MEAN DISTANCES
TRAVELED BY INDIVIDUALS

MONTHS	OBSERVATIONS	MEANS (METERS)
MARCH	33	2.41
APRIL	47	2.58
MAY	213	2.93
JUNE	77	3.06
JULY	50	1.84
AUGUST	57	1.17

CONSIDERATIONS ON THE RELEVANCE OF IMMATURE STAGES TO THE SYSTEMATICS OF TRICHOPTERA

G.B. WIGGINS

SUMMARY

This paper is a response to the view (Schmid, 1979) that knowledge of the immature stages is not essential in developing a sound classification and phylogeny for the Trichoptera.

Examples are given where immature stages have been critical in revealing relationships in Trichoptera. Higher classification of the order is outlined in an historical context, showing that data from immature stages have been significant in its development. An integrated data base combining the widest possible array of characters from all stages is held to be the most useful in classification and phylogeny; and while some types of data tend to be more useful than others – depending upon whether the systematic objective is species discrimination, classification or phylogeny, the view that data from any restricted source such as the imaginal stage are *a priori* more correct or informative than those from other sources such as the immature stages is not consistent with current developments in systematics. Moreover, only an integrated data base provides insight into the ecological factors governing evolution.

'The actual state of the classification of Trichoptera shows that the study of the immature stages is not essential ... Over the years I made many important changes in the classification of Trichoptera and these were made exclusively on the basis of imaginal characters. None of these changes has since been contradicted by the study of immature stages. I certainly made many mistakes, such as placing *Lepania* in the Moropsychini ...' (Schmid, 1979: 54).

'... Therefore, evidence from larvae does not support the interpretation based on adults (Schmid, 1968) that *Imania* and *Moselyana* are closely related and that these genera should be grouped with the Apataniinae. In terms of functional taxonomy, inclusion of *Moselyana*, *Imania,* and *Manophylax* in the Apataniinae adds no new characters to the diagnosis for the subfamily and nullifies those that did exist.' (Wiggins, 1973a: 30).

'Assignment of *Lepania cascada* to the Apataniinae (Moropsychini) was based primarily on characters of the genitalia (Schmid, 1968), and in the absence of any knowledge of the immature stages of either group. Evidence presented here from larval characters in *Lepania* reveals that the relationship between *Lepania* and *Moropsyche* is so distant that the two genera cannot be considered members of the same subfamily (p.21) . . . Available evidence shows that the genus *Lepania* cannot be assigned to the Apataniinae, as previously proposed, but rather that this genus has more in common with the limnephilid subfamily Goerinae.' (p. 29) (Wiggins, 1973b).

'. . . *Pedomoecus* was referred to the Dicosmoecinae (Schmid, 1955) as was *Rossiana* (Schmid, 1968); perhaps these assignments will prevail. But until the larval morphology in all of these genera can be interpreted consistently with that of the adults, I prefer to regard their position as uncertain.' (Wiggins, 1977: 183).

INTRODUCTION

Recently, the opinion has been advanced that knowledge of the immature stages is not essential, not even necessary, in developing a sound classification and phylogeny for the Trichoptera (Schmid, 1979, in press). The basis for this view is that the imaginal stages are a richer source of morphlogical characters than the immatures, and that the present classification, held to be generally satisfactory, has been built over the past century or so by authors using mostly imaginal characters. This opinion has been developed at some length to counter the view recently stated explicitly by me (Wiggins, 1977:4), and held in one form or another by a number of students of the order, that data from larval morphology are essential in assessing systematic relationships of Trichoptera and for advancing hypotheses concerning their phylogeny.

The issue as stated (perhaps more aptly in current idiom, non-issue) is surely of minor importance. Moreover, classification and phylogeny serve whatever purpose those constructing them wish to achieve; and the data used in their construction will be chosen accordingly. Not surprisingly, systematists interested solely in attributes of the imagines will see little need to incorporate data from the immature stages. For others who believe that classification and phylogeny should serve as a basis for generalization about the entire organism throughout its life cycle, the necessity to incorporate data from immature as well as imaginal stages is clear. Thus, conclusion is largely determined by premise. But I am sufficiently concerned at the restrictive and retrogressive viewpoint represented by my colleague's comments that I take this opportunity to outline the case for a holistic or integrated approach to trichopteran systematics.

Briefly, my view is that data from the immature stages have often enough revealed relationships in family-group and higher levels in the Trichoptera that they can indeed be regarded as essential for efficient progress in systematics; that data from the immature stages are valuable in establishing generality and hence validity of genus-group taxa; and that diagnostic larval morphology at the species level holds the key to admission of the behavioural repertoire of larvae into the systematic process. This support for data from immature stages diminishes not at all the importance I attach to imaginal characters; I believe the two sets of data should be used together and that trichopteran systematists must develop both data bases. Moreover, the predictive and generalizing functions of classification require initial data from the immature stages for maximum effectiveness.

EXAMPLES OF CONTRIBUTIONS FROM IMMATURE STAGES

Cited here are several cases in which knowledge of the immature stages has been critically important in revealing or resolving questions of affinity and classification.

The limnephilid genus *Imania*[1] was assigned to the Apataniinae upon its original description by Martynov (1935), then to the Dicosmoecinae by Schmid (1955a), and was returned to the Apataniinae by Schmid (1968), along with *Moselyana*. The first analysis of data from larval stages (Wiggins, 1973a, see quotation above) revealed marked discordance between these genera and the Apataniinae, resulting in their placement as *incertae sedis*. Again because of discordant evidence from the larval stages, two additional genera previously assigned to the Dicosmoecinae, *Pedomoecus* (Schmid, 1955a) and *Rossiana* (Schmid, 1968), were placed as *incertae sedis* (Wiggins, 1977, see quotation above). No satisfactory solution to this problem in classification has yet been advanced. I contend only that reliance solely upon data from imaginal stages led to a classification that was inadequate and confusing, and that this weakness was identified because data from larval stages were introduced.

Further support for the importance of larval data is seen in recent treatment of the classification of the eastern Palaearctic genus *Archithremma*, assessed from adults alone as a primitive goerid (Schmid, 1979:57); but after study of the larva, which does not show typical goerid characteristics, *Archithremma* was reassigned as Limnephilidae *incertae sedis* (Levanidova and Schmid, in press).

'A ma connaissance, les rares cas qui font exception à cette situation générale sont les petits genres relictes et très particuliers qui se sont détachés de la base du tronc des Dicosmoecines: *Allomyia*[1] Banks, *Moselyana* Denning, *Pedomoecus* Ross et *Rossiana* Denning. Wiggins (1977) sur la base des caractères larvaires, les a détachés de la sous-famille des Dicosmoecines où je les avais inclus (1955) sur la base des caractères imaginaux, pour en faire une catégorie 'incertae sedis'. Depuis, les genres *Manophylax* Wiggins et *Archithremma* Martynov ont été découverts ou redécouverts et également inclus dans ce groupe d''incertae sedis genera'. L'initiative de Wiggins est certainement justifiée. Elle représente un incontestable progrès dans nos connaissances de ces genres, mais pas

[1] By strict application of the rule of nomenclatorial priority the correct name for species previously placed in *Imania* Martynov 1935 must be *Allomyia* Banks 1916 (Fischer, 1973; Schmid, 1980).

pour leur classification[2], car leur statut d' 'incertae sedis' est aussi insatisfaisant pour l'esprit que leur inclusion dans la sous-famille des Dicosmoecines. Le cas de ces genres est des plus ambigus.' (Schmid, in press)

The idea that only in rare cases have immature stages revealed weaknesses in classification hardly agrees with the additional examples outlined below. But apart from that, the attempt to evaluate data from immature stages on some ratio of past successes misses entirely the point that most systematists concerned with corroborating interpretation of their imaginal data have seen for a long time.

Originally placed in the Dicosmoecinae (Schmid, 1955a), the western Nearctic species *Lepania cascada* Ross was later transferred to the Apataniinae (Schmid, 1968). Discovery and analysis of the larva (Wiggins, 1973b, see quotation above) revealed that this species was in fact a member of the Goerinae(idae).

Analysis of larval characters of *Phryganopsyche* (formerly *Phryganopsis*) first revealed that this genus was not a member of the Phryganeidae, to which it had always been assigned, and the new family Phryganopsychidae was created (Wiggins, 1959).

Larval characters were instrumental in demonstrating relationships of the aberrant *Yphria californica* (Banks) as a subfamily of the Phryganeidae (Wiggins, 1962).

The south-eastern Nearctic *Pseudogoera singularis* Carpenter retained an uncertain position in the Goeridae (Flint, 1966) for nearly 40 years until discovery of the larva demonstrated that it belonged in the Odontoceridae (Wallace and Ross, 1971).

Familial classification for genera of the Polycentropodidae s.1, Psychomyiidae, and related groups has been a problem of long-standing in the Trichoptera. Ulmer (1951) considered this problem at length, stressing repeatedly the importance of data from immature stages; he erected the subfamily Pseudoneureclipsinae for the genus *Pseudoneureclipsis* in which larvae are aberrant. Lepneva (1956) dealt with another aspect by analyzing immature stages, concluding with elevation of the subfamily Ecnominae to family level. The Neotropical genus *Austrotinodes,* originally assigned on imaginal evidence to the Psychomyiidae (Schmid, 1955b), then suspected to belong to the Ecnominae(idae) (Schmid, 1958) was clearly shown to be a member of the Ecnomidae when larvae were discovered (Flint, 1973). Flint's (1971) reservation that immature stages are an essential component for a satisfactory resolution of the relationships in all of these groups, criticized by Schmid (1979) as an example of unnecessary caution, arises from the same concern that Ulmer voiced two decades earlier.

---

[2] Another issue, here tangential, is the assertion (Schmid, in press) that assignment of these genera as *incertae sedis* does not represent progress in their classification because an unspecified position is as unsatisfactory as their inclusion in the Dicosmoecinae (or Apataniinae). I believe that admission *pro tem* of uncertain position for genera is a far more responsible systematic choice than assigning them on incomplete evidence to groups where they distort existing congruity.

A new genus assigned to the Hydroptilidae, *Petrotrichia,* from the Seychelles Islands (Ulmer, 1910), was shown to belong to the Helicopsychidae when larvae were discovered (Marlier, 1978).

The Nearctic genera *Neothremma* and *Farula* have had an uncertain history in classification. Both have always been placed among the Limnephilidae. *Farula,* assigned to the Dicosmoecinae (Schmid, 1955), was later recognized as a close relative of *Neothremma* in the Neophylacinae and transferred to that subfamily (Schmid, 1968). Later, in a summary of the Neophylacinae (Schmid, 1980:96): '*Neothremma* Banks et *Farula* Milne sont très isolés et n'entrent que difficilement dans la description ci-dessus. Ils sont placés ici tentativement et il est possible qu'ils soient un jour déplacés parmi les Goérides.' In fact, larval and pupal morphology and case-making behaviour show strikingly that these two genera are closely related, and that the affinity of both is with the Asiatic family Uenoidae (Wiggins, 1977). Imaginal data are consistent with this relationship (Wiggins, in prep.).

In many of these cases evidence was also found in the imagines that was consistent with relationships suggested by data from the larvae; but the significance of that evidence was revealed because the larvae were studied.

HISTORICAL CONSIDERATIONS

Review of the development of familial and higher level taxa in the Trichoptera shows that present composition and understanding of family groups and their phyletic relationships was not derived solely from imagines but that data and insight from immature stages have been significant.

Entomologists know that most of the early development in identification and classification of insects was based on the imaginal stage. Trichoptera were no exception, and it was a fortunate coincidence that the more readily available adult insects were the most fruitful sources of the systematic characters that led to definition of individual species and to a classification by which growing numbers of these species could be comprehended. The result, not unexpected, is that much of the early classification of Trichoptera was based on imaginal characters. From the single genus *Phryganea* recognized by Linnaeus in 1758, comprising 17 species not all of which were even Trichoptera, knowledge of imaginal Trichoptera increased so substantially that by 1884 when the additional supplement to Robert McLachlan's monumental *Monographic revision and synopsis of the Trichoptera of the European fauna* was published, some 500 species were known from Europe alone. McLachlan was little involved in the study of immature stages, nor was much information known at that time, although he cited what was available. Through consistent use of genitalic and other characters, McLachlan brought a level of precision to the study of imaginal Trichoptera not previously achieved, making a major advance in the systematics of the order. However, even with this advance, the higher classification remained essentially as Kolenati had proposed in 1859. Within this system, McLachlan recognized seven families and identified in certain families sections of related genera. This classification (Table I) represents the level of understanding of the order before organized knowledge of the larval stages was available.

Additional families were recognized by Wallengren (1891). Several of these were elevations of McLachlan's sections, others were not. Perhaps it is significant that most of those families Wallengren knew as larvae were accepted (Beraeidae, Molannidae, Psychomyiidae, and Glossosomatidae), while those for which he did not know larvae were not accepted (Chimarrhidae, Apataniidae, Agrypniidae). The true relationship of *Chimarra* with *Philopotamus* and *Wormaldia*, not appreciated by either McLachlan or Wallengren, was seen by Morton (1888) from larval morphology. But Wallengren retained McLachlan's Philopotamus section of the Hydropsychidae, a section which included genera of the Philopotamidae, Polycentropodidae, and Arctopsychidae, and it is noteworthy that he did not know larvae of *Wormaldia, Neureclipsis, Polycentropus*, or *Arctopsyche*. Separation of the Polycentropodidae and Philopotamidae as distinct families was done by Ulmer (1906, 1907), who was deeply involved in the systematics of the immature stages of Trichoptera. Ulmer elevated another of McLachlan's sections to family level as the Calamoceratidae, and recognized as subfamilies others that later became families – Helicopsychinae, Lepidostomatinae, Brachycentrinae, and Goerinae. The long history of these latter groups as subfamilies of the Sericostomatidae was finally corrected by Ross (1944) with their elevation to family status; each was shown to have phyletic affinity with different families (Ross, 1956, 1967, 1978), and much of the evidence was drawn from the larval stage.

Not widely appreciated, however, is that the work of Klápalek (1888 etc.), Ulmer (1906 etc.), and Siltala (1907 etc.), the principal founders of a systematic data base for immature Trichoptera, showed marked disagreement with Kolenati's higher divisions Aequipalpia and Inaequipalpia (Table I). Their studies reveal appreciation of fundamental differences in morphology and case-making behaviour of larvae, and in characteristics of the eggs which was the foundation for a new alignment of families, finally formalized in nomenclature with Martynov's (1924, 1930) suborders Annulipalpia and Integripalpia. This interpretation is confirmed by Martynov's own comment (1930:5, footnote): '... It is well known that the larval phases in these suborders differ very greatly, but I have found that the imagines also present marked differences in many respects. The old division of the order Trichoptera into the suborders Aequipalpia and Inaequipalpia is quite artificial.'

The suborders Annulipalpia and Integripalpia have been further subdivided into superfamilies Rhyacophiloidea, Hydropsychoidea, and Limnephiloidea (Ross, 1967); immature stages were also involved in the definition of these categories. Views differ concerning assignment of the Rhyacophiloidea – to the Annulipalpia where Martynov originally placed these families (Schmid, 1980), or to the Integripalpia (Ross, 1967). Refinements in this higher classification may lie ahead[3], but there appears to be a broad consensus that the system in its present general form is soundly based.

[3] A comment by Ulmer (1957:463) on Ross' (1956) phylogeny of families is of interest: 'Ich glaube, dass durch Ausscheiden der Rhyacophilidae, Glossosomatidae und Hydroptilidae aus den Annulipalpia ein Fortschritt erzielt ist.'

Table I. Classification of Trichoptera from McLachlan, 1884.

Division I	Inaequipalpia	
	Family	Phryganeidae
	Family	Limnephilidae Limnephilinae Apataniinae
	Family	Sericostomatidae section of *Sericostoma* section of *Silo* section of *Brachycentrus* section of *Lepidostoma*
Division II	Aequipalpia	
	Family	Leptoceridae section of *Beraea* section of *Molanna* section of *Odontocerum* section of *Leptocerus* section of *Calamoceras*
	Family	Hydropsychidae section of *Oestropsis* section of *Macronema* section of *Hydropsyce* section of *Philopotamus* section of *Tinodes*
	Family	Rhyacophilidae section of *Chimarrha* section of *Rhyacophila* section of *Agapetus*
	Family	Hydroptilidae

Following what might be termed the classic period of family designation outlined above, several small families were established: Stenopsychidae (Martynov, 1924), Arctopsychidae (Martynov, 1924), Thremmatidae (Martynov, 1935), Philorheithridae (Mosely, 1936), Limnocentropodidae (Kitagamiidae, Tsuda, 1936), Ecnomidae (Lepneva, 1956), and Phryganopsychidae (Wiggins, 1959). In the two latter families, immature stages contributed substantially to the evidence validating their recognition. Larvae of the Limnocentropodidae were known before that family was created. The literature indicates that the others were established without explicit reference to immature stages, but most of them have been subsequently corroborated as distinct and valid families by knowledge of the immature stages. Only the Arctopsychidae are not generally accepted as a family, and are placed as a subfamily of Hydropsychidae by a number of workers, mainly non-European. The principal reason supporting subfamilial status is that the immature stages are so similar to Hydropsychidae s.s. that to recognize the Arctopsychidae as a family is to open the way for subfamilies in other groups to be elevated with equal if not greater justification (e.g. subfamily Yphriinae of the Phryganeidae). Thus immature stages have shown a level of importance in higher classification such that their role in these systematic decisions cannot be treated as unnecessary.

Growing knowledge of southern hemisphere Trichoptera has opened a new period of growth for family-level taxa, with early contributions by Mosely and Kimmins (1953), Ross (1967 etc.), and Riek (1968 etc.). More recently, work by Neboiss (1977 etc.) on imagines in Australia, Cowley (1976, 1978) on immatures in New Zealand, and Flint (1967 etc.) on both imaginal and immature stages in South America is yielding much new information. It remains to be seen what family level classification and relationships will be resolved from all of this, and how data from the immature stages will figure in their justification. A new South American family, Anomalopsychidae, proposed by Flint at this Symposium has been partly substantiated by data from immature stages.

INTEGRATED DATA BASE

I believe that the systematic data base for Trichoptera is undergoing an integrating phase, beginning in an organized form with the works of Ulmer and Siltala early in the present century. This phase is well exemplified by Ulmer's Sunda Islands studies (1951, 1955, 1957), where data from immature stages are constantly being incorporated into the diagnoses of new and existing taxa along with data from the imagines. It is my impression that most systematists of the Trichoptera concur in this integrating phase as a logical and necessary development in further testing and consolidating hypotheses of evolutionary relationships, and in strengthening the basis of classification for more effective use in prediction and generalization. This phase is leading, it seems to me, to a standard for the order by which taxa lacking diagnostic data for immature stages are incompletely defined. That is not to say that contributions solely concerned with imaginal stages are not important; nor that contributions concerned with immature stages are in any way superior. Quite simply, systematic knowledge of Trichoptera is reaching a level where partial taxonomic diagnoses are no longer sufficient. Systematists of all persuasions contribute to this integrated character base. Continuing analysis of taxonomic categories in the light of new data from immature stages is by this view an important part of the refinement and consolidation of the classification of Trichoptera. Much recent taxonomic work is directed to this end, e.g. in publications by Botosaneanu, Flint, Marlier, Smith, I.D. Wallace, Wiggins, and others.

One insight emerging from this integrating data base is that there is a high level of congruence between imaginal and larval systematic data from the generic level upwards. Were this not so, the systematics of Trichoptera would indeed be chaotic. Because of this important central fact, trichopteran systematists can take considerable reassurance that most of the established taxonomic groupings within the order reflect monophyletic groups. Since the larval and adult stages of endopterygote insects have evolved in response to different biological requirements, congruence in characters derived from separate data sets provides an effective check on each set. Instances have arisen where the two data sets are not congruent, e.g. some groups in the Hydroptilidae (Ulmer, 1957), and undoubtedly others have still to be discovered; even after closer study anomalies such as these may not be resolvable with present knowledge (e.g. Michener, 1977), but in each case incongruity is a sign that the characters and interpretation ought to be reviewed. Workers categorically denying that data from immature stages are essential in classification shield their work from this additional check. There is little justification for attaching *a priori* importance to data from imaginal stages and for insisting that imaginal stages hold all the data necessary for obtaining a sound classification and phylogeny because these characters are more numerous than those from larval stages. Systematists are concerned with congruency in as many widely different characters as it is possible to detect, and more particularly with congruency in derived characters. Whether these come from imaginal or immature stages is not important. What is important is whether the characters are reliable indicators of the evolutionary history of particular taxa. Parallelism is a problem for all systematic data. I believe that tests for congruency and parallelism involving data from both imaginal and immature stages are more rigorous, and hence more valuable in the systematic process, than tests based on either data set alone.

The integrated character base developing in the systematics of Trichoptera is used in different ways at different taxonomic levels; some types of data tend to be more useful than others, depending upon whether the objective is species discrimination, classification, or phylogeny. For most systematic work at the species level, morphological characters of imagines remain the most incisive because there are more readily detectable and precise characters in adults than in other stages. Initial diagnosis of species and formal designation of types must of course be based on the imaginal stage. Diagnostic larval characters at the species level are indispensable to ecology but their ultimate importance in systematics is that they open the way for precise information on case-making behaviour, oviposition and egg characteristics, habitat, etc. to be provided for the integrated data base.

At generic levels, imaginal data again provide much of the data for designating precise limits and for detecting character transformations. Characters from immature stages are of value in broadening the generality, and hence validity, of the genus; the behavioural and biological components built by accumulating species level data show the biological distinctness (or lack) of a genus.

It is at the family level (incl. infra- and supra-) that data from immature stages have proven to be particularly valuable. The general facies of larval and pupal stages in established families and subfamilies is clear and distinctive. As preceding examples demonstrate, family or subfamily affinity of a genus or its uniqueness, is illuminated through immature stages. This view reduces in no way the value of imaginal data at family levels. It affirms only that family groups in Trichoptera are elegantly confirmed by an amalgamation of data from imaginal and immature data. Taken together, the morphological and behavioural components from the immature stages provide important insight into character transformations.

403

'Evolution of the family units within the order is apparent primarily in the structures of the larvae and pupae and many of their associated habits. Evolution of generic units within the families is apparent more often in adult structures.' (Ross, 1956:6)

A corollary to Ross' principle (1956 above) is that essential evidence for admission of a genus to a particular family is to be found in the immature stages. For example, Schmid (1979, 1980) may be proven correct in the claim that *Phylocentropus* belongs in the Hyalopsychidae and that it was misplaced in the Dipseudopsidae (Ross and Gibbs, 1973); but until larvae of unquestioned hyalopsychids are discovered, an important component of the evidence is lacking — at least for those concerned with the holistic approach to classification.

GOERIDAE AND THE GENUS GOEREILLA

My paper on *Goereilla baumanni* Denning (Wiggins, 1976), providing new information on immature stages and biology of this interesting species, has been criticized (Schmid, 1979, in press) for drawing phylogenetic inferences without recourse to the full range of imaginal characters available. That paper was intended only to draw attention to the phylogenetic implications of the new data presented on immature stages; nonethless, the criticism is justified, although perhaps the shortcoming is no more a transgression against an integrated character base than works by my colleague (e.g. Schmid, 1968). Although the phylogeny of the Goeridae lies outside the scope of this paper, I will say that the definition of the group remains a fundamental problem. When genera such as *Archithremma* are primitive Goeridae on the basis of adults (Schmid, 1979), but Limnephilidae *incertae sedis* on the basis of larvae (Levanidova and Schmid, in press), the importance of immature stages in defining family-group taxa is very well illustrated. Furthermore, it matters little whether this group is recognized as Goeridae, or Goerinae of the Limnephilidae. This question should be resolved not on the basis of how distinct the most disparate members of the group are from the Limnephilidae (Schmid, 1980), but rather on whether there are diagnostic characters, consistent with established family-level distinctions in Trichoptera, to separate the Limnephilidae from those goerids most closely related to them.

OBJECTIVES IN TRICHOPTERAN SYSTEMATICS

The objectives of systematics include advancement of understanding of the evolution and biology of Trichoptera, and development of a classification based upon this understanding that serves for identification, communication, generalization, and prediction. Even allowing for the varying degrees of emphasis placed on these objectives by different workers, the view that imaginal data are alone sufficient for their realization is retrogressive. I have attempted to show that the present systematic base for Trichoptera is increasingly a synthesis of data from both imaginal and immature stages. Moreover, this integrated data base provides, through access by functional morphology of the immature stages, direct connections with the fields of behavioural and evolutionary ecology (e.g. J.B. Wallace, 1975 etc.; Wiggins and Mackay, 1978). An outstanding example is the general hypothesis of trichopteran evolution proposed by Ross (1956), stating that both the origin of Trichoptera and their diversification into major lineages at family (and many subfamilial) levels occurred in cool lotic habitats, and that warm lotic and lentic sites were invaded only after this diversification had taken place. Thus, the varied paths to exploitation of energy resources by Trichoptera were established through behavioural-morphological interactions in immature stages inhabiting cool running waters; net-spinning, case-making, sedentary, and motile modes of life were ecological adjustments (Mackay and Wiggins, 1979) leading to establishment of the major groups now recognized taxonomically as families and subfamilies. The synthesis of data from imaginal and immature stages underlying such an hypothesis is drawn from 100 years of accumulating observations. Insights at this level of generality for Trichoptera are not available from imaginal stages alone.

REFERENCES

Cowley, D.R., N.Z. Jl Zool. 3:99–109, 1976.
———, N.Z. Jl Zool. 5:639–750, 1978.
Fischer, F.C.J., Trichopterorum Catalogus. Amsterdam: Ned. Ent. Vereen.,Vol. XV, 1973.
Flint, O.S., Proc. U.S. Natn. Mus. 118(3530):373–389, 1966.
———, Beitr. Neotr. Fauna V(1):45–68, 1967.
———, Amazoniana III (1):7–43, 1971.
———, Proc. Biol. Soc. Wash. 86:127–142, 1973.
Klápalek, F., Arch. Naturw. LandDurchforsch. Böhm. 6:1–64, 1888.
Kolenati, F.A., Wien. Ent. Mschr. 15–23, 56–59, 1859.
Lepneva, S.G., Rev. Ent. USSR 35:8–27, 1956. (In Russian).
Levanidova, I.M. and Schmid, F., Considerations on *Archithremma ulachensis* Martynov (Trichoptera, Limnephilidae). (In press).
Mackay, R.J. and Wiggins, G.B., A. Rev. Ent. 24:185–208, 1979.
Marlier, G., *In* Proc. 2nd Int. Symp. Trichoptera, ed. M.I. Crichton, pp. 31–54, Junk, The Hague, 1978.
Martynov, A.B., Pract. Entomology, 5, LXVII and 388 pp., 1924 (In Russian).
———, Proc. Zool. Soc. Lond., Part. 1(V):5–112, 1930.
———, Trudy Zool. Inst., Leningr. 2:205–395, 1935.
McLachlan, R., A Monographic Revision and Synopsis of the Trichoptera of the European Fauna. First Add. Suppl. 1884.
Michener, C.D., Syst. Zool. 26:32–56, 1977.
Morton, K.J., Entomologist's mon. Mag. 25:269, 1888.
Mosely, M.E., Proc. Zool. Soc. Lond. 1936:395–424, 1936.
——— and Kimmins, D.E., The Trichoptera (Caddis-flies) of Australia and New Zealand. Brit. Mus. (Nat. Hist.), 1953.
Neboiss, A., Mem. Natn. Mus. Vict. 38:1–208, 1977.
Riek, E.F., J. Aust. ent. Soc. 7:109–114, 1968.
Ross, H.H., Bull. Ill. St. Nat. Hist. Surv. 23, 1944.
———, Evolution and Classification of the Mountain Caddisflies. Urbana: University Illinois Press, 1956.

————, Ann. Rev. Ent. 12:169–206, 1967.

————, Proc. 2nd Symp. Trichoptera, ed. M.I. Crichton, pp. 1–6, Junk The Hague, 1978.

———— and Gibbs, D.G., J. Ga. Ent. Soc. 8:312–316, 1973.

Schmid, F., Mitt. Schweiz. ent. Ges. 28, 1955a.

————, Mem. Soc. Vaudoise Sci. Nat. 11:117–160, 1955b.

————, Mitt. Zool. Mus. Berl. 34:183–217, 1958.

————, Naturaliste Can. 95:637–698, 1968.

————, Bull. Ent. Soc. Can. 11:48–57, 1979.

————, Agric. Can. Publ. 1692, 1980.

————, Esquisse pour une classification et une phylogénie des Goérides (Trichoptera). (In press).

Siltala, A.J., Zool. Jb. Suppl. 9:309–626, 1907.

Tsuda, M., Annotnes Zool. Jap. 15:394–398, 1936.

Ulmer, G., Notes Leyden Mus. 28:1–116, 1906.

————, Trichoptera, *In* Genera Insect. 60, 1907.

————, Trans. Linn. Soc. Lond., 2d Ser. Zool. 14:41–54, 1910.

————, Arch. Hydrobiol. Suppl. 19:1–528, 1951.

————, Arch. Hydrobiol. Suppl. 21:408–608, 1955.

————, Arch. Hydrobiol. Suppl. 23:109–470, 1957.

Wallace, J.B., Ann. Ent. Soc. Am. 68:463–472, 1975.

———— and Ross, H.H., Ann. Ent. Soc. Am. 64:891–894, 1971.

Wallengren, H.D.J., Andra Afdelningen. Svenska Ak. Hand., 24:1–173, 1891.

Wiggins, G.B., Can. Ent. 91:745–757, 1959.

————, Can. J. Zool. 40:879–891, 1962.

————, Contr. Life Sci. Div. R. Ont. Mus. 94, 1973a.

————, Contr. Life Sci. Div. R. Ont. Mus. 91, 1973b.

————, *In* Proc. Ist. Int. Symp. Trichoptera, ed. H. Malicky, pp. 7–19, Junk The Hague, 1976.

————, Larvae of the North American Caddisfly Genera (Trichoptera). Toronto: University of Toronto Press, 1977.

———— and Mackay, R.J., Ecology 59:1211–1220, 1978.

DISCUSSION

Nielsen: You were kind enough to send me some *Apatania tavala* larvae. They are definitely *A.* larvae, but they build typical goërine cases. On the other hand, *Lithax* larvae are typical goërine larvae, but they build cases like those of, for example, *Apatania muliebris* and *A. zonella*. This is a phylogenic puzzle to me. The position of *Ecnominae;* in the structure of the female genitalia *Ecnomus* shows a striking similarity to *Tinodes* but, also, a typical polycentropid characteristic. The larva builds a polycentropid-type net. The plesiomorphy of the female *Ecnomus* too, is a puzzle to me.

Moretti: I agree with Dr Wiggins of the absolute necessity of including the aquatic stages, so that there may be the widest possible range of data for classification.

Botosaneanu: I agree with almost all that Prof. Wiggins has said. I am convinced that all sources of information are important for sound systematics — morphology, number of larval stages, larval and pupal morphology, male, female, ethology and, let us say, physiology, biochemistry or karyology; and, as the speaker wrote 'some types of data tend to be more useful than others, depending on whether the systematic objective is species discrimination, classification or phylogeny'. Indeed, this is the most important point! I have only one remark. I am not sure that in the paper by Schmid (1979) the aim was to demonstrate that 'the immature stages are not necessary for these aspects of systematics (i.e. a sound classification and phylogeny of Trichoptera)'. I remember that in the last lines of this paper F. Schmid expressed the hope that both larval and imaginal systematics would be used in the future, with integrated and 'convergent' results.

Wiggins: Whatever else may have been the aim of the paper (Schmid, 1979), there is certainly the statement that knowledge of the immature stages is not essential for obtaining a sound classification and phylogeny of Trichoptera. However, this issue is so clouded by exception and qualification that not only is the aim of that paper unclear, but also approaches to trichopteran systematics are confused. My remarks here are an attempt to set forth the approach I believe to be important – an attempt I should have considered unnecessary before that paper appeared. Your final remark is, in fact, not borne out by the paper.

QUATERNARY SUB-FOSSIL TRICHOPTERA LARVAE FROM A SITE IN THE ENGLISH LAKE DISTRICT

B. WILKINSON

SUMMARY

Trichoptera larval sclerites in two Late-Glacial cores from Low Wray Bay, Lake Windermere were identified by comparison with modern larvae. Apparent changes in climate and environment during the period under investigation were noted from changes in the fauna.

INTRODUCTION

Fossil Trichoptera are abundant in Quaternary deposits, predominantly as discrete larval sclerites. In this paper I report on a sequence of Trichoptera fossils from Low Wray Bay, Windermere, Cumbria an extensively investigated site (Coope and Pennington, 1977; Howell, 1971; Pennington 1943, 1977.).

Insect remains from two cores taken by Mrs W. Tutin of the University of Leicester were extracted from 5 cm samples and the Coleoptera identified and published (Coope, 1977). Trichoptera form much of the residual insect fauna.

The stratigraphy is illustrated in Fig. 1, below.

Larval fossils only were found, represented by the frontoclypeus, gena, labrum, pronota, mesonota, legs and jaws. Of these, the most easily identifiable sclerite was the frontoclypeus. Where possible sclerites were identified by direct comparison with specimens of final instar larvae and exuviae (loaned by several colleagues), or in their absence with drawings in Hickin (1967), and Lepneva (1964, 1966). All available species, both British and European, were considered.

RESULTS

Polycentropodidae

The most common species was *Polycentropus flavomaculatus* Pictet. The frontoclypeus has a highly distinctive outline and markings (Fig. 3). A similar species *P. kingi* McL. is broader and less angular. A single specimen of *Cyrnus*

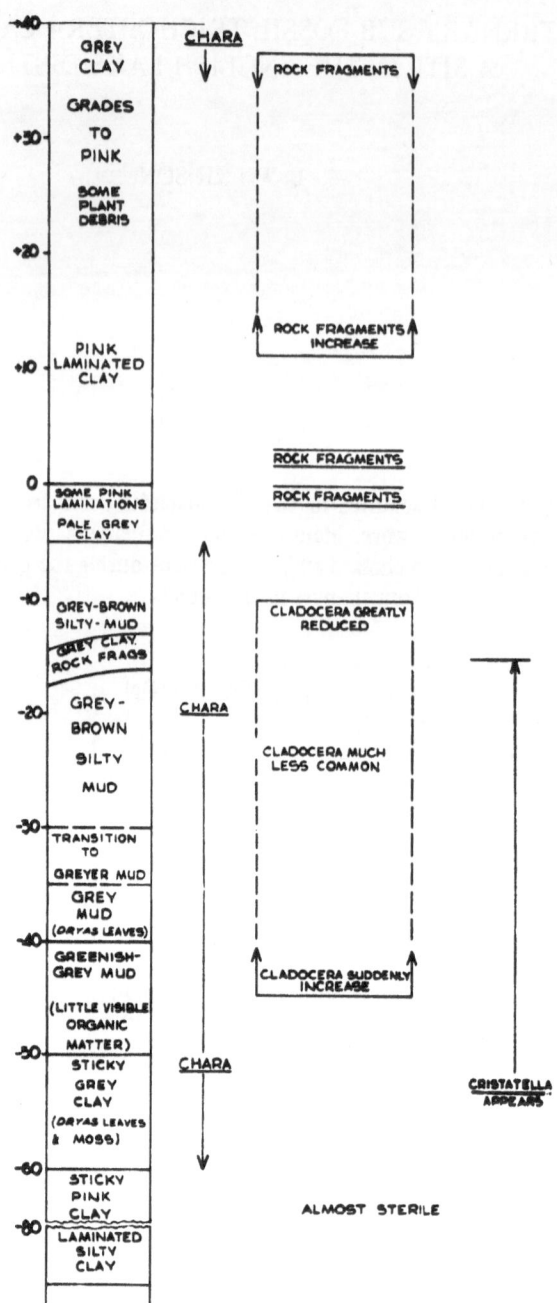

Fig. 1. Geological succession. After Coope (1977)

410

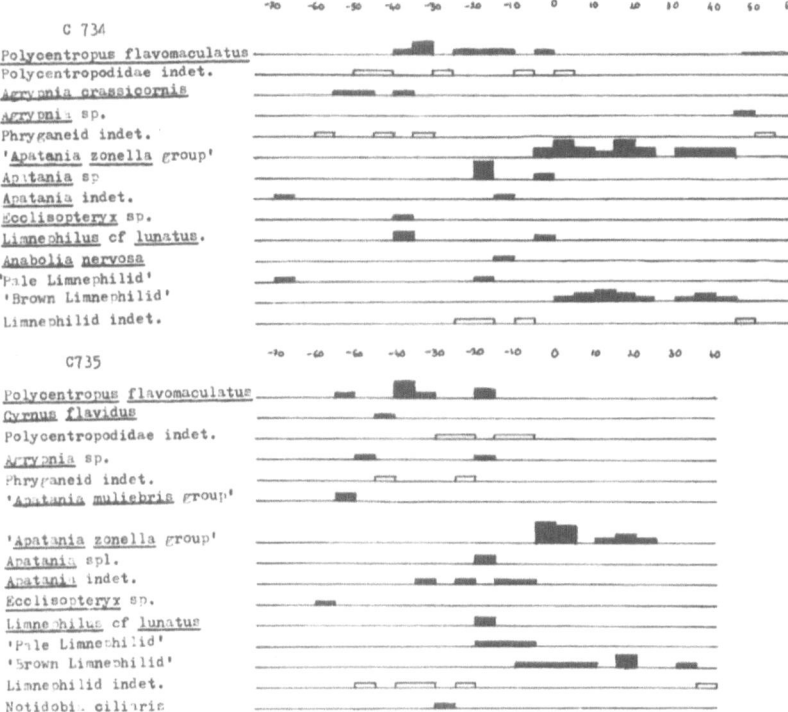

Fig. 2. Species found in the cores.

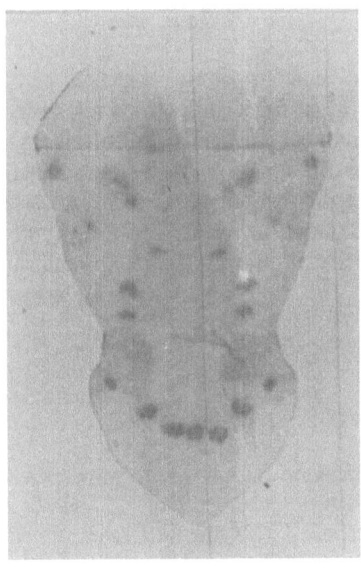

Fig. 3. Polycentropus flavomaculatus, × 40

411

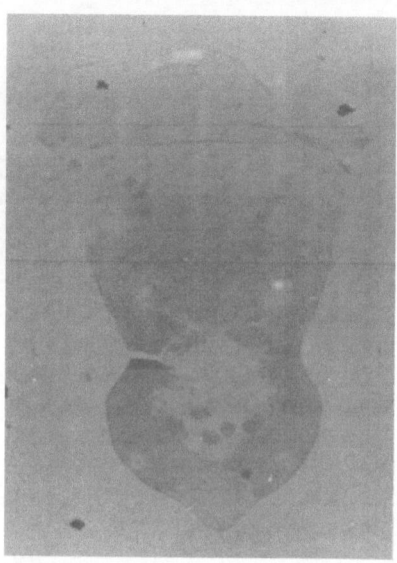

Fig. 4. Cyrnus flavidus, × 40

flavidus McL. was found (Fig. 4); note the central white fleck and the white spots around the four laterally placed setigerous pores. In addition numerous fragments of Polycentropodid pronota and gena were found in horizons devoid of the frontoclypeus.

Phryganeidae

The fossils comprised two forms of *Agrypnia* and some unidentifiable fragments. One form, identifiable only to genus, (Fig. 5), possessed a dark stripe down the frontoclypeus reaching the anterior margin. This is referred to as *Agrypnia* sp.. The other (Fig. 6) had a pronounced 'V shaped' fleck that did not reach the anterior margin. Modern larval descriptions (Bray, 1967; Lepneva, 1966) leave *A. crassicornis* McL. as the only possible candidate. This species has a disjunct distribution on the continent, and in Britain is found only in Malham Tarn. This does not necessarily suggest a continuous occupation of this region.

Limnephilidae

Three forms of *Apatania* were found. One form bears a reddish-brown fronto-clypeus, similar to Icelandic specimens of *A. zonella* Zett. given to me by Dr Gisláson; I have labelled this the '*Apatania zonella* group' (Fig. 7). A second form resembles *A. muliebris* McL. specimens lent to me by Dr G.N. Philipson, having a pale central fleck; these I have labelled the '*A. muliebris*

412

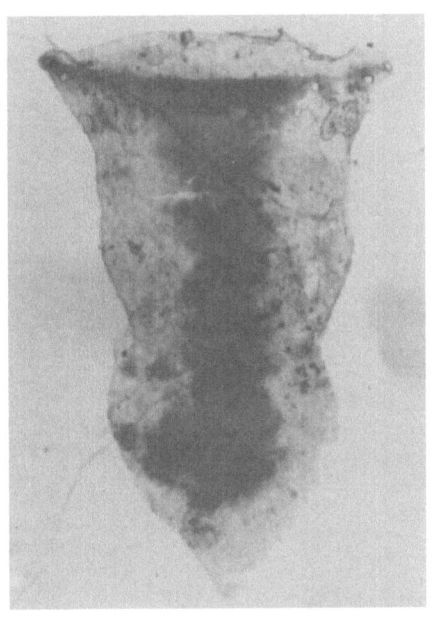

Fig. 5. Agrypnia sp., × 40

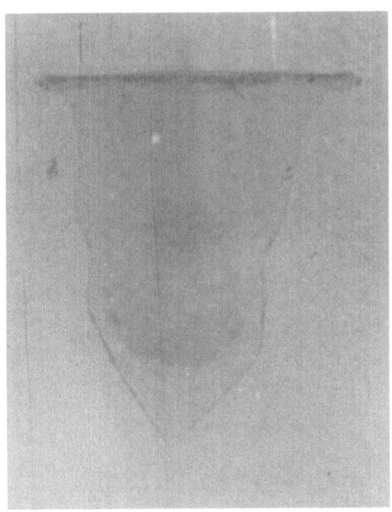

Fig. 6. Agrypnia crassicornis, × 40

413

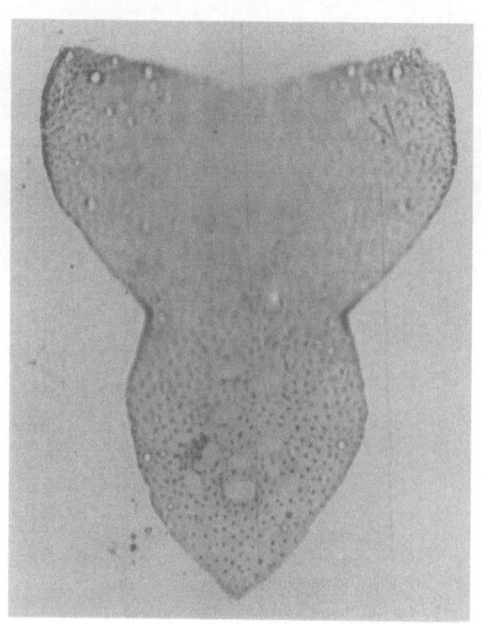

Fig. 7. Apatania zonella group, × 100

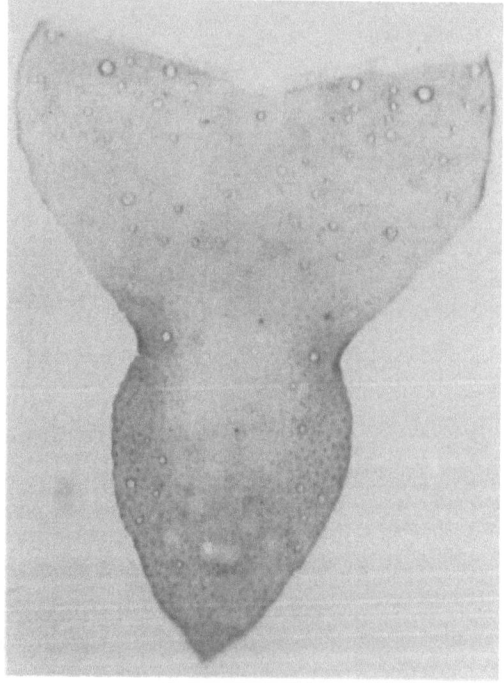

Fig. 8. Apatania muliebris group, × 100

414

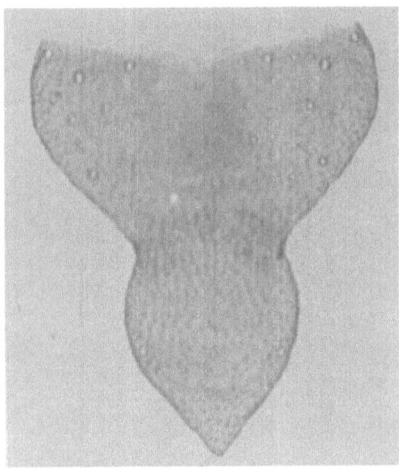

Fig. 9. Apatania sp. 1, × 100

group' (Fig. 8). The third form does not resemble any of the British species of *Apatania*; I have called it *Apatania* sp 1. (Fig. 9).

Ecclisopteryx (Fig. 10) is represented by two specimens bearing a close but not exact resemblance to *E. guttulata* Pictet. The possibility of their being *Drusus* is precluded by their outlines and patterns of setigerous pores.

One species of *Limnephilus* was recovered, (Fig. 11), similar to *L. lunatus* Curtis. I have seen insufficient numbers of *Limnephilus* species to make a confident identification.

Anabolia nervosa Curtis was easily identified because of the characteristic markings on its frontoclypeus (Fig. 12).

Two remaining Limnephilids could not be identified beyond family level. Working names were used for these species. The 'Pale Limnephilid' (Fig. 13) was represented by five frontoclypeal apotomes. The similar but distinctive 'Brown Limnephilid' (Fig. 14) was found as twenty frontoclypeal apotomes and numerous other sclerites, all a characteristic chestnut brown colour. None of the modern species examined resembled either of these Limnephilids.

A Sericostomatid was also found. The frontoclypeus lacked the ridges noted by Wallace (1977) in *Sericostoma* and resembled a specimen of *Notidobia ciliaris* L. lent to me by him. I have therefore assigned this specimen to *Notidobia ciliaris*, a common species, in the absence of more precise information on other Sericostomatids (Fig. 15).

Fig. 2 shows the occurrence of Trichoptera in the two cores. The stratigraphical levels are taken from the usual datum 0cms, selected at a sharp, easily recognised lithological change, present in all cores from this site. The numbers of individuals at each level are calculated from the maximum number of any diagnostic sclerite of the same instar. Since the samples are volumetrically identical, and if it is assumed that deposition rate was constant, any variation in the number of individuals may well represent a change in the abundance of the species.

415

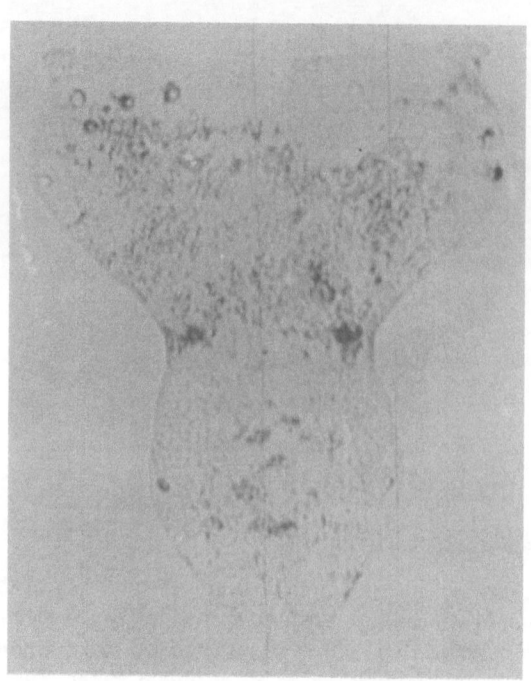

Fig. 10. Ecclisopteryx sp., × 100

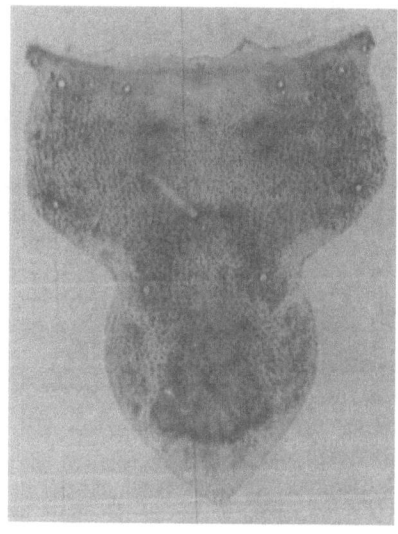

Fig. 11. Limnephilus of *lunatus*, × 40

416

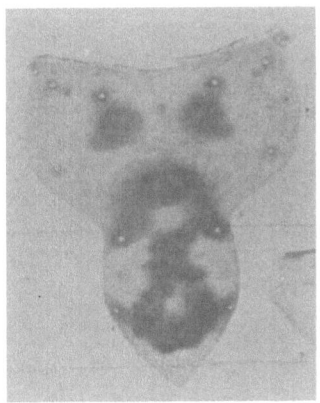

Fig. 12. Anabolia nervosa, × 40

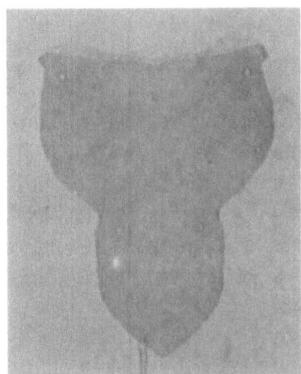

Fig. 13. Pale Limnephilid, × 40

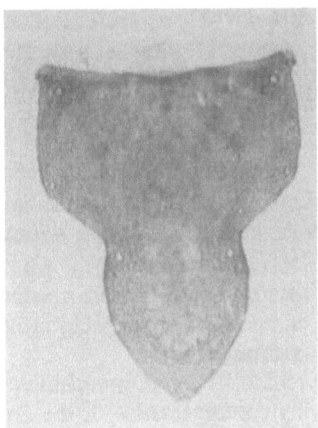

Fig. 14. Brown Limnephilid, × 40

417

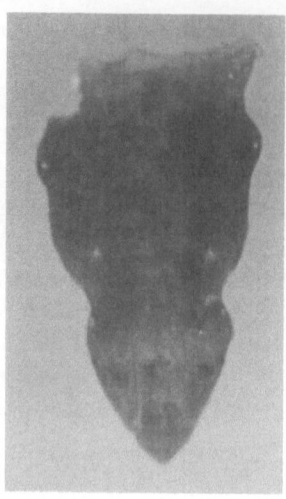

Fig. 15. Notidobia ciliaris, × 40

Trichoptera sclerites are transported only a short way, unlike pollen or even adult Trichoptera which may be blown great distances, so the ecological development of the neighbourhood may be inferred both locally and regionally. An example of a local effect is the washing of the flowing water species *Ecclisopteryx* into a lake deposit. Regional environmental alterations are indicated by the drastic change at Ocms. Local differences can even be seen between the two cores. For example *Agrypnia crassicornis* is found in only one of them, this could reflect its original patchy occurrence. The faunal change at Ocms appears imprecise in C734 in contrast to that in C735, and also there is a difference in the position of the change indicating the beginning of the Post-Glacial period. Because of such discrepancies separate diagrams were drawn. In spite of these differences there is a broad agreement between the two cores. The levels below Ocms are characterised by an increasingly diverse fauna, indicating a temperate climate, little difference from that at the present day, as only temperate species are found. Both diagrams show a marked change in the fossil assemblage at about Ocms, corresponding with the lithological change. All of the previously found species disappear, and in the ensuing 40 cms, only two Trichoptera species occur, the '*Apatania zonella* group' and the 'Brown Limnephilid'. These levels have an arctic Coleopteran fauna e.g. *Amara alpina* Payk. and *Bembidion fellmani* Mnh. (Coope, 1977) and flora (Pennington, 1977). The lithology is a pink laminated clay with layers of fine gravel, indicating glaciolacustrine conditions with little vegetation nearby. This environmental picture should provide a clue to the habitat and identity of the 'Brown Limnephilid'.

The renewed presence of different Limnephilids and Phryganeids at the very top corresponds with the appearance of the Post-Glacial organic clays. No attempt has been made at this stage to follow the development of the Trichoptera fauna into the Post-Glacial.

418

REFERENCES

Coope, G.R., Phil. Trans. Roy. Soc. B 280: 313–340, 1977.
——, and Pennington, W., Phil. Trans. Roy. Soc. B 280: 337–339, 1977.
Hickin, N.E., Caddis Larvae. ed. Hutchinson, London, 1967.
Howell, F.T., Geological Journal. 7: 329–334, 1971.
Lepneva, S.G., Fauna of the U.S.S.R. Trichoptera, Israel Program for Scientific Translations, Jerusalem, Vol. II, No. 1, 1970, 2, 1971.
Pennington, W., New Phytologist. 42: 27, 1943.
——, Phil. Trans. Roy. Soc. B 280: 247–271, 1977.
Wallace, I.D., Freshwater Biology 7: 93–98, 1977.

DISCUSSION

Harrison: Is the faunal discontinuity in Trichoptera and Coleoptera also represented in the floral evidence?

Wilkinson: Yes, it is.

Moretti: 1) Do you find the diatom frustules on the larvae clypeus? 2) or apodeme remains on the inner surface of the clypeus? 3) What period of the Quaternary does your material come from?

Wilkinson: 1) No, the magnification that I used was not sufficient. 2) No, the larvae were disarticulated and the smaller sclerites could not be found. 3) Between 10,000 and 15,000 years ago.

Higler: I find this a fascinating study which contributes as much to our knowledge of zoogeography as to larval morphology. However, as you base your identification primarily on the frontoclypeal shape, I doubt whether you are able to identify at species level as you did with *Limnephilus* species. I suggest the use of numerical taxonomy methods as practised by Buch for *Rhyacophila*. He uses the measurements derived from the chaetography of the frontoclypeus.

Wilkinson: The identification of *Limnephilus* species are by no means certain. The number of specimens found do not enable statistical studies, and the chaetography of *Limnephilus* is too constant to help with identification of the species.

A PROPOSED SETAL NOMENCLATURE AND HOMOLOGY
FOR LARVAL TRICHOPTERA

N.E. WILLIAMS AND G.B. WIGGINS

SUMMARY

Recently there has been an increasing use of larval chaetotaxy in trichopteran taxonomy and systematics, which places added importance on adoption of a single, comprehensive system of setal nomenclature, in particular for abdominal segments.

This paper re-examines the definitions of tactile and proprioceptor setae, and sensory pits. The history of the taxonomic and systematic use of these structures and previous numbering systems are discussed. A new nomenclature is proposed for the primary setae and pits of all body parts; its relationship to past systems proposed for Trichoptera is discussed and homology with equivalent Lepidopteran structures is proposed. Examples are given of differences in setal arrangement in various caddisfly taxa.

Recently there has been an increase in the use of larval chaetotaxy by trichopteran taxonomists and systematists. For example Hiley (1972, 1976) has made considerable use of setal characters in taxonomic studies of British Sericostomatidae (*sensu lato*) and Limnephilidae; his key to the limnephilid larvae refers to the numbers positions and colours of the setae of the femora in particular. Gíslason (1979) used Hiley's terminology in his key and descriptions of Icelandic larvae. Resh (1976) used some setal characters to define *Ceraclea* species. Wallace (1980) used numbers and positions of setae in British limnephilids, particularly those of the first and ninth abdominal segments and the femora. Many chaetotaxal characters were introduced in keys to North American families and genera (Wiggins, 1977). Thus, with the increasing initiative in the use of setal characters, we feel that it is important at this time to re-examine the already established setal nomenclature and to complete and standardize the system before a proliferation of terminologies takes place.

The first step in this procedure is to ensure an understanding of the basic structures referred to in chaetotaxy. Siltala (1907) gave detailed definitions of setae ('Borsten') and spines ('Dornen') and these concur with the general entomological use of the terms. Setae are those cuticular appendages which have alveoli, while spines are those without, although various forms are found within each category. For example, spurs are included within the definition of setae. Siltala further distinguished 'true setae', long in comparison with their diameter and usually dark; and 'little setae', ('Borstchen'), short, pale and thin such as often occur on the doral surface of coxae or on intersegmental membranes. Those on the membrane are the proprioceptors (Chapman, 1975), serving to inform the animal of the position of one body segment with respect to the next. His 'true setae' are the tactile setae.

Sensory pits were not as clearly defined by Siltala, but their identity was clarified by Nielsen (1942) when he suggested that they could be classified according to Imms (1925) as coeloconic sense organs and that they might be modified setae. The external process of a coeloconic sensilla is sunken into a pit and its function is generally considered to be chemosensory. A sensory pit should not be confused with the alveolus remaining after a seta breaks off.

Primary setae and pits are those occurring in the first instar. Since the number of these is the same in virtually all Trichoptera (Siltala, 1907), and there are no secondary proprioceptors or sensory pits, all of these structures can be readily distinguished from one another, given moderate practice combined with a knowledge of their usual positions and arrangement. When this is done, all three types become more useful in identification and systematics. For example, Siltala (1907) first proposed investigating the degree of specialization of various groups by looking at the resemblance of chaetotaxy of mature larvae to that of first instar larvae. He considered the Rhyacophilidae to be the most primitive family since their primary setae remain virtually unchanged through all larval instars and secondary setae are little developed. He also developed a key to first instar larvae for twelve families, based mainly on setal characters.

Unfortunately this auspicious start at trichopteran chaetotaxy was largely ignored for the next forty years. The second, and only other, major contribution was made by Nielsen (1936, 1937, 1942, 1943, 1948). He enlarged greatly (1942) on Siltala's discussion of the placement of setae, commenting on the homology of these structures in the seventeen species of 9 families which he studied, and giving arabic numbers to the setae of the labrum (1–6), genae (7–18), maxillae (1–9), and anal claws (1–8). Later (1943), Nielsen used the chaetotaxy of the legs as one character in support of separating the subfamily Ecclisopteryginae (Limnephilidae); and in 1948 he proposed for the Hydroptilidae a system of letters and numbers to describe the positions of the dorsal thoracic setae, e.g., M1 – first marginal seta; ml – first premarginal seta; L – lateromarginal; A – angular; O – oral surface. Nielsen's setal nomenclature (particularly the head numbers) has been adopted partially by many authors, but none has since pursued chaetotaxal studies with his thoroughness, and many have failed entirely to recognize the taxonomic and systematic value of this approach.

In 1956, Ross designated setal areas of the thoracic nota as sa 1, 2 and 3 for the anterior, posterior and anterolateral positions. This terminology has been generally adopted and extended to the setae of the abdomen (Wiggins, 1977). Flint (1960) appears to have been the first person to number the setae of the frontoclypeus (1–6). No further numbering systems have since been proposed.

It now seems in order to establish a simple, standard and complete system which deviates as little as possible from the terminology in current use. In the various systems proposed for larvae of the order Lepidoptera, where chaetotaxy has been the major taxonomic tool for many years, setal nomenclature has suffered from a proliferation of complex systems, including those of Müller (1886), Dyar (1896), Dampf (1910), Fracker (1915), Heinrich (1916), Ripley (1923), Forbes (1923), Gerasimov (1935) and Hinton (1946). The resulting confusion in this order has influenced our decision to designate setae simply by arabic numbers beginning at 1 for each body part. Sensory pits are similarly numbered but prefixed by the letter P.

Figs. 1–4 illustrate the locations of the primary setae and pits of the head, thorax, legs and abdomen numbered according to the system proposed here. A rhyacophilid (*Rhyacophila acropedes* Banks) has been chosen for these basic diagrams since its chaetotaxy is relatively simple, i.e., free of secondary setae and close to that of its lepidopteran relatives.

Tactile setae of the labrum and head (Fig. 1) are numbered according to Nielsen (1942) and Flint (1960) 1 to 6 and 1 to 18, with the five proprioceptors of the head numbered 19 to 23. The sensory pits are numbered P1 to P19. Table 1 describes their usual locations.

The setae and pits of each thoracic and abdominal segment (Figs. 2, 3) are believed to be homologous and therefore the same numbers have been applied to equivalent structures on each segment. Proprioceptors again have the higher numbers 13 to 21. It should be noted however, that not all setae and pits are present on every segment. For example, setae 7, 8 and 9 are missing from thoracic segments II and III (Fig. 2), while seta 14 is usually missing from all abdominal segments (Fig. 3). Pit P1 and seta 18 are usually present only on the prothorax (Fig. 2).

Ross' setal area designations for the thoracic nota are incorporated for the most part into the new system, since the primary setae of each area are few. Sa 1 corresponds to seta 1, sa 2 to setae 2, 3 and 4, and sa 3 to setae 5 and 6. Since Nielsen's terminology for setae of these segments has never attained general use, no confusion should result here.

On the legs, setal numbers begin at 1 on each segment, commencing with tactile setae of the posterior face.

This system was developed as one of the initial stages in a systematic study of larval chaetotaxy currently under way. There seems a strong possibility that chaetotaxy will prove as useful in the taxonomy and systematics of larval Trichoptera as it has in Lepidoptera where much of the larval diagnosis is based on setal characters, particularly those of the abdomen. We have studied larval Lepidoptera from families considered to be among the more primitive. On the basis of the positions, lengths and associations of primary setae and pits we are proposing an hypothesis of homology for the majority of these

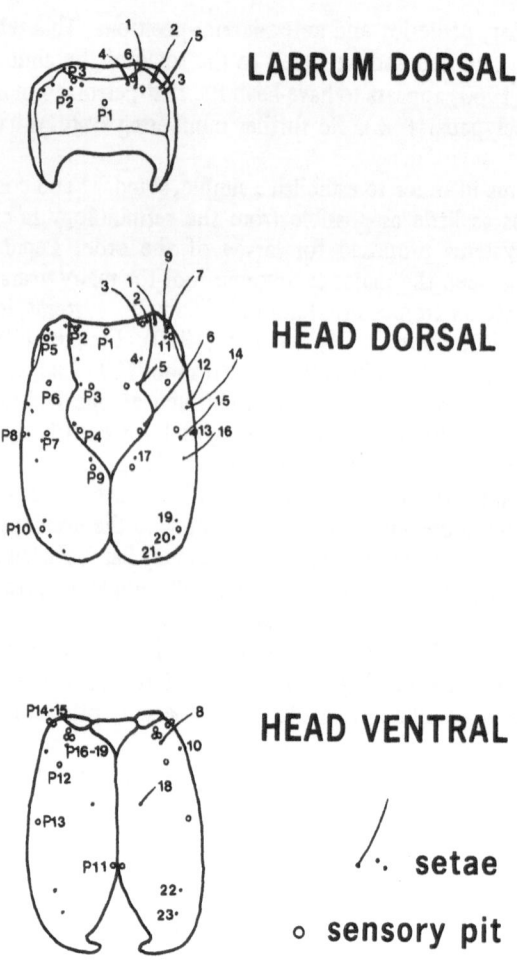

LABRUM DORSAL

HEAD DORSAL

HEAD VENTRAL

setae

o sensory pit

Fig. 1. Setae and sensory pits of the labrum and head of a 5th instar *Rhyacophila* larva.

structures found on larvae in the two orders (Table I). Although further study may well modify opinions concerning some of these, homology seems particularly good on the head, where the twenty-three trichopteran setae can be equated to the twenty-three lepidopteran setae with only two or three uncertain cases. On the thorax and abdomen twelve of the fourteen (thirteen) tactile setae of each group can be equated, while there are more differences in proprioceptors and sensory pits. The chaetotaxy of the two sister groups can now be accurately compared and the Lepidoptera will therefore be useful in determining primitive characters in the Trichoptera.

Our studies have begun to reveal useful differences in setal arrangement in various taxa. Fig. 5 illustrating the second and third abdominal segments of species in several families shows that although the positions of certain setae

424

Table I Proposed nomenclature for trichopteran setae and sensory pits and Trichoptera/ Lepidoptera homology.

Body Part	Seta/ Pit Number	Comments	Lepidopteran Equivalent
LABRUM	1		M3
	2		LA3
	3	numbers after Nielsen (1942)	LA1
	4		M1
	5		LA2
	6		M2
	P1	single, central	
	P2	near seta 4	
	P3	anterior margin	
HEAD	1		A1
	2		C1
	3	numbers after Flint (1960)	C2
	4		F1
	5		AF1
	6		AF2
	7		SO1
	8		SO?
	9		O1
	10	numbers after Nielsen (1942)	SO2?
	11		A2
	12		O2
	13		L1
	14		A3
	15		P3
	16		P2
	17		P1
	18		O3
	19		V1
	20	dorsal proprioceptors	V2
	21		V3
	22	ventral proprioceptors	G1
	23		G2
	P1	single, central	Fa
	P2	posterior to seta 2	
	P3	near tentorial pit	AFa
	P4	posterior to tentorial pit	
	P5	anterior — near seta 9	Ob
	P6	near seta 11	Aa
	P7	posterior to seta 14	Pa
	P8	lateral, near seta 13	La
	P9	posterior, near seta 17	Pb
	P10	with dorsal proprioceptors	Va
	P11	near seta 18	Oa
	P12	posterior to seta 8	SOa
	P13	near ventral proprioceptors	Ga
	P14		
	P15	between setae 7 and 8	
	P16		
	P17	anterior ventral	
	P18		
	P19		
FEMUR	1	dorsal, median	
	2	dorsal, distal	
	3	posterior, distal	
	4	posterior ventral,	
	5	anterior, ventral, proximal	
	6	anterior, distal	
	P1	anterior, ventral, proximal	
TIBIA	1	posterior, ventral, distal	
	2	posterior, median, distal	
	3	posterior, dorsal, distal	
	4	anterior, ventral, distal	
	5	anterior, median, distal	
	6	anterior, dorsal, distal	
	P1	dorsal	

Body Part	Seta/ Pit Number	Comments	Lepidopteran Equivalent
THORAX AND ABDOMEN	1	sa1	D1
	2		D2
	3	sa2	
	4		
	5	sa3	SD1
	6		SD2
	7		L1
	8		L2
	9		L3
	10		SV1
	11		SV2
	12		V1
	13		MD1
	14		MSD1
	15		MSD2
	16	proprioceptors	MV3
	17		MV1?
	18		MV2?
	19	on trochantin (episternum)	
	20		
	21		MXD1
	22	prothorax only	XD1
	23		XD2
			XD6
	P1	near seta 1	
	P2	near seta 2	
LEGS-COXA	1	posterior, dorsal	
	2	posterior, ventral, distal	
	3	anterior, dorsal, proximal	
	4	anterior, dorsal, distal	
	5	anterior, median	
	6	anterior median	
	7	anterior, ventral, distal	
	8	posterior, dorsal, proximal	
	9	posterior, median, proximal	
	10	anterior, median, proximal	
	11	anterior, dorsal, proximal	
	12	" " "	
	P1	dorsal	
TROCHANTER	1	posterior, median	
	2	posterior, ventral, distal	
	3	" " "	
	4	anterior, median	
	5	anterior, ventral, distal	
	6	" " "	
	7	anterior, ventral, median	
	8	anterior, dorsal, median	
	P1	posterior, median	
	P2	" "	
	P3	" "	
	P4	" "	
	P5	anterior, median	
	P6	" "	
TARSUS	1	posterior, ventral, distal	
	2	dorsal, distal	
	3	dorsal, distal	
	4	anterior, ventral, distal	
	P1	dorsal	
CLAW	1		

*Lepidopteran setal names are in accordance with Hinton (1946) The letters refer to setal areas e.g. AF — adfrontal, P-posterior.

425

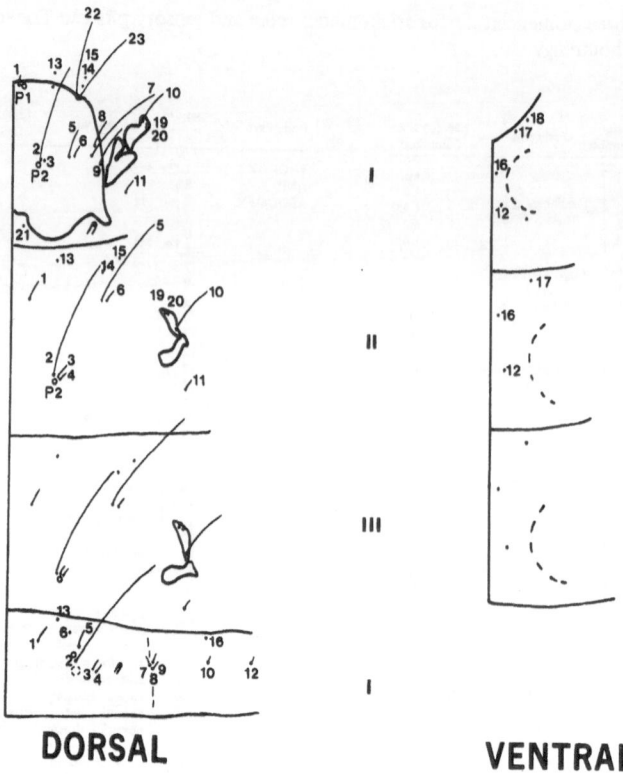

DORSAL   **VENTRAL**

Fig. 2. Setae and sensory pits of the thorax and first abdominal segment. The symbol ⌐ indicates the position of a spiracle and the symbol ◇ indicates the base of a gill in this and subsequent figures.

vary, they maintain associations with other setae, a pit or the spiracle. This has proved useful in locating specific setae, particularly on specimens with dense secondary setation. Seta 5 for example can usually be found just dorsal to the spiracle.

Setae are sometimes missing. The lateral line setae 7, 8 and 9 are all present in *Rhyacophila*, but one of these is missing in members of most other families. Seta 17 may be present or absent as well.

In some taxa there is also variation in setal position between segments, e.g. setae 7 and 8 in *Phryganopsyche* and *Anabolia* (Fig. 2 D, E), seta 12 in *Anabolia*.

In conclusion, we would like to make a plea for more complete and accurate illustration in all works involving description of larvae. A quick sketch of characters thought to be diagnostic at the time may be of little use in later years as further study of a taxonomic group is made. The time spent in clearing, preparing slides, and drawing accurately and completely any specimen to be described in the literature not only prevents frustration for future researchers, but also contributes to better observation and often

MID-DORSAL

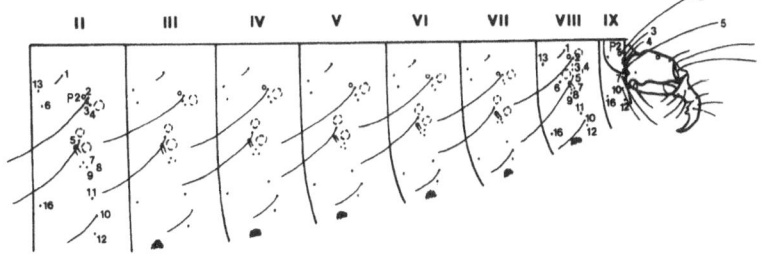

Fig. 3. Setae and sensory pits of the abdomen.

interesting discoveries. We have observed, for example, previously undescribed structures on the abdominal segments of many Hydropsychoidea and Rhyacophiloidea (Fig. 5). These appear to be small patches of minute spines (?) modified into various rippled or comb-like shapes. They would have gone unnoticed if the specimens had not been cleared, mounted on slides and every detail thus exposed included in drawings. We hope therefore, that our setal nomenclature and homology will encourage chaetotaxal studies and help to make available new systematic data.

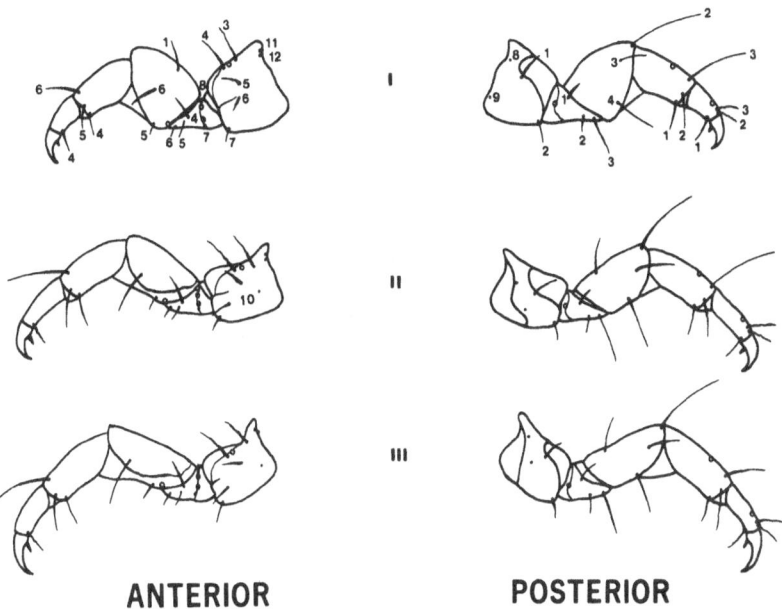

ANTERIOR   **POSTERIOR**

Fig. 4. Setae and sensory pits of the legs.

427

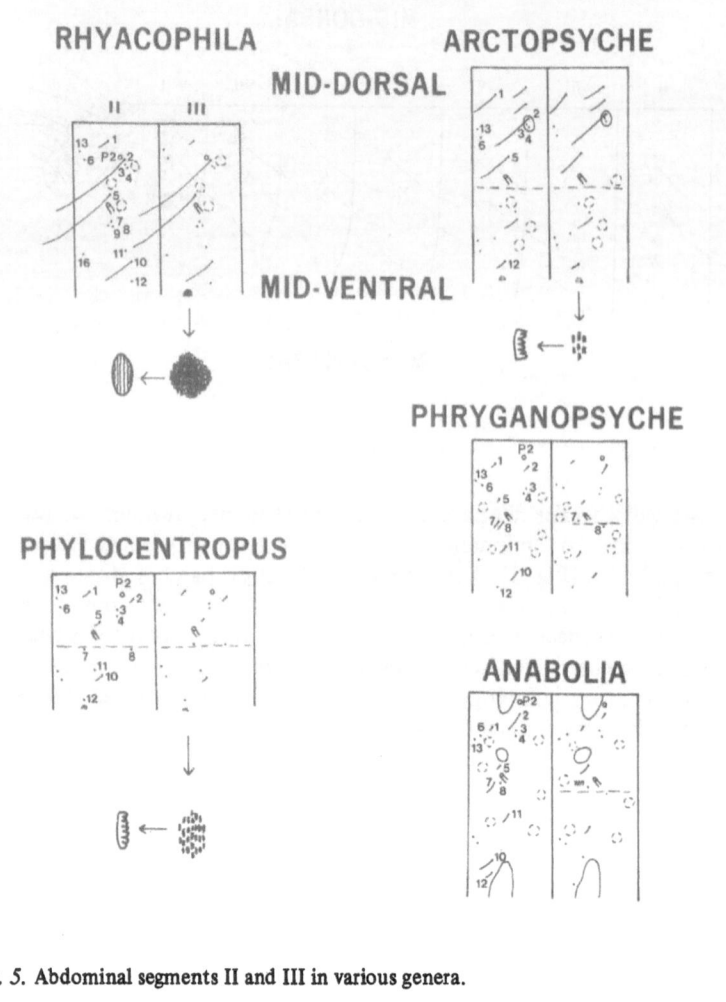

Fig. 5. Abdominal segments II and III in various genera.

REFERENCES

Chapman, R.F., The Insects, Structure and Function. (3rd Ed.) English Universities Press Ltd., 1975.
Dampf, A., Zool. Jb. Suppl. 12: 513–608, 1910.
Dyar, H.G., J.N.Y. Ent. Soc. 4: 92–93, 1896.
Flint, O.S., Entomologica Am. 40: 1–119, 1960.
Forbes, W.T.M., Mem. Cornell Univ. Agric. Exp. Stn. 68: 1–729, 1923.
Fracker, S.B., Illinois Biol. Monogr. 2: 1–169, 1915.
Gerasimov, A.M., Zool. Anz. 112: 177–194, 1935.
Gislason, G.M., Ent. Scand. 10: 161–176, 1979.
Heinrich, C., Proc. Ent. Soc. Wash. 20: 27–38, 1916.
Hiley, P.D., Entomologist's Gaz. 23: 105–119, 1972.
———, Syst. Ent. 1: 147–167, 1976.
Hinton, H.E., Trans. R. Ent. Soc. London. 97: 1–37, 1946.
Imms, A.D., A General Textbook of Entomology., Methuen and Co., London, Ltd., 1925.

428

Müller, W., Zool. Jb. (Syst.) 1: 417–678, 1886.

Nielsen, A., Zool. Anz. 113: 255–266, 1936.

————, Arch. Hydrobiol. 31: 253–263, 1937.

————, Arch. Hydrobiol. Suppl. 17: 255–631, 1942.

————, Vidensk. Meddr. dansk naturh. Foren. 107: 105–120, 1943.

————, K. Danske Vidensk. Selsk. Skr. Biolog. 5: 1–200, 1948.

Resh, V.H., Ann. Ent. Soc. Am. 69: 1039–1061, 1976.

Ripley, L.B., Illinois Biol. Monogr. 8: 1–169, 1923.

Ross, H.H., Evolution and Classification of the Mountain Caddisflies. Urbana: University of Illinois Press, 1956.

Siltala, A.J., Zool. Jb. Suppl. 9: 309–626, 1907.

Wallace, I.D., Freshwat. Biol. 10: 171–189, 1980.

Wiggins, G.B., Larvae of the North American Caddisfly Genera (Trichoptera). University Toronto Press, 1977.

DISCUSSION

Smith: Did you look at *Rhyacophila* larvae other than those with gills.

Williams: No, I only looked at *R. acropedes*.

SOME ASPECTS OF THE LIFE HISTORY AND FEEDING ECOLOGY OF *DOLOPHILODES DISTINCTUS* (WALKER) IN TWO ONTARIO STREAMS

D. WILLIAMS AND N.E. WILLIAMS

SUMMARY

Dolophilodes distinctus (Philopotamidae) is a net-spinning species widespread in eastern North America. Data are presented from populations in two rivers: Duffin Creek, a highly productive, hard water stream in southern Ontario, and the Root River, a less productive, brown water stream in northern Ontario. In Duffin Creek, the species appears to have at least a biroltine life cycle with one emergence in late June and another in August. Maximum larval densities in this stream occur in August and this coincides with the maximum amount of fine particulate organic matter (FPOM), which is known to be the species' principal food source, in the habitat. The gut contents of all 5 instars in August are compared with those for June when there is much less FPOM in the stream. The results are compared with existing data in the literature and comments are made on the methods currently used to analyze the size composition of small ingested food particles. Locations of the feeding nets in the streams are given together with descriptions of characteristic net clusters, containing a mixture of instars, at preferred sites. The net structure is discussed with regard to its functioning in low and high productivity streams. Aquatic invertebrates found living on the feeding nets are listed.

INTRODUCTION

Wallace et al. (1977) contend that the presence of net-spinning Trichoptera in running water increases the efficiency of the ecosystem by tightening the cycling (or spiralling) of edible organic material that passes through it. Caddisflies do this by retarding the passage of, and storing, organic materials as they pass downstream. The fact that different species feed on different parts of a wide spectrum of particle sizes and types (Williams and Hynes, 1973; Wallace, 1975) must mean that where they coexist the efficiency is increased further still.

A variety of net types is found within the four net-spinning families and each has been shown to function differently according to its design and characteristic location (microdistribution) in the habitat (e.g. Nielsen, 1942;

Philipson, 1953; Brickenstein, 1955; Sattler, 1962, 1963; Wallace et al. 1976). Differences between the functioning of nets of the same species in streams of differing character have been little studied, however. This study compares the nets constructed by two populations of the philopotamid *Dolophilodes distinctus*, one occurring in a low productivity stream and the other occurring in a high productivity stream. Data on the life cycle and diet of the species in the latter location are also given.

THE STUDY SITES

Duffin Creek is a highly productive, hard water stream in southern Ontario. At the sampling site (43°58'N; 79°05'W at 180 m elevation) the stream is about 5 m wide and has a series of riffles each about 15 m long. Depth on the riffles ranges from 20 cm in midsummer to 60 cm during spring runoff, with corresponding current velocities of between 25 and 110 cm/s. The substrate consists of mixed limestone gravels with some sand and clay patches underlain by clay at a depth of 20 cm. Flat cobbles, up to 20 cm in diameter, are frequently found on the surface.

The Root River is a low productivity, brown water stream in northern Ontario. It is about 12 m wide at the sampling site (46°35'N; 84°15'W at 240 m elevation) and also consists of a fairly uniform series of alternating riffles and pools. Depth on the riffle areas is only about 12 cm in midsummer when the river flows over only part of the channel bed. Current velocity on the riffles at this time is approximately 30–40 cm/s and the bed consists of a deep layer of mixed granite gravels with some larger cobbles imbedded near the surface.

LIFE HISTORY

D. distinctus is a species widespread in eastern North America (Wiggins, 1977). Its life cycle in Duffin Creek has been deduced from the occurrence of larvae on baskets of clean gravel that were set out in the stream each month for one year, and from adults caught in small emergence cages.

Separation into larval instars was not easy in this species due to the large variation in widths of head capsules in any size class. This was ultimately overcome by measuring several hundred specimens and plotting a size-frequency histogram; five reasonably well-defined peaks resulted.

The species appears to have at least a bivoltine life cycle with emergence taking place almost continually from late June to late August resulting from two summer generations in quick succession (Figs 1 and 2). Larval densities on the colonization baskets dropped during the fall and winter but this may reflect decreased mobility of the larvae at this time. The presence of some fourth instar larvae in midwinter might suggest a winter emergence, as has been shown to occur in the species elsewhere (Ross, 1944). Maximum larval densities in the stream occurred in August. This coincided with the maximum amount of fine particulate organic matter (FPOM), which is known to be the species' principle food source (Wallace and Malas, 1976), in the habitat.

432

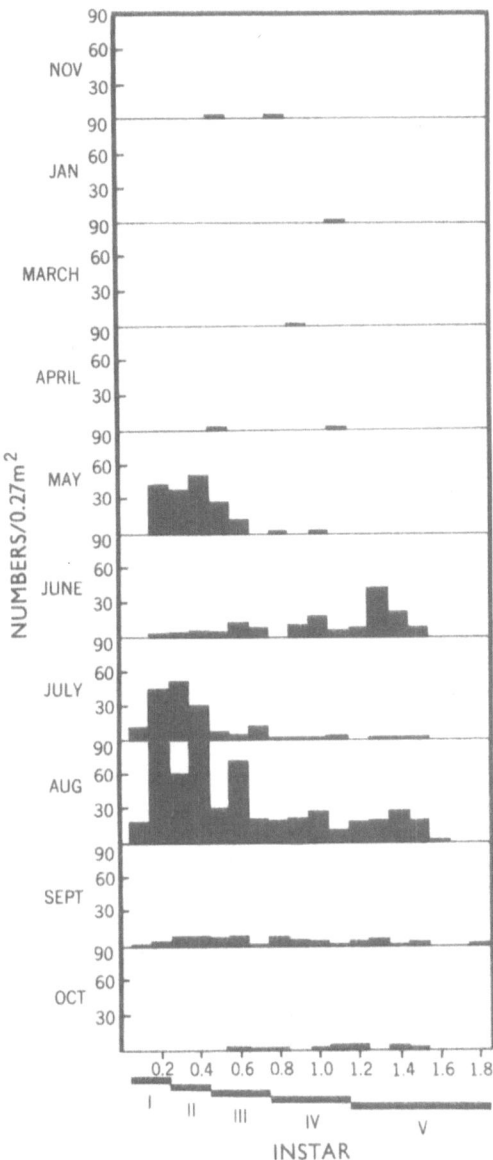

Fig. 1. Size distribution of larvae in Duffin Creek from November 1977 to October 1978.

LARVAL FOOD

Analyses of gut contents were made after the material had been dissected out and mounted on 0.45 μm filters according to the method of Coffman et al. (1971). 200 items from each gut were then identified along a transect of the

433

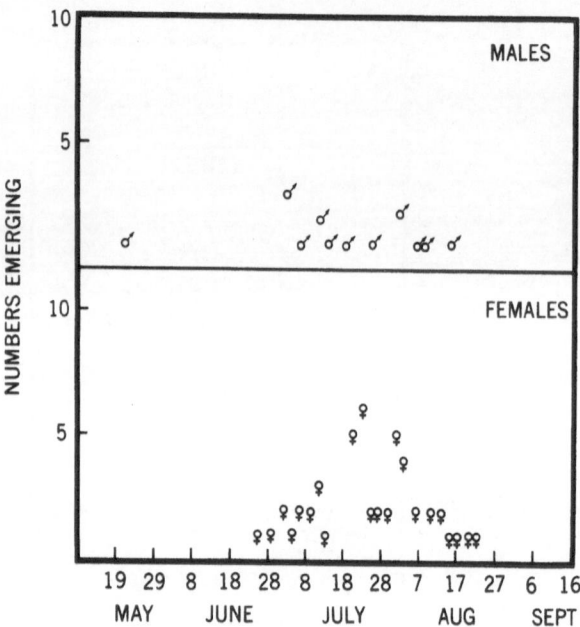

Fig. 2. Emergence period of adults in Duffin Creek in 1979.

filter, using a micrometer eyepiece, and the approximate volume of each was then calculated.

The gut contents of three larvae of all five instars, from Duffin Creek, for June and August 1979 (the summer months when the quantity of FPOM in the stream was low and high, respectively) are compared in Fig. 3. Items were divided into three main categories: FPOM, diatoms and other, the latter consisting mostly of large fragments (up to 0.5 mm in length) of terrestrial plant tissue (see also Mecom, 1972). FPOM was by far the most important material eaten in both months and, in general, the percentage composition for each instar was similar for both months. In both months, the first three instars had eaten proportionately more diatoms and other plant material than the fourth and fifth instars. Further, Fig. 4 shows that the size range of diatoms eaten, as exemplified by *Diatoma*, was similar for small and large larvae.

The size composition of the FPOM ingested was similar between June and August for each instar pair (Fig. 5). In both months, there seems to be an indication of a shift in preference to larger-sized particles with increase in larval size, with fifth instars feeding regularly on a range of particles up to 50.0 μm in diameter. This is perhaps not surprising as the mesh opening size of the larval capture net increases with instar (Wallace and Malas, 1976). Unlike the findings of these latter authors, very few particles below 2.5 μm diameter were found in the guts.

The Coffman/Cummins method of preparation of gut contents for analysis was not deemed totally satisfactory for the material eaten by *D. distinctus*, even though it has been used in previous studies on the species (e.g. Wallace and Malas, 1976; Wallace et al., 1977). The difficulty lies in the fact that a

434

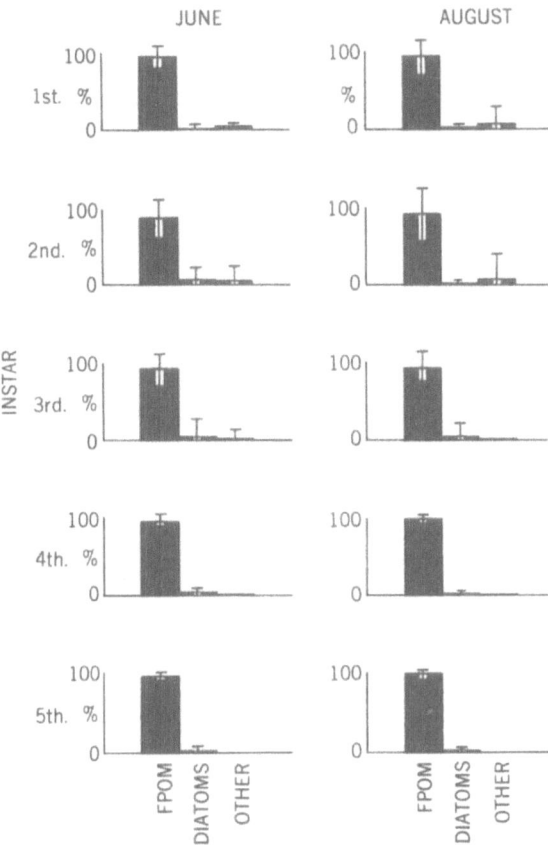

Fig. 3. Larval gut contents during June and August. Data are expressed as percentage composition of main components (95% confidence limits, based on three replicate samples, are given).

mixture of very small particle sizes, many of which may be in an aggregated form, is eaten by the animal. The particles are then compressed in the gut and subjected to at least partial breakdown during digestion. The Coffman/ Cummins method involves resuspending these particles, by stirring, prior to settling on filter paper. Obviously, the extent and intensity of stirring may affect the final size composition of the typically amorphous FPOM. However, had the size distribution of particles on the filter paper reflected entirely the method of slide preparation, one would have expected to see no differences in the size distribution of particles between the various instars. Some vestige of the relationship to the true quantitative nature of the particles ingested must, therefore, have been measured, but any bias due to the preparation technique may have gone unnoticed. The analysis obtained in this and similar studies should be accepted, consequently, with this shortcoming in mind.

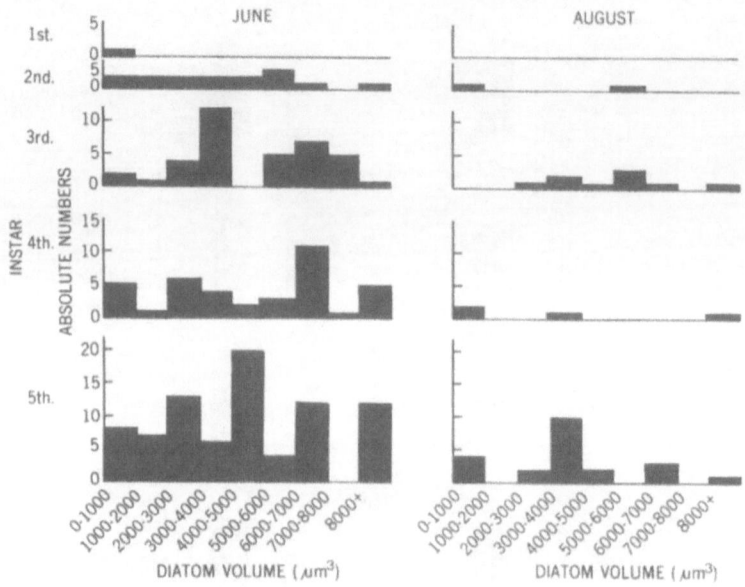

Fig. 4. Size analysis of *Diatoma* taken in larval guts (numbers represent the combined totals of three guts).

LARVAL FEEDING NETS

The distribution of feeding nets of *D. distinctus* when collected in mid-summer in both rivers was far from random. In Duffin Creek they were most often found in areas of reduced current (less than 30 cm/s) particularly under, or on the lower edge of, small (usually less than 30 cm diameter) rocks. In the colonization baskets, larvae built nets in the interstices of the gravel particles (3.2 cm diameter), where the current was reduced, and their distribution was often highly clumped. In the Root River, the feeding nets were most common underneath and attached to the edges of small rocks at the junction of the heads of riffles and upstream pools. At these preferred sites, the nets invariably occurred in tight clusters, presumably in response to heavy competition for optimal current regimes. Close examin-ation of net clusters from both rivers revealed that they consisted of as many as two dozen individual nets attached side by side, each containing a larva (Fig. 6). Table I gives examples of the typical composition of in-stars found in single net clusters from both rivers. Clearly, all instars live side by side with, in many instances, smaller instars making use of the nets of the larger larvae as attachment sites for their own nets. As the table shows, other taxa, particularly chironomid larvae, are intimately associated with the outside of these net clusters, and most probably benefit from eating the detritus trapped there. Wallace and Malas (1976) recorded finding smaller nets attached to larger ones but did not find any larvae in the former. They concluded that the smaller structures represented the feeding nets of the

436

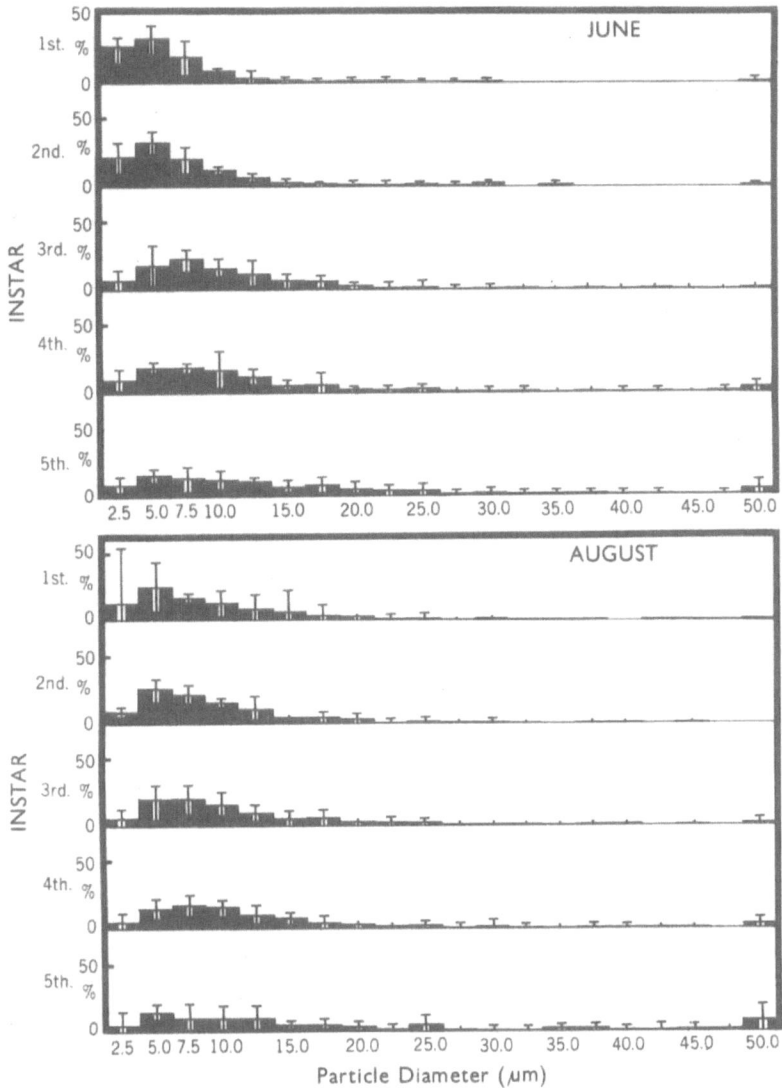

Fig. 5. Particle size distribution of FPOM in larval guts (95% c.l., based on three replicate samples, are given).

preceding larval instars. In general, the size and shape of the larval feeding nets agreed with the descriptions given by Wallace and Malas (1976), for example, fifth instar nets were between 26 and 50 mm long and about 5 mm in diameter.

The structure of individual nets was determined after they had been warmed at 45°C in 5% potassium hydroxide for 24 hours to remove some of the encrusting organic matter. In the Root River, the net structure was much

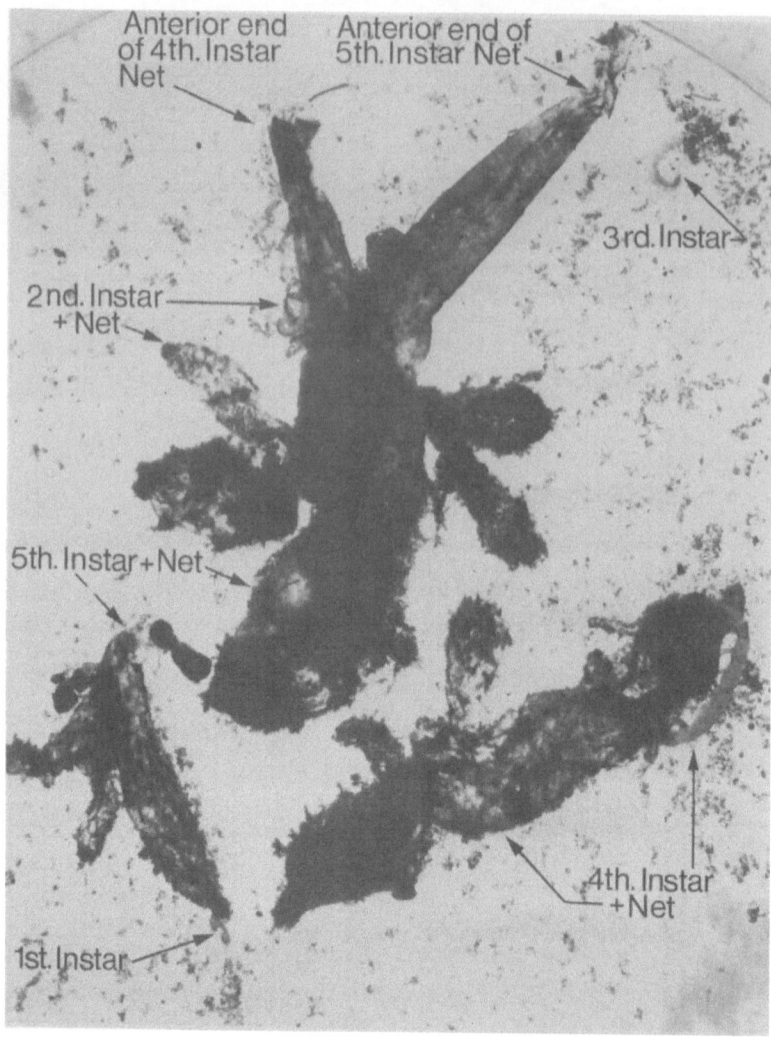

Fig. 6. Partially dissected net cluster from Root River showing component parts and larval occupants.

as described by Wallace and Malas (1976) for populations from Georgia and North Carolina. The anterior one third of the net consisted of several (usually three) layers of differing construction. The innermost layer was a tightly woven mesh of threads, laid down in two directions roughly at right angles to one another, with virtually no spaces between the threads. On top of this was another layer in which bundles of three to six threads were laid down together, again with the bundles running roughly at right angles to one another. The third, outer, layer consisted of more widely-spaced multi-directional fibres. Such a dense meshing of threads in this section of the net

438

Table 1. Size composition of larvae in typical net clusters together with associated fauna (midsummer).

Duffin Creek

Instar	No. of Individuals	Associated Fauna	
1	1	Chironomidae —	*Polypedilum* sp.
2	1		*Eukiefferiella* sp.
3	2		*Corynoneura* sp.
4	2		*Heterotrissocladius* sp.
5	5	Ephemeroptera —	*Ephemerella* sp.

Root River

Instar	No. of Individuals	Associated Fauna	
1	6	Chironomidae —	*Polypedilum* sp.
2	2		*Eukiefferiella* sp.
3	9		*Corynoneura* sp.
4	1		*Thienemanniella* sp.
5	2		*Tanytarsus* sp.
		Ephemeroptera —	*Leptophlebiidae*
		Trichoptera	— *Symphytopsyche* sp.
		Nematoda	
		Oligochaeta	— Naididae

rules out, as Wallace and Malas suggested, any filtering function and probably serves to strengthen the neck of the net near its point of attachment to the substrate.

The posterior two thirds of the net was composed of the very regularly arranged threads, with mesh openings of virtually identical dimensions (e.g. 5th instar 1 x 6 μm), described by Wallace and Malas. In the Root River nets, only one layer was present, usually, but it was sometimes overlain in places by another regular layer or patches of irregularly arranged fibres. By way of contrast, in Duffin Creek, the posterior two thirds of each net consisted of many layers of regularly arranged threads (frequently more than eight layers) some of which lacked the double-stranded support threads recorded in previous studies. Often, a layer of irregularly arranged threads occurred outermost.

It is interesting to speculate on why there was such a difference in the number of layers of mesh in the posterior sections of the nets from the two rivers. Perhaps, because of the greater amount of food, in the form of FPOM, in Duffin Creek, the larvae add new linings of regular mesh to the inside of the net as the outer layers become plugged. Because of the very small mesh size, at some point (presumably after only three or four layers) the net must necessarily cease to function as a filter and merely acts as a tube through which water and FPOM flow. Some FPOM probably settles on the inside wall of the net due to the reduced current resulting from the restricted posterior net opening, and the larva must then brush these particles off with its highly specialized labrum. The fact that the innermost layer of mesh in this region of the net was invariably clean, upon dissection, would tend to support this supposition. In the less-productive Root River, sparcity of FPOM would not necessitate the production of periodic new net linings, thus keeping the net functioning as the ultrafine filter unit for which it seems to have been designed originally. One might speculate, therefore, that the genus *Dolophilodes* evolved in flowing waters of comparatively low productivity.

ACKNOWLEDGEMENTS

We thank C. Filkin, K. Moore and S. Cohen for technical assistance, and the Natural Sciences and Engineering Research Council of Canada for financial support.

REFERENCES

Brickenstein, C., Abh. Bayer. Akad. Wiss. N.F. 69: 1–44, 1955.
Coffman, W.P.; Cummins, K.W. and Wuycheck, J.C., Arch. Hydrobiol. 68: 232–276, 1971.
Mecom, J.O., Oikos 23: 401–407, 1972.
Nielsen, A., Arch. Hydrobiol. Suppl. 17: 255–631, 1942.
Philipson, G.N., Proc. R. Ent. Soc. Lond. A. 28: 17–23, 1953.
Ross, H.H., Bull. Ill. Nat. Hist. Surv. 23: 1–326, 1944.
Sattler, W., Arch. Hydrobiol. 58: 125–135, 1962.
———, Arch. Hydrobiol. 59: 26–60, 1963.
Wallace, J.B., Ann. Ent. Soc. Am. 68: 463–472, 1975.
——— and Malas, D., Can. J. Zool. 54: 1788–1802, 1976.
———; Webster, J.R. and Woodal, W.R., Arch. Hydrobiol. 79: 506–532, 1977.
———; Woodall, W.R. and Staats, A.A., Ann. Ent. Soc. Am. 69: 149–154, 1976.
Wiggins, G.B., Larvae of the North American Caddisfly Genera (Trichoptera). Univ. Toronto Press, pp 401, 1977.
Williams, N.E. and Hynes, H.B.N., Oikos 24: 73–84, 1973.

OBSERVATIONS AND A THEORY ON THE BUILDING STYLE CHANGE IN THE *POTAMPHYLAX LATIPENNIS* (CURT.) NEB. – TRICHOPTERA, LIMNEPHILIDAE

H. ZINTL

SUMMARY

A spontaneous change takes place in the style of the case during the 4th or 5th instar. Observations and experimental results suggest that the tendency to build a new style decreases slowly and in oscillations. The 3rd/4th and 4th/5th moults probably have an influence on this decreasing tendency.

INTRODUCTION

It is known that some larvae of the Limnephilidae, Sericostomatidae, Lepidostomatidae and Brachycentridae families change their case-building material, and in consequence their style, in certain instars and, strangely, the instar during which the change takes place may vary within the same species; in *Micrasema longulum* (Brachyc.)(Bohle, 1974) it occurs at the end of the 1st or the beginning of the 2nd instar, in *Lepidostoma hirtum* (Lepidost.) (Hansell, 1972) during the 3rd or at the start of the 4th instar, in *Potamophylax latipennis* (Limnephil.)(Zintl, 1974) in the 4th or at the beginning of the 5th instar. Furthermore, it has been noticed that some *Micrasema longulum*, *Crunoecia irrorata* (Sericostomatidae)(Sattler, 1957) and *Potamophylax latipennis* larvae retrogress to their former style at least once before accomplishing a complete style change. These results make it unlikely that the style change is triggered by a simple hormone mechanism, such as that responsible for wandering and spinning in caterpillars before pupation (Bollenbacher et al., 1978).

Therefore, in recent years, my investigations have been confined to the purely behavioural analysis of *Potamophylax latip.* Previous data are included in the present evaluations and the methods have, already, been described (Zintle, 1976).

RESULTS

The *Potamophylax latip.* larva changes its style in such a way that it ceases to construct the dorsal and ventral rows of its case with panels cut from dead leaves and substitutes them with small particles. Initially the larva may use either tiny fragments of leaves or grains of sand, but, finally, it builds only with sand grains. In order to understand the style change, it is essential to establish the time relationships between this and the 3rd/4th and 4th/5th moults.

Distinct maxima were observed in the 1973 and 1976 samples after the 4th/5th moult and, in 1976, also before that moult. The period of the 4th instar in these two years was not unduly spread over time and was medium compared to the brief duration of the same instar in 1968 and the very long ones of 1969 and 1970 (Fig. 1). If the time-interval between the 3rd/4th moult and the style change is plotted against the time-interval between the 3rd/4th and 4th/5th moults, the resulting correlation can be expressed as a regression line (Fig. 2). With respect to style change before the 4th/5th moult only the 1968 samples failed to show a significant correlation (Fig. 3, below). Only 4 data were available for 1973, but in 1967 and 1976, when the

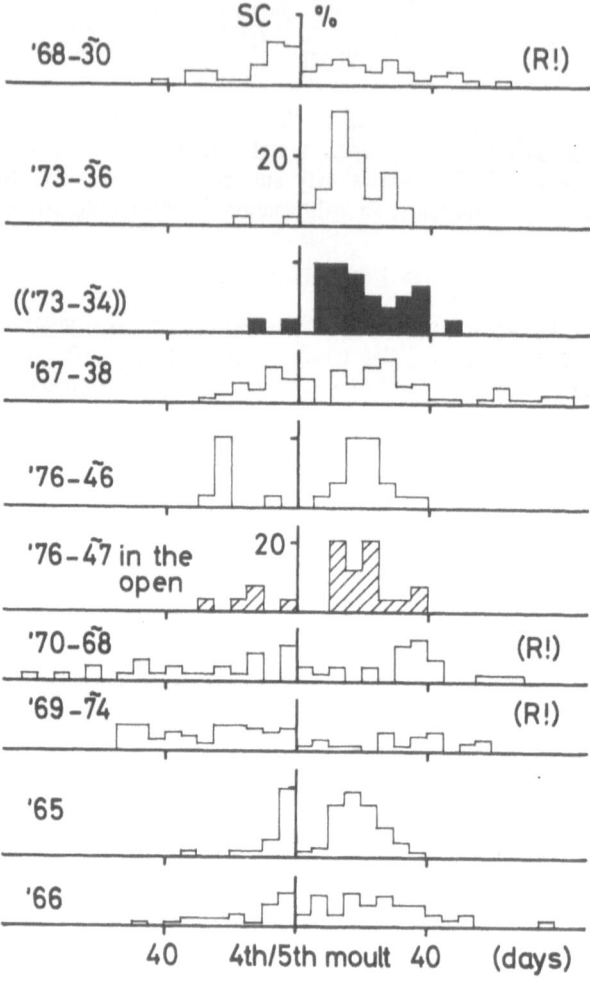

Fig. 1. Style change before and after the 4th/5th moult related to this moult. Data gathered from 5 days periods. N = 27 to 116 per year. From top to bottom ascending duration of the 4th instar (median beside the year). Black: sample with only leaf environment. Hatched: sample in the open. '65 and '66 length of the 4th instar unknown. R!: frequent relapses to the panel style.

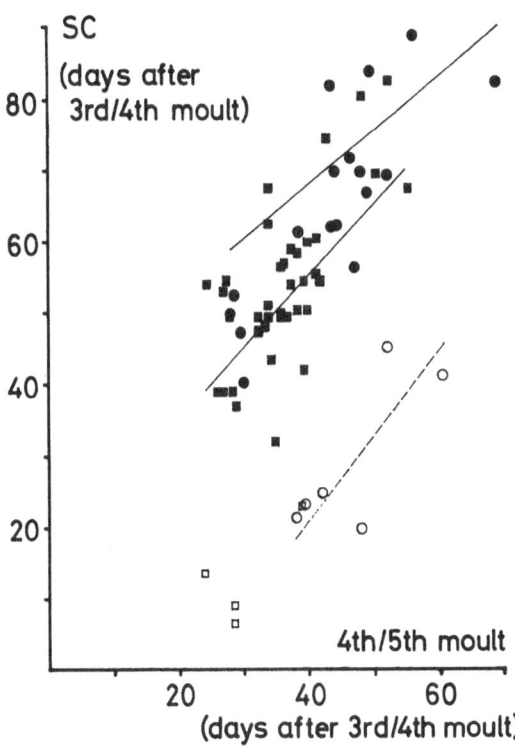

Fig. 2. Relationship between the length of the 4th instar and the point of time of the style change. Squares: sample '73; circles: sample '76. Open signs: style changes before the 4th/5th moult; filled signs: style changes after the 4th/5th moult.

duration of the instar was medium, good correlations ($r_{SP} = 0.8$) were achieved. However, with the numerous long 4th instars of 1970, the correlation coefficient decreased to 0.6. Although significant correlations ($\alpha < 0.01$) had been revealed in the style change after the 4th/5th moult, it was not until 1976 that a coefficient of 1.0 was reached. It decreased only slightly to 0.8 in 1970. The correlation was higher for larvae kept in a sandless environment in 1973 than it was for those in the normal sand-and-leaves environment. A possible explanation for this result is that, in these experimental conditions, only when the state of readiness to style change is very high does the change become manifest. No correlation could be found for the style change either before or after the 4th/5th moult in the open 1976 samples and this is somewhat puzzling. These significant correlations offer the first indication that, at least for larvae maintained in the laboratory, there is a relationship between the style change and the 4th/5th moult. The fact that style changes were almost entirely absent before that moult in both environments in 1973 (Fig. 1) warrants further investigation.

The question now was whether there were periods during the 4th instar in which the style change before the 4th/5th moult was inhibited. To examine this problem, more than 15 4th instar groups of different durations,

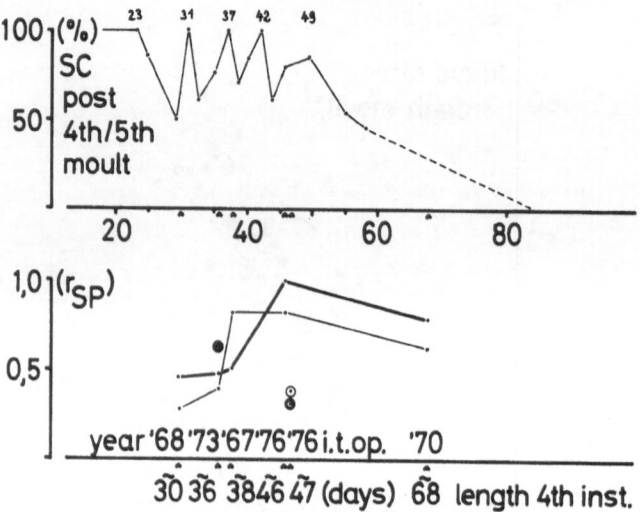

Fig. 3. Above: Distribution of the style changes with respect to the periods before and after the 4th/5th moult depending on the length of the 4th instar. Number of data used for the points (from left to right): 14, 23, 28, 8, 13, 26, 6, 14, 13, 7, 8, 5, 14, 9, 13. Further explanations in the text.

Below: Coefficients of correlation (Spearman) of the relationship 'period between the style change and the 3rd/4th moult' to 'period between the 4th/5th and the 3rd/4th moult'. Thinly drawn curve: Style changes before the 4th/5th moult; thickly drawn curve: Style changes after the 4th/5th moult. Circle with cross: '73, only leaf environment. Circles with centre: '76, in the open. Abscissa: year and median of the length of the 4th instar.

each two days apart, were evaluated across all the years and the style change percentage after the moult recorded for each group (Fig. 3). This figure shows that style change may take place in the 4th instar and that not only must a minimum time interval be surpassed in the 4th instar before the style change can occur, but that when this time interval is exceeded, the style change is inhibited at certain periods. Owing to the method of data sampling across the years, the position of the peaks represent orders of magnitude and are only indicative. It could not, therefore, be ascertained whether the periods between the 4th instar durations connected with style change inhibition were equal and, if they were, whether they would have been different for the various larvae.

An analysis of the retrogression to the former style, observed in a reasonably high number of larvae in 1968, 1969 and 1970, was considered useful. Nineteen percent of the 1968 larvae which had already started the style change retrogressed and added one or several leaf panels and so only accomplished a definitive style change when they made a second attempt. The majority of the larvae in this sample began the style change with particles of leaves (Table I).

The larvae which begin the style change with particles of leaves use mainly sand grains later; those which start the style change with sand do not retrogress (Table II). The starting of the style change with particles of leaves, therefore, seems to be an expression of a style change readiness

444

Table I. Type of the building material used during the start of the style change

| | | number of larvae | % starting SC with | | |
			particles of leaves	mixed particles	sand
Ascholding	'60	41	00	85	15
laboratory	'60	127	11	13	76
"	'68	79	70	9	21
"	'70	65	25	24	51
"	'73	45	20	11	69
"	'76	26	23	8	69
in the open	'76	26	23	15	62

Table II. Type of the building material used during the start of the style change and frequency of following relapses into the panel style

	SC with	no relapse	relapse following
	part. of leav.	41	13
'68	mixed part.	6	4
	sand	15	0
	part. of leav.	12	4
'70	mixed part.	11	5
	sand	32	1

which is still low. Therefore, both the preliminary style changes and those which were definitive at the first attempt are shown in relationship to a time scale after the 3rd/4th moult in Fig. 4. Reading from top to bottom, one sees that the pattern 4th/5th moult closely followed by a style change, style change further into the 5th instar, style change in the middle of the 4th instar, 4th/5th moult closely followed by a style change, and so on, tends to repeat itself. There is no style change in the 4th instar when this is shorter than 27 days; when it exceeds 27 days, style changes do not take place at intervals of 2 to 4 days.

On one occasion two larvae underwent a style change at the same time; in one it was definitive, in the other only temporarily. This could indicate variations in the increase of the style-change readiness in larvae at the same stage of the 4th instar. It is now obvious that the retrogression phenomenon

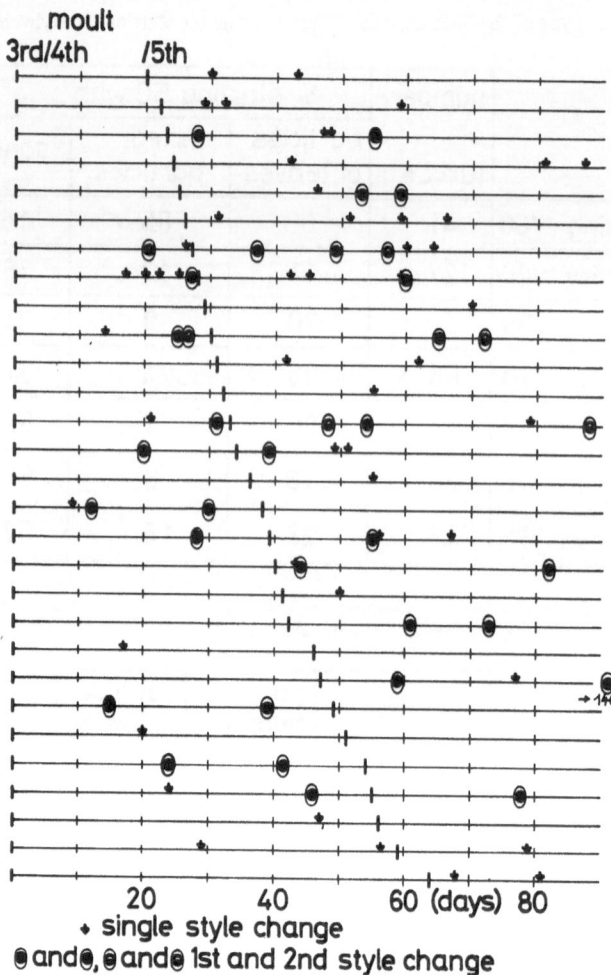

Fig. 4. Distribution of the definitive and the preliminary style changes to the periods before and after the 4th/5th moult. Further explanations in the text.

is not limited to short instars, as was previously suspected. The rather long period which often divides the temporary and definitive style changes indicates that, at least in some larvae, the change in readiness is a slow process. A similar behavioural pattern was exhibited by larvae extracted from their cases, some of which re-added panels for as long as 35 days after the beginning of the style change, thereby reverting to their former style (Zintl, 1976).

HYPOTHESES

Perhaps, it is a little early to formulate a theory on the style change, but the results suggest certain hypotheses.

Style change means the ability to build with panels is lost and the readiness towards this type of building decreases to zero in the nervous system. Not only does this happen slowly, as mentioned above, but the regression in the panel-building tendency would seem to occur in a diminishing oscillation. If the assumption of such an oscillation is not accepted, there is no way of explaining the sequence: firstly, a style change; secondly, the addition of one or several panels and, thirdly, a definitive style change. It must, also, be assumed that when the 'common building tendency' arises in the nervous system, it coincides with a tendency to build with leaf panels and that the readiness to this is at first temporary and below the threshold level, then transitorily above it and, finally, permanently below it.

REFERENCES

Bollenbacher, W.E.; H. Zvenk, A.K. and Kumaran, L.I. Gilbert, Gen. and Comp. Endocrinology 34: 169–179, 1978.
Hansel, M.H., J. Zool., Lond. 167: 179–192, 1972.
Sattler, W., Ber. Limnol. FluBstat. Freudenthal 7: 18–22, 1957.
Zintl, H., In Proc. First. Int. Symp. Trichoptera, ed. H. Malicky, Junk The Hague, 1974.

DISCUSSION

Nielsen: In 1974, I pointed out that a period in which the larva builds a flat case of dry beech leaves is characteristic for *Potamophylax nigricornis*. I have since found such case in *P. latipennis* in a small forest stream on the island of Sjaelland. However, *P. latipennis* also lives in faster-flowing streams and there it never has a flat case. This is quite natural, since it has no access to beech leaves.

NAME INDEX

Adams, H.R., 8, 10
Adlmannseder, A., 16
Aisa, E., 1 ff.
Aldrich, F.A., 156
Allan, J.D., 375, 381
Anderson, N.H., 44, 45, 299, 300, 373
Angelier, E., 256
Augustin, C.I.

Badcock, R.M., 5 ff., 16, 18, 73, 111, 345
Bajkov, A., 68, 69, 74
Barnard, K.H., 269
Barnard, P.C., 207, 208
Barton, D.R., 240, 244
Baudoin, J., 3, 213, 217
Bautista, I., 300, 303 ff.
Beck, W.M., 229
Benzecri, J.P. 349, 362
Berner, L., 378, 381
Beron, P., 146
Betten, C., 264
Bicchierai, M.C., XVI, 193 ff.
Blickle, R.L., 40, 44, 45
Bohle, H.V., 441
Bollenbacher, W.E., 441, 447
Boon, P.J., 87, 92, 329, 335, 349, 353, 355, 357, 363
Botosaneanu, L., XVI, 11 ff., 21 ff., 31 ff., 38, 50, 55, 87, 92, 139, 144, 146, 147, 163, 175, 178, 180, 199, 206, 207, 208, 210, 217, 256, 278, 314, 315, 318, 363, 375, 381, 402, 406
Bournaud, M., XVI, 114, 138, 230, 347 ff.
Bouvet, Y., XVI, 157, 160, 161, 162
Branch, H.E., 366, 373

Bray, R.P., 412
Brickenstein, C., 432, 440
Briden, J.C., 263, 264
Brindle, A., 355, 363
Brunskill, G.J., 68, 69, 74
Bueno Soria, J., 33 ff.
Bullock, T.H., 111, 116
Buresch, I., 144, 146
Bykhovskaya-Pavlovskaya, I.E., 3

Cairns, J., 156
Campbell, P., 68, 74, 285, 286, 290
Canton, S.P., 39 ff.
Carlsson, M., 299, 300
Caspers, N., 363
Chapin, J.W., 156
Chapman, R.F., 422, 428
Chavanon, G., 363
Christophersen, J., 117
Christovic', G., 139, 146
Cianficconi, F., XVI, 12, 17, 18, 19, 165, 181, 182, 186, 187, 189, 193, 198, 199 ff., 237, 242, 247, 256
Clark, K.L., 290, 291
Coffman, W.P., 440
Coineau, J., 25, 29
Colby, J.A., 380, 381
Collardeaux, C., 111, 116
Collardeaux-Roux, C., 111, 116
Connel, J.H., 150, 156
Coob, D.G., 67 ff.
Coope, G.R., 409, 410, 418, 419
Corallini-Sorcetti, C., 169, 211, 213 ff., 226, 243 ff.
Corbet, P.S., 154, 156
Cowley, D.R., 402, 405
Crichton, M.I., 47 ff., 104, 108, 109, 138, 155, 156, 159, 160, 161, 162,

449

453

SUBJECT INDEX

465

471

*taxa other than Trichoptera